Computability and Models

THE UNIVERSITY SERIES IN MATHEMATICS

Recent volumes in the series:

COMPUTABILITY AND MODELS: Perspectives East and West
S. Barry Cooper and Sergey S. Goncharov

CONTEMPORARY GEOMETRY (J.-Q. Zhong Memorial Volume)
Edited by Hung-Hsi Wu

ELLIPTIC COHOMOLOGY
Charles B. Thomas

AN INTRODUCTION TO ALGEBRAIC NUMBER THEORY
Takashi Ono

LINEAR CONTROL SYSTEMS: with solved problems and MATLAB examples
Branislav Kisačanin and Gyan C. Agarwal

MATRIX THEORY: A Second Course
James M. Ortega

PROBABILITY MEASURES ON SEMIGROUPS: Convolution Products, Random Walks, and Random Matrices
Göran Högnäs and Arunava Mukherjea

RECURRENCE IN TOPOLOGICAL DYNAMICS: Furstenberg Families and Ellis Actions
Ethan Akin

A SHORT INTRODUCTION TO INTUITIONISTIC LOGIC
Grigori Mints

TOPICS IN NUMBER THEORY
J. S. Chahal

VARIATIONS ON A THEME OF EULER: Quadratic Forms, Elliptic Curves, and Hopf Maps
Takashi Ono

Computability and Models

Perspectives East and West

S. Barry Cooper
University of Leeds
Leeds, United Kingdom

Sergey S. Goncharov
Novosibirsk State University
Novosibirsk, Russia

Springer Science+Business Media, LLC

Library of Congress Cataloging-in-Publication Data

Cooper, S. B. (S. Barry)
Computability and models: perspectives east and west/S. Barry Cooper and Sergey S. Goncharov.
p. cm.
Includes bibliographical references and index.
ISBN 978-1-4613-5225-9 ISBN 978-1-4615-0755-0 (eBook)
DOI 10.1007/978-1-4615-0755-0

1. Recursion theory. I. Goncharov, S. S. II. Title.

QA9.6 .C68 2002
511.3'5—dc21

2002031660

ISBN 978-1-4613-5225-9

Originally published by Kluwer Academic / Plenum Publishers in 2003
Softcover reprint of the hardcover 1st edition 2003

Contributing Authors

Marat Arslanov is Professor, and Head of Chair of Algebra and Logic, at Kazan State University. He is a Correspondent-Member of the Academy of Science of the Republic of Tatarstan.

Serikzhan Badaev is Professor at the Kazakh National University in Almaty, and organiser of the International Conference and Workshop on *Computability and Models* in Almaty, 2002.

Vasco Brattka is DFG Research Fellow at the Computer Science Department of the University of Hagen. He was co-organiser of the Dagstuhl Seminar on *Computability and Complexity in Analysis* 2001.

Barry Cooper is Professor of Mathematical Logic at the University of Leeds, and founder of the *Recursive Function Theory Newsletter.* He is co-ordinator of the INTAS-RFBR Research Project *Computability and Models.*

William Gasarch is Professor of Computer Science at the University of Maryland. He currently edits the SIGACT News Book Review Column.

Sergey Goncharov is a Corresponding Member of the Russian Academy of Sciences, and Professor and Dean of the Department of Mechanics and Mathematics at Novosibirsk State University. He is also Deputy Director of the Institute of Discrete Mathematics and Informatics.

Eberhard Herrmann, of the Humboldt University of Berlin, has a strong international reputation, initially based on his ground-breaking proof of the undecidability of the theory of the lattice of computably enumerable sets.

Sanjay Jain is Professor of Computer Science at the National University of Singapore. He is well known for his work in inductive inference, and is a coauthor of the MIT Press book *Systems That Learn.*

Iskander Kalimullin is an innovative young researcher from Kazan State University.

Steffen Lempp is Professor at the University of Wisconsin at Madison. A past editor of the *Journal of Symbolic Logic,* and former member of the Executive Committee of the A.S.L., he is an editor of the AK Peters Lecture Notes in Logic, and of the de Gruyter Series in Logic and Its Applications.

Angsheng Li is Full Professor at the Institute of Software, Chinese Academy of Sciences, P.R. of China.

Charles McCoy is Van Vleck Assistant Professor at the University of Wisconsin at Madison, as a prelude to entering the Notre Dame seminary.

Andrei Morozov is Leading Research Fellow at the Sobolev Institute of Mathematics of Siberian Branch of Russian Academy of Sciences and Professor at the Novosibirsk State University. He is a well-known speaker at international conferences.

Piergiorgio Odifreddi is Professor of Mathematical Logic at the University of Turin, and a celebrated author, interviewer and broadcaster within Italy. He is also the author of the two-volume major work *Classical Recursion Theory,* and the editor of the AK Peters collection of articles *Kreiseliana : About and Around Georg Kreisel.*

Roland Omanadze is Professor at the I. Vekua Institute of Applied Mathematics at Tbilisi State University, and is the world leading expert on the theory of Q -reducibility.

Sergey Podzorov is a Senior Research Fellow at the Sobolev Institute of Mathematics of the Siberian Branch of the Russian Academy of Sciences.

Victor Selivanov is Professor at the A. P. Ershov Institute of Informatics Systems, Siberian Division of Russian Academy of Sciences, and Head of Chair of Informatics and Discrete Mathematics at the Novosibirsk Pedagogical University.

Reed Solomon is Van Vleck Assistant Professor at the University of Wisconsin at Madison, and will take up a tenure-track appointment at the University of Connecticut in 2003.

Boris Solon is Professor of Mathematics, and Head of Chair of Higher Mathematics, at Ivanovo State University, and a leading expert on enumeration reducibility.

Andrea Sorbi is Professor of Mathematical Logic at the University of Siena, and editor of *Complexity, Logic and Recursion Theory.* He is the co-ordinator of the INTAS Research Project *Computability in Hierarchies and Topological Spaces.*

Frank Stephan is a Heisenberg fellow of the Deutsche Forschungsgemeinschaft (DFG), at the University of Heidelberg and an expert in Inductive Inference.

Preface

Science involves *descriptions* of the world we live in. It also depends on nature exhibiting what we can best describe as a high *algorithmic content.* The theme running through this collection of papers is that of the interaction between descriptions, in the form of formal theories, and the algorithmic content of what is described, namely of the *models* of those theories. This appears most explicitly here in a number of valuable, and substantial, contributions to what has until recently been known as 'recursive model theory' – an area in which researchers from the former Soviet Union (in particular Novosibirsk) have been pre-eminent. There are also articles concerned with the computability of aspects of familiar mathematical structures, and — a return to the sort of basic underlying questions considered by Alan Turing in the early days of the subject — an article giving a new perspective on computability in the real world. And, of course, there are also articles concerned with the classical theory of computability, including the first widely available survey of work on quasi-reducibility.

The contributors, all internationally recognised experts in their fields, have been associated with the three-year INTAS-RFBR Research Project "Computability and Models" (Project No. 972-139), and most have participated in one or more of the various international workshops (in Novosibirsk, Heidelberg and Almaty) and other research activities of the network. Although based on just eight research centres – Almaty, Heidelberg, Ivanovo, Kazan, Leeds, Novosibirsk, Siena and Turin – the project has acted as a focus for researchers from all over Europe and beyond. This has been an exciting and rewarding experience for everybody involved, and has helped transform the fragmented European scene of ten or more years ago (so vividly described by George Odifreddi in his entertaining introduction to this volume) into the lively community of researchers we now see developing.

The articles which follow approach this important and growing area of research from many different angles. The authors were encouraged to provide *readable* introductions to their research. All have responded either with timely surveys of work inadequately covered elsewhere, or with interesting and important new results, with clear pointers to the wider context. All articles have been rigorously refereed, and revised accordingly. We wish to express our gratitude to the contributors, and hope that the reader will be able to benefit from the expertise and enthusiasm expressed in their articles.

We also wish to acknowledge with thanks the support given by the EU, through INTAS, and by the RFBR, without which this volume would not have been possible.

BARRY COOPER

SERGEY GONCHAROV

Contents

Introduction xv
Piergiorgio Odifreddi

Truth-Table Complete Computably Enumerable Sets 1
Marat M. Arslanov
1. Introduction 1
2. A Truth-Table Completeness Criterion 3

References 9

Completeness and Universality of Arithmetical Numberings 11
Serikzhan Badaev, Sergey Goncharov, Andrea Sorbi
1. Basic Notions and Preliminaries 12
2. Complete Numberings and Completions 19
3. Universal Numberings of Finite Families 40
4. Open Questions 43

References 43

Algebraic Properties of Rogers Semilattices of Arithmetical Numberings 45
Serikzhan Badaev , Sergey Goncharov, Sergey Podzorov, Andrea Sorbi
1. The Cardinality of Rogers Semilattices 46
2. Intervals in Rogers Semilattices 50
3. Empty Intervals. Minimal Elements and Minimal Covers of Rogers Semilattices 71

References 76

Isomorphism Types and Theories of Rogers Semilattices of Arithmetical Numberings 79
Serikzhan Badaev, Sergey Goncharov, Andrea Sorbi
1. Distinguishing Rogers Semilattices of Finite Families 80

2. Distinguishing Rogers Semilattices of Finite Families from Rogers Semilattices of Infinite Families 80
3. Weak Distributivity 81
4. Distinguishing Rogers Semilattices of Infinite Families 84
5. Isomorphism Types 85

References 90

Computability over Topological Structures 93
Vasco Brattka
1. Introduction to Computable Analysis 93
2. Representations and Computable Functions 95
3. Topologically Admissible Representations 99
4. Real Number Representations 102
5. Computable Metric Spaces 104
6. Computable Subsets of Metric Spaces 107
7. Computable Multi-Valued Operations 110
8. Recursive Closure Schemes 113
9. Recursive Operations over Structures 118
10. A Stability Theorem for Perfect Structures 121
11. Topological and Metric Structures 125
12. Recursive Sets over Structures 129

References 133

Incomputability In Nature 137
S. Barry Cooper, Piergiorgio Odifreddi
1. An Historical Parallel 137
2. Matter and Mathematical Definability 140
3. Finitism in a Universe with Algorithmic Content 143
4. The Inseparability of Algorithmic Content, Complexity and Incomputability 146
5. A Closer Look at the Turing Model 147
6. Scientifically Presenting the Universe 151
7. What the Turing Model Delivers 153

References 160

Gems in the Field of Bounded Queries 161
William Gasarch
1. Definitions 162
2. The Complexity of C_n^A 168
3. The Complexity of $\#_n^A$ 172
4. The Complexity of ODD_n^A 175
5. Do More Queries Help? 179
6. Does Allowing Divergence Help? 182
7. Does Order Matter? 188
8. Acknowledgments 193

References 193

Finite End Intervals in Definable Quotients of $\mathcal{E}$ 197
Eberhard Herrmann

1. Examples of Congruence Relations on $\mathcal{E}$, Finite End Intervals 199
2. The Congruence Relation $\approx_{ms} \vee \approx_{\mathcal{QM}}$ 202
3. Basic Constructions 204
4. Finite End Intervals in $\mathcal{E}_{/\approx^\circ}$ 210

References 214

A Tour of Robust Learning 215
Sanjay Jain, Frank Stephan
1. Introduction 215
2. Function Identification 218
3. Bārzdiņš' Conjecture 220
4. Uniformly Robust Learning 225
5. Robust Learning and Consistency 229
6. Hyperrobust Learning 232
7. Team-Learning and the Union-Theorem 240
8. Learning Aided by Context 241
9. Conclusion 244

References 245

On Primitive Recursive Permutations 249
Iskander Kalimullin
1. P.t.c.e Sets and P.r.e. Sets 251
2. P.t.c.e. Sets, P.r.e. Sets and Permutations 252
3. A Presentation of Computable Permutations via Three Polynomial Time Computable Permutations 253
4. Generating Group Terms 255

References 258

On Self-Embeddings of Computable Linear Orders 259
Steffen Lempp, Andrei S. Morozov, Charles F. D. McCoy, D. Reed Solomon
1. Background and Theorems 260
2. The Proof of Theorem 4 262

References 265

Definable Relations on the Computably Enumerable Degrees 267
Angsheng Li
1. Structure and Hierarchies 268
2. Definable Relations 274
3. Continuity in the C.E. Degrees 279
4. Automorphisms 281
5. Defining $\mathcal{E}$ in $\mathcal{E}_n$ 282

References 285

Quasi-Degrees of Recursively Enumerable Sets 289
Roland Sh. Omanadze
1. Some Classes of Recursively Enumerable Sets and Quasi-reducibility 291
2. Complexity Properties of Recursively Enumerable Sets and Quasi-degrees 295

3. On the Upper Semilattices of Recursively Enumerable Quasi-degrees 303
4. Relations Between Degrees 310
5. Complete Degrees 314

References 316

Positive Structures 321
Victor Selivanov
1. Introduction 321
2. Preliminaries 323
3. General Facts 325
4. Some Paricular Topics 330
5. Applications 338
6. Conclusion 346

References 346

Local Properties of the Non-Total Enumeration Degrees 351
Boris Solon

References 369

Piergiorgio Odifreddi

Introduction

"East is East and West is West, and never the twain shall meet"

says Rudyard Kipling in *The Ballad of East and West.* It certainly looked so to most recursion theorists in the late seventies, except for a vague knowledge of the fact that Post's Problem had been solved independently by the Soviet mathematician Muchnik in 1956, and the North American Richard Friedberg in 1957.

In his book, *Theory of Recursive Functions and Effective Computability*, Hartley Rogers called the result "the Friedberg-Muchnik Theorem", ironically preferring the alphabetical to the priority order! Except for the references to this theorem, Muchnik is mentioned only once (in an exercise) by Rogers, and never by Bob Soare in *Recursively Enumerable Sets and Degrees*, despite the fact that he had obtained a number of other results: first and foremost, what everybody (except me) still calls "the Friedberg Splitting Theorem".

From Rogers' book one got the impression that Recursion Theory was a totally Western enterprise, with practically no Eastern contribution besides Muchnick's. At least, this was the impression I had formed when I landed in Urbana in 1978, to study with Carl Jockusch. An impression that was to change very quickly, thanks to him and Marat Arslanov, who had arrived a few days earlier than I. He was met at the airport by Carl who, after unsuccessfully trying to communicate with him in English or German, asked whether he spoke any language besides Russian, and got the answer: "Oh yes! Tatar."

Jockusch was the foremost expert on what are now called "strong reducibilities", and knew that much work had been done in the Soviet Union. In particular, he was aware of Marchenkov's new solution to Post's Problem and Ershov's characterization of the structure of the m-degrees. However, the original sources were not easy to handle, because of a two-fold problem: they were either Russian originals, or English translations of abstracts without proofs.

During the year I spent in Urbana I wrote a long paper on "Strong Reducibilities", with a lot of input from Carl. I handed him my only copy, handwritten in pencil. After returning the first half of it with his comments, he suggested that I expand the treatment to other strong reducibilities. When I asked him whether the second part was not enough, he asked surprised: “Which second part?” For a few days we thought we would never see it again, but it eventually re-emerged from his maximum entropy desk in the very colourful office he was sharing with Ken Appel. The latter had just proved his big theorem, and the University was happily stamping all its mail with: “Four Colours Suffice!”. By coincidence Ken had the same birthday as Carl and I, and that year we all celebrated together.

As for my paper, I sent it to the *Bulletin of the American Mathematical Society*, without knowing that the New Series surveys were not supposed to be submitted by postdocs, but rather requested from luminaries. Luckily, however, the paper was accepted, on condition that a general introduction explained the purposes of Recursion Theory and its relationships to the rest of mathematics. This was done “with a little help from my friends”, Richard Shore and Bob Soare in particular, and the publication earned me an unexpected check of a thousand dollars from the American Mathematical Society, for reasons that I did not attempt to clarify (fearing a mistake). Less personally, the paper also played a certain role in making some Eastern work known to a Western audience.

That was the most I could do from a distance. To proceed further it was necessary to go to the Soviet Union, which I did in 1982, after having studied Russian (and other things, of course) in Los Angeles for a year. My flight from Moscow to Novosibirsk was forced to stop in Omsk because of a snow storm, and I was taken to some hotel by Inturist. A couple of hours later, since the weather had improved, I was thrown out of bed and taken to my destination. Obviously, nobody was expecting me anymore and I had to wait for the rest of the night sitting on an airport chair. In the morning Serghei Goncharov came to pick me up, introduced himself and asked me how I had slept. I did not dare start off with a complaint, and as a result I was immediately whisked to the weekly seminar. After the greetings of Yuri Ershov, whom I had already met in Los Angeles, the seminar started and I immediately and happily fell asleep.

The seminar in Novosibirsk was to logicians what Picadilly Circus is to tourists: if you want to meet someone, you just wait there and sooner or later he will come. I thus met most of the Soviet recursion theorists whose work I had quoted in “Strong Reducibilities”: in particular Alexander Degtev, Sergei Denisov, and Victor Selivanov, who explained to me their results and helped me translate their “Eastern-style” proofs into “Western-style” language.

This was not always easy or straightforward, because they usually exploited structural properties of the r.e. sets to derive degree-theoretical ones, instead of using direct constructions. Their approach was more informative but less flexible, and they did miss a number of results because of it. A typical case is Degtev’s structural proof of the existence of a singular r.e. tt-degree. When Rod Downey found a direct construction, he was able to adapt it to derive a wealth of theorems on the structure of r.e. m-degrees inside r.e. tt-degrees.

Some of the Soviets have later been converted, at least partially, to our way of working. For example, Selivanov once flattered me by saying that when he saw my treatment of his structural results on the m-degrees of index sets he thought: "Finally I understand what I did!" And Arslanov published a book on degrees which followed closely, at times perhaps even too much, the first (and otherwise unpublished) draft of *Classical Recursion Theory*.

The greatest difficulty I encountered was the "translation" of Ershov's characterization of the structure of the m-degrees, which he had obtained in the framework of his Theory of Enumerations. I had to develop everything from scratch, and the result of this work is now Chapter VI of my book. Once again, the direct proof could be adapted to derive similar results in other settings, in particular for the polynomial time honest m-degrees.

Incidentally, although this had no effect on my work, one of the mathematicians I met in Novosibirsk was Efim Zelmanov. He was then a brilliant young Jew, who at the time of my visit was not allowed to attend the 1983 Warsaw International Congress of Mathematicians and give one of the invited talks. He did attend the 1994 Zürich Congress, where he was awarded a Fields Medal for his solution of a restricted version of the first Burnside conjecture in group theory.

In Novosibirsk I saw again Arslanov, whom I also visited in Kazan. While in Urbana we could not speak at ease, because none of us was comfortable with English, but this time I had learned enough Russian to appreciate his wonderful sense of humour. For example, we were once asked which of the definitions that we had both given of "effective hypersimplicity" was better, and he immediately answered: "George's is more general, but mine is more important". The structure of differences of r.e. sets, whose study he had independently initiated with his students Bukaraev and Ishmuchametov, is now a favourite in the West.

In Moscow I made a point of meeting Sergei Marchenkov, who treated me very nicely and offered me a wonderful lunch, but refused to talk about Recursion Theory. At the time I could not understand how he could have lost interest in such a beautiful field, to the point of not even remembering his own results, but I'm afraid time has taught me the lesson. I did not meet Muchnik, probably because he was away, but I remember a surreal visit to Vladimir Ouspensky. He decided the priority order in which we all had to enter and leave the room, and greeted me with a: "I hope you're not a shitty set theorist!" When I asked him what was "shitty" in Set Theory, he told me that it was a perfect image of capitalism: you start with nothing, and accumulate by building on power ...

In Tbilisi I was hosted by Georgi Kobzev and Roland Omanadze, who told me of their surprise in opening one day the *Bulletin of the American Mathematical Society* and finding their names and results quoted in the West. Kobzev was then the world expert on η-hyperhypersimple sets and had done marvels with them, but he missed Marchenkov's solution to Post's Problem. He recalled with a sad look that he could have done it very straightforwardly, putting together facts that he knew very well, if only he had thought of doing it. But he didn't.

When I published *Classical Recursion Theory*, I summed up in the Preface the set of events here described as follows: "I had learned that the Soviets were doing work that the Westerners did not know much about, and they themselves were largely unaware of what people did in the West. I found this an odd situation, and decided I would go to the Soviet Union to bridge the gap, at least in my knowledge."

At the end of my two years in the Soviet Union the bridge had indeed been built, at least in my knowledge. I cannot judge whether, or how much, my papers and books have influenced others, but it is a fact that towards the mid eighties a number of people, led by Richard Shore, started a fruitful work in areas of Recursion Theory that had previously been a Soviet monopoly. In my chapter for the *Handbook of Computability Theory* I have reported on the progress made on the problems I had stated in "Strong Reducibilities". And I hope that the relevant chapters of my book show how Recursion Theory must now be thought of as a collective Eastern and Western enterprise.

East and West have thus met, at least in our field, but this does not prove Kipling wrong. Indeed, it is too often forgotten that the negative, well know *incipit* quoted above was immediately corrected by the following positive verses, which can be taken as a poetic summary of the prosaic report in which I have indulged, and for which I beg forgiveness:

> *"But there is neither East nor West, Border, nor Breed, nor Birth,*
> *When two strong men [or schools] stand face to face, tho' they*
> *come from the ends of the Earth."*

Computability and Models

TRUTH-TABLE COMPLETE COMPUTABLY ENUMERABLE SETS

Marat M. Arslanov*
Kazan State University
Department of Mathematics
Kremlevskaja 18
Kazan 420008
Russia
Marat.Arslanov@ksu.ru

Abstract We prove a truth-table completeness criterion for computably enumerable sets.

Keywords: Computably enumerable sets, truth-table reducibility, completeness criterion, fixed-point free functions

1. Introduction

There exist two kinds of completeness criteria for computably enumerable (c.e.) sets for several reducibilities. The first of them describes Turing complete c.e. sets by means of properties of productiveness of their complements, and the second describes Turing complete c.e. sets by means of diagonally noncomputable functions or, equivalently, fixed-point free functions, which are reducible to these sets. The author in his earlier papers [Ars81, Ars87, Ars89], see also [Soa87], worked out second kind criteria for Turing (T-), weak truth-table (wtt-) and m-reducibilities. For the latter the m-reducibility of a function f to a set A is defined as follows: $f \leq_m A$ if and only if there are computable functions a, b and g such that for all $x \in \omega$, $f(x) = a(x)$, if $g(x) \in A$, and $f(x) = b(x)$ otherwise. (The natural definition of m-reducibility of the function f to the set A via m-reducibility of the graph f to the set A is not suitable for this purpose, since if $\{\langle x, f(x)\rangle \mid x \in \omega\} \leq_m A$ and A is c.e., then the function f is computable, and by the recursion theorem has fixed points.)

The question of the existence of a second kind completeness criterion for truth-table (tt-) reducibility remained open. V.K. Bulitko in his paper [Bul91,

*The authors research was partially supported by Russian Foundation of Basic Research, Project 99-01-00830, and RFBR-INTAS, Project 97-91-71991.

Computability and Models: Perspectives East and West
Edited by Cooper and Goncharov, Kluwer Academic/Plenum Publishers, 2003

Theorems 11 and 12] published the following tt-completeness criterion for c.e. sets: a c.e. set is tt-complete if and only if there exists a function $f \leq_{tt} A$ such that $\forall x(W_x \neq W_{f(x)})$, and f is majorized by a computable function. Unfortunately, the proposed proof of this result was incorrect, and later Bulitko withdrew this claim (see [Bul93]).

Below we first show that Bulitko's criterion does not hold in fact. Indeed, let us consider the example of a wtt-complete but not tt-complete set, as constructed by Lachlan [La75]. We will follow the construction from [Odi89].

The desired c.e. set A is defined using an auxiliary computable function f, where $f(0) = 0$, and a sequence of intervals $I_x = \{z \mid f(x) \leq z < f(x+1)\}$. By the construction, if $x \in K_{s+1} - K_s$, where K is the creative set, then the least integer $m \in I_x \cap \bar{A}_s$ is enumerated into A (as well as some other numbers). By the construction such an integer always exists. Now let $t(e) = g(m)$, where m is the least integer from the $h(e)$–th interval $I_{h(e)}$ such that $m \notin A$. Here g is a computable function such that $W_{g(x)} = \{x\}$, and the computable function $h(x)$ we define using an auxiliary c.e. set $T(x, e)$, which depends on x and e.

If $x \in K$ and there is a stage s and a number y such that $W_{e,s} = \{y\}$, and $y = \mu t\{t \in I_x \,\&\, t \notin A_s\}$, then $T(x, e) = \{x\}$. Otherwise $T(x, e) = \emptyset$.

Now let $p(x, e)$ be the following computable function: $W_{p(x,e)} = T(x, e)$. By the recursion theorem there exists a computable function $h(e)$ such that for any e, $W_{h(e)} = W_{p(h(e),e)}$. Therefore, for any e, $h(e) \in K \Leftrightarrow h(e) \in W_{h(e)} \Leftrightarrow \exists s(W_{e,s} = \{y\})$, where $y = \mu x(x \in I_{h(e)} \cap \bar{A}_s)$.

It is clear that $t \leq_{tt} A$ and the function t is majorized by the computable function $\max\{g(x) \mid x \in I_{h(e)}\}$. We also have $\forall e(W_{t(e)} \neq W_e)$. Indeed, suppose, that for an integer e we have the equalities $W_e = W_{t(e)} = W_{g(m)} = \{m\}$, where m is the least integer from the $h(e)$-th interval $I_{h(e)}$ such that $m \notin A$.

If $h(e) \notin K$ then for some s we have $W_{e,s} = \{m\}$, since $W_e = \{m\}$,and, by definition of h, at a stage $s' > s$ the number $h(e)$ must be enumerated in K. If $h(e) \in K$, then, since for some s we will have $W_{e,s} = \{m\}$, by the construction of A we enumerate from $I_{h(e)}$ exactly this number m. A contradiction.

We constructed a function $f \leq_{tt} A$ such that $\forall e(W_e \neq W_{f(e)})$ and the function f is majorized by the computable function $\max\{g(x) \mid x \in I_{h(e)}\}$. But, nevertheless, the set A is not tt-complete.

2. A Truth-Table Completeness Criterion

Below we show that the conditions of Bulitko describe the weak truth-table ($wtt-$) complete c.e. sets. Probably this claim can be obtained via a careful analysis of Bulitko's original construction. But we give an independent simple proof of this result, which uses the recursion theorem.

Theorem 1. *A c.e. set A is wtt-complete if and only if there is a computable function $f \leq_{tt} A$ such that for any x, $W_x \neq W_{f(x)}$ and f is majorized by a computable function g.*

Proof. ($\Longleftarrow$) For this direction the claim immediately follows from my wtt-completeness criterion. (See, for instance, [Odi89].)

($\Longrightarrow$) Now suppose that the creative set $K = \{x \mid x \in W_x\}$ is wtt-reducible to the set A by a partial-computable functional Φ, i.e. for any x, $K(x) = \Phi^A(x)$ and the *use*-function $\varphi(x)$ (supremum of all questions posed to the oracle during the computation of $\Phi^A(x)$) is a computable function.

Using the recursion theorem we define the following computable function $n(x)$:

$$W_{n(x)} = \begin{cases} \omega, & \text{if } \exists s(\Phi_s^{A_s}(n(x)) \downarrow = 0 \;\& \\ & \qquad\qquad \mid A_s\lceil\varphi(n(x)) \mid \in W_{x,s}), \\ \emptyset, & \text{otherwise.} \end{cases}$$

Here $\mid X \mid$ means the cardinality of the set X.

Let g be such an increasing computable function, such that $W_{g(x)} = \{x\}$ for all x. Define $f(x) = g(\mid \lceil\varphi(n(x)) \mid)$. Therefore, for all x we have: $W_{f(x)} = \{\mid A\lceil\varphi(n(x)) \mid\}$. It is clear, that for any x, $f(x) \leq g(\varphi(n(x)))$ and $f \leq_{tt} A$.

Suppose that there exists an integer x_0 such that $W_{x_0} = W_{f(x_0)}$.

If $W_{n(x_0)} = \emptyset$, then $h(x_0) \notin K$ and, since $K = \Phi^A$, there exists a stage s, such that $\Phi_s^{A_s}(n(x_0)) \downarrow \;\& \; W_{x_0,s} = \{\mid A_s\lceil\varphi(n(x_0)) \mid\}$. But then by the definition of the function $n(x)$ we have $W_{n(x_0)} = \omega$, which is a contradiction.

Therefore, $W_{n(x_0)} \neq \emptyset$. It follows that $W_{n(x_0)} = \omega$ and $h(x_0) \in W_{n(x_0)}$. Therefore $n(x_0) \in K$.

Now by definition of $n(x_0)$ we have the following:
$\exists s(\Phi_s^{A_s}(n(x_0)) \downarrow = 0 \;\& \; W_{x_0,s} = \mid A_s\lceil\varphi(n(x_0)) \mid \in W_{x,s})$.
Since $K(x_0) = \Phi^A(n(x_0)) = 1$, then $\mid A_s\lceil\varphi(n(x_0)) \mid \neq \mid A\lceil\varphi(n(x_0)) \mid$, where $\mid A\lceil\varphi(n(x_0)) \mid$ is the single element of $W_{f(x_0)}$. Therefore, $W_{x_0} \neq W_{f(x_0)}$. This contradiction completes the proof. □

Below we prove a new truth-table completeness criterion for c.e. sets A. It also uses a fixed-point free function, which is in some sense (we will give below the exact definition) "truth-table reducible" to the set A. The truth-table reducibility of the function f to set A we define by the same way, as it was done for m-reducibility of function f to the set A (this definition was given at the beginning of this paper).

We first show that the m-completeness criterion of c.e. sets from [Ars87] holds even if we suppose that functions a and b from the definition of m-reducibility of function f to set A are partial-computable. Namely, we accept the following definition:

Definition 1. *Let $f : \omega \to \omega$ be a total function and $A \subseteq \omega$. We define $f \leq_R A$, if there exist partial-computable functions a and b, and a (total) computable function g such that*

$$f(x) = \begin{cases} a(x), & \textit{if } g(x) \in A, \\ b(x), & \textit{if } g(x) \notin A. \end{cases}$$

(In particular, we have dom $a(x) \supset \{x \mid g(x) \in A\}$, and dom $b(x) \supset \{x \mid g(x) \notin A\}$.)

As before we have the following

Theorem 2. *A c.e. set A is m-complete if and only if there exists a function $f \leq_R A$ such that $\forall x(W_x \neq W_{f(x)})$.*

The proof of this theorem is quite similar to the proof of an analogous theorem from [Ars87] and we do not give it here.

We may also define R-reducibility of a set A to a set B in this way. Let $A \leq_R B$ if there exist partial-computable functions a,b and a computable function g such that

$$A(x) = \begin{cases} a(x), & \text{if } g(x) \in B, \\ b(x), & \text{if } g(x) \notin B. \end{cases}$$

We now have the following

Theorem 3. *If A is a c.e. set then $K \leq_R A \to K \leq_m A$.*
(Here $K = \{x : x \in W_x\}$ is the creative set.)

Proof. We first show that there exists a fixed-point free function f which is m-reducible to the set K. Indeed, let h be a computable productive function for $\omega - K$: for any x, $W_x \subseteq \omega - K \to h(x) \in \omega - K - W_x$. Let

$$f(x) = \begin{cases} a, & \text{if } h(x) \in K, \\ b(x), & \text{if } h(x) \notin K. \end{cases}$$

Here by definition $W_a = \emptyset, W_{b(x)} = \{h(x)\}$ for any x. Then for any x, $W_x \neq W_{f(x)}$. Indeed, if $W_x = W_{f(x)}$ for some x, then suppose that $h(x) \in K$ and $f(x) = a$. Then $W_{f(x)} = \emptyset \subset \omega - K$ and $h(x) \in \omega - K$. A contradiction.

If $h(x) \notin K$ and $f(x) = b(x)$, then $W_x = \{h(x)\} \subset \omega - K$ and $h(x) \in \omega - K - W_x$. Since in this case we have $h(x) \in W_x$, we again obtain a contradiction. Therefore, $\forall x(W_x \neq W_{f(x)})$.

Now suppose that

$$K(x) = \begin{cases} c(x), & \text{if } q(x) \in A, \\ d(x), & \text{if } q(x) \notin A, \end{cases}$$

where c and d are partial-computable functions, and q is a computable function.

We will show that in this case A is creative. Let n be a computable function such that for any x, $\Phi_{\Phi_x(n(x))} \equiv \Phi_{n(x)}$. We define the following computable function g. Let $\{W_{x,s}\}$ and A_s be fixed enumerations of c.e. sets W_x and A accordingly.

Define

$$\Phi_{g(x)}(y) = \begin{cases} b(y), & \text{if } \exists s\{q(h(y)) \in W_{x,s} \& d(h(y)) \downarrow = 0\}, \\ a, & \text{if } \exists s\{q(h(y)) \in W_{x,s} \& d(h(y)) \downarrow = 1\}, \\ a, & \text{if } \exists s\{q(h(y)) \in A_s \& c(h(y)) \downarrow = 1\}, \\ b(y), & \text{if } \exists s\{q(h(y)) \in A_s \& c(h(y)) \downarrow = 0\}, \\ \uparrow, & \text{otherwise.} \end{cases} \tag{1}$$

If in (1) several of first four strings hold simultaneously, then we choose the string with the least such stage s. Here for any x, $W_a = \emptyset$, $W_{b(x)} = \{h(x)\}$, and h is the computable productive function for $\omega - K$.

It is clear, that for any x, $\Phi_{\Phi_{g(x)}(n(g(x)))} \equiv \Phi_{n(g(x))}$, and

$$W_{n(g(x))} = \begin{cases} W_{\Phi_{g(x)}(n(g(x)))}, & \text{if } \Phi_{g(x)}(n(g(x))) \downarrow, \\ \emptyset, & \text{otherwise.} \end{cases}$$

Now suppose that $W_x \subseteq \omega - A$.

If $\Phi_{g(x)}(n(g(x))) \downarrow$, then we have one of the following three possibilities.

a) $\Phi_{g(x)}(n(g(x))) = b(n(g(x))) \& q(h(n(g(x))) \in A \& c(h(n(g(x)))) = 0$. Then $W_{f(n(g(x)))} = W_{b(n(g(x)))} = W_{\Phi_{g(x)}(n(g(x)))} = W_{n(g(x))}$, a contradiction.

b) $\Phi_{g(x)}(n(g(x))) = b(n(g(x))) \& q(h(n(g(x))) \in W_x \subseteq \bar{A} \& d(h(n(g(x)))) = 0$. Then again $W_{f(n(g(x)))} = W_{b(n(g(x)))} = W_{\Phi_{g(x)}(n(g(x)))} = W_{n(g(x))}$, a contradiction.

c) $\Phi_{g(x)}(n(g(x))) = a \& q(h(n(g(x))) \in W_x \subseteq \bar{A} \& d(h(n(g(x)))) = 1$. Then $W_{f(n(g(x)))} = W_a = W_{\Phi_{g(x)}(n(g(x)))} = W_{n(g(x))}$, a contradiction.

d) $\Phi_{g(x)}(n(g(x))) = a \& q(h(n(g(x))) \in A \& c(h(n(g(x)))) = 1$. Then again $W_{f(n(g(x)))} = W_a = W_{\Phi_{g(x)}(n(g(x)))} = W_{n(g(x))}$, a contradiction.

Therefore, $\Phi_{g(x)}(n(g(x))) \uparrow$. But then it follows from (1) that $q(h(n(g(x)))) \notin A \cup W_x$, i. e. $q(h(n(g(x)))) \in ((\omega - A) - W_x)$. We proved that A is creative set with a computable productive function $q(h(n(g(x))))$. Therefore, $K \leq_m A$. □

It follows that for R-reducibility we again have the following:

Theorem 4. *Let A be a c.e. set. Then $K \leq_R A$ if and only if there exists a function $f \leq_R A$ such that $\forall x(W_x \neq W_{f(x)})$.*

Indeed, we proved above part ($\Longrightarrow$) of this claim. Part ($\Longleftarrow$) follows from theorem 2, since $K \leq_m A \rightarrow K \leq_R A$.

In general the implication $A \leq_R B \rightarrow A \leq_m B$ does not hold for all c.e. sets A and B. In this connection it is interesting to note that the part $K \leq_R A \rightarrow K \leq_m A \vee K \leq_m \omega - A$ holds for all sets $A \in \omega$ (i. e. not only for c.e. sets, as shown above). This was proved by I. Kalimullin.

Proposition 5. *(Kalimullin) Let $K \leq_R A$. Then for all sets $A \subseteq \omega$ either $K \leq_m A$, or $K \leq_m \omega - A$.*

Proof. Let

$$K(x) = \begin{cases} a(x), & \text{if } p(x) \notin A, \\ b(x), & \text{if } p(x) \in A, \end{cases}$$

where a and b are partial-computable function, and p is a computable function. Here again $K = \{x \mid x \in W_x\}$.

Define

$$W_{h(n,m,x)} = \begin{cases} \omega, & \text{if } a(x) \downarrow= 0 \& n \in K \vee b(x) \downarrow= 0 \& m \in K \vee \\ & \qquad a(x) \downarrow= 0 \& b(x) \downarrow= 0, \\ \emptyset, & \text{otherwise.} \end{cases}$$

There exists a computable function f such that $W_{f(n,m)} = W_{h(n,m,f(n,m)}$. Define $B = \{\langle n, m \rangle \mid a(f(n,m)) \downarrow= 0\}, C = \{\langle n, m \rangle \mid b(f(n,m)) \downarrow= 0\}$.

$B \cup C = \omega$, since if $\langle n, m \rangle \notin B \cup C$, then $W_{f(n,m)} = \emptyset$, therefore $f(n,m) \notin K$, a contradiction. $B \cap C = \emptyset$, since if $\langle n, m \rangle \in B \cap C$, then $W_{f(n,m)} = \omega$. Therefore $f(n,m) \in K$, and we again obtain a contradiction.

We have following properties of sets B and C:

a) If $\langle n, m \rangle \in B$, then $n \in K \Leftrightarrow p(f(n,m)) \in A$;

b) If $\langle n, m \rangle \in C$, then $m \in K \Leftrightarrow p(f(n,m)) \notin A$.

Below we prove the property $a)$. For $b)$ the proof is similar.

$(\Longrightarrow)$ $n \in K \rightarrow W_{f(n.m)} = \omega \rightarrow f(n,m) \in K \rightarrow p(f(n,m)) \in A$.

$(\Longleftarrow)$ $p(f(n,m)) \in A \rightarrow b(f(n,m)) \downarrow= 1 \rightarrow f(n.m) \in K \rightarrow W_{f(n.m)} \neq \emptyset \rightarrow n \in K$.

There are two possible cases.

Case 1. $\forall n \exists m(\langle n,m\rangle \in B$. Then there is a computable function $m(n)$ such that $\langle n, m(n)\rangle \in B$ for each n. Then $n \in K \Leftrightarrow p(f(n,m(n)) \in A$, i. e. $K \leq_m A$.

Case 2. $\forall m \exists n(\langle n,m\rangle \in C$. Then there is a computable function $n(m)$ such that $\langle n(m), m\rangle \in C$ for each m. Then $m \in K \Leftrightarrow p(f(n(m),m) \notin A$, i. e. $K \leq_m \omega - A$. □

Below we prove the following truth-table completeness criterion for c.e. sets:

Theorem 6. *A c.e. set A is tt-complete if and only if there exist computable functions* a, b *and* g *such that for all* x $W_x \neq W_{f(x}$, *where*

$$f(x) = \begin{cases} a(x), & \textit{if } A \models \sigma_{g(x)}, \\ b(x), & \textit{if } A \not\models \sigma_{g(x)}. \end{cases}$$

Proof. $(\Longrightarrow)$ If a c.e. set A is tt-complete, then define computable functions a, b and g as follows: for any x, $W_x \neq \emptyset \leftrightarrow A \models \sigma_{g(x)}, W_{a(x)} = \emptyset, W_{b(x)} = \omega$. Let

$$f(x) = \begin{cases} a(x), & \text{if } A \models \sigma_{g(x)}, \\ b(x), & \text{if } A \not\models \sigma_{g(x)}. \end{cases}$$

It is clear that $f \leq_{tt} A$, and for any x, $W_x \neq W_{f(x)}$.

$(\Longleftarrow)$ Now suppose that there exists a function f such that $\forall e(W_{f(e)} \neq W_e)$, and for any x, $f(x) = a(x)$, if $A \models \sigma_{g(x)}$, and $f(x) = b(x)$, if $A \not\models \sigma_{g(x)}$, where a, b and g are computable functions.

Repeatedly using the recursion theorem, for each e we define below pairwise disjoint sets of indices $M_e = \{x < i_1 < i_2 < \ldots < i_{2^n}\}$, where $n = \mid \sigma_{g(x)} \mid$. Let $x_1, \ldots, x_n$ and $x_1, \ldots, x_m, m \geq n$ list all variables for $\sigma_{g(x)}$ and $\sigma_{g(x)}, \sigma_{g(i_1)}, \ldots, \sigma_{g(i_{2^n})}$, accordingly. Now for each e we define a tt-condition τ_e with an arbitrary oracle X as follows:

$$\tau_e(X(x_1), X(x_2), \ldots, X(x_m)) =$$

$$[\sigma_{g(x)}(X(x_1), X(x_2), \ldots, X(x_n)) \Longleftrightarrow \sigma_{g(i_k)}(X(x_1), X(x_2), \ldots, X(x_m))],$$

where k is the number of n-tuples $X(x_1), X(x_2), \ldots, X(x_n)$ in a fixed enumeration of all n-tuples, consisting of 0s and 1s.

To meet the condition $e \in K \Leftrightarrow A \models \tau_e$ we sequentially define, using the recursion theorem, c.e. sets $W_{i_1}, W_{i_2}, \dots, W_{i_{2^n}}, W_x$.

While $e \notin K_s$, compute $(b_1, \dots, b_m) = (A_s(x_1), \dots, A_s(x_m))$ and find its number k in the enumeration of all m-tuples, consisting of 0s and 1s. Then check whether $A_s \models \tau_e$. If "not", then do nothing. If "yes", then consider following two cases.

Case 1. $A_s \models \sigma_{g(e)}$. In this case define $W_{i_k} = W_{a(i_k)}$. (Therefore, $f(i_k) = b(i_k) \neq a(i_k)$, and $A \not\models \sigma_{g(i_k)}$.)

Case 2. $A \not\models \sigma_{g(e)}$. In this case define $W_{i_k} = W_{b(i_k)}$. (Therefore, $f(i_k) = a(i_k) \neq b(i_k)$, and $A \models \sigma_{g(i_k)}$.)

Therefore, if $A_s\lceil\{x_1, \dots, x_n\} = A\lceil\{x_1, \dots, x_n\}$, then $A \models (\sigma_{g(x)} \Leftrightarrow \sigma_{g(i_k)})$, hence $A \not\models \tau_e$.

Further, if at a stage $t > s$ we obtain $(b_1, \dots, b_m) =$ $(A_t(x_1), \dots, A_t(x_m)) \neq (A_s(x_1), \dots, A_s(x_m))$, then define $W_{i_{k'}}$ as in the previous case. Here k' is the number of the m-tuple $(A_t(x_1), \dots, A_t(x_m))$, and so on.

At last, if at a stage s' we obtain $e \in K_{s'} - K_{s'-1}$, then we define W_x as above, forcing the following condition to be satisfied:

$$A \models \sigma_{g(x)} \Leftrightarrow A_{s'-1} \not\models \sigma_{g(x)}.$$

In particular, we force $(A_{s'-1}(x_1), \dots, A_{s'-1}(x_m))$ to change.

Further, at any stage $s'' > s'$ such that $A_{s''} \models \sigma_{g(x)}$ and $(A_{s''}(x_1), \dots, A_{s''}(x_m))$ is a new k''-th m-tuple, we define $W_{i_{k''}}$ so that

$$A \models \sigma_{g(i_{k''})} \Leftrightarrow A_{s''} \models \sigma_{g(x)}.$$

Therefore, at the end of the construction we have:

$$e \notin K \rightarrow (A \models \sigma_{g(x)} \Leftrightarrow A \not\models \sigma_{g(i_k)}),$$

$$e \in K \rightarrow (A \models \sigma_{g(x)} \Leftrightarrow A \models \sigma_{g(i_k)}),$$

where $(A(x_1), \dots, A_{s''}(x_m))$ is k-th n-tuple. Therefore, $K \leq_{tt} A$. □

Note that theorem 6 can be reformulated as follows:

Theorem 7. *A c.e. set A is tt-complete if and only if there exist a set $B \leq_{tt} A$ and a function $f \leq_m B$ such that for any x, $W_x \neq W_{f(x)}$.*

Proof. For the proof it is enough to check that theorems 6 and 7 are equivalent.

Indeed, suppose that we have all the conditions of theorem 6. Let $B = \{x \mid A \models \sigma_{g(x)}\}$. It is clear that $B \leq_{tt} A$. We have

$$f(x) = \begin{cases} a(x), & \text{if } x \in B, \\ b(x), & \text{if } x \notin B. \end{cases}$$

Therefore, $f \leq_m B$.

Now suppose that we have all the conditions of the theorem 7, i.e. there exist computable functions a, b, g, h and a c.e. set B such that for all x, $W_x \neq W_{f(x)}$, where

$$f(x) = \begin{cases} a(x), & \text{if } g(x) \in A, \\ b(x), & \text{if } g(x) \notin A. \end{cases}$$

We have also $x \in B \Leftrightarrow A \models \sigma_{h(x)}$.

But then

$$f(x) = \begin{cases} a(x), & \text{if } A \models \sigma_{h(g(x))}, \\ b(x), & \text{if } A \not\models \sigma_{h(g(x))}. \end{cases}$$

Therefore, functions a, b and f satisfy the conditions of the theorem 6. □

Obviously, we can consider the properties of the function f from theorem 6 as a new definition of truth-table reducibility of the function f to the set A. Let us denote this reducibility by tt^*. Therefore, $f \leq_{tt^*} A$, if there exist computable functions a, b and g such that $f(x) = a(x)$, if $A \models \sigma_{g(x)}$, and $f(x) = b(x)$, if $A \not\models \sigma_{g(x)}$. Clearly, $f \leq_{tt^*} A \rightarrow f \leq_{tt} A$. It follows from theorems 1 and 6 that $f \leq_{tt} A \not\rightarrow f \leq_{tt^*} A$. This can also be easily checked directly. It is easy also to check that $f \leq_{tt} A \not\rightarrow f \leq_{wtt} A$. But, obviously, if $f \leq_{tt} A$ and f is majorized by some computable function then $f \leq_{wtt} A$.

In conclusion we note that, in our opinion, for all "Post reducibilities" there are completeness criteria which can be described using a fixed-point free function f which is reducible to A. Here for strong ("tabular") reducibilities the reducibility of function f to the set A is defined by the scheme described in this paper, and for reducibilities which are "close" to Turing reducibility (in V.Bulitko [Bul91] they are so-called "sub-Turing reducibilities") the reducibility is defined in the usual sense.

References

[Ars81] M.M. Arslanov. On some generalizations of fixed-point theorem. *Izvestija VUZov. Matem.* **5**,1981, 9–16.

[Ars87] M.M. Arslanov. *Local theory of degrees of unsolvability and Δ_2^0-sets*. Kazan: Kazan State University, 1987.

[Ars89] M.M. Arslanov. Completeness in arithmetical hierarchy and fixed-points. *Algebra i Logika* **28**,1989, 3–17.

[Bul91] V.K. Bulitko. On ways of characterization of complete sets. *Izvestija AN SSSR, seria matem.* **55**,1991, 227–253.

[Bul93] V.K. Bulitko. A letter to editors. *Izvestija AN SSSR, seria matem.* **57**,1993, 207.

[La75] A.H. Lachlan. wtt-complete sets are not necessarily tt-complete. *Proc. Amer. Math. Soc.* **48**,1975, 429–434.

[Odi89] P. Odifreddi. *Classical Recursion Theory (Volume I)*. North–Holland Publishing Co., Amsterdam.

[Soa87] R. I. Soare. *Recursively Enumerable Sets and Degrees*. Perspectives in Mathematical Logic, Omega Series. Springer–Verlag, Heidelberg, 1987.

COMPLETENESS AND UNIVERSALITY OF ARITHMETICAL NUMBERINGS

Serikzhan Badaev*
Kazakh National University
Almaty, Kazakhstan
badaev@math.kz

Sergey Goncharov†
Institute of Discrete Mathematics and Informatics
Novosibirsk State University
Novosibirsk, Russia
gonchar@math.nsu.ru

Andrea Sorbi‡
University of Siena
Siena, Italy
sorbi@unisi.it

Abstract We investigate completeness and universality notions, relative to different oracles, and the interconnection between these notions, with applications to arithmetical numberings. We prove that principal numberings are complete; completeness is independent of the oracle; the degree of any incomplete numbering is meet-reducible, uniformly complete numberings exist. We completely characterize which finite arithmetical families have a universal numbering.

Keywords: Numberings; arithmetical hierarchy; complete numberings; computable numberings; computable function; oracle; universal numberings.

*Partial funding provided by grant INTAS-RFBR *Computability and Models* no. 97-139 and by grant PICS-541-Kazakhstan.
†Partial funding provided by grant INTAS-RFBR *Computability and Models* no. 97-01-71001; and by grant of "University of Russia".
‡Partial funding provided by grant INTAS-RFBR *Computability and Models* no. 97-139; work done under the auspices of GNSAGA.

1. Basic Notions and Preliminaries

As is known, the s_n^m -theorem and the fixed point property hold for the universal numbering of the family of all partial computable functions, and these properties play an important role in the development of computability theory. The existence of universal numberings for different families of constructive objects as well as the study of the properties of such numberings have attracted the attention of many researchers (see, for instance, the monograph of Yu.L. Ershov [5] and the papers of A. Lachlan [8], A.I. Malt'sev [11] and H. Rogers [20]).

The notion of a complete numbering introduced by A.I. Malt'sev [9] became a very productive notion closely connected with both the fixed point property and the universality of numberings. Our paper is devoted to the study of these fundamental notions of completeness and universality for the class of arithmetical numberings.

For unexplained terminology and notations relative to computability theory, our main references are the textbooks of A.I. Mal'tsev [10], H. Rogers [13] and R. Soare [14]. For the main concepts and notions of the theory of numberings we refer to the book of Yu.L. Ershov [5].

A surjective mapping α of the set $\mathbf{N}$ of natural numbers onto a nonempty set $\mathcal{A}$ is called a *numbering* of $\mathcal{A}$. The collections of all numberings of $\mathcal{A}$ will be denoted by $\mathrm{Num}(\mathcal{A})$. Suppose that $\mathcal{A}$ is a family of objects that admit constructive descriptions. By this we mean that one can define a language $\mathcal{L}$ (henceforth identified with a corresponding set of "well formed formulas") and an interpretation of (fragments of) this language via an onto partial mapping $i\colon\ \mathcal{L} \longrightarrow \mathcal{A}$. For any object $a \in \mathcal{A}$, each formula Φ of $\mathcal{L}$ such that $i(\Phi) = a$ is interpreted as a "description" of a. Suppose further that $G\colon\ \mathbf{N} \longrightarrow \mathcal{L}$ is a Gödel numbering.

Following [12], we propose:

Definition 1.1. A numbering α of $\mathcal{A}$ is called *computable in $\mathcal{L}$ with respect to i* if there exists a computable mapping f such that $\alpha = i \circ G \circ f$.

An example to illustrate the previous definition may be in order at this point. A Turing machine can be viewed as a finite sequence of symbols from a certain alphabet Σ. Thus the set of Turing machines can be identified with a particular "set of formulas" $\mathcal{L} \subseteq \Sigma^*$. Let i be the mapping associating to each $M \in \mathcal{L}$ the partial function $i(M)$ computed by M, and let G be the standard Gödel numbering of the Turing machines. Let $G(x) = M_x$. Then a numbering α of the set of partial computable functions is computable in $\mathcal{L}$ with respect to i according to Definition 1.1 if for some computable function f, one has $\alpha(x) = i(M_{f(x)})$. This coincides with the familiar notion of a computable numbering of the family of partial computable functions, as studied in the classical theory of numberings, see e.g. [5].

It is immediate to see that Definition 1.1 does not depend on the choice of the Gödel numbering G. Hence, via identification of $\mathcal{L}$ with $\mathbf{N}$ through some fixed Gödel numbering, the above definition states that α is computable if there is some computable function f from $\mathbf{N}$ to $\mathcal{L}$ such that $\alpha = i \circ f$.

Definition 1.1 has a wide scope of applications, based upon suitable choices of $\mathcal{L}$ and i. Throughout this paper we will confine ourselves to families of arithmetical sets and relations. As language $\mathcal{L}$ we take in this case the collection of arithmetical first–order formulas, and i will be a mapping associating to each formula the corresponding set or relation defined by that formula in the standard model $\mathfrak{N}$ of Peano arithmetic. Restriction to this particular case leads to the following definition, see [12]:

Definition 1.2. A numbering α of a family $\mathcal{A}$ of Σ^0_{n+1}– sets, with $n \geq 0$, is called Σ^0_{n+1}– *computable* if there exists a computable function f from $\mathbf{N}$ into the Σ_{n+1}– formulas of Peano arithmetic such that, for every m,

$$\alpha(m) = \{x \in \mathbf{N} \mid \mathfrak{N} \models f(m)(\overline{x})\}$$

(where the symbol $\overline{x}$ stands for the numeral for x). The set of Σ^0_{n+1}– computable numberings of $\mathcal{A}$ will be denoted by $\mathrm{Com}^0_{n+1}(\mathcal{A})$.

In other words, a Σ^0_{n+1}– computable numbering is just a computable numbering in $\mathcal{L}$ with respect to i – in the sense of Definition 1.1, where $\mathcal{L}$ and i are as above specified – of a family of Σ^0_{n+1}– sets.

Computable numberings of families of sets which are first–order definable in the standard model of Peano arithmetic are called *arithmetical numberings*. A family $\mathcal{A}$ for which $\mathrm{Com}^0_{n+1}(\mathcal{A}) \neq \emptyset$ will be called Σ^0_{n+1}– *computable*.

We now revise some of the basic definitions of the theory of numberings. In order to avoid trivial cases, we will always deal in the following with families $\mathcal{A}$ that contain at least two elements, i.e. $|\mathcal{A}| > 1$.

We will use the symbols $\langle\cdot,\cdot\rangle$ to denote the usual computable pairing function between $\mathbf{N}^2$ and $\mathbf{N}$, with computable projections π_0, π_1 (hence $\pi_i(\langle x_0, x_1\rangle) = x_i$ and $\langle \pi_0(x), \pi_1(x)\rangle = x$). Likewise, the symbol $\langle\cdot,\cdot,\cdot\rangle$ will denote the usual computable bijection of $\mathbf{N}^3$ onto $\mathbf{N}$, with computable projections $\pi^3_0, \pi^3_1, \pi^3_2$, etc.

Definition 1.3. Let $\alpha \in \mathrm{Num}(\mathcal{A})$, $\beta \in \mathrm{Num}(\mathcal{B})$, where $\mathcal{A}$ and $\mathcal{B}$ are any given families. We say that α is *reducible* to β (in symbols, $\alpha \leqslant \beta$) if $\alpha = \beta \circ f$ for some computable function f. If $\alpha \leqslant \beta$ and $\beta \leqslant \alpha$ then $\mathcal{A} = \mathcal{B}$ and we say that α and β are *equivalent* (in symbols, $\alpha \equiv \beta$)numberings of $\mathcal{A}$. The equivalence class of a numbering α is called the *degree* of α, denoted by $\deg(\alpha)$. The set of all degrees of the elements of $\mathrm{Num}(\mathcal{A})$ will be denoted by $L(\mathcal{A})$.

It is straightforward to see that this notion of reducibility on numberings of $\mathcal{A}$ is a pre-ordering relation on $\mathrm{Num}(\mathcal{A})$, and induces a partial ordering relation (still denoted by the symbol $\leqslant$) on $L(\mathcal{A})$, where $\deg(\alpha) \leqslant \deg(\beta)$ if $\alpha \leqslant \beta$. Further, $L(\mathcal{A})$ is an upper semilattice: If $\alpha, \beta \in \mathrm{Num}(\mathcal{A})$ then the least upper bound of $\deg(\alpha)$ and $\deg(\beta)$ is easily seen to be $\deg(\alpha \oplus \beta)$ where

$$\alpha \oplus \beta(x) \leftrightharpoons \begin{cases} \alpha(y) & \text{if } x = 2y, \\ \beta(y) & \text{if } x = 2y+1. \end{cases}$$

The numbering $\alpha \oplus \beta$ is called the *join of* α and β.

We are interested in the substructure of $L(\mathcal{A})$ obtained by restricting the universe to the degrees of elements of $\mathrm{Com}^0_{n+1}(\mathcal{A})$, if $\mathcal{A}$ is Σ^0_{n+1}–computable.

Definition 1.4. Let $\mathcal{R}^0_{n+1}(\mathcal{A}) \leftrightharpoons \langle \mathrm{Com}^0_{n+1}(\mathcal{A})/_{\equiv}, \leqslant \rangle$.

It is easy to see that $\mathcal{R}^0_{n+1}(\mathcal{A})$ is an upper semilattice with the least upper bound operation still given by $\deg(\alpha) \vee \deg(\beta) = \deg(\alpha \oplus \beta)$. In fact $\mathcal{R}^0_{n+1}(\mathcal{A})$ is an ideal of $L(\mathcal{A})$. $\mathcal{R}^0_{n+1}(\mathcal{A})$ is called the *Rogers semilattice* of the family $\mathcal{A}$.

1.1 Relativized reducibility

The above defined reducibility provides a way of "translating" numberings into other numberings. Different kinds of "translations" can be devised, though. For instance, when dealing with computable numberings of c.e. sets, it is often useful to compare numberings with respect to the reducibility provided by $\mathbf{0}'$–computable functions, i.e. α is reducible to β if there exists a $\mathbf{0}'$–computable function f such that $\alpha = \beta \circ f$. We now provide a general setting for some extended notions of reducibility between numberings.

Given a set X, let the corresponding boldface letter $\mathbf{X}$ denote the Turing degree of X, i.e. $\mathbf{X} = \deg_T(X)$. We will follow the usual notation $\mathbf{0}^{(n)} = \deg_T(\emptyset^{(n)})$, and we let $\emptyset^{(0)} = \emptyset$ and $\mathbf{0}^{(0)} = \mathbf{0}$. Since a partial function is X–computable if and only if is Y–computable for *all* $Y \in \mathbf{X}$, for a given Turing degree $\mathbf{X}$ we will say that a partial function φ is said to be $\mathbf{X}$–*computable*, if φ is X–computable for some $X \in \mathbf{X}$. Consequently, for every set X we will often write $\varphi_e^{\mathbf{X}}$ to denote φ_e^X, in the standard listing of the partial functions which are computable with oracle X.

Well known notions from classical computability theory are relativized in a obvious way. For instance, an $\mathbf{X}$–*maximal* set is a set M which is c.e. in X, and for every set $V \supseteq M$ which is c.e. in X we have that either $V \setminus M$ is finite or $\mathbf{N} \setminus V$ is finite. In these cases, one gets the classical notion by taking $X = \emptyset$.

Definition 1.5. Let $\alpha \in \mathrm{Num}(\mathcal{A})$, $\beta \in \mathrm{Num}(\mathcal{A})$, and let $X \subseteq \mathbf{N}$. We say that α is $\mathbf{X}$–*reducible* to β (in symbols, $\alpha \leqslant_{\mathbf{X}} \beta$) if there exists an $\mathbf{X}$–computable function f such that $\alpha = \beta \circ f$. We say that the numberings α and β of $\mathcal{A}$ are $\mathbf{X}$–*equivalent* (in symbols, $\alpha \equiv_{\mathbf{X}} \beta$) if $\alpha \leqslant_{\mathbf{X}} \beta$ and $\beta \leqslant_{\mathbf{X}} \alpha$. The set $\mathrm{Num}(\mathcal{A})$ of numberings of $\mathcal{A}$ is partitioned by $\equiv_{\mathbf{X}}$ into equivalence classes.

The equivalence class of a numbering $\alpha \in \mathrm{Num}(\mathcal{A})$ under $\equiv_{\mathbf{X}}$ will be denoted by $\deg_{\mathbf{X}}(\alpha)$. The set $L^{\mathbf{X}}(\mathcal{A}) \leftrightharpoons \mathrm{Num}(\mathcal{A})/_{\equiv_{\mathbf{X}}}$ is a partially ordered set, in fact an upper semilattice, with the least upper bound operation still induced by the join of numberings. Its partial ordering relation will still be denoted by the symbol $\leqslant_{\mathbf{X}}$.

Restriction to $\mathrm{Com}^0_{n+1}(\mathcal{A})$ gives:

Definition 1.6. Let $\mathcal{A}$ be a Σ^0_{n+1}–computable family, and let X be any set. Define

$$\mathcal{R}^{0,\mathbf{X}}_{n+1}(\mathcal{A}) \leftrightharpoons \langle \mathrm{Com}^0_{n+1}(\mathcal{A})/\equiv_{\mathbf{X}}, \leqslant_{\mathbf{X}} \rangle.$$

The following definitions provide suitable relativizations of two important notions of the theory of numberings.

Definition 1.7. We say that a numbering α of a family $\mathcal{A}$ is $\mathbf{X}$–*universal* in $\mathrm{Com}^0_{n+1}(\mathcal{A})$ if $\alpha \in \mathrm{Com}^0_{n+1}(\mathcal{A})$ and $\beta \leqslant_{\mathbf{X}} \alpha$ for all numberings $\beta \in \mathrm{Com}^0_{n+1}(\mathcal{A})$.

Remark 1.1. The notion of a *principal numbering of* a family $\mathcal{A}$, [5], coincides with the notion of a numbering which is $\mathbf{0}$– universal in $\mathrm{Com}^0_1(\mathcal{A})$.

Definition 1.8. A numbering $\alpha \in \mathrm{Com}^0_{n+1}(\mathcal{A})$ is said to be $\mathbf{X}$–*minimal* if there exists no $\beta \in \mathrm{Com}^0_{n+1}(\mathcal{A})$ such that $\beta <_{\mathbf{X}} \alpha$. In other words, $\deg_{\mathbf{X}}(\alpha)$ is a minimal element in the poset $\mathcal{R}^{0,\mathbf{X}}_{n+1}(\mathcal{A})$.

We shall usually omit to mention X and $\mathbf{X}$ in our notations if X is a computable set.

For a Σ^0_{n+1}–computable family $\mathcal{A}$, the following are immediate:

- $\mathcal{R}^{0,\mathbf{X}}_{n+1}(\mathcal{A})$ is an upper semilattice, called the $\mathbf{X}$–*Rogers semilattice of* $\mathcal{A}$. Similarly to what happens for $\mathcal{R}^0_{n+1}(\mathcal{A})$, the least upper bound of $\deg_{\mathbf{X}}(\alpha)$ and $\deg_{\mathbf{X}}(\beta)$ is $\deg_{\mathbf{X}}(\alpha \oplus \beta)$.

- If $Y \leq_T X$, then every degree of $\mathcal{R}^{0,\mathbf{X}}_{n+1}(\mathcal{A})$ is union of degrees from $\mathcal{R}^{0,\mathbf{Y}}_{n+1}(\mathcal{A})$:

$$\deg_{\mathbf{X}}(\alpha) = \bigcup \{\deg_{\mathbf{Y}}(\beta) : \beta \in \deg_{\mathbf{X}}(\alpha)\}.$$

- If $Y \leq_T X$ and $|\mathcal{R}^{0,\mathbf{Y}}_{n+1}(\mathcal{A})| = 1$ then $|\mathcal{R}^{0,\mathbf{X}}_{n+1}(\mathcal{A})| = 1$.

- $|\mathcal{R}^{0,\mathbf{0}^{(n+2)}}_{n+1}(\mathcal{A})| = 1$, as the relation "$\alpha(m) = \alpha(n)$" is $\mathbf{0}^{(n+2)}$– decidable for $\alpha \in \mathrm{Com}^0_{n+1}(\mathcal{A})$.

- If α is a numbering which is $\mathbf{0}^{(m)}$– universal in $\mathrm{Com}^0_{n+1}(\mathcal{A})$, then it is $\mathbf{0}^{(k)}$– universal in $\mathrm{Com}^0_{n+1}(\mathcal{A})$ for every $k \geq m$.

1.2 Some results on Σ^0_{n+1}–computable numberings

Although straightforward, the following theorem and its immediate corollary give a useful characterization of Σ^0_{n+1}–computable numberings. The theorem is stated in the more general case of numberings of arithmetical relations rather than arithmetical sets.

Theorem 1.1 (S.Goncharov and A.Sorbi, [12]). *Let $\mathcal{A}$ be a family of Σ^0_{n+1}–subsets of $\mathbf{N}^k$, $k \geq 1$. Then a numbering α: $\mathbf{N} \longrightarrow \mathcal{A}$ is Σ^0_{n+1}–computable if and only if the set $\{(\overrightarrow{x}, m) \mid \overrightarrow{x} \in \alpha(m)\}$ is $\mathbf{0}^{(n)}$–computably enumerable.*

Applied to families of subsets of $\mathbf{N}$, the theorem gives:

Corollary 1.1.1 ([12]). *A numbering α of a family $\mathcal{A}$ of Σ^0_{n+1}–subsets of $\mathbf{N}$, is Σ^0_{n+1}–computable if and only if the relation "$x \in \alpha(y)$" is Σ^0_{n+1}.*

For the case $n = 0$, this is exactly the classical notion of computable numbering of a family of c.e. sets.

The following result is an extension to the Σ^0_{n+2}–case of a useful criterion introduced for Σ^0_2–computable numberings by Goncharov and Sorbi, [12].

Lemma 1.1. *A numbering α of a family $\mathcal{A}$ of Σ^0_{n+2}–sets is Σ^0_{n+2}–computable if and only if there exists a $\mathbf{0}^{(n)}$–computable function $f(m, s)$ such that, letting $B_{m,s} = D_{f(m,s)}$ for every $m, s \in \mathbf{N}$ (where D_u denotes the finite set with canonical index u), the following holds, for every m:*

$$\alpha(m) = \underline{\lim}_{\,s} B_{m,s}.$$

(Where, given a family $\{X_s\}_{s\in\mathbf{N}}$ of subsets of $\mathbf{N}$, we let

$$\underline{\lim}_{\,s} X_s \leftrightharpoons \{x \mid \exists t \forall s \geq t (x \in X_s)\}).$$

Proof. Let $\alpha \in \mathrm{Com}^0_{n+2}(\mathcal{A})$. Since the set $\{\langle x, m\rangle \mid x \in \alpha(m)\}$ is Σ^0_{n+2}, let $\{B^s \mid s \in \omega\}$ be a $\mathbf{0}^{(n)}$–computable sequence of finite sets such that

$$x \in \alpha(m) \Leftrightarrow (\exists t)(\forall s > t)(\langle x, m\rangle \in B^s).$$

Then take

$$B_{m,s} \leftrightharpoons \{x \mid \langle x, m\rangle \in B^s\}.$$

The converse is trivial. □

Theorem 1.2. *Let $\mathcal{A}$ be a Σ^0_{n+1}–computable family, $n \geq 0$, and let α be a numbering of $\mathcal{A}$. Then the following statements are equivalent:*

(i) α is Σ^0_{n+1}–computable;

(ii) α is reducible to some Σ^0_{n+1}–computable numbering of $\mathcal{A}$;

(iii) α is $\mathbf{0}^{(n)}$– reducible to some Σ^0_{n+1}–computable numbering of $\mathcal{A}$;

(iv) α is reducible to a numbering universal in $\mathrm{Com}^0_{n+1}(\Sigma^0_{n+1})$;

(v) α is $\mathbf{0}^{(n)}$– reducible to a numbering universal in $\mathrm{Com}^0_{n+1}(\Sigma^0_{n+1})$.

Proof. It is easy to see that

$$\text{(i)} \Rightarrow \text{(ii)} \Rightarrow \text{(iii)} \Rightarrow \text{(i)} :$$

To show that (iii) $\Rightarrow$ (i) just notice that if $\alpha = \beta \circ f$, where β is a Σ^0_{n+1}–computable numbering and f is some $\mathbf{0}^{(n)}$– computable function, then

$$x \in \alpha(y) \Leftrightarrow \exists z(f(y) = z \;\&\; x \in \beta(z)),$$

which gives that the relation "$x \in \alpha(y)$" is Σ^0_{n+1}, hence, by Corollary 1.1.1, α is Σ^0_{n+1}– computable.

Let us now show that

$$\text{(i)} \Rightarrow \text{(iv)} \Rightarrow \text{(v)}.$$

To show that (i) $\Rightarrow$ (iv), it is enough to show that there exists a numbering β universal in $\mathrm{Com}^0_{n+1}(\Sigma^0_{n+1})$. By the relativized Universal Function Theorem, let $K(x, y)$ be some partial $\mathbf{0}^{(n)}$– computable function that is universal for the unary partial $\mathbf{0}^{(n)}$– computable functions. Then the numbering

$$\beta(x) \leftrightharpoons \{y \mid K(x, y) \downarrow\}$$

is a Σ^0_{n+1}– computable numbering of the family of all Σ^0_{n+1}– sets. If now ν is any Σ^0_{n+1}– computable numbering, then by Post's Theorem, the relation "$x \in \alpha(y)$" is the domain of some partial $\mathbf{0}^{(n)}$– computable function $\varphi(x, y)$. By the relativized s^m_n– Theorem there exists a computable function φ such that $\varphi^{\emptyset^{(n)}}_{f(x)}(y) = \varphi(x, y)$. Hence $\nu = \beta \circ f$. Since ν is arbitrary, it follows that the numbering β is universal in $\mathrm{Com}^0_{n+1}(\Sigma^0_{n+1})$.

The rest of the proof is now trivial since, exactly as in the proof of (iii) $\Rightarrow$ (i), one can easily show that (v) $\Rightarrow$ (i) . □

Badaev and Goncharov, [3], have solved the problem of the cardinality of the set of minimal elements in $\mathcal{R}^0_{n+2}(\mathcal{A})$, for any infinite family $\mathcal{A}$. The following theorem will have several applications throughout the paper.

Theorem 1.3 (S.Badaev and S.Goncharov, [3]). *For every n, if $\mathcal{A}$ is an infinite Σ^0_{n+2}–computable family, then $\mathcal{R}^0_{n+2}(\mathcal{A})$ has infinitely many minimal elements.*

Proof. Let $\mathcal{A}$ be an infinite Σ^0_{n+2}– family, and let $\alpha \in \mathrm{Com}^0_{n+2}(\mathcal{A})$. Take any maximal set M. Assume that

$$\overline{M} = \{m_0 < m_1 < \cdots\}.$$

For every $A \in \mathcal{A}$ define

$$\alpha_{M,A}(x) \leftrightharpoons \begin{cases} \alpha(i) & \text{if } x = m_i, \\ A & \text{if } x \in M. \end{cases}$$

It is possible to show:

- $\alpha_{M,A} \in \mathrm{Com}^0_{n+2}(\mathcal{A})$;
- $\alpha_{M,A}$ is minimal. To see this, assume that $\gamma = \alpha_{M,A} \circ f$, for some computable function f. Let $R \leftrightharpoons \mathrm{range}(f)$, and let g be a computable function such that $M = \mathrm{range}(g)$. Since $\mathcal{A}$ is infinite, $R \cap \overline{M}$ is infinite, and thus by maximality of M it follows that $\overline{R} \cap \overline{M}$ is finite. Finally, let k be such that $\gamma(k) = A$. Towards showing that $\alpha_{M,A} \leqslant \gamma$ define, for every $x \in R \cup M$,

$$h(x) \leftrightharpoons \begin{cases} k & \text{if } \exists y(g(y) = x \;\&\; \forall z < y(f(z) \neq x)), \\ \mu y\,(f(y) = x) & \text{otherwise.} \end{cases}$$

  Then, for every $x \in R \cup M$ we have $\alpha_{M,A}(x) = \gamma(h(x))$. Since $\overline{R \cup M}$ is finite, one can suitably extend h to a computable function reducing $\alpha_{M,A}$ to γ, as desired.
- If $B \in \mathcal{A}$ and $A \neq B$ then $\alpha_{M,A} \not\leqslant \alpha_{M,B}$ (thus there are infinitely many minimal elements in $\mathcal{R}^0_{n+2}(\mathcal{A})$, as $\mathcal{A}$ is infinite). Suppose by contradiction that $\alpha_{M,A} = \alpha_{M,B} \circ h$, via some computable function h. Since $\alpha_{M,B}$ maps every element of M to B, we have that $h[\alpha^{-1}_{M,A}(A)] \subseteq \overline{M}$. On the other hand, if $x \in h[\alpha^{-1}_{M,A}(A)]$, then $\alpha_{M,B}(x) = A$, therefore $\overline{M} \setminus h[\alpha^{-1}_{M,A}(A)]$ is infinite. But $M \subseteq \alpha^{-1}_{M,A}(A)$, hence $h[M] \subseteq \overline{M}$ and thus $h[M]$ is finite by maximality of M. Finally if $y \in h^{-1}[h[M]]$ then $\alpha_{M,B}(h(y)) = A$, so $\overline{M} \setminus h^{-1}[h[M]]$ is infinite, and thus as before $h^{-1}[h[M]] \setminus M$ is finite. As $M \subseteq h^{-1}[h[M]]$ and $h[M]$ is finite, this would imply that M is computable, a contradiction.
- $\alpha_{M,A} \equiv_{\mathbf{0}'} \alpha$, for every A.

□

Remark 1.2. Inspection of the proof of the previous theorem shows that if we define $\alpha_{M,A}$ starting from a $\mathbf{0}^{(i)}$– maximal set M, with $i \leq n$, then for every A the numbering $\alpha_{M,A}$ is $\mathbf{0}^{(i)}$– minimal; if $A \neq B$ then $\alpha_{M,A} \not\leqslant_{\mathbf{0}^{(i)}} \alpha_{M,B}$; and finally $\alpha_{M,A} \equiv_{\mathbf{0}^{(i+1)}} \alpha$.

Remark 1.3. In the case of a finite family $\mathcal{A}$, the construction of the previous theorem gives us *atoms* in the Rogers semilattice $\mathcal{R}^0_{n+2}(\mathcal{A})$.

More precisely, if $\mathcal{A} \subseteq \Sigma^0_{n+2}$ is a finite family, and some element $B \in \mathcal{A}$ different from A has infinitely many α– indices then the numbering $\alpha_{M,A}$ is not minimal and for every numbering $\beta \in \mathrm{Com}^0_{n+2}(\mathcal{A})$, the fact that $\beta \leqslant \alpha_{M,A}$ implies that either β is reducible to any numbering of $\mathcal{A}$ or $\beta \equiv \alpha_{M,A}$.

2. Complete Numberings and Completions

In this section we develop the theory of complete numberings in the context of arithmetical numberings. Both the classical notion of reducibility as well as some relativized versions are taken into account. We also introduce the notion of a uniformly complete numbering.

2.1 Complete numberings: The basic facts

Complete numberings were introduced by A.I. Mal'tsev, [10]. For a thorough investigation of complete numberings, see [5].

Definition 2.1. A numbering α of an abstract nonempty family $\mathcal{A}$ is called *complete with respect to* $a \in \mathcal{A}$, if for every partial computable function φ there exists a total computable function f such that, for every x,

$$\alpha(f(x)) = \begin{cases} \alpha(\varphi(x)) & \text{if} \quad \varphi(x)\downarrow, \\ a & \text{otherwise.} \end{cases}$$

We say in this case that f *α– extends φ with respect to a.* The element a is said to be a *special object for α.*

The following definition is a bit less demanding, [5]:

Definition 2.2. A numbering α is *precomplete* if for every partial computable function φ there is a computable function f such that, for every x, if $\varphi(x)\downarrow$ then $\alpha(\varphi(x)) = \alpha(f(x))$. We say that f *α– extends φ.*

Remark 2.1. The previous definitions are given in terms of partial computable functions of one variable. Of course they could equivalently be given in terms of functions of several variables. If for instance α is precomplete and $\lambda x, y\, \varphi(x, y)$ is a partial computable function of two variables then there exists a computable function $\lambda x, y\, f(x, y)$ such that $\alpha(\varphi(x, y)) = \alpha(f(x, y))$, for all x, y such that $\varphi(x, y)$ is defined. To see this, given φ, just consider a computable function g that α– extends the partial computable function $\lambda u\, \varphi(\pi_0(u), \pi_1(u))$. Then one can take $f(x, y) \leftrightharpoons g(\langle x, y\rangle)$.

It is immediate to see that every complete numbering is precomplete. We have:

Theorem 2.1 (Yu.L.Ershov, [5]). *Let α be a numbering of a family $\mathcal{A}$. Then the following statements are equivalent:*

(1) α *is precomplete;*

(2) there exists a computable function h *such that for every e,* $\varphi_{h(e)}$ *is total and for all* x*,*

$$\varphi_e(x)\downarrow\Rightarrow \alpha(\varphi_{h(e)}(x)) = \alpha(\varphi_e(x));$$

(3) (The Uniform Fixed Point Theorem) there exists computable function g *such that for every e,*

$$\varphi_e(g(e))\downarrow\Rightarrow \alpha(g(e)) = \alpha(\varphi_e(g(e))).$$

Proof. (1) $\Rightarrow$ (2). Let α be precomplete. To show (2), we can argue as follows. By the Universal Function Theorem, let U be a unary universal partial computable function, for instance let $U(\langle e, x\rangle) = \varphi_e(x)$. Let f be a computable function such that $\alpha(U(z)) = \alpha(f(z))$, for every $z \in \mathrm{dom}(U)$. By the s^m_n–Theorem, let h be a computable function such that $\varphi_{h(e)}(x) = f(\langle e, x\rangle)$. It is easy to see that h is the desired function.

(2) $\Rightarrow$ (3). To show the Uniform Fixed Point property (3), by the s^m_n–Theorem let f be a computable function such that $\varphi_{f(e)}(x) = \varphi_e(\varphi_x(x))$ for all e, x, and by (2) let h be a computable function such that for every e $\varphi_{h(e)}$ is total and for every x,

$$\varphi_{f(e)}(x)\downarrow\Rightarrow \alpha(\varphi_{h(e)}(x)) = \alpha(\varphi_{f(e)}(x)).$$

Define $g(e) = \varphi_{h(e)}(h(e))$ for all e. It is easy to see that if $\varphi_e(g(e))\downarrow$ then $\alpha(g(e)) = \alpha(\varphi_e(g(e)))$.

(3) $\Rightarrow$ (1). Let g be a computable function such that for every e,

$$\varphi_e(g(e))\downarrow\Rightarrow \alpha(g(e)) = \alpha(\varphi_e(g(e))).$$

For a given partial computable function φ, define the function $\psi(x, y)$,

$$\psi(x,y) = \begin{cases} \varphi(x) & \text{if } \varphi(x)\downarrow, \\ \uparrow & \text{otherwise.} \end{cases}$$

By the s^m_n–Theorem there exists a computable function f such that $\varphi_{f(x)} = \lambda y\, \psi(x, y)$, for all x. It is easy to verify that the computable function $g \circ f$ α–extends the partial function φ.

□

Remark 2.2. In [9], A.I. Mal'tsev considered numberings with the uniform fixed point property (3) and called them "complete" numberings. Later in [10], he defined "complete" numbering just as in Definition 2.1 and renamed "precomplete" numberings the numberings with the uniform fixed point property.

The definition of "precomplete numberings" as in Definition 2.2 was introduced by Yu.L. Ershov in [5].

Theorem 2.2 (A.I.Mal'tsev, [10]). *Let α, β be numberings of a family $\mathcal{A}$. If $\alpha \equiv \beta$, and α is complete with respect to a special object a (precomplete), then β is complete as well with respect to a (precomplete, respectively). In fact, if α and β are precomplete and $\alpha \equiv \beta$ then α and β are computably isomorphic, i.e. there exists a computable permutation p of* **N** *such that $\beta = \alpha \circ p$.*

Proof. This is a well known fact of the theory of numberings, see e.g. [5]. We give however a sketch of the proof. We first show that if $\alpha \equiv \beta$, and α is complete with respect to the special object a, then β is complete with respect to a. (The same argument will show that if α is precomplete so is β.) If $\alpha = \beta \circ h$, $\beta = \alpha \circ k$, where h, k are computable functions, and α is complete with respect to a, then one can argue that β is complete with respect to a as follows. If φ is partial computable, then let f α–extend $k \circ \varphi$ with respect to a. It easily follows that $h \circ f$ β–extends φ with respect to A.

Now, if $\alpha = \beta \circ h$ and $\beta = \alpha \circ k$ and α and β are precomplete, then by a standard back–and–forth argument one can produce a computable permutation p such that $\alpha = \beta \circ p$. For this, we can define by stages a computable approximation $\{p^s\}_{s \in \mathbf{N}}$ to p, so that, for every s, p^s is a partial function with finite domain, and $p^s \subseteq p^{s+1}$. We start up with defining $p^0 \risingdotseq \emptyset$. At stage $s = 2t + 1$ we arrange, if $t \notin \mathrm{dom}(p^{s-1})$, that $t \in \mathrm{dom}(p^s)$, by defining $p^s \risingdotseq p^{s-1} \cup \{(t, y)\}$, for a suitable number y. At stage $s = 2t + 2$ we arrange, if not already achieved at any of the previous stages, that $t \in \mathrm{range}(p^s)$, by defining $p^s \risingdotseq p^{s-1} \cup \{(y, t)\}$, for suitable number y. In order to choose a correct number y at each stage so as to achieve that p is 1–1 and $\alpha = \beta \circ p$, one can use the fact that if ν is precomplete, then from any finite set $F = \{n_1, \ldots, n_k\}$ of numbers such that $\nu(n_1) = \cdots = \nu(n_k)$ we can uniformly find $n \notin F$ such that $\nu(n) = \nu(n_1)$. Indeed, if f is a computable function such that $\varphi_{f(e)}$ is total and ν–extends φ_e (such a function exists by (2) of Theorem 2.1), then by the Recursion Theorem let e be such that

$$\varphi_e(x) = \begin{cases} n_1 & \text{if } \varphi_{f(e)}(0) \notin F, \\ \max F + 1 & \text{otherwise.} \end{cases}$$

Then the number

$$n \risingdotseq \begin{cases} \varphi_{f(e)}(0) & \text{if } \varphi_{f(e)}(0) \notin F, \\ \max F + 1 & \text{otherwise.} \end{cases}$$

is the desired number. Using this, given any m one can uniformly enumerate an infinite set X of numbers such that $\nu(m) = \nu(n)$ for every $n \in X$. □

Finally, the degrees of complete numberings satisfy a distinguished structural property in both the semilattices $L(\mathcal{A})$ and $\mathcal{R}^0_{n+1}(\mathcal{A})$. We first need the following definition:

Definition 2.3. A numbering $\alpha \in \mathrm{Num}(\mathcal{A})$ is called *splittable* if it is equivalent to the join of some pair of incomparable numberings of subfamilies of $\mathcal{A}$. Likewise α is called $\mathbf{X}$– *splittable* if it is $\mathbf{X}$– equivalent to the join of some pair of numberings of subfamilies of $\mathcal{A}$ that are incomparable with respect to $\leqslant_{\mathbf{X}}$.

We recall that an element of an upper semilattice is called *splittable* if it is the least upper bound of two incomparable elements of that semilattice. Note that if $\alpha \in \mathrm{Num}(\mathcal{A})$ is not splittable then the degree $\deg(\alpha)$ is not a splittable element in any upper semilattice of numberings of $\mathcal{A}$, but the converse is not necessarily true.

Theorem 2.3 (Yu.L.Ershov, [5]). *Precomplete numberings are not splittable.*

Proof. Suppose that α is precomplete and splittable. By Theorem 2.2 we may assume that $\alpha = \alpha_0 \oplus \alpha_1$. Let φ and ψ be partial computable functions such that

(1) if $\varphi_e(x)$ is even, then $\varphi(e,x) = x$;

(2) if $\varphi_e(x)$ is odd then $\psi(e,u) = x$ and $\varphi(e,x) = u$ for some u, and, for every e, $\lambda u\ \psi(e,u)$ is total if there exist infinitely many numbers x such that $\varphi_e(x)$ is odd.

(To achieve this, one can use a 1–1 enumeration of the graph of the universal function $K(e,x)$. Whenever a triple $\langle e,x,y\rangle$ appears in this enumeration such that y is even, then we enumerate $\langle e,x,x\rangle$ in the graph of φ; whenever a triple $\langle e,x,y\rangle$ appears such that y is odd, then we enumerate $\langle e,u,x\rangle$ in the graph of ψ and $\langle e,x,u\rangle$ in the graph of φ, where u is the least number such that $\psi(e,u)\uparrow$.) By precompleteness of α let $f(e,x)$ be a computable function that α– extends φ, and by the Recursion Theorem, let e be such that $\varphi_e(x) = f(e,x)$, so that φ_e is total. Finally let

$$A = \{x \mid \varphi_e(x) \text{ odd}\}.$$

If A is finite, then for almost all x we have $x = \varphi(e,x)$, thus

$$\begin{aligned}\alpha(x) &= \alpha(\varphi(e,x)) = \alpha(f(e,x)) \\ &= \alpha(\varphi_e(x)) \\ &= \alpha_0\left(\frac{\varphi_e(x)}{2}\right)\end{aligned}$$

from which we get that $\alpha \leqslant \alpha_0$.

If, on the contrary, A is infinite, then for every u, $\psi(e,u)$ is defined, $u = \varphi(e,\psi(e,u))$ and $\varphi_e(\psi(e,u)) = \varphi_e(x)$, for some x such that $\varphi_e(x)$ is odd. Hence

$$\begin{aligned}\alpha(u) &= \alpha(\varphi(e,\psi(e,u)))\\ &= \alpha(f(e,\psi(e,u)))\\ &= \alpha(\varphi_e(\psi(e,u))).\end{aligned}$$

But for some x, $\psi(e,u) = x$ and $\varphi_e(x)$ is odd; then

$$\alpha(u) = \alpha_1(\frac{\varphi_e(\psi(e,u)) - 1}{2})$$

which implies that $\alpha \leqslant \alpha_1$. □

For more on precomplete numberings see [5].

2.2 The completion operator and its properties

In the theory of numberings there is a well known and powerful construction, due to Ershov [5], which allows, given any numbering α of a family $\mathcal{A}$, to find a complete numbering of $\mathcal{A}$ with respect to any special object a. This construction is defined as follows.

Definition 2.4. Let $U(x)$ be a unary universal partial computable function, for instance let $U(\langle e,x\rangle) = \varphi_e(x)$. Given numbering α of a family $\mathcal{A}$ and $a \in \mathcal{A}$, define

$$\alpha_a^{\mathbf{0}}(x) \leftrightharpoons \begin{cases}\alpha(U(x)) & \text{if} \quad U(x)\downarrow,\\ a & \text{otherwise.}\end{cases}$$

(The superscript **0** denotes that the completion is defined through a function which is universal for the class of all partial computable functions, no oracle needed. Suitable generalizations to the case of functions computable via oracles will be provided later.)

Clearly, since range$(U) = \mathbf{N}$, $\alpha_a^{\mathbf{0}}$ is a numbering of the whole family.

Theorem 2.4 (Yu.L. Ershov). *For every α, the numbering $\alpha_a^{\mathbf{0}}$ is complete with respect to a.*

Proof. (See [5]) Let φ_e be any partial computable function. By the s_n^m–Theorem let s be a computable function such that

$$U(\langle s(e),x\rangle) = \varphi_{s(e)}(x) = U(\varphi_e(x)).$$

We have

$$\alpha_a^{\mathbf{0}}(\varphi_e(x)) = \begin{cases}\alpha(U(\varphi_e(x))) & \text{if } U(\varphi_e(x))\downarrow,\\ a & \text{otherwise.}\end{cases}$$

On the other hand, $U(\varphi_e(x)) = U(\langle s(e), x\rangle)$, and as

$$\alpha_a^0(\langle s(e), x\rangle) = \begin{cases} \alpha(U(\langle s(e), x\rangle)) & \text{if } U(\langle s(e), x\rangle)\downarrow, \\ a & \text{otherwise,} \end{cases}$$

it follows that $\alpha_a^0(\varphi_e(x)) = \alpha_a^0(\langle s(e), x\rangle)$. Hence the computable function $\lambda x\ \langle s(e), x\rangle$ α_a^0–extends φ_e with respect to a. □

Let us now turn to the case of Σ_{n+1}^0–computable families. We will go back to the general setting in Section 1.1, where we propose a relativized notion of completion. The basic question in which we are interested now is the following: If $\alpha \in \mathrm{Com}_{n+1}^0(\mathcal{A})$, is $\alpha_a^0 \in \mathrm{Com}_{n+1}^0(\mathcal{A})$? In order to avoid incomputability, in the case of families of Σ_1^0–sets we usually need to take a to be the least element $\perp$ of $\mathcal{A}$ under inclusion, if any such element exists. This is justified by the fact that if $\mathcal{A}$ is a Σ_1^0–computable family and $\mathcal{A}$ has the least element $\perp$ under $\subseteq$, then for every $\alpha \in \mathrm{Com}_1^0(\mathcal{A})$ we have that $\alpha_\perp^0 \in \mathrm{Com}_1^0(\mathcal{A})$.

To see this, one just needs to observe that

$$x \in \alpha_\perp^0(y) \Leftrightarrow x \in \perp \vee [U(y)\downarrow \ \&\ x \in \alpha(U(y))]$$

for every x, y. Thus the relation "$x \in \alpha_\perp^0(y)$" is Σ_1^0.

On the other hand, the following theorem shows that for every Σ_{n+2}^0–computable numbering α of any family $\mathcal{A} \subseteq \Sigma_{n+2}^0$ and for *any* $A \in \mathcal{A}$, the numbering α_A^0 is still Σ_{n+2}^0–computable. This, together with the fact that the completion operator is well behaved with respect to the reducibility $\leqslant$ (see next theorem, item (2)), shows that the mapping $\alpha \mapsto \alpha_A^0$ induces an operator on the semilattice $\mathcal{R}_{n+2}^0(\mathcal{A})$.

Theorem 2.5. *Let $\mathcal{A}$ be a family of Σ_{n+2}^0–sets; let $\alpha, \beta \in \mathrm{Com}_{n+2}^0(\mathcal{A})$ and pick $A \in \mathcal{A}$. Then*

(1) $\alpha_A^0 \in \mathrm{Com}_{n+2}^0(\mathcal{A})$;

(2) if $\alpha \leqslant \beta$ then $\alpha_A^0 \leqslant \beta_A^0$.

(3) $\alpha \leqslant \alpha_A^0$;

(4) $\alpha_A^0 \equiv_{0'} \alpha$;

(5) $\alpha_A^0 \equiv \alpha$ if and only if α is complete with respect to A.

Proof. Let $\alpha, \beta \in \mathrm{Com}_{n+2}^0(\mathcal{A})$, and let $A \in \mathcal{A}$.

(1) For every x, y we have

$$x \in \alpha_A^0(y) \Leftrightarrow \exists z(U(y) = z \ \&\ x \in \alpha(z)) \vee (U(y)\uparrow \ \&\ x \in A).$$

Since the relations "$x \in \alpha(z)$", "$x \in A$", are Σ^0_{n+2}, and the relations "$U(y) = z$", "$U(y) \uparrow$" are $\mathbf{0}'$–computable, it follows that the relation "$x \in \alpha^0_A(y)$" is Σ^0_{n+2}.

Remark 2.3. The same argument shows also that the ternary relation, in x, i, y, "$x \in \alpha^0_{\alpha(i)}(y)$" is Σ^0_{n+2}.

(2) Suppose that $\alpha \leqslant \beta$ via some computable function f, i.e. $\alpha = \beta \circ f$. Let e be such that $\varphi_e = f \circ U$. Then $f(U(x))) = U(\langle e, x \rangle)$. Define $g(x) = \langle e, x \rangle$. It easily follows that

$$\alpha^0_A = \beta^0_A \circ g.$$

Indeed, $U(x) \downarrow$ if and only if $U(\langle e, x \rangle) \downarrow$, thus if $U(x) \uparrow$ then $\alpha^0_A(x) = \beta^0_A(g(x)) = A$; otherwise

$$\begin{aligned} \alpha^0_A(x) &= \alpha(U(x)) \\ &= \beta(f(U(x))) \\ &= \beta(U(g(x))) = \beta^0_A(g(x)). \end{aligned}$$

Remark 2.4. Notice that the proof is uniform: One can effectively find, independently from A, an index for g starting from any index for f.

(3) Let e be an index of the identity function, i.e. $\varphi_e(x) = x$. Then

$$\alpha^0_A(\langle e, x \rangle) = \alpha(U(\langle e, x \rangle)) = \alpha(x).$$

Hence $\alpha \leqslant \alpha^0_A$ via the function $\lambda x\ \langle e, x \rangle$. (Notice that this function does not depend on either α or A.)

(4) By construction, $\alpha^0_A \leqslant_{\mathbf{0}'} \alpha$, thus by (3) we have that $\alpha^0_A \equiv_{\mathbf{0}'} \alpha$.

(5) As observed in Theorem 2.2, if $\beta \equiv \alpha$ and α is complete with respect to some object, then β is complete with respect to the same special object as α. It follows that if α is not complete with respect to A then $\alpha < \alpha^0_A$.

On the other hand, suppose that α is complete with respect to A. Let f be a computable function that α–extends U with respect to A. Therefore $\alpha^0_A(x) = \alpha(f(x))$ for every x and therefore $\alpha^0_A \leqslant \alpha$.

□

Notice that the previous theorem remains true for α^0_A in the classical case when $\mathcal{A} \subseteq \Sigma^0_1$, provided that $\mathcal{A}$ has least set $\bot$ with respect to inclusion, and we take $A = \bot$.

Corollary 2.5.1. *If α is a numbering which is universal in $\mathrm{Com}^0_{n+2}(\mathcal{A})$, then α is complete with respect to any member of $\mathcal{A}$.*

Proof. If α is universal in $\mathrm{Com}^0_{n+2}(\mathcal{A})$ then, by (1) and (3) of Theorem 2.5, $\alpha \equiv \alpha^0_A$, for every A. It follows that α is complete with respect to any $A \in \mathcal{A}$. □

Corollary 2.5.1 gives us easily natural examples of numberings which are complete with respect to every element. This should be compared with Denisov and Lavrov, [4], who showed for the first time that one can have numberings which are complete with respect to more than one element.

Corollary 2.5.2. *For every $\alpha \in \mathrm{Com}^0_{n+2}(\mathcal{A})$, the numbering α^0_A is not splittable. In particular, the greatest degree of $\mathcal{R}^0_{n+2}(\mathcal{A})$, if any, is not splittable.*

Proof. Immediate, by Theorem 2.3 and Corollary 2.5.1. □

Corollary 2.5.3. *Let $\alpha \in \mathrm{Com}^0_{n+2}(\mathcal{A})$, and let $A \in \mathcal{A}$. Then any creative set is $m-$reducible both to the α^0_A-index set of any $B \in \mathcal{A}$ different from A and to the α^0_A-index set of all the elements of $\mathcal{A}$ different from A. In particular, the α^0_A-index set of A is productive.*

Proof. The properties of the numbering α^0_A proposed in the corollary hold, indeed, for any complete numbering with the special object A (see A.I. Mal'tsev, [10]).

Let C be a creative set, and let b be such that $\alpha^0_A(b) = B$ and $B \neq A$. Define a partial computable function φ as follows:

$$\varphi(x) \rightleftharpoons \begin{cases} b & \text{if } x \in C, \\ \uparrow & \text{otherwise.} \end{cases}$$

Let f be a computable function that α^0_A-extends φ with respect to A. Then

$$\alpha^0_A(f(x)) = \begin{cases} B & \text{if } x \in C, \\ A & \text{otherwise.} \end{cases}$$

Therefore, for every x,

$$x \in C \Leftrightarrow \alpha^0_A(f(x)) = B \Leftrightarrow \alpha^0_A(f(x)) \neq A$$

i.e. the function f $m-$reduces the set C to the sets $\{x \mid \alpha^0_A(x) = B\}$ and $\{x \mid \alpha^0_A(x) \neq A\}$. □

Corollary 2.5.3 has a number of useful consequences. We recall that α is a *Friedberg numbering* if for every x, y we have that $\alpha(x) = \alpha(y)$ if and only if $x = y$, and α is a *positive* numbering if the relation in x, y "$\alpha(x) = \alpha(y)$" is c.e. Note that for Friedberg and positive numberings, any index set is c.e. Corollary 2.5.3 implies that Friedberg and positive numberings, as well as the minimal numberings which one can build by the method introduced by Badaev and Goncharov (see Theorem 1.3) are not complete with respect to any element. In the last case, the claim follows from the fact that one of the index sets contains a maximal set, while the other index sets are included in the complement of this maximal set; thus no index set can be productive.

The following theorem gives an elegant and rather unexpected proof of the fact that in $\mathcal{R}^0_{n+2}(\mathcal{A})$ any non complete numbering is meet–reducible.

Theorem 2.6. *If* $\alpha \in \mathrm{Com}^0_{n+2}(\mathcal{A})$ *and* $A, B \in \mathcal{A}$, *with* $A \neq B$, *then, in* $\mathcal{R}^0_{n+2}(\mathcal{A})$, $\deg(\alpha)$ *is the greatest lower bound of* $\deg(\alpha^0_A)$ *and* $\deg(\alpha^0_B)$.

Proof. By Theorem 2.5, we have that $\alpha \leqslant \alpha^0_A$ and $\alpha \leqslant \alpha^0_B$.

Let now f and g be computable functions which reduce a numbering γ to α^0_A and α^0_B respectively. Thus, for every x,

$$\gamma(x) = \alpha^0_A(f(x)) = \begin{cases} \alpha(U(f(x))) & \text{if } U(f(x)) \downarrow, \\ A & \text{otherwise} \end{cases}$$

and

$$\gamma(x) = \alpha^0_B(g(x)) = \begin{cases} \alpha(U(g(x))) & \text{if } U(g(x)) \downarrow, \\ B & \text{otherwise.} \end{cases}$$

Since $A \neq B$ then $U(f(x)) \downarrow$ or $U(g(x)) \downarrow$. Let $\varphi_v = \lambda x U(f(x))$ and let $\varphi_w = \lambda x U(g(x))$ and define

$$h(x) \leftrightharpoons \begin{cases} U(f(x)) & \text{if } \exists s(\varphi_{v,s}(x) \downarrow \ \& \ \varphi_{w,s}(x) \uparrow), \\ U(g(x)) & \text{otherwise.} \end{cases}$$

So h is computable and we have $\gamma = \alpha \circ h$, hence $\gamma \leqslant \alpha$. □

Notice that γ need not be a numbering of the whole family in the previous proof.

Corollary 2.6.1. *If* $\alpha \in \mathrm{Com}^0_{n+2}(\mathcal{A})$ *and there exists a pair of distinct objects of* $\mathcal{A}$ *with respect to which* α *is not complete, then* $\deg(\alpha)$ *is meet–reducible in* $\mathcal{R}^0_{n+2}(\mathcal{A})$.

Proof. Immediate by Theorem 2.5 and Theorem 2.6. □

Remark 2.5. Theorem 2.6 and Corollary 2.6.1 remain true if one works in the upper semilattice $\mathrm{Num}(\mathcal{A})$ instead of $\mathrm{Com}^0_{n+2}(\mathcal{A})$, as γ need not be a Σ^0_{n+2}–computable numbering of $\mathcal{A}$ in the proofs of Theorem 2.6 and Corollary 2.6.1.

2.3 Relativization of the completion operator

We now go back to the relativized version $\leqslant_{\mathbf{X}}$ of reducibility on numberings, introduced in Section 1.1, and provide a suitable generalization of the completion operator. We generalize the notion of a complete numbering, by allowing functions that are computable relatively to oracles.

We begin with some general definitions, relative to numberings of any abstract family.

Definition 2.5. Let $X \subseteq \mathbf{N}$. A numbering α of a set $\mathcal{A}$ is called **X**–*complete with respect to the special object* $a \in \mathcal{A}$ if for every partial **X**–computable function $\varphi(x)$ there exists some total **X**–computable function $f(x)$ such that

$$\alpha(f(x)) = \begin{cases} \alpha(\varphi(x)) & \text{if } \varphi(x)\downarrow, \\ a & \text{otherwise.} \end{cases}$$

We say that f α–*extends* φ *with respect to* a.

The definition of an **X**–precomplete numbering is given accordingly.

We begin with noting that **X**–complete numberings satisfy a relativized version of the fixed point property stated in Theorem 2.1, which now reads as follows:

Theorem 2.7. *If α is an **X**–precomplete numbering, then for every partial **X**–computable function φ one can uniformly find a number n such that if $\varphi(n)\downarrow$ then $\alpha(n) = \alpha(\varphi(n))$.*

Proof. Assume that α is an **X**–precomplete numbering. Given any index of a partial **X**–computable function φ, one can uniformly find an index z of an **X**–computable function f such that $\alpha(f(x)) = \alpha(\varphi(\varphi_x^{\mathbf{X}}(x)))$, for every x such that $\varphi(\varphi_x^{\mathbf{X}}(x))\downarrow$. Take now $n = f(z)$, and argue as in the proof of Theorem 2.1. □

The relativized version of Theorem 2.2 reads as follows:

Theorem 2.8. *For every set $X \subseteq \mathbf{N}$ and any pair of numberings α, β, if α and β are **X**–equivalent and α is **X**–complete with respect to a then β is **X**–complete with respect to a. In fact in this case α and β are **X**–computably isomorphic (i.e. $\alpha = \beta \circ p$, for some **X**–computable permutation p. This holds also if α and β are just both precomplete).*

Proof. The proof is a direct relativization of the proof of Theorem 2.2. To show for instance that if α and β are **X**–equivalent and α is **X**–complete with respect to a then β is **X**–complete with respect to a, one can argue as follows. Let $\alpha = \beta \circ h$ and $\beta = \alpha \circ k$ where h, k are **X**–computable, and let φ be any partial **X**–computable function. Then $k \circ \varphi$ is partial **X**–computable, and there exists an $\mathbf{X}$– computable function f that α– extends $k \circ \varphi$ with respect to a. Then $h \circ f$ is $\mathbf{X}$– computable and β– extends φ with respect to a. □

Finally we give the relativized version of Theorem 2.3.

Theorem 2.9. *If α is $\mathbf{X}$-precomplete, then α is not $\mathbf{X}$– splittable.*

Proof. The proof of this fact is a straightforward relativization of the proof of Theorem 2.3. Let $\alpha = \alpha_0 \oplus \alpha_1$ be $\mathbf{X}$– precomplete and let $K^{\mathbf{X}}(e, x) \rightleftharpoons \varphi_e^{\mathbf{X}}(x)$ be the universal partial $\mathbf{X}$– computable function for the unary partial $\mathbf{X}$– computable functions. Starting from $K^{\mathbf{X}}$, one defines suitable partial $\mathbf{X}$– computable functions $\varphi(e, x)$ and $\psi(e, s)$ satisfying properties (1) and (2) stated in the proof of Theorem 2.3. Then we apply the relativized Recursion Theorem to find e such that $\varphi_e^{\mathbf{X}}(x) = f(e, x)$, where f is an $\mathbf{X}$– computable function that α– extends φ. As in the proof of Theorem 2.3, it follows that either $\alpha \leqslant_{\mathbf{X}} \alpha_0$ or $\alpha \leqslant_{\mathbf{X}} \alpha_1$. □

We now want to relativize the completion operator which we have introduced in the previous section. We need for this to fix some preliminary notations and terminology. First of all, for every $X \subseteq \mathbf{N}$, let $\{\varphi_e^X \mid e \in \omega\}$ be the standard Kleene–numbering of the partial functions which are computable with oracle X. (As anticipated in Section 1.1 we will write $\varphi_e^{\mathbf{X}}$ for φ_e^X.) The relativized Universal Function Theorem justifies the following definition.

Definition 2.6. For every arbitrary subset X of $\mathbf{N}$ define

$$U^{\mathbf{X}}(\langle e, x\rangle) \rightleftharpoons \varphi_e^{\mathbf{X}}(x)$$

for all $e, x \in \mathbf{N}$. The function $U^{\mathbf{X}}$ is partial $\mathbf{X}$– computable.

Let now α be a numbering of a family $\mathcal{A}$, and let $a \in \mathcal{A}$.

Definition 2.7. Define

$$\alpha_a^{\mathbf{X}}(x) \rightleftharpoons \begin{cases} \alpha(U^{\mathbf{X}}(x)) & \text{if } \; U^{\mathbf{X}}(x)\downarrow, \\ a & \text{otherwise.} \end{cases}$$

Since range$(U^{\mathbf{X}}) = \mathbf{N}$, $\alpha_a^{\mathbf{X}}$ is a numbering of the whole family.

Notice that if X is computable, then we may assume $U^{\mathbf{X}}$ to be the function U of Definition 2.4, and thus in this case $\alpha_a^{\mathbf{X}}$ is exactly the numbering $\alpha_a^{\mathbf{0}}$, which we have defined in Definition 2.4.

We would like now to point out some properties of the completion operator which hold for abstract numberings. We remind the reader that we always consider numbered families that contain at least two elements. Moreover, if $\mathcal{A}$ is a family of sets with least element under inclusion then this element will be denoted by $\perp$.

The following theorem follows along the lines of Theorem 2.5 and Theorem 2.6.

Theorem 2.10. *Let $\mathcal{A}$ be any family, let $a, b \in \mathcal{A}$, and let α, β be any numberings of $\mathcal{A}$. Then the following statements hold for every subsets $X, Y \subseteq \mathbf{N}$:*

(1) if $Y \leq_T X$, then the numbering $\alpha_a^{\mathbf{X}}$ is $\mathbf{Y}$–complete with respect to the special object a;

(2) if $\alpha \leqslant_{\mathbf{X}} \beta$ then $\alpha_a^{\mathbf{X}} \leqslant \beta_a^{\mathbf{X}}$.

(3) $\alpha \leqslant \alpha_a^{\mathbf{X}}$;

(4) $\alpha_a^{\mathbf{X}} \equiv_{\mathbf{X}'} \alpha$;

(5) $\alpha \equiv_{\mathbf{X}} \alpha_a^{\mathbf{X}}$ if and only if α is $\mathbf{X}$–complete with respect to a;

(6) if $\beta \leqslant_{\mathbf{X}} \alpha_a^{\mathbf{X}}$ then $\beta \leqslant \alpha_a^{\mathbf{X}}$;

(7) if $Y \leq_T X$ then $\alpha_a^{\mathbf{X}'} \equiv (\alpha_a^{\mathbf{X}'})_b^{\mathbf{Y}}$. In particular $\alpha_a^{\mathbf{X}'}$ is $\mathbf{Y}$–complete with respect to any special element b;

(8) if $a \neq b$ then for every γ such that $\gamma \leqslant_{\mathbf{X}} \alpha_a^{\mathbf{X}}$ and $\gamma \leqslant_{\mathbf{X}} \alpha_b^{\mathbf{X}}$ one has $\gamma \leqslant_{\mathbf{X}} \alpha$.

Proof. Let $\mathcal{A}$ and a be given. The proof of this theorem consists for the most part a straightforward relativization of Theorems 2.5 and 2.6. We will be however a bit more detailed than necessary, because this will enable us to make some interesting remarks on the proof. Finally, let $Y \leq_T X$.

(1) Let $\varphi = \varphi_e^{\mathbf{Y}}$ be any partial $\mathbf{Y}$– computable function. Then the function $U^{\mathbf{X}}(\varphi(x))$ is partial $\mathbf{X}$–computable. By the relativized s_n^m–Theorem let s be a computable function such that

$$U^{\mathbf{X}}(\langle s(e), x\rangle) = \varphi_{s(e)}^{\mathbf{X}}(x) = U^{\mathbf{X}}(\varphi_e^{\mathbf{Y}}(x)).$$

We have

$$\alpha_a^{\mathbf{X}}(\varphi_e^{\mathbf{Y}}(x)) = \begin{cases} \alpha(U^{\mathbf{X}}(\varphi_e^{\mathbf{Y}}(x))) & \text{if } U^{\mathbf{X}}(\varphi_e^{\mathbf{Y}}(x))\downarrow, \\ a & \text{otherwise.} \end{cases}$$

On the other hand, $U^{\mathbf{X}}(\varphi_e^{\mathbf{Y}}(x)) = U^{\mathbf{X}}(\langle s(e), x\rangle)$, and as

$$\alpha_a^{\mathbf{X}}(\langle s(e), x\rangle) = \begin{cases} \alpha(U^{\mathbf{X}}(\langle s(e), x\rangle)) & \text{if } U^{\mathbf{X}}(\langle s(e), x\rangle)\downarrow, \\ a & \text{otherwise,} \end{cases}$$

it follows that $\alpha_a^{\mathbf{X}}(\varphi_e^{\mathbf{Y}}(x)) = \alpha_a^{\mathbf{X}}(\langle s(e), x\rangle)$.

Remark 2.6. Notice that the function that $\alpha_a^{\mathbf{X}}$–extends φ with respect to a is the function $g(x) = \langle s(e), x\rangle$, i.e. a computable function! Furthermore, the proof is uniform in any index of φ.

(2) Let $\alpha \leqslant_{\mathbf{X}} \beta$ via some $\mathbf{X}$–computable function f, i.e. $\alpha = \beta \circ f$. Let e be such that $\varphi_e^{\mathbf{X}} = f \circ U^{\mathbf{X}}$. Then $f(U^{\mathbf{X}}(x))) = U^{\mathbf{X}}(\langle e, x\rangle)$. Define $g(x) = \langle e, x\rangle$. As in the proof of (2) of Theorem 2.5, it easily follows that $\alpha_a^{\mathbf{X}} = \beta_a^{\mathbf{X}} \circ g$.

Remark 2.7. Notice that the function g is computable! As in the previous item, the proof is uniform, i.e. an index for g can be uniformly found from any index of f.

(3) Let e be an index of the identity function, i.e. $\varphi_e^{\mathbf{X}}(x) = x$. Then, as $U^{\mathbf{X}}(\langle e, x\rangle)$ is always defined,

$$\alpha_a^{\mathbf{X}}(\langle e, x\rangle) = \alpha(U^{\mathbf{X}}(\langle e, x\rangle)) = \alpha(x).$$

Hence $\alpha \leqslant \alpha_a^{\mathbf{X}}$ via the function $\lambda x\ \langle e, x\rangle$.

(4) By definition, we immediately have $\alpha_a^{\mathbf{X}} \leqslant_{\mathbf{X}'} \alpha$. Since $\alpha \leqslant \alpha_a^{\mathbf{X}}$ by the previous item, we get $\alpha_a^{\mathbf{X}} \equiv_{\mathbf{X}'} \alpha$.

(5) Due to (3), claim (5) can be equivalently formulated as follows: $\alpha_a^{\mathbf{X}} \leqslant_{\mathbf{X}} \alpha$ if and only if α is $\mathbf{X}$–compete with respect to a. To see this, first assume that $\alpha_a^{\mathbf{X}} \leqslant_{\mathbf{X}} \alpha$. Then, by (3), $\alpha_a^{\mathbf{X}} \equiv_{\mathbf{X}} \alpha$, and thus by Theorem 2.8 α is $\mathbf{X}$–complete with respect to a. Conversely, suppose that α is $\mathbf{X}$–complete with respect to a. Let f be an $\mathbf{X}$–computable function that α–extends $U^{\mathbf{X}}$ with respect to a. Then $\alpha_a^{\mathbf{X}} = \alpha \circ f$, thus $\alpha_a^{\mathbf{X}} \leqslant_{\mathbf{X}} \alpha$.

(6) Let f be an $\mathbf{X}$–computable function such that $\beta = \alpha_a^{\mathbf{X}} \circ f$. Then, by definition of $\alpha_a^{\mathbf{X}}$, we have, for every x,

$$\beta(x) = \alpha_a^{\mathbf{X}}(f(x)) = \begin{cases} \alpha(U^{\mathbf{X}}(f(x))) & \text{if } U^{\mathbf{X}}(f(x))\downarrow \\ a & \text{otherwise.} \end{cases}$$

Let e be such that $U^{\mathbf{X}} \circ f = \varphi_e^{\mathbf{X}}$. Then $U^{\mathbf{X}}(f(x)) = U^{\mathbf{X}}(\langle e, x\rangle)$ and thus, for every x,

$$\beta(x) = \alpha_a^{\mathbf{X}}(\langle e, x\rangle).$$

Then the function $g(x) = \langle e, x\rangle$ reduces β to $\alpha_a^{\mathbf{X}}$.

Remark 2.8. Notice that the proof is uniform, i.e. there is some computable function g such that if $\varphi_u^{\mathbf{X}}$ reduces β to $\alpha_a^{\mathbf{X}}$, then the computable function (with no oracle!) $\varphi_{g(u)}$ reduces β to $\alpha_a^{\mathbf{X}}$.

(7) By definition of completion we have

$$(\alpha_a^{\mathbf{X}'})_b^{\mathbf{Y}}(x) = \begin{cases} \alpha_a^{\mathbf{X}'}(U^{\mathbf{Y}}(x)) & \text{if } U^{\mathbf{Y}}(x)\downarrow, \\ b & \text{otherwise} \end{cases}$$

$$= \begin{cases} \alpha(U^{\mathbf{X}'}(U^{\mathbf{Y}}(x))) & \text{if } U^{\mathbf{X}'}(U^{\mathbf{Y}}(x))\downarrow, \\ a & \text{if } U^{\mathbf{Y}}(x)\downarrow \text{ and } U^{\mathbf{X}'}(U^{\mathbf{Y}}(x))\uparrow \\ b & \text{otherwise.} \end{cases}$$

Since $\text{range}(U^{\mathbf{X}'}) = \mathbf{N}$, there exists x_0 such that $U^{\mathbf{X}'}(x_0)\downarrow$ and $\alpha(U^{\mathbf{X}'}(x_0)) = b$. Consider the total $\mathbf{X}'$–computable function $h(x)$ defined as follows

$$h(x) = \begin{cases} U^{\mathbf{Y}}(x) & \text{if } U^{\mathbf{Y}}(x)\downarrow, \\ x_0 & \text{if } U^{\mathbf{Y}}(x)\uparrow. \end{cases}$$

Obviously, $(\alpha_a^{\mathbf{X}'})_b^{\mathbf{Y}}(x) = \alpha_a^{\mathbf{X}'}(h(x))$. Therefore $(\alpha_a^{\mathbf{X}'})_b^{\mathbf{Y}} \leqslant_{\mathbf{X}'} \alpha_a^{\mathbf{X}'}$. On the other hand, $\alpha_a^{\mathbf{X}'} \leqslant (\alpha_a^{\mathbf{X}'})_b^{\mathbf{Y}}$ by (3), thus $(\alpha_a^{\mathbf{X}'})_b^{\mathbf{Y}} \equiv_{\mathbf{X}'} \alpha_a^{\mathbf{X}'}$. By (6) it follows that $(\alpha_a^{\mathbf{X}'})_b^{\mathbf{Y}} \leqslant \alpha_a^{\mathbf{X}'}$. On the other hand, by (3) we have $\alpha_a^{\mathbf{X}'} \leqslant (\alpha_a^{\mathbf{X}'})_b^{\mathbf{Y}}$. Hence $(\alpha_a^{\mathbf{X}'})_b^{\mathbf{Y}} \equiv \alpha_a^{\mathbf{X}'}$. The claim now follows by item (5).

(8) Let us show that if $\gamma \leqslant_{\mathbf{X}} \alpha_a^{\mathbf{X}}$ and $\gamma \leqslant_{\mathbf{X}} \alpha_b^{\mathbf{X}}$ (via, say, $\mathbf{X}$–computable functions f and g) then $\gamma \leqslant_{\mathbf{X}} \alpha$. We have, for every x,

$$\gamma(x) = \alpha_a^{\mathbf{X}}(f(x)) = \begin{cases} \alpha(U^{\mathbf{X}}(f(x))) & \text{if } U^{\mathbf{X}}(f(x))\downarrow, \\ a & \text{otherwise} \end{cases}$$

and

$$\gamma(x) = \alpha_b^{\mathbf{X}}(g(x)) = \begin{cases} \alpha(U^{\mathbf{X}}(g(x))) & \text{if } U^{\mathbf{X}}(g(x))\downarrow, \\ b & \text{otherwise.} \end{cases}$$

Since $a \neq b$, obviously we have that $U^{\mathbf{X}}(f(x))\downarrow$ or $U^{\mathbf{X}}(g(x))\downarrow$. Let $\varphi_v^{\mathbf{X}} = \lambda x U^{\mathbf{X}}(f(x))$ and $\varphi_w^{\mathbf{X}} = \lambda x U^{\mathbf{X}}(g(x))$ and define

$$h(x) \rightleftharpoons \begin{cases} U^{\mathbf{X}}(f(x)) & \text{if } \exists s(\varphi_{v,s}^{\mathbf{X}}(x)\downarrow \ \&\ \varphi_{w,s}^{\mathbf{X}}(x)\uparrow), \\ U^{\mathbf{X}}(g(x)) & \text{otherwise.} \end{cases}$$

So we have $\gamma = \alpha \circ h$, where h is an $\mathbf{X}$–computable function. Hence $\gamma \leqslant_{\mathbf{X}} \alpha$. (As remarked for Theorem 2.6, γ need not be a numbering of the whole family in the previous argument.)

□

We now list several useful corollaries of the previous theorem, for numberings of some fixed family.

Corollary 2.10.1. *If* $Y \leq_T X$ *then* $(\alpha_a^{\mathbf{X}})_a^{\mathbf{Y}} \equiv \alpha_a^{\mathbf{X}}$.

Proof. By (1), (3), (5) and (6) of Theorem 2.10. □

Corollary 2.10.2. $\alpha \leqslant \alpha_a^{\mathbf{0}^{(n)}}$. *Moreover,* α *is* $\mathbf{0}^{(n)}$*– complete with respect to* a *if and only if* $\alpha_a^{\mathbf{0}^{(n)}} \equiv_{\mathbf{0}^{(n)}} \alpha$.

Proof. By (3) and (5) of Theorem 2.10. □

Corollary 2.10.3. *The numbering* $\alpha_a^{\mathbf{0}^{(n)}}$ *is* $\mathbf{0}^{(i)}$*– complete with respect to* a *for all* $i \leq n$.

Proof. By (1) of Theorem 2.10. □

Corollary 2.10.4. *If* $\beta \leqslant_{\mathbf{0}^{(n)}} \alpha_a^{\mathbf{0}^{(n)}}$ *then* $\beta \leqslant \alpha_a^{\mathbf{0}^{(n)}}$.

Proof. By (6) of Theorem 2.10. □

Corollary 2.10.5. $(\alpha_a^{\mathbf{0}^{(n+1)}})_b^{\mathbf{0}^{(n)}} \equiv \alpha_a^{\mathbf{0}^{(n+1)}}$*, all* a, b*. In particular,* $\alpha_a^{\mathbf{0}^{(n+1)}}$ *is* $\mathbf{0}^{(i)}$*– complete with respect to each element of the family and for all* $i \leq n$.

Proof. By (1) and (7) of Theorem 2.10. □

Finally we give the relativized analogues of Corollary 2.5.2 and Corollary 2.5.3.

Corollary 2.10.6. *For every numbering* α *and every set* X*, the numbering* $\alpha_a^{\mathbf{X}}$ *is not* $\mathbf{X}$*– splittable.*

Proof. It follows from Theorem 2.9. □

Corollary 2.10.7. *The set of all* $\alpha_a^{\mathbf{X}}$*– indices of the special object* a *is* $\mathbf{X}$*– productive.*

Proof. The proof is a straightforward relativization of the proof of Corollary 2.5.3. Let e be such that $\alpha_a^{\mathbf{X}}(e) \neq a$. Choose any $\mathbf{X}$– creative set C, and define

$$\varphi(x) \coloneqq \begin{cases} e & \text{if } x \in C, \\ \uparrow & \text{otherwise.} \end{cases}$$

Then φ is partial $\mathbf{X}$– computable, and by $\mathbf{X}$– completion, there is some $\mathbf{X}$– computable function f that $\alpha_a^{\mathbf{X}}$– extends φ with respect to a. Then it is easy to see that the function f m– reduces C to the complement of the $\alpha_a^{\mathbf{X}}$– index set $\{x \mid \alpha_a^{\mathbf{X}}(x) = a\}$. □

2.4 Computability of completions

We now go back to consider in details Σ^0_{n+1}–computable families. Let $\alpha \in \mathrm{Com}^0_{n+1}(\mathcal{A})$. We are interested in completions of α which are still in $\mathrm{Com}^0_{n+1}(\mathcal{A})$.

Theorem 2.11. *Let* $\alpha \in \mathrm{Com}^0_{n+1}(\mathcal{A})$. *Then*

(1) $\alpha_A^{\mathbf{0}^{(m)}} \in \mathrm{Com}^0_{n+1}(\mathcal{A})$ *for every* $m < n$ *and each* $A \in \mathcal{A}$;

(2) $\alpha_A^{\mathbf{0}^{(n)}} \in \mathrm{Com}^0_{n+1}(\mathcal{A})$ *if and only if* $\mathcal{A}$ *has the least set* $\perp$ *and* $A = \perp$;

(3) $\alpha_A^{\mathbf{0}^{(m)}} \notin \mathrm{Com}^0_{n+1}(\mathcal{A})$ *for every* $m > n$ *and each* $A \in \mathcal{A}$.

Proof. Let $\alpha \in \mathrm{Com}^0_{n+1}(\mathcal{A})$.

(1) Let $A \in \mathcal{A}$ and $m < n$. By definition of completion it follows that $x \in \alpha_A^{\mathbf{0}^{(m)}}(y)$ if and only if

$$\exists z(U^{\mathbf{0}^{(m)}}(y) = z \;\&\; x \in \alpha(y)) \;\vee\; (U^{\mathbf{0}^{(m)}}(y)\uparrow \;\&\; x \in A).$$

Since $U^{\mathbf{0}^{(m)}}$ is a partial $\mathbf{0}^{(m)}$–computable function, it is immediate to check that the relation "$x \in \alpha_A^{\mathbf{0}^{(n)}}(y)$" is Σ^0_{n+1}.

(2) Let $A = \perp \in \mathcal{A}$. In this case

$$x \in \alpha_\perp^{\mathbf{0}^{(n)}}(y) \Leftrightarrow x \in \perp \;\vee\; \exists z(U^{\mathbf{0}^{(n)}}(y) = z \;\&\; x \in \alpha(y)).$$

Hence $\alpha_\perp^{\mathbf{0}^{(n)}} \in \mathrm{Com}^0_{n+1}(\mathcal{A})$.

Conversely, let $\alpha_A^{\mathbf{0}^{(n)}} \in \mathrm{Com}^0_{n+1}(\mathcal{A})$, and suppose that A is not the least element of $\mathcal{A}$. Hence there is some $B \in \mathcal{A}$ such that $A \not\subseteq B$. Let $a \in A \setminus B$, and let b be such that $\alpha_A^{\mathbf{0}^{(n)}}(b) = B$. Let

$$C \rightleftharpoons \{x \mid a \in \alpha_A^{\mathbf{0}^{(n)}}(x)\}.$$

Obviously, $C \in \Sigma^0_{n+1}$, and thus the partial function φ defined as follows,

$$\varphi(x) = \begin{cases} b & \text{if } x \in C, \\ \uparrow & \text{otherwise,} \end{cases}$$

is $\mathbf{0}^{(n)}$–computable. Since $\alpha_A^{\mathbf{0}^{(n)}}$ is $\mathbf{0}^{(n)}$–complete with respect to A, let f be a $\mathbf{0}^{(n)}$–computable function that $\alpha_A^{\mathbf{0}^{(n)}}$–extends φ with respect to A, i.e.

$$\alpha_A^{\mathbf{0}^{(n)}}(f(x)) = \begin{cases} B & \text{if } x \in C, \\ A & \text{otherwise.} \end{cases}$$

By the relativized version of the Fixed Point Theorem (see Theorem 2.7) let e be such that $\alpha_A^{\mathbf{0}^{(n)}}(e) = \alpha_A^{\mathbf{0}^{(n)}}(f(e))$. Hence

$$e \in C \Leftrightarrow a \in \alpha_A^{\mathbf{0}^{(n)}}(e)$$

and on the other hand, by properties of f,

$$e \in C \Leftrightarrow \alpha_A^{\mathbf{0}^{(n)}}(e) = B.$$

This yields a contradiction.

(3) Let $A \in \mathcal{A}$ and $m > n$. Recall that we consider only nontrivial families, i.e. $|\mathcal{A}| > 1$. Let $B, C \in \mathcal{A}$, $B \not\subseteq C$ and choose a number $b \in B \setminus C$.

Assume now that $\alpha_A^{\mathbf{0}^{(m)}} \in \mathrm{Com}_{n+1}^0(\mathcal{A})$ and consider the set $X \leftrightharpoons \{x \mid b \in \alpha_A^{\mathbf{0}^{(m)}}(x)\}$. X is clearly $\mathbf{0}^{(n+1)}$–computable. Inequality $m > n$ implies that X is also $\mathbf{0}^{(m)}$–computable. The subfamilies $\alpha_A^{\mathbf{0}^{(m)}}[X]$ and $\alpha_A^{\mathbf{0}^{(m)}}[\mathbf{N} \setminus X]$ are disjoint and contain B and C respectively. Denote by x_0 and x_1 $\alpha_A^{\mathbf{0}^{(m)}}$–indices of B and C respectively, and define a function $f(x)$ as follows

$$f(x) \leftrightharpoons \begin{cases} x_1 & \text{if } x \in X, \\ x_0 & \text{otherwise.} \end{cases}$$

By Theorem 2.7, $\alpha_A^{\mathbf{0}^{(m)}}(f(x_2)) = \alpha_A^{\mathbf{0}^{(m)}}(x_2)$ for some x_2. We have

$$b \in \alpha_A^{\mathbf{0}^{(m)}}(x_2) \Leftrightarrow x_2 \in X$$

and

$$b \in \alpha_A^{\mathbf{0}^{(m)}}(f(x_2)) \Leftrightarrow x_2 \in \mathbf{N} \setminus X,$$

a contradiction. Therefore, $\alpha_A^{\mathbf{0}^{(m)}}$ is not a Σ_{n+1}^0–computable numbering.

□

Corollary 2.11.1. *Let α be $\mathbf{0}^{(m)}$–universal in $\mathrm{Com}_{n+1}^0(\mathcal{A})$. Then:*

(1) if $m < n$ then α is $\mathbf{0}^{(m)}$–complete with respect to every element of $\mathcal{A}$;

(2) if $m \leqslant n$ and $\mathcal{A}$ has the least element $\perp$ then α is $\mathbf{0}^{(m)}$–complete with respect to $\perp$;

(3) α is not $\mathbf{0}^{(n)}$–complete with respect to any non-least element of $\mathcal{A}$;

(4) α is not $\mathbf{0}^{(m)}$–complete with respect to any $A \in \mathcal{A}$ if $m > n$.

Proof. Let A be any element of $\mathcal{A}$. By Theorem 2.10, $\alpha_A^{\mathbf{0}^{(m)}}$ is Σ^0_{n+1}-computable numbering if $m < n$ or if $m \leq n$ and $A = \bot$. Since α is $\mathbf{0}^{(m)}$–universal, it follows that $\alpha_A^{\mathbf{0}^{(m)}} \leqslant_{\mathbf{0}^{(m)}} \alpha$. Then by Corollary 2.10.2, α is $\mathbf{0}^{(m)}$–complete with respect to A. So, claims (1) and (2) are proved.

Let now either $m = n$ and $A \neq \bot$ or $m > n$. By Theorem 2.11, numbering $\alpha_A^{\mathbf{0}^{(m)}}$ is not Σ^0_{n+1}-computable. Therefore, $\alpha_A^{\mathbf{0}^{(m)}} \nleqslant_{\mathbf{0}^{(m)}} \alpha$ because all Σ^0_{n+1}-computable numberings of $\mathcal{A}$ are $\mathbf{0}^{(m)}$–reducible to α. Corollary 2.10.2 in this case implies that α is not $\mathbf{0}^{(m)}$–complete with respect to A. □

We now ask whether α is $\mathbf{Y}$–complete with respect to a if $Y \leq_T X$ and α is $\mathbf{X}$–complete with respect to a. This question naturally arises from (1) of Theorem 2.10. In particular, we have seen that $\alpha_A^{\mathbf{0}^{(m)}} \in \mathrm{Com}^0_{n+1}(\mathcal{A})$ if $m < n$, and $\alpha_A^{\mathbf{0}^{(m)}}$ is $\mathbf{0}^{(i)}$–complete with respect to A, for all $i \leq m$. It is then natural to ask, given any object of the family, whether it is possible to find numberings that are $\mathbf{0}^{(i)}$–complete, for some numbers $i < n$, with respect to that object, but not $\mathbf{0}^{(j)}$–complete, for different numbers $j < n$. The following theorem gives an answer to these questions.

Theorem 2.12. *Let $\mathcal{A}$ be any non-trivial Σ^0_{n+1}–computable family with $n \geq 1$. Then*

(1) for every $A \in \mathcal{A}$ and every set $I \subseteq \{0, 1, \ldots, n-1\}$, there exists a numbering $\alpha \in \mathrm{Com}^0_{n+1}(\mathcal{A})$ such that α is $\mathbf{0}^{(i)}$–complete with respect to A if and only if $i \in I$;

(2) if $\mathcal{A}$ has least set $\bot$, and if it has a Σ^0_{n+1}–computable numbering which is not $\mathbf{0}^{(n)}$–complete with respect to $\bot$ then for every set $I \subseteq \{0, 1, \ldots, n\}$, there exists a numbering $\alpha \in \mathrm{Com}^0_{n+1}(\mathcal{A})$ such that α is $\mathbf{0}^{(i)}$–complete with respect to $\bot$ if and only if $i \in I$.

Proof. We will consider only the case when $\mathcal{A}$ is an infinite family. The case with a finite $\mathcal{A}$ can be proved almost by the same way due to Remark 1.3. Let us prove (1). So, let $\mathcal{A}$ be any infinite Σ^0_{n+1}–computable family, $A \in \mathcal{A}$, and $I \subseteq \{0, 1, \ldots, n-1\}$. We prove the claim by induction on the cardinality k of the set $C_I = \{0, 1, \ldots, n-1\} \setminus I$.

For $k = 0$, i.e. $I = \{0, 1, \ldots, n-1\}$, by Corollary 2.10.3 and (1) of Theorem 2.11 it suffices to take $\alpha = \beta_A^{\mathbf{0}^{(n-1)}}$, for any numbering $\beta \in \mathrm{Com}^0_{n+1}(\mathcal{A})$.

Suppose that the claim is true of k, and let $I \subseteq \{0, 1, \ldots, n-1\}$ be such that the cardinality of C_I is $k+1$. Let j_0 be the least element of C_I. By induction, there exists a numbering $\beta \in \mathrm{Com}^0_{n+1}(\mathcal{A})$ such that β is $\mathbf{0}^{(i)}$–complete with respect to A if and only if $i \in I \cup \{j_0\}$.

We distinguish two cases:

Case 1. If $j_0 = 0$ then take any maximal set M, choose $B \in \mathcal{A}$ different from A, and apply the Badaev–Goncharov construction ([3] or see here Theorem 1.3 and Remark 1.2) to produce a minimal numbering $\gamma \leftrightharpoons \beta_{M,B}$. It follows from Theorem 1.3 that $\beta_{M,B} \equiv_{\mathbf{0}'} \beta$. Thus by Theorem 2.8 γ is $\mathbf{0}^{(i)}$–complete with respect to A if and only if β is $\mathbf{0}^{(i)}$–complete with respect to A for every $i \neq 0$. On the other hand, γ is not complete with respect to A, since by Corollary 2.5.3 in this case the γ– index set of A should be productive, whereas it is contained in the complement of the maximal set M.

Case 2. Assume now that $j_0 > 0$. In this case take a $\mathbf{0}^{(j_0)}$– maximal set M; choose again $B \in \mathcal{A}$. Apply the Badaev–Goncharov construction (see Theorem 1.3) to produce a numbering $\gamma' \leftrightharpoons \beta_{M,B}$. We have that $\beta_{M,B} \equiv_{\mathbf{0}^{(j_0+1)}} \beta$.

By Theorem 2.10(5) it follows that for every $i > j_0$ the numbering γ' is $\mathbf{0}^{(i)}$–complete with respect to A if and only if β is $\mathbf{0}^{(i)}$–complete with respect to A. Moreover, since M is $\mathbf{0}^{(j_0)}$–maximal, it follows also that γ' is not $\mathbf{0}^{(j_0)}$–complete since in this case, by Corollary 2.10.7, the γ'– index set for A should be $\mathbf{0}^{(j_0)}$–productive.

Now, define $\gamma \leftrightharpoons (\gamma')_A^{\mathbf{0}^{(j_0-1)}}$. By Corollary 2.10.3, we have that γ is $\mathbf{0}^{(i)}$–complete with respect to A, for every $i < j_0$. But $\gamma \equiv_{\mathbf{0}^{(j_0)}} \gamma'$ and then for every $i \geq j_0$, γ is $\mathbf{0}^{(i)}$–complete with respect to A if and only if γ' is $\mathbf{0}^{(i)}$–complete with respect to A.

Putting things together, we have that γ is $\mathbf{0}^{(i)}$–complete with respect to A if and only if $i \in I$, as desired.

The proof for the case $\perp \in \mathcal{A}$ is similar, starting from the observation (see (2) of Theorem 2.11) that $\alpha_\perp^{\mathbf{0}^{(n)}}$ is $\mathbf{0}^{(i)}$–complete with respect to $\perp$ for every $i \leq n$. □

Remark 2.9. For every $n \geq 2$, there exists an infinite family $\mathcal{A}_n \subseteq \Sigma_n^0$ with $\perp \in \mathcal{A}_n$ such that all the numberings of $\mathrm{Com}_n^0(\mathcal{A})$ are $\mathbf{0}^{(n)}$–equivalent. In particular, all these numberings are $\mathbf{0}^{(n)}$–complete with respect to $\perp$.

The existence of such families $\mathcal{A}_n$, $n \geq 2$, follows by relativized versions of the construction of S.Badaev and S.Goncharov, [3], of an infinite family $\mathcal{A}_1$ of c.e. sets with least set under inclusion such that $|\mathcal{R}_1^0(\mathcal{A}_1)| = 1$.

2.5 Uniformly X– complete numberings

We now consider a useful although natural strengthening of the notion of completeness.

Definition 2.8. We say that a numbering β of a set $\mathcal{A}$ is *uniformly* **X**– *complete* if there exists a total **X**–computable function $h(i, m, x)$ such that for every

i, m, x

$$\beta(h(i,m,x)) = \begin{cases} \beta(\varphi_i^{\mathbf{X}}(x)) & \text{if } \varphi_i^{\mathbf{X}}(x)\downarrow, \\ \beta(m) & \text{otherwise.} \end{cases}$$

We already know (see Theorem 2.10 (7)) that for every family $\mathcal{A}$, for every numbering α of $\mathcal{A}$ and for every $a \in \mathcal{A}$, the numbering $\alpha_a^{\mathbf{X}'}$ is complete with respect to any object of the family. In fact, one can prove:

Theorem 2.13. *For every set X and for every numbering α of some family $\mathcal{A}$, and for each $a \in \mathcal{A}$, the numbering $\alpha_a^{\mathbf{X}'}$ is uniformly $\mathbf{X}$–complete.*

Proof. Let us fix a set X, a family $\mathcal{A}$, a numbering $\alpha \in \mathrm{Num}(\mathcal{A})$, and $a \in \mathcal{A}$. Define the partial $\mathbf{X}'$–computable function $H(i,m,x)$ as follows

$$H(i,m,x) = \begin{cases} U^{\mathbf{X}'}(\varphi_i^{\mathbf{X}}(x)) & \text{if } \varphi_i^{\mathbf{X}}(x)\downarrow, \\ U^{\mathbf{X}'}(m) & \text{otherwise.} \end{cases}$$

By the s_n^m–Theorem let $s(i,m)$ be a computable function such that for every i, m, x, $H(i,m,x) = \varphi_{s(i,m)}^{\mathbf{X}'}(x)$. Let $h(i,m,x) = \langle s(i,m), x\rangle$ for all i, m, x. Now we show that the function h satisfies Definition 2.8 for the numbering $\alpha_a^{\mathbf{X}'}$, i.e.

$$\alpha_a^{\mathbf{X}'}(h(i,m,x)) = \begin{cases} \alpha_a^{\mathbf{X}'}(\varphi_i^{\mathbf{X}}(x)) & \text{if } \varphi_i^{\mathbf{X}}(x)\downarrow, \\ \alpha_a^{\mathbf{X}'}(m) & \text{otherwise.} \end{cases}$$

Recall that $U^{\mathbf{X}'}(\langle e, y\rangle) = \varphi_e^{\mathbf{X}'}(y)$ for all e, y and

$$\alpha_a^{\mathbf{X}'}(h(i,m,x)) = \begin{cases} \alpha(U^{\mathbf{X}'}(h(i,m,x))) & \text{if } U^{\mathbf{X}'}(h(i,m,x))\downarrow, \\ a & \text{otherwise} \end{cases}$$

for all i, m, x.

Case 1. Assume that $\varphi_i^{\mathbf{X}}(x)\downarrow$ and $U^{\mathbf{X}'}(\varphi_i^{\mathbf{X}}(x))\downarrow$. In this case $H(i,m,x)$ is defined, $H(i,m,x) = U^{\mathbf{X}'}(\varphi_i^{\mathbf{X}}(x))$, and

$$\begin{aligned} \alpha_a^{\mathbf{X}'}(h(i,m,x)) &= \alpha(U^{\mathbf{X}'}(h(i,m,x))) = \alpha(\varphi_{s(i,m)}^{\mathbf{X}'}(x)) \\ &= \alpha(H(i,m,x)) = \alpha(U^{\mathbf{X}'}(\varphi_i^{\mathbf{X}}(x))) \\ &= \alpha_a^{\mathbf{X}'}(\varphi_i^{\mathbf{X}}(x)). \end{aligned}$$

Case 2. Next assume that $\varphi_i^{\mathbf{X}}(x)\downarrow$ and $U^{\mathbf{X}'}(\varphi_i^{\mathbf{X}}(x))\uparrow$. In this case $H(i,m,x)\uparrow$ and, hence, $U^{\mathbf{X}'}(h(i,m,x))\uparrow$ since

$$U^{\mathbf{X}'}(h(i,m,x)) = U^{\mathbf{X}'}(\varphi_{s(i,m)}^{\mathbf{X}}(x)) = U^{\mathbf{X}'}(H(i,m,x)).$$

This implies that $\alpha_a^{\mathbf{X}'}(h(i,m,x)) = a$.

On the other hand, from

$$\alpha_a^{\mathbf{X}'}(\varphi_i^{\mathbf{X}}(x)) = \begin{cases} \alpha(U^{\mathbf{X}'}(\varphi_i^{\mathbf{X}}(x))) & \text{if } U^{\mathbf{X}'}(\varphi_i^{\mathbf{X}}(x)) \downarrow, \\ a & \text{otherwise} \end{cases}$$

we have that $\alpha_a^{\mathbf{X}'}(\varphi_i^{\mathbf{X}}(x)) = a$.

Notice that Cases 1 and 2 show that if $\varphi_i^{\mathbf{X}}(x) \downarrow$ then $\alpha_a^{\mathbf{X}'}(\varphi_i^{\mathbf{X}}(x)) = \alpha_a^{\mathbf{X}'}(h(i,m,x))$.

Case 3. Now assume that $\varphi_i^{\mathbf{X}}(x) \uparrow$. Let us show that $\alpha_a^{\mathbf{X}'}(h(i,m,x)) = \alpha_a^{\mathbf{X}'}(m)$. Consider the following two subcases.

Subcase 3*a*. $U^{\mathbf{X}'}(m) \uparrow$.

This subcase immediately yields $\alpha_a^{\mathbf{X}'}(m) = a$. On the other hand, $U^{\mathbf{X}'}(m) \uparrow$ implies $H(i,m,x) \uparrow$ and, just as in Case 2, we have that $U^{\mathbf{X}'}(h(i,m,x)) \uparrow$. Therefore, $\alpha_a^{\mathbf{X}'}(h(i,m,x)) = a$.

Subcase 3*b*. $U^{\mathbf{X}'}(m) \downarrow$.

In this subcase we have

$$\alpha_a^{\mathbf{X}'}(m) = \alpha(U^{\mathbf{X}'}(m)) = \alpha(H(i,m,x))$$

since $H(i,m,x) = U^{\mathbf{X}'}(m)$. Moreover,

$$\alpha(H(i,m,x)) = \alpha(U^{\mathbf{X}'}(h(i,m,x))) = \alpha_a^{\mathbf{X}'}(h(i,m,x)).$$

□

Restriction to arithmetical families gives:

Corollary 2.13.1. *For every numbering $\alpha \in \mathrm{Com}_{n+3}^0(\mathcal{A})$ there exists a Σ_{n+3}^0-computable uniformly complete numbering β such that $\alpha \leqslant \beta$.*

Proof. Let $A \in \mathcal{A}$. A simple calculation shows that $\alpha_A^{\mathbf{0}'} \in \mathrm{Com}_{n+3}^0(\mathcal{A})$. Thus, by the previous theorem, it suffices to take $\beta = \alpha_A^{\mathbf{0}'}$. □

This implies also:

Corollary 2.13.2. *For every numbering $\alpha \in \mathrm{Com}_{n+3}^0(\mathcal{A})$ there exists a numbering $\beta \in \mathrm{Com}_{n+3}^0(\mathcal{A})$ such that $\alpha \leqslant \beta$ and β is complete with respect to every element $B \in \mathcal{A}$.*

Proof. Immediate by the previous corollary. □

3. Universal Numberings of Finite Families

We now take a closer look at numberings of finite arithmetical families. Clearly, for every n each finite family $\mathcal{A}$ of Σ^0_{n+1}–sets has Σ^0_m–computable numberings for all $m \geq n+1$.

It is a well known fact of the theory of numberings that every finite family of c.e. sets has a principal numbering, see e.g. [5]. By relativizing this result we obtain:

Theorem 3.1. *For every n, each finite family $\mathcal{A}$ of Σ^0_{n+1}–sets has a numbering α which is $\mathbf{0}^{(n)}$–universal in* $\mathrm{Com}^0_{n+1}(\mathcal{A})$.

Proof. The following proof is based on ideas of Lachlan, [8]. Let $\mathcal{A}$ be a finite family of Σ^0_{n+1}–sets, say $\mathcal{A} = \{A_1, \dots, A_n\}$. Then it is possible to find finite sets $F_1, \dots, F_n$ such that, for all $1 \leq i, j \leq n$,

$$F_i \subseteq F_j \Leftrightarrow A_i \subseteq A_j;$$
$$F_i \subseteq A_j \Leftrightarrow A_i \subseteq A_j.$$

Assume first that $\mathcal{A}$ has a least element $\bot$, say $A_1 = \bot$. Assume $F_1 = \emptyset$. Let $\mathfrak{C}$ be the set of all chains, i.e. strictly increasing sequences $F_{i_1} \subset \cdots \subset F_{i_k}$, of finite sets from the list $F_1, \dots, F_n$, and for every chain $C \in \mathfrak{C}$ let F_C denote the maximal element in the chain, and let A_C denote the set of the family corresponding to F_C (i.e. $A_C = A_i$ if and only if $F_C = F_i$). It is not difficult (see [5] for details) to construct a Σ^0_{n+1}–computable numbering α such that, for every e,

- there exists $C \in \mathfrak{C}$ such that $F_C \subseteq W_e^{\mathbf{0}^{(n)}}$ and $\alpha(e) = A_C$;
- if $W_e^{\mathbf{0}^{(n)}} \in \mathcal{A}$ then $W_e^{\mathbf{0}^{(n)}} = \alpha(e)$.

Let now $\rho(e) = W_e^{\mathbf{0}^{(n)}}$. If $\beta \in \mathrm{Com}^0_{n+1}(\mathcal{A})$ then since ρ is a universal numbering in the class $\mathrm{Com}^0_{n+1}(\Sigma^0_{n+1})$ (consisting of the Σ^0_{n+1}–computable numberings of the family of all Σ^0_{n+1}–sets), there exists a computable function f such that $\beta = \rho \circ f$, but then $\beta = \alpha \circ f$, thus giving that α is universal. So in this case there exists a numbering which is universal in $\mathrm{Com}^0_{n+1}(\mathcal{A})$.

Suppose now that $\mathcal{A}$ does not have a least member. Let

$$\mathcal{A}_0 \leftrightharpoons \mathcal{A} \cup \{\emptyset\}$$

and by the above argument let α_0 be universal in $\mathrm{Com}^0_{n+1}(\mathcal{A}_0)$. Let f be a $\mathbf{0}^{(n)}$–computable function such that

$$\mathrm{range}(f) = \{x \mid \alpha_0(x) \neq \emptyset\}.$$

Then $\alpha = \alpha_0 \circ f \in \mathrm{Com}^0_{n+1}(\mathcal{A})$. Let now $\beta \in \mathrm{Com}^0_{n+1}(\mathcal{A})$, and define

$$\beta_0(x) \leftrightharpoons \begin{cases} \emptyset & \text{if } x = 0, \\ \beta(x-1) & \text{otherwise} \end{cases}$$

and let h be a computable function such that $\beta_0 = \alpha_0 \circ h$. Hence, for every x

$$\begin{aligned} \beta(x) &= \beta_0(x+1) \\ &= \alpha_0(h(x+1)). \end{aligned}$$

But $h(x+1) \in \mathrm{range}(f)$. Let

$$k(x) \leftrightharpoons \mu\, y\,(f(y) = h(x+1)).$$

It follows that $\beta(x) = \alpha_0(f(k(x)))$, i.e. $\beta = \alpha \circ k$. Since k is $\mathbf{0}^{(n)}$–computable, it follows that $\beta \leqslant_{\mathbf{0}^{(n)}} \alpha$, hence α is $\mathbf{0}^{(n)}$–universal in $\mathrm{Com}^0_{n+1}(\mathcal{A})$. □

Thus, as already remarked, every finite family $\mathcal{A}$ of Σ^0_1–sets has a numbering which is universal in $\mathrm{Com}^0_1(\mathcal{A})$. The situation is different if we consider Σ^0_{n+2}–computable finite families. We are able in fact to give a characterization of the Σ^0_{n+2}–computable finite families $\mathcal{A}$ possessing a numbering which is $\mathbf{0}^{(n)}$–universal in $\mathrm{Com}^0_{n+2}(\mathcal{A})$.

Theorem 3.2. *Let $\mathcal{A} \subseteq \Sigma^0_{n+2}$ be a finite family. Then the following statements are equivalent:*

(1) there exists a numbering of $\mathcal{A}$ which is universal in $\mathrm{Com}^0_{n+2}(\mathcal{A})$;

(2) $\mathcal{A}$ has a numbering which is $\mathbf{0}^{(n)}$– universal in $\mathrm{Com}^0_{n+2}(\mathcal{A})$;

(3) $\mathcal{A}$ contains a least element $\perp$ under inclusion.

Proof. (1) $\Rightarrow$ (2) is evident.

(2) $\Rightarrow$ (3). Let α be a $\mathbf{0}^{(n)}$–universal numbering in $\mathrm{Com}^0_{n+2}(\mathcal{A})$. Assume that the partially ordered set $\langle \mathcal{A}, \subseteq \rangle$ has no least element, and denote by $A_0, A_1, \ldots, A_k$ all its minimal elements. Clearly $k \geq 1$. For every $i \leq k$, let a_i stand for some α– index of A_i, i.e. $\alpha(a_i) = A_i$. As in the proof of the previous theorem, for every $i \leq k$ choose a finite set F_i such that $F_i \subseteq A_i$ and $F_i \nsubseteq A_j$ for all $j \neq i, j \leq k$. We will assume that $i+1$ denotes 0 if $i = k$. It is evident that for every $i \leq k$ the set $Q_i \leftrightharpoons \{x \mid F_i \subseteq \alpha(x)\}$ is $\mathbf{0}^{(n+1)}$–computably enumerable. By the relativized Reduction Theorem, there exist $\mathbf{0}^{(n+1)}$–computably enumerable pairwise disjoint sets $R_0, R_1, \ldots, R_k$ such that $R_i \subseteq Q_i$ for all $i \leq k$ and $\bigcup_{i \leq k} R_i = \bigcup_{i \leq k} Q_i$. It is clear that the sets $R_0, R_1, \ldots, R_k$

form a partition of $\mathbf{N}$ into $\mathbf{0}^{(n+1)}-$computable sets. For every x, find i such that $x \in R_i$ and define

$$f(x) \leftrightharpoons \begin{cases} a_{i+1} & \text{if } \varphi_x^{\mathbf{0}^{(n)}}(x)\downarrow \text{ and } \varphi_x^{\mathbf{0}^{(n)}}(x) \in R_i, \\ a_i & \text{otherwise} \end{cases}$$

The function f is clearly $\mathbf{0}^{(n+1)}-$computable.

Define a numbering β of the family $\{A_0, A_1, \dots, A_k\}$ by $\beta \leftrightharpoons \alpha \circ f$. By Theorem 1.2, β is $\Sigma^0_{n+2}-$computable. By construction of f, we have that $\beta \not\leqslant_{\mathbf{0}^{(n)}} \alpha$. Define a numbering γ as follows. Let $b_0, b_1, \dots, b_m$ be any list of $\alpha-$indices of the elements of $\mathcal{A}$. Let $\gamma(x) = \alpha(b_x)$ if $x \leq m$, and let $\gamma(m+y+1) = \beta(y)$ for all y.

Obviously, $\gamma \in \mathrm{Com}^0_{n+2}(\mathcal{A})$ and $\gamma \not\leqslant_{\mathbf{0}^{(n)}} \alpha$. Therefore, the numbering α can not be $\mathbf{0}^{(n)}-$universal in $\mathrm{Com}^0_{n+2}(\mathcal{A})$. Contradiction.

(3) $\Rightarrow$ (1). Let $\mathcal{A}$ contain a least element $\perp$ under inclusion. By Theorem 3.1, the family $\mathcal{A}$ has a numbering α which is $\mathbf{0}^{(n+1)}-$universal in $\mathrm{Com}^0_{n+2}(\mathcal{A})$. We will prove that $\alpha_\perp^{\mathbf{0}^{(n+1)}}$ is universal in $\mathrm{Com}^0_{n+2}(\mathcal{A})$.

By Theorem 2.11(2), the completion $\alpha_\perp^{\mathbf{0}^{(n+1)}}$ of α is $\Sigma^0_{n+2}-$computable. By Theorem 2.10(3), $\alpha \leqslant \alpha_\perp^{\mathbf{0}^{(n+1)}}$, and since α is $\mathbf{0}^{(n+1)}-$universal in $\mathrm{Com}^0_{n+2}(\mathcal{A})$ it follows that $\alpha \equiv_{\mathbf{0}^{(n+1)}} \alpha_\perp^{\mathbf{0}^{(n+1)}}$.

Therefore, for every $\beta \in \mathrm{Com}^0_{n+2}(\mathcal{A})$ we have $\beta \leqslant_{\mathbf{0}^{(n+1)}} \alpha_\perp^{\mathbf{0}^{(n+1)}}$, and hence, by Theorem 2.10(6), $\beta \leqslant \alpha_\perp^{\mathbf{0}^{(n+1)}}$. So, $\alpha_\perp^{\mathbf{0}^{(n+1)}}$ is a universal in $\mathrm{Com}^0_{n+2}(\mathcal{A})$. □

Corollary 3.2.1. *Let $\mathcal{A}$ be a non-trivial finite family of $\Sigma^0_{n+2}-$sets. Then*

(1) if $\mathcal{A}$ has the least set $\perp$ then for all m, $\mathcal{A}$ possesses numberings which are $\mathbf{0}^{(m)}-$universal in $\mathrm{Com}^0_{n+2}(\mathcal{A})$;

(2) if $\mathcal{A}$ does not contain the least set under inclusion then $\mathcal{A}$ has a numbering $\mathbf{0}^{(m)}-$universal in $\mathrm{Com}^0_{n+2}(\mathcal{A})$ if and only if $m \geq n+1$.

Proof. It immediately follows from Theorems 3.1, 3.2. □

Corollary 3.2.2. *If $\mathcal{A}$ is a finite family of Σ^0_{n+2} sets, then for every $i \leq n$, $\mathcal{R}^{0,\mathbf{0}^{(i)}}_{n+2}(\mathcal{A})$ has greatest element if and only if $\mathcal{A}$ has a least element with respect to inclusion.*

Proof. By Theorem 3.2. □

Corollary 3.2.3. *If a finite family $\mathcal{A} \subseteq \Sigma^0_{n+1}$ contains $\perp$ then the numberings of $\mathcal{A}$ which are universal in $\mathrm{Com}^0_{n+1}(\mathcal{A})$ are all $\mathbf{0}^{(i)}$–complete with respect to $\perp$ for every $i \leq n$.*

Proof. Immediate from Theorem 3.2 and Corollary 3.2.1. □

As to infinite families, for every n there are infinite Σ^0_{n+1}–computable families $\mathcal{A}$ without numberings which are universal (even $\mathbf{0}^{(n)}$–universal) in $\mathrm{Com}^0_{n+1}(\mathcal{A})$, and there exist infinite Σ^0_{n+1}–computable families $\mathcal{A}$ with numberings which are universal in $\mathrm{Com}^0_{n+1}(\mathcal{A})$. This will be shown in more details in Theorem 1.4 and Theorem 1.5, [2].

4. Open Questions

We conclude this overview of complete numberings and completions by listing some open problems.

We know from Remark 1.2 that the minimal numberings constructed by Badaev and Goncharov in Theorem 1.3 can not be complete. So the following question is quite natural:

Question 1. *Prove or disprove that no Σ^0_{n+2}–computable minimal numberings of any non–trivial family $\mathcal{A}$ can be complete.*

As a consequence of Theorem 2.10 and Corollary 2.10.1, we may ask:

Question 2. *Is it true that*

$$((\alpha_a^{\mathbf{0}^{(n)}})_b^{\mathbf{0}^{(n)}})_a^{\mathbf{0}^{(n)}} \equiv (\alpha_a^{\mathbf{0}^{(n)}})_b^{\mathbf{0}^{(n)}}.$$

In particular, is it true that

$$((\alpha_a^{\mathbf{0}})_b^{\mathbf{0}})_a^{\mathbf{0}} \equiv (\alpha_a^{\mathbf{0}})_b^{\mathbf{0}})?$$

We know that no complete numbering is splittable (see Theorem 2.3). One may ask:

Question 3. *Let $\alpha \in \mathrm{Com}^0_{n+2}(\mathcal{A})$ be a numbering of a non–trivial family $\mathcal{A}$ and suppose that α is not complete with respect to $A \in \mathcal{A}$. Does there exist a numbering β such that*

$$\alpha < \beta < \alpha_A^{\mathbf{0}}?$$

Question 4. *Let $\alpha \in \mathrm{Com}^0_{n+2}(\mathcal{A})$ be a numbering of a non–trivial family $\mathcal{A}$, and assume that α is not complete with respect to $A \in \mathcal{A}$. Does there exist a non–splittable numbering β such that*

$$\alpha < \beta < \alpha_A^{\mathbf{0}}?$$

References

[1] S.A. Badaev, S.S. Goncharov, *On Rogers semilattices of families of arithmetical sets.* Algebra i Logika, 2001, vol. 40, no. 5, pp. 507–522 (Russian); Algebra and Logic, 2001, vol. 40, no. 5, pp. 283–291 (English translation).

[2] S.A. Badaev, S.S. Goncharov, A. Sorbi, *Arithmetical numberings: completeness and universality.* This volume.

[3] S.S. Goncharov, S.A. Badaev, *Families with one-element Rogers semilattice.* Algebra i Logika, 1998, vol. 37, no. 1, pp. 36–62 (Russian); Algebra and Logic, 1998, vol. 37, no. 1, pp. 21–34 (English translation).

[4] S.D. Denisov, I.A. Lavrov, *Complete numerations with an infinite number of special elements.* Algebra i Logika, 1970, vol. 9, pp. 503–509 (Russian); Algebra and Logic, 1970, vol. 9, pp. 301–304 (English translation).

[5] Yu.L. Ershov, *On inseparable pairs.* Algebra i Logika, 1970, vol. 9, no. 6, pp. 661–666 (Russian); Algebra and Logic, 1970, vol. 9, no. 6, pp. 396–399 (English translation).

[6] Yu.L. Ershov,*Theory of Numberings.* Nauka, Moscow, 1977 (Russian).

[7] S.S. Goncharov, A. Sorbi, *Generalized computable numerations and non-trivial Rogers semilattices.* Algebra i Logika, 1997, vol. 36, no. 6, pp. 621–641 (Russian); Algebra and Logic, 1997, vol. 36, no. 6, pp. 359–369 (English translation).

[8] A.H. Lachlan, *Standard classes of recursively enumerable sets.* Zeit. Mat. Log. Grund. Math., 1964, vol. 10, pp. 23–42.

[9] A.I. Mal'tsev, *Constructive algebras, I* Uspekhi Mat. Nauk, 1961, vol. 16, no. 3, pp. 3–60 (Russian); in [13], pp. 248–214 (English translation).

[10] A.I. Mal'tsev, *Sets with complete numberings.* Algebra i Logika, 1963, vol. 2, no. 2, pp. 4–29 (Russian); in [13], pp. 287–312 (English translation).

[11] A.I. Mal'tsev, *Towards a theory of computable families of objects.* Algebra i Logika, 1963, vol. 3, no. 4, pp. 5–31 (Russian); in [13], pp. 353–378 (English translation).

[12] A.I. Mal'tsev, *Algorithms and Recursive Functions.* Nauka, Moscow, 1965 (Russian) ; Wolters-Noordoff Publishing, Groningen, 1970 (English translation).

[13] A.I. Mal'tsev, *The Metamathematics of Algebraic Systems.* North—Holland, Amsterdam, 1971.

[14] H. Rogers, *Gödel numberings of partial computable functions.* J. Symbolic Logic, 1958, v.23, n.3, pp.49–57.

[15] H. Rogers, Jr. *Theory of Recursive Functions and Effective Computability.* McGraw-Hill, New York, 1967.

[16] R.I. Soare *Recursively Enumerable Sets and Degrees.* Springer-Verlag, Berlin Heidelberg, 1987.

ALGEBRAIC PROPERTIES OF ROGERS SEMILATTICES OF ARITHMETICAL NUMBERINGS *

Serikzhan Badaev
Kazakh National University
Almaty, Kazakhstan
badaev@math.kz

Sergey Goncharov
Institute of Discrete Mathematics and Informatics
Novosibirsk State University,
Novosibirsk, Russia
gonchar@math.nsu.ru

Sergey Podzorov
Institute of discrete mathematics and Informatics
Novosibirsk State University
Novosibirsk, Russia
podz@math.nsu.ru

Andrea Sorbi
University of Siena
Siena, Italy
sorbi@unisi.it

Abstract We investigate initial segments and intervals of Rogers semilattices of arithmetical families. We prove that there exist intervals with different algebraic properties; the elementary theory of any Rogers semilattice at arithmetical level $n \geq 2$ is hereditarily undecidable; the class of all Rogers semilattices of a fixed level $n \geq 2$ has an incomplete theory.

* All authors were partially supported by grant INTAS-RFBR *Computability and Models* no. 97-139 and grant INTAS *Computability in Hierarchies and Topological Spaces* no. 00-499.

Keywords: Numberings; reducibility of numberings; arithmetical hierarchy; Rogers semilattice; initial segments; intervals; hereditarily undecidable theory; the lattice of c.e. sets.

One of the main tasks of the theory of numberings is the study and the characterization of algebraic and elementary properties of the Rogers semilattices of the families of numberings under investigation. The study of Rogers semilattices was started by H. Rogers, [20]. In the theory of numberings, see [5], the notion of a Rogers semilattice plays a fundamental role. Rogers semilattices are used to classify properties of computable numberings for different families.

For the basic notions and notations relative to arithmetical numberings the reader is referred to [2]. For unexplained terminology and notations relative to computability theory, our main references are the textbooks of A.I. Mal'tsev [10], H. Rogers [13] and R. Soare [14]. For the main concepts and notions of the theory of numberings we refer to the book of Yu.L. Ershov [5].

This paper intends to continue the program of research initiated by S. Badaev and S. Goncharov in [2]. In particular we address some of the possible questions on how complicated the Rogers semilattice $\mathcal{R}^0_{n+1}(A)$, of a given Σ^0_{n+1}–computable family, can be.

1. The Cardinality of Rogers Semilattices

The question of the cardinality of the Rogers semilattices of Σ^0_1– families was settled by Khutoretsky, [14].

Theorem 1.1 (A. Khutoretsky). *Let $\mathcal{A}$ be a Σ^0_1–computable family. If the Rogers semilattice $\mathcal{R}^0_1(\mathcal{A})$ contains at least two distinct elements then it is infinite.*

Proof. See [14]. □

Indeed, Khutoretsky shows that it is possible to embed a linear ordering of type ω into $\mathcal{R}^0_1(\mathcal{A})$ above any non-greatest element of $\mathcal{R}^0_1(\mathcal{A})$. Subsequently Badaev, [1], was able to prove that in every nontrivial Rogers semilattice $\mathcal{R}^0_1(\mathcal{A})$, for every non-greatest element one can embed a chain with no endpoints containing that element, and for every non-minimal element one can embed a chain with no endpoints containing that element.

The following result, [22], shows that $\mathcal{R}^0_1(\mathcal{A})$ is never a lattice, if $\mathcal{R}^0_1(\mathcal{A})$ is not trivial.

Theorem 1.2 (V. Selivanov). *Let $\mathcal{A}$ be a Σ^0_1–computable family of c.e. sets. If $\mathcal{R}^0_1(\mathcal{A})$ contains at least two elements then it is not a lattice.*

Proof. See [22]. □

For Σ^0_1– families, both Theorem 1.1 and Theorem 1.2 answer questions posed by Ershov, [6]. The corresponding questions for Σ^0_{n+2}– computable families have been answered by Goncharov and Sorbi, [12].

Theorem 1.3 (S. Goncharov, A. Sorbi). *If a Σ^0_{n+2}– computable family $\mathcal{A}$ contains at least two elements then the Rogers semilattice $\mathcal{R}^0_{n+2}(\mathcal{A})$ is infinite and is not a lattice.*

Proof. See [12]. The proof consists of two cases, distinguishing whether we deal with a finite family or an infinite Σ^0_{n+2}– computable family $\mathcal{A}$. We notice that if $\mathcal{A}$ is infinite, then the claim follows also from Badaev and Goncharov's result [3] and from Theorem 1.3, [2], which reads that for every infinite Σ^0_{n+2}– computable family $\mathcal{A}$ the corresponding Rogers semilattice $\mathcal{R}^0_{n+2}(\mathcal{A})$ has infinitely many minimal elements. Thus $\mathcal{R}^0_{n+2}(\mathcal{A})$ is obviously infinite and not a lattice since no pair of distinct minimal elements can have greatest lower bound.

The original proof of this claim given in [12] is however quite different. We wish to sketch this proof here since it employs a method for constructing numberings with some desired property which promises to have applications also in different contexts.

So, let $\mathcal{A}$ be a Σ^0_{n+2}– computable infinite family. Let $\alpha \in \mathrm{Com}^0_{n+2}(\mathcal{A})$, and let a, b be such that $\alpha(a) \neq \alpha(b)$. We will construct a family $\{\alpha_i\}_{i\in\mathbf{N}}$ of Σ^0_{n+2}– computable numberings of $\mathcal{A}$, such that for no $\beta \in \mathrm{Com}^0_{n+2}(\mathcal{A})$ do we have $\beta \leqslant \alpha_i, \alpha_j$ if $i \neq j$. Notice that this implies that $\mathcal{R}^0_{n+2}(\mathcal{A})$ is infinite and not a lattice.

By Lemma 1.1, [2], let $\{B_{k,t}\}_{k,t\in\mathbf{N}}$ be a $\mathbf{0}^{(n)}$– computable sequence of finite sets (given according to their canonical indices) such that for every k, $\alpha(k) = \underline{\lim}\,_t B_{k,t}$. In constructing the numberings α_i, $i \in \mathbf{N}$, first of all we will ensure that each α_i is a numbering of a subfamily of $\mathcal{A}$ by arranging that for every x there exists k such that $\alpha_i(x) = \underline{\lim}\,_t B_{k,t}$. Then we aim to satisfy the following requirements, for every i, j, k, l, y, with $i \neq j$:

$$P_{i,j,k,l} : \varphi_k, \varphi_l \text{ total } \Rightarrow (\exists x)(\alpha_i(\varphi_k(x)) \neq \alpha_j(\varphi_l(x))$$
$$Q_{i,y} : (\exists x)(\alpha(y) = \alpha_i(x)).$$

We observe that satisfaction of all $Q_{i,y}$'s ensures that each α_i is a numbering of the whole family $\mathcal{A}$.

In the course of the construction below, at step t, for every i, k we define a finite set $D^i_{k,t}$ so that eventually

$$\alpha_i(k) \leftrightharpoons \underline{\lim}\,_t D^i_{k,t}.$$

We will arrange that the sequence $\{D^i_{k,t}\}_{k,t\in\mathbf{N}}$ is $\mathbf{0}^{(n)}$– computable. Then by Lemma 1.1, [2], each α_i is Σ^0_{n+2}– computable.

The construction is by stages. Suppose that $D^i_{k,t} = B_{x,t}$, and at stage $t+1$, we define $D^i_{k,t+1} = B_{x',t+1}$ for $x' \neq x$. Then we say that at $t+1$ we *redefine* D^i_k; if we do not redefine D^i_k at stage $t+1$ then it is understood that $D^i_{k,t+1} = B_{x,t+1}$.

We arrange the requirements in an effective priority listing

$$R_0 < R_1 < \dots ,$$

in which each $P_{i,j,k,l}$, $i \neq j$, and each $Q_{i,y}$ appears exactly once.

At stage t we define also the value of a restraint function $r(h,t)$ for every requirement R_h, and we declare some requirements temporarily satisfied or temporarily unsatisfied.

Stage 0) Define $D^i_{k,0} \leftrightharpoons B_{k,0}$, for every $i, k \in \mathbf{N}$. Let $r(h,0) \leftrightharpoons -1$, for every $h \in \mathbf{N}$. All requirements are *temporarily unsatisfied.*

Stage $t+1$) We say that a requirement R_h *requires attention at* $t+1$, if R_h is temporarily unsatisfied at the end of stage t, and there is a number x such that $\varphi_{k,t}(x), \varphi_{l,t}(x) > r(h,t)$ if $R_h = P_{i,j,k,l}$.

Choose the least h such that R_h requires attention.

- if $R_h = P_{i,j,k,l}$, then choose some such x and define

$$D^i_{\varphi_k(x),t+1} \leftrightharpoons B_{a,t+1}$$
$$D^j_{\varphi_l(x),t+1} \leftrightharpoons B_{b,t+1}.$$

  (Notice that if we never redefine $D^i_{\varphi_k(x)}$ and $D^j_{\varphi_l(x)}$ at any later stage then

$$\alpha_i(\varphi_k(x)) = \underline{\lim}_t D^i_{\varphi_k(x),t} = \underline{\lim}_t B_{a,t} = \alpha(a),$$
$$\alpha_j(\varphi_l(x)) = \underline{\lim}_t D^j_{\varphi_l(x),t} = \underline{\lim}_t B_{b,t} = \alpha(b),$$

  hence $\alpha_i(\varphi_k(x)) \neq \alpha_j(\varphi_l(x))$.) Set

$$r(h', t+1) = \max\{\varphi_k(x), \varphi_l(x)\}$$

  for all $h' > h$.

- If $R_h = Q_{i,y}$ then choose $x = r(h,t)+1$ and define

$$D^i_{x,t+1} \leftrightharpoons B_{y,t+1}.$$

  Set $r(h', t+1) = x$, for all $h' > h$.

Whatever the case, declare R_h *temporarily satisfied* at the end of stage $t+1$, declare $R_{h'}$ *temporarily unsatisfied* at the end of stage $t+1$, for all $h' > h$.

This ends the construction. Notice that $r(h,t)$ and the function assigning to each triple i,k,t a number x such that $D^i_{k,t} = B_{x,t}$ are in fact computable, so $\{D^i_{k,t}\}_{k,t\in\mathbf{N}}$ is $\mathbf{0}^{(n)}$–computable. For the verification, a standard inductive argument shows that each requirement requires attention only finitely many times, and thus for every i,k we redefine D^i_k only finitely often. We pause only to point out where the assumption that $\mathcal{A}$ is infinite is needed in this argument. Consider $R_h = P_{i,j,k,l}$, and suppose that t_0 is the least stage such that no $R_{h'}$ with $h' < h$ requires attention at any $t \geq t_0$. Thus $r(h) = \lim_t r(h,t)$ exists, being $r(h) = r(h,t_0)$, and, starting from stage t_0, R_h is temporarily unsatisfied until it eventually requires attention. Now, if $\beta \in \mathrm{Com}^0_{n+2}(\mathcal{A})$ is such that $\beta = \alpha_i \circ \varphi_k$ and $\beta = \alpha_j \circ \varphi_l$ (with φ_k, φ_l total) then since $\mathcal{A}$ is infinite we have that both φ_k and φ_l have infinite ranges, so that eventually we find x such that $\varphi_k(x), \varphi_l(x) > r(h)$. At this point R_h requires attention (for the last time), we act accordingly on R_h, $D^i_{\varphi_k(x)}$ and $D^j_{\varphi_l(x)}$ are never again redefined, and thus, as already remarked, our action ensures that $\beta \neq \alpha_i \circ \varphi_k$ or $\beta \neq \alpha_j \circ \varphi_l$, and therefore the requirement R_h is eventually met. (Notice that a requirement of the form $P_{i,j,k,l}$ can be met even if it is eventually always temporarily unsatisfied, for instance if φ_k and φ_l are not both total.)

Remark 1.1. Notice that the relation in i,x,y "$x \in \alpha_i(y)$" is itself Σ^0_{n+2}.

Let us consider the case in which $\mathcal{A}$ is finite, say $\mathcal{A} = \{A_0, \dots, A_k\}$. If C is a proper c.e. subset of $\mathbf{N}$ then define

$$\alpha^C(x) \leftrightharpoons \begin{cases} A_{x+2} & \text{if } x < k-1, \\ A_0 & \text{if } x \geq k-1 \text{ and } x+1-k \in C \\ A_1 & \text{if } x \geq k-1 \text{ and } x+1-k \notin C. \end{cases}$$

If now $\beta = \alpha^C \circ f$, where f is computable, then one easily shows that $\beta \equiv \alpha^D$, where $D = f^{-1}[C]$. Hence every principal ideal of the c.e. m–degrees is isomorphic to an ideal of $R^0_{n+2}(\mathcal{A})$. The result then follows in this case from the fact that the semilattice of the c.e. m–degrees is not a lattice, and that it has infinite ideals, see [7] and [17]. □

Apparently, classical computability may seem very close to Σ^{-1}_n–computability (the classes Σ^{-1}_n refer to the finite levels of the Ershov hierarchy of the Δ^0_2 sets), since classical computability is definable by means of Σ-formulas of first order arithmetic, while Σ^{-1}_n–computability is definable by means of Boolean combinations of these formulas.

Unfortunately, the methods employed in Theorems 3.1 – 3.3 do not seem to apply to the case of Σ^{-1}_n–computability. In this regard, both Ershov's questions – on cardinality of Rogers semilattices and on whether or not a Rogers semilattice is always a lattice – are still open. It looks appropriate to raise these questions here (see also [2]). Consider computable numberings of families of Σ^{-1}_{n+1} sets in the sense of Definition 1.1, [2]:

Question 5. *Let $\mathcal{A} \subseteq \Sigma^{-1}_{n+1}$ be such that $\mathcal{R}^{-1}_{n+1}(\mathcal{A})$ contains at least two elements. Is $\mathcal{R}^{-1}_{n+1}(\mathcal{A})$ infinite?*

Question 6. *Let $\mathcal{A} \subseteq \Sigma^{-1}_{n+1}$ be such that $\mathcal{R}^{-1}_{n+1}(\mathcal{A})$ contains at least two elements. Is $\mathcal{R}^{-1}_{n+1}(\mathcal{A})$ a lattice?*

2. Intervals in Rogers Semilattices

We first introduce some preliminary notations and remarks. If α and β are Σ^0_{n+1}–computable numberings of a family $\mathcal{A}$, then we denote by $[\alpha, \beta]$ the closed interval of $\mathcal{R}^0_{n+1}(\mathcal{A})$,

$$[\alpha, \beta] \leftrightharpoons \{\deg(\gamma) \mid \deg(\alpha) \leqslant \deg(\gamma) \leqslant \deg(\beta)\}.$$

We also let

$$(\alpha, \beta) \leftrightharpoons \{\deg(\gamma) \mid \deg(\alpha) < \deg(\gamma) < \deg(\beta)\}.$$

Moreover, we will denote by $\hat{\beta}$ the principal ideal of $\mathcal{R}^0_{n+1}(\mathcal{A})$,

$$\hat{\beta} \leftrightharpoons \{\deg(\gamma) \mid \deg(\gamma) \leqslant \deg(\beta)\}.$$

Finally, we let $\hat{\hat{\beta}} \leftrightharpoons \hat{\beta} \setminus \{\deg(\beta)\}$.

In the rest of the paper, the following Definition 2.1 and Lemma 2.2 will play fundamental role.

Definition 2.1. Let β be a numbering of any family $\mathcal{A}$, and let C be any nonempty $\mathbf{X}$– c.e. set. For every $\mathbf{X}$– computable function f such that range$(f) = C$, define $\beta_{C,\mathbf{X}} \leftrightharpoons \beta \circ f$. If C is c.e. we simply write β_C instead of $\beta_{C,\mathbf{0}}$.

Then $\beta_{C,\mathbf{X}}$ is a numbering of some subfamily $\mathcal{A}_0 \subseteq \mathcal{A}$. Clearly, for every pair of $\mathbf{X}$– computable function f and g such that range$(f) =$ range$(g) = C$, we have

$$\beta \circ f \equiv_{\mathbf{X}} \beta \circ g.$$

Thus, in contexts in which we are interested in comparing numberings with respect to $\leqslant_{\mathbf{X}}$, we are justified in writing $\beta_{C,\mathbf{X}}$ without mentioning any particular $\mathbf{X}$– computable function f such that range$(f) = C$.

The following lemma holds:

Lemma 2.1. *For every triple of sets X, Y, Z such that $X \oplus Y \leqslant_T Z$ and for every pair of numberings β, γ we have:*

(1) if A is $\mathbf{X}$– c.e. then $\beta_{A,\mathbf{X}} \leqslant_{\mathbf{X}} \beta$;

(2) *if A is* $\mathbf{X}$*–c.e., B is* $\mathbf{Y}$*–c.e. then the following are equivalent:*

(a) $\beta_{A,\mathbf{X}} \leqslant_{\mathbf{Z}} \beta_{B,\mathbf{Y}}$;

(b) *there exists a partial* $\mathbf{Z}$*–computable function* φ *satisfying the conditions:* $\mathrm{dom}(\varphi) \supseteq A$, $\varphi(A) \subseteq B$ *and for all* $x \in A$, $\beta(x) = \beta(\varphi(x))$;

(c) *there exists a* $\mathbf{Z}$*–c.e. equivalence relation* η *such that*

$$\forall x, y((x, y) \in \eta \,\&\, x \neq y \Rightarrow x, y \in A \cup B),$$
$$\forall x, y((x, y) \in \eta \Rightarrow \beta(x) = \beta(y)),$$
$$\forall x \in A \exists y \in B((x, y) \in \eta);$$

(3) *if A is* $\mathbf{X}$*–c.e., B is* $\mathbf{Y}$*–c.e., and* $A \subseteq B$ *then* $\beta_{A,\mathbf{X}} \leqslant_{\mathbf{Z}} \beta_{B,\mathbf{Y}}$;

(4) *if A is* $\mathbf{X}$*–c.e., B is* $\mathbf{Y}$*–c.e. then* $\beta_{A\cup B,\mathbf{Z}} \equiv_{\mathbf{Z}} \beta_{A,\mathbf{X}} \oplus \beta_{B,\mathbf{Y}}$;

(5) *if A is* $\mathbf{X}$*–c.e., B is* $\mathbf{Y}$*–c.e., and if* $\beta_{A,\mathbf{X}} \leqslant_{\mathbf{Z}} \beta_{B,\mathbf{Y}}$*, then* $\beta_{B,\mathbf{Y}} \equiv_{\mathbf{Z}} \beta_{A\cup B,\mathbf{Z}}$;

(6) *if* $\gamma \leqslant_{\mathbf{X}} \beta$ *then there exists an* $\mathbf{X}$*–c.e. set A such that* $\gamma \equiv_{\mathbf{X}} \beta_{A,\mathbf{X}}$;

(7) *if* $\gamma \leqslant_{\mathbf{X}} \beta$*, and* $\gamma \equiv_{\mathbf{X}} \beta_{A,\mathbf{X}}$*, for some* $\mathbf{X}$*–c.e. A, then for every* α *such that* $\gamma \leqslant_{\mathbf{X}} \alpha \leqslant_{\mathbf{X}} \beta$ *there exists an* $\mathbf{X}$*–c.e. set B with* $A \subseteq B$ *and* $\alpha \equiv_{\mathbf{X}} \beta_{B,\mathbf{X}}$;

(8) *if* $\gamma \leqslant_{\mathbf{X}} \beta$ *via some* $\mathbf{X}$*–computable function f, then* $\gamma \leqslant_{\mathbf{Z}} \beta_{A,\mathbf{Y}}$*, for every* $\mathbf{Y}$*–c.e. set A such that* $A \supseteq \mathrm{range}(f)$;

(9) *if A and B are* $\mathbf{X}$*–c.e. and for some* $a \in B$, $\beta[A] = \{\beta(a)\}$ *then* $\beta_{A\cup B,\mathbf{X}} \equiv_{\mathbf{X}} \beta_{B,\mathbf{X}}$;

(10) *if A and B are* $\mathbf{X}$*–c.e., A is finite and* $\beta[A] \subseteq \beta[B]$ *then we have* $\beta_{A\cup B,\mathbf{X}} \equiv_{\mathbf{X}} \beta_{B,\mathbf{X}}$.

Proof. We now show the various items, one by one.

(1) Immediate.

(2) Let f and g be an $\mathbf{X}$–computable function and a $\mathbf{Y}$–computable function, respectively, such that $A = \mathrm{range}(f)$, $B = \mathrm{range}(g)$.

(a) ⇒ (b). Assume that $\beta_{A,\mathbf{X}} \leqslant_{\mathbf{Z}} \beta_{B,\mathbf{Y}}$ via some $\mathbf{Z}$–computable function h. Then $\beta \circ f = \beta \circ g \circ h$. Define

$$\varphi(x) = \begin{cases} g(h(\mu y\,(x = f(y)))) & \text{if } x \in A, \\ \uparrow & \text{otherwise.} \end{cases}$$

Then φ is the desired partial $\mathbf{Z}$–computable function. (Notice that $\mathrm{dom}(\varphi) = A$.)

(b) $\Rightarrow$ (c). Let η be the $\mathbf{Z}$–c.e. equivalence relation generated by the set of pairs

$$\{(x, \varphi(x)) \mid x \in A\}.$$

It is easy to see that η satisfies (c).

(c) $\Rightarrow$ (a). Let η satisfy the conditions in (c); let $\eta = \bigcup_s \eta^s$ where $\{\eta^s\}_{s \in \mathbf{N}}$ is a $\mathbf{Z}$–computable sequence of finite sets approximating η. Define a function h as follows:

$$h(x) \leftrightharpoons \pi_0(\mu s((f(x), g(\pi_0(s))) \in \eta^{\pi_1(s)})).$$

It is now easy to see that h is a $\mathbf{Z}$–computable function reducing $\beta_{A,\mathbf{X}}$ to $\beta_{B,\mathbf{Y}}$.

(3) This item follows from (2b) taking φ to be the identity embedding of A into B, i.e.

$$\varphi(x) \leftrightharpoons \begin{cases} x & \text{if } x \in A, \\ \uparrow & \text{otherwise.} \end{cases}$$

Notice that φ is in this case partial $\mathbf{X}$–computable.

(4) Claim (3) implies that $\beta_{A,\mathbf{X}} \leqslant_{\mathbf{Z}} \beta_{A\cup B,\mathbf{Z}}$ and $\beta_{B,\mathbf{Y}} \leqslant_{\mathbf{Z}} \beta_{A\cup B,\mathbf{Z}}$. Therefore, $\beta_{A,\mathbf{X}} \oplus \beta_{B,\mathbf{Y}} \leqslant_{\mathbf{Z}} \beta_{A\cup B,\mathbf{Z}}$.

To prove that $\beta_{A\cup B,\mathbf{Z}} \leqslant_{\mathbf{Z}} \beta_{A,\mathbf{X}} \oplus \beta_{B,\mathbf{Y}}$ and let f_1 be an $\mathbf{X}$–computable function, f_2 a $\mathbf{Y}$–computable function and f a $\mathbf{Z}$–computable function whose ranges are A, B and $A \cup B$ respectively. By the Reduction Theorem there exist $\mathbf{Z}$–c.e. sets A_1 and B_1 such that $A_1 \subseteq A$, $B_1 \subseteq B$, $A_1 \cap B_1 = \emptyset$ and $A_1 \cup B_1 = A \cup B$. Then $f^{-1}(A_1) \cup f^{-1}(B_1) = \mathbf{N}$ and $f^{-1}(A_1) \cap f^{-1}(B_1) = \emptyset$. Hence, $f^{-1}(A_1), f^{-1}(B_1)$ are $\mathbf{Z}$–computable sets. For $i = 1, 2$, let $g_i(x) \leftrightharpoons \mu y(f_i(y) = f(x))$ and define

$$g(x) \leftrightharpoons \begin{cases} 2g_1(x) & \text{if } x \in f^{-1}(A_1), \\ 2g_2(x) + 1 & \text{if } x \in f^{-1}(B_1) \end{cases}.$$

It is easy to see that $\beta_{A\cup B,\mathbf{Z}}$ is reducible to $\beta_{A,\mathbf{X}} \oplus \beta_{B,\mathbf{Y}}$ by the $\mathbf{Z}$–computable function g.

(5) Immediately from (4).

(6) A numbering $\gamma \leqslant_{\mathbf{X}} \beta$ can be considered of the form $\gamma = \beta_{A,\mathbf{X}}$, where $A = \mathrm{range}(f)$, and f is any $\mathbf{X}$–computable function that reduces γ to β.

(7) This follows from (5) and (6).

(8) Let $\gamma = \beta \circ f$, where f is some $\mathbf{X}$–computable function, and let $A \supseteq \mathrm{range}(f)$ be any $\mathbf{Y}$–c.e. set, with say $A = \mathrm{range}(g)$ and g is $\mathbf{Y}$–computable. Then

$$h(x) \rightleftharpoons \mu y\,(g(y) = f(x))$$

is the desired $\mathbf{Z}$–computable function that reduces γ to $\beta_{A,\mathbf{Y}}$.

(9) Let η be the least $\mathbf{X}$–c.e. equivalence relation containing the set $\{(x, a) \mid x \in A\}$. Then by (2), $\beta_{A,\mathbf{X}} \leqslant_{\mathbf{X}} \beta_{B,\mathbf{X}}$. Thus by (5), $\beta_{A\cup B,\mathbf{X}} \equiv_{\mathbf{X}} \beta_{B,\mathbf{X}}$, and statement (9) is proved.

(10) This is a consequence of (9).

□

Particularly important is the case in which $X = Y = \emptyset$. In this case the previous lemma gives:

Lemma 2.2. *For every pair A, B of c.e. sets and for every pair of numberings α, β, we have*

(1) The following are equivalent:

(a) $\beta_A \leqslant \beta_B$;

(b) there is a partial computable function φ satisfying $\mathrm{dom}(\varphi) \supseteq A$, $\varphi(A) \subseteq B$ and for all $x \in A$, $\beta(x) = \beta(\varphi(x))$;

(c) there exists a c.e. equivalence relation η such that

$$\begin{aligned}&\forall x, y((x,y) \in \eta \;\&\; x \neq y \Rightarrow x, y \in A \cup B),\\&\forall x, y((x,y) \in \eta \Rightarrow \beta(x) = \beta(y)),\\&\forall x \in A \exists y \in B((x,y) \in \eta);\end{aligned}$$

(2) if $A \subseteq B$ then $\beta_A \leqslant \beta_B$;

(3) if $\beta_A \leqslant \beta_B$, then $\beta_B \equiv \beta_{A\cup B}$;

(4) if $\alpha \leqslant \beta$ then $\alpha = \beta_C$ for some c.e. set C;

(5) if $\alpha \leqslant \beta$, and $\alpha \equiv \beta_C$, for some c.e. set C, then for every γ such that $\alpha \leqslant \gamma \leqslant \beta$ there exists a c.e. set D with $C \subseteq D$ and $\gamma \equiv \beta_D$;

(6) $\beta_{A\cup B} \equiv \beta_A \oplus \beta_B$;

(7) let U be any set; if $\gamma \leqslant_{\mathbf{U}} \beta$ via some $\mathbf{U}$–computable function f, then $\gamma \leqslant_{\mathbf{U}} \beta_C$, for every c.e. set C such that $C \supseteq \mathrm{range}(f)$.

(8) *if for some* $a \in B$, $\beta[A] = \{\beta(a)\}$ *then* $\beta_{A \cup B} \equiv \beta_B$;

(9) *if* A *is finite set and* $\beta[A] \subseteq \beta[B]$ *then* $\beta_{A \cup B} \equiv \beta_B$.

Proof. The statements follow immediately from the previous lemma. □

We will consider mainly intervals and ideals of Rogers semilattices $\mathcal{R}^0_m(\mathcal{A})$ with $m \geq 2$. It should be mentioned that that the principal ideals of $\mathcal{R}^0_1(\mathcal{A})$, in the case of finite families $\mathcal{A}$ of c.e. sets, were completely described by Yu.L. Ershov and I.A. Lavrov in [8]. V.V. V'jugin studied initial segments of $\mathcal{R}^0_1(\mathcal{A})$ for some infinite families $\mathcal{A} \subseteq \Sigma^0_1$ (see [15] and [25]). A first insight into the ideals of $\mathcal{R}^0_m(\mathcal{A})$, for $m \geq 2$, is provided by the following theorem.

Theorem 2.1 (S.S.Goncharov). *Let* $\mathfrak{B}_{k+1}$ *be the finite Boolean algebra with* $k+1$ *atoms. Then*

(1) *for every infinite* Σ^0_{n+2}*–computable family* $\mathcal{A}$, $\mathcal{R}^0_{n+2}(\mathcal{A})$ *contains an element* $\deg(\beta)$ *such that the subsemilattice* $\hat{\beta}$ *is isomorphic to* $\mathfrak{B}_{k+1} \setminus \{0\}$;

(2) *for every finite* Σ^0_{n+2}*–computable family* $\mathcal{A}$, $\mathcal{R}^0_{n+2}(\mathcal{A})$ *contains an element* $\deg(\beta)$ *such that the subsemilattice* $\hat{\beta}$ *is isomorphic to* $\mathfrak{B}_{k+1}$.

Proof. Let $R_0, R_1, \ldots, R_k$ be infinite computable sets such that

$$\bigcup_{i \leq k} R_i = \mathbf{N} \qquad \text{and} \qquad R_i \cap R_j = \emptyset \text{ if } i \neq j.$$

For every $i \leq k$ let f_i be a computable 1–1 function such that $\text{range}(f_i) = R_i$. We will consider the Boolean algebra $\mathfrak{B}_{k+1}$ as the subalgebra of the Boolean algebra of all subsets of $\mathbf{N}$, generated by $R_0, R_1, \ldots, R_k$.

(1) Let $\mathcal{A}$ be an infinite Σ^0_{n+2}–computable family, and let $A_0, \ldots, A_k$ be distinct elements of $\mathcal{A}$. Choose $\alpha \in \text{Com}^0_{n+2}(\mathcal{A})$, and let M be a maximal set. For every $i \leq k$, let $M_i \leftrightharpoons f_i[M]$, and let

$$R_i \setminus M_i \leftrightharpoons \{m^i_0 < m^i_1 < m^i_2 < \ldots\}.$$

Define a numbering β of $\mathcal{A}$ as follows: For every $x \in \mathbf{N}$, find $i \leq k$ such that $x \in R_i$, and let

$$\beta(x) \leftrightharpoons \begin{cases} A_i & \text{if } x \in M_i, \\ \alpha(j) & \text{if } x = m^i_j. \end{cases}$$

Following the proof of Theorem 1.3, [2], one easily shows that for every $i \leq k$, β_{R_i} is a minimal Σ^0_{n+2}–computable numbering of $\mathcal{A}$, and $\beta_{R_i} \not\leq \beta_{R_j}$ if $i \neq j$. Lemma 2.2(6) implies that

$$\beta \equiv \beta_{R_0 \cup R_1 \cup \ldots R_k} \equiv \beta_{R_0} \oplus \beta_{R_1} \oplus \cdots \oplus \beta_{R_k}.$$

We will now prove that $\hat{\beta}$ is isomorphic to $\mathfrak{B}_{k+1} \setminus \{0\}$. It is sufficient to show that for every $\gamma \in \mathrm{Com}^0_{n+2}(\mathcal{A})$, if $\gamma \leqslant \beta$ then

$$\gamma \equiv \beta_{R_{i_0}} \oplus \beta_{R_{i_1}} \oplus \cdots \oplus \beta_{R_{i_s}}$$

for some unique sequence $0 \leq i_0 < i_1 < \ldots < i_s \leq k$.

So, let $\gamma \leqslant \beta$ and choose by Lemma 2.2(4) some c.e. set Q such that $\gamma \equiv \beta_Q$. Denote by I_0 and I_1 the sets

$$I_0 \leftrightharpoons \{i \mid i \leq k \;\&\; Q \cap (R_i \setminus M_i) \text{ is infinite}\}$$

and

$$I_1 \leftrightharpoons \{0, 1, \ldots, k\} \setminus I_0,$$

respectively. Note that $I_0 \neq \emptyset$ since $\mathcal{A}$ is infinite. Define

$$Q_0 \leftrightharpoons \bigcup_{i \in I_0} (R_i \setminus M_i) \cap \overline{Q},$$
$$Q_1 \leftrightharpoons \bigcup_{i \in I_1} (R_i \setminus M_i) \cap Q,$$
$$Q_2 \leftrightharpoons \bigcup_{i \in I_1} (Q \cap M_i),$$
$$Q_3 \leftrightharpoons \bigcup_{i \in I_0} ((Q \cup Q_0) \cap R_i).$$

Since M is maximal, it follows that Q_0 is finite. Then, by Lemma 2.2(9),

$$\beta_{Q \cup Q_0} \equiv \beta_Q.$$

We have $Q \cup Q_0 = Q_1 \cup Q_2 \cup Q_3$. The set Q_1 is finite by choice of I_1, therefore, by Lemma 2.2(9), $\beta_Q \equiv \beta_{Q_2 \cup Q_3}$. Since $\beta[Q \cap M_i] \subseteq \{A_i\}$ for all $i \leq k$ it follows by Lemma 2.2(6) and (8) that

$$\beta_Q \equiv \beta_{Q_3} \equiv \beta_{R_{i_0}} \oplus \beta_{R_{i_1}} \oplus \cdots \oplus \beta_{R_{i_s}},$$

where $i_0, i_1, \ldots, i_s$ are all the elements of I_0. Thus, the ideal $\hat{\beta}$ of the semilattice $\mathcal{R}^0_{n+2}(\mathcal{A})$ is isomorphic to $\mathfrak{B}_{k+1} \setminus \{0\}$.

For uniqueness it is sufficient to show that, for $i \leq k$,

$$\beta_{R_i} \not\leqslant \beta_{R_0} \oplus \ldots \oplus \beta_{R_{i-1}} \oplus \beta_{R_{i+1}} \oplus \ldots \oplus \beta_{R_k}.$$

By Lemma 2.2(6) it is sufficient to show that $\beta_{R_i} \not\leqslant \beta_{R'_i}$ where $R'_i = \bigcup_{j \neq i} R_j$. Let $\beta \circ f_i = \beta \circ g \circ h$ where g and h are computable functions and $\mathrm{range}(g) = R'_i$.

Then $(g \circ h)[M] \subseteq \bigcup_{j \neq i}(R_j \setminus M_j)$ and $(g \circ h)[M]$ is finite since $\bigcup_{j \neq i}(R_j \setminus M_j)$ is immune. But this implies that M is computable, contradicting the choice of M. Thus, the ideal $\hat{\beta}$ of the semilattice $\mathcal{R}^0_{n+2}(\mathcal{A})$ is isomorphic to $\mathfrak{B}_{k+1} \setminus \{0\}$.

(2) If $\mathcal{A}$ is finite then, as in Theorem 1.3, we can easily show that $\mathcal{R}^0_{n+2}(\mathcal{A})$ contains any principal ideal of the c.e. $m-$degrees. It follows from a result of Lachlan, [17], that there exists a c.e. set such that the principal ideal of $m-$degrees generated by the $m-$degree of this set is isomorphic to $\mathfrak{B}_{k+1}$. □

We are now going to exhibit a great variety of intervals and ideals of $\mathcal{R}^0_{n+2}(\mathcal{A})$, for every $n \geq 0$. We still need some definitions and notations.

Definition 2.2. *Let ε denote the poset of all c.e. subsets of ω with respect to inclusion. As is well known, ε is a bounded (i.e. with least and greatest element) distributive lattice.*

Let ε^* denote the bounded distributive lattice obtained by dividing ε modulo the ideal of all finite sets.

If X is any c.e. set, then ε_X denotes the principal filter of ε generated by X, i.e. the collection of all c.e. supersets of X: Clearly ε_X is still a bounded distributive lattice. Likewise, ε^*_X denotes the bounded distributive lattice obtained by dividing ε_X modulo the finite sets.

The following three lemmas play a key role in understanding the constructions in the proofs of several theorems in the sequel of the paper.

Lemma 2.3. *Let α, β be $\Sigma^0_{n+2}-$computable numberings of a family $\mathcal{A}$. Then the mapping $\psi_1 : \varepsilon \longrightarrow \mathcal{R}^0_{n+2}(\mathcal{A})$ defined by $\psi_1(C) \leftrightharpoons \deg(\alpha \oplus \beta_C)$ for every $C \in \varepsilon$, induces an isomorphism from ε^* onto $[\alpha, \alpha \oplus \beta]$ if and only if*

(1) for every infinite $C \in \varepsilon$, $\beta_C \nleqslant \alpha$;

(2) for every $C_1, C_2 \in \varepsilon$ if $C_1 \setminus C_2$ is infinite then $\beta_{C_1} \nleqslant \beta_{C_2}$.

Proof. The conditions (1) and (2) are clearly necessary. As to sufficiency, let (1) and (2) hold. Clearly, by (2) and (9) of Lemma 2.2, if $C_1 \subseteq^* C_2$ then $\alpha \oplus \beta_{C_1} \leqslant \alpha \oplus \beta_{C_2}$.

Moreover, we can use (5) and (6) of Lemma 2.2 to show that the mapping induced by ψ_1 is onto. In fact, let γ be a numbering of $\mathcal{A}$ such that $\alpha \leqslant \gamma \leqslant \alpha \oplus \beta$ and let C_1 be the set of all even numbers. Then $\alpha \equiv (\alpha \oplus \beta)_{C_1}$. By Lemma 2.2 (5), there exists a c.e. superset C_2 of the set C_1 such that $\gamma \equiv (\alpha \oplus \beta)_{C_2}$. Then by Lemma 2.2 (6), $(\alpha \oplus \beta)_{C_2} \equiv (\alpha \oplus \beta)_{C_1} \oplus (\alpha \oplus \beta)_{C_2 \setminus C_1}$. Therefore $\gamma \equiv \alpha \oplus \beta_D$ where $D = \{x : 2x + 1 \in C_2 \setminus C_1\}$.

We are only left to show that if $C_1 \setminus C_2$ is infinite then $\alpha \oplus \beta_{C_1} \nleqslant \alpha \oplus \beta_{C_2}$. For the sake of a contradiction, suppose otherwise, and let $C_1 \setminus C_2$ be infinite and $\alpha \oplus \beta_{C_1} \leqslant \alpha \oplus \beta_{C_2}$. Then $\beta_{C_1} \leqslant \alpha \oplus \beta_{C_2}$. Let f be a computable function such that range$(f) = C_1$, and let g be a computable function that reduces β_{C_1} to $\alpha \oplus \beta_{C_2}$. Define

$$C_3 \leftrightharpoons \{f(x) \mid g(x) \text{ even}\}$$
$$C_4 \leftrightharpoons \{f(x) \mid g(x) \text{ odd}\}.$$

Then $C_1 = C_3 \cup C_4$ and $C_3 \cap C_4 = \emptyset$. Clearly $\beta_{C_3} \leqslant \alpha$. If C_3 were infinite, then by condition (1) we would get $\beta_{C_3} \nleqslant \alpha$. Thus C_3 is finite. But then $C_1 =^* C_4$ and thus, being $\beta_{C_4} \leqslant \beta_{C_2}$, we would have $\beta_{C_1} \equiv \beta_{C_4} \leqslant \beta_{C_2}$, contradicting (2). □

Lemma 2.4. *Let β be a Σ^0_{n+2}–computable numbering of an infinite family $\mathcal{A}$ such that every set of $\mathcal{A}$ has infinitely many β-indices. Let $\psi_2 : \varepsilon \longrightarrow \mathcal{R}^0_{n+2}(\mathcal{A})$ be a mapping defined as follows: $\psi_2(C) \leftrightharpoons \deg(\beta_C)$, $C \in \varepsilon$. Then*

(1) ψ_2 induces an epimorphism of upper semilattices from the semilattice $\langle \varepsilon^* \setminus \{\bot\}, \cup^*, \subseteq^* \rangle$ onto the ideal $\hat{\beta}$ of the semilattice $\mathcal{R}^0_{n+2}(\mathcal{A})$ if and only if for every infinite $C \in \varepsilon$, β_C is a numbering of the whole family;

(2) ψ_2 induces an isomorphism between $\langle \varepsilon^* \setminus \{\bot\}, \cup^*, \subseteq^* \rangle$ and $\mathcal{R}^0_{n+2}(\mathcal{A})$ if and only if for every pair of infinite sets $C_1, C_2 \in \varepsilon$, if $C_1 \setminus C_2$ is infinite then $\beta_{C_1} \nleqslant \beta_{C_2}$.

Proof. Let $\mathcal{A}$ and β be chosen as in statement of the lemma.

(1) Necessity is obvious (if the condition does not hold then ψ_2 is not well defined). For sufficiency we only need to show that ψ_2 preserves the binary operation and induces a mapping onto $\hat{\beta}$. This follows immediately from (4) and (6) of Lemma 2.2 (notice that ψ_2 is well defined by (9) of lemma 2.2).

(2) Sufficiency is evident. Let us prove necessity. Let ψ_2 induce an isomorphism of $\langle \varepsilon^* \setminus \{\bot\}, \cup^*, \subseteq^* \rangle$ onto $\mathcal{R}^0_{n+2}(\mathcal{A})$ and assume that $\beta_{C_1} \leqslant \beta_{C_2}$ for some pair of infinite c.e. sets C_1, C_2 such that $C_1 \setminus C_2$ is infinite. Then by (3) of Lemma 2.2, $\beta_{C_2} \equiv \beta_{C_1 \cup C_2}$. Therefore, $\psi_2[C_2] = \psi_2[C_1 \cup C_2]$ but $C_2 \neq^* C_1 \cup C_2$. This is in contradiction with the injectivity of the mapping induced by ψ_2 on $\varepsilon^* \setminus \{\bot\}$.

□

Lemma 2.5. *Let $\mathcal{A}$ be a finite family of Σ^0_{n+2}–sets, and suppose that for some $\beta \in \mathrm{Com}^0_{n+2}(\mathcal{A})$ and some finite set F we have that $\beta[F] = \mathcal{A}$. Let $\psi_2 : \varepsilon \longrightarrow \mathcal{R}^0_{n+2}(\mathcal{A})$ be defined as in the previous lemma. Then*

(1) ψ_2 induces an epimorphism of upper semilattices from $\langle \varepsilon^*_F, \cup^*, \subseteq^* \rangle$ onto the ideal $\hat{\beta}$ of the semilattice $\mathcal{R}^0_{n+2}(\mathcal{A})$;

(2) *ψ_2 induces an isomorphism between $\langle \varepsilon_F^*, \cup^*, \subseteq^* \rangle$ and $\mathcal{R}_{n+2}^0(\mathcal{A})$ if and only if for every pair of sets $C_1, C_2 \in \varepsilon_F$, if $C_1 \setminus C_2$ is infinite then $\beta_{C_1} \nleqslant \beta_{C_2}$.*

Proof. Let $\mathcal{A}$, β, and F be chosen as in statement of lemma.

(1) The mapping ψ_2 is onto by (4) and (9) of Lemma 2.2 and preserves the binary operation by (6) of the same lemma. (Notice that $\varepsilon_F^* = \varepsilon^*$).

(2) Similar to the proof of (2) in Lemma 2.4.

□

Theorem 2.2. *Let $\mathcal{A}$ be a Σ_{n+2}^0–computable family. If $\alpha \in \mathrm{Com}_{n+2}^0(\mathcal{A})$ is not $\mathbf{0}'$–universal in $\mathrm{Com}_{n+2}^0(\mathcal{A})$ then there exists a numbering $\beta \in \mathrm{Com}_{n+2}^0(\mathcal{A})$ such that the following conditions hold:*

- *The interval $[\alpha, \alpha \oplus \beta]$ is isomorphic to ε^*;*
- *the ideal $\hat{\beta}$ is isomorphic to $\varepsilon^* \setminus \{\bot\}$ if $\mathcal{A}$ is infinite, and $\hat{\beta}$ is isomorphic to ε^* if $\mathcal{A}$ is finite family.*

Proof. Let $\mathcal{A}$ and α be given as in the statement of the theorem. We will construct the numbering β by stages. Let $\gamma \in \mathrm{Com}_{n+2}^0(\mathcal{A})$ be such that $\gamma \nleqslant_{\mathbf{0}'} \alpha$, and let a_1, a_2 be two numbers such that $\gamma(a_1) \neq \gamma(a_2)$. Let $\{\epsilon_t\}_{t \in \mathbf{N}}$ be a standard listing of all c.e. equivalence relations. Firstly we will consider the case when $\mathcal{A}$ is infinite family.

The requirements. In view of Lemma 2.3 and Lemma 2.4, we will build a numbering β of $\mathcal{A}$ so as to satisfy the following requirements, for every p, q, t:

$$P : \beta \leqslant_{\mathbf{0}'} \gamma$$
$$Q_p : W_p \text{ infinite } \Rightarrow \gamma \leqslant_{\mathbf{0}'} \beta_{W_p}$$
$$R_{t,p,q} : W_p \setminus W_q \text{ infinite } \Rightarrow \beta_{W_p} \nleqslant \beta_{W_q} \text{ via } \epsilon_t$$

where we say that $\beta_{W_p} \nleqslant \beta_{W_q}$ *via* ϵ_t if ϵ_t does not witness the reduction of β_{W_p} to β_{W_q} in the sense of (1c) of Lemma 2.2.

We will satisfy P by building a $\mathbf{0}'$–computable function h such that $\beta = \gamma \circ h$. This feature of β (i.e. $\beta \leqslant_{\mathbf{0}'} \gamma$) will be used in the verification.

For every p we will satisfy Q_p by building, whenever W_p is infinite, a $\mathbf{0}'$–computable function f_p such that range$(f_p) \subseteq W_p$ and $\gamma = \beta \circ f_p$. Thus, by (7) of Lemma 2.2, $\gamma \leqslant_{\mathbf{0}'} \beta_{W_p}$, which implies that β_{W_p} is a numbering of the whole family, and $\beta_{W_p} \nleqslant \alpha$, otherwise $\gamma \leqslant_{\mathbf{0}'} \alpha$.

Finally, if $W_p \setminus W_q$ is infinite then we will satisfy $R_{t,p,q}$ by choosing in the interval $(k(t), k(t+1))$ – where k is a suitable increasing $\mathbf{0}'$–computable function – numbers x, y with $(x, y) \in \epsilon_t$, and letting $\beta(x) \neq \beta(y)$.

Together with the numbering β, and the functions h, f_p, during the construction we will also define the values of a counter function $\lambda t\, s(p,t)$, for every number p. In fact, the definition of β will follow automatically from h by letting $\beta^0 \rightleftharpoons \{\langle 0, \gamma(h(0))\rangle\}$ and

$$\beta^{t+1} \rightleftharpoons \beta^t \cup \{\langle x, \gamma(h(x))\rangle \mid k(t) < x \leq k(t+1)\}$$

for all t.

The construction. By stages:

Stage 0) Define

$$k(0) = h(0) \rightleftharpoons 0, f_p^0 \rightleftharpoons \emptyset, s(p,0) \rightleftharpoons p$$

for every $p \in \mathbf{N}$ and go to next stage.

Stage $t+1$). Go through Steps 1, 2), 3) in order.

Step 1) Check whether

$$\exists x_0 \exists x_1 \exists y_0 \exists y_1 [(\forall i \leq 1)\, (k(t) < x_i, y_i \\ \& \ |\{x_0, x_1, y_0, y_1\}| = 4 \ \& \ (x_i, y_i) \in \epsilon_t)] \quad \text{(A)}$$

If (A) does not hold then define $k(t+1) \rightleftharpoons k(t)+1$. If (A) holds then fix such numbers x_0, x_1, y_0, y_1 so that, say, $\langle x_0, x_1, y_0, y_1\rangle$ is the least such quadruple, and define

$$k(t+1) \rightleftharpoons \max\{x_0, x_1, y_0, y_1\} + 1.$$

Move to Step 2).

Step 2) Check if the following condition $B(p)$ holds, for any $p \leq t$:

$$B(p): \qquad \exists z (k(t) < z \leq k(t+1) \ \& \ z \in W_p).$$

If $B(p)$ holds of some p, then fix p_0 such that $B(p_0)$ holds and

$$\forall p \leq t (B(p) \ \& \ p \neq p_0 \Rightarrow (s(p_0,t) < s(p,t) \\ \vee \ (s(p_0,t) = s(p,t) \ \& \ p_0 < p)).$$

Let z_0 be the least z for which $B(p_0)$ holds. Choose the least number $u_0 \notin \mathrm{dom}(f_{p_0}^t)$. Fix the least $i_0 \in \{0,1\}$ such that $z_0 \notin \{x_{i_0}, y_{i_0}\}$. Go to Step 3).

Step 3) We distinguish the following cases:

(a) If (A) is not true and $B(p)$ fails for every $p \leq t$ then choose the least number $y \notin \mathrm{range}(h^t)$ and let

$$h^{t+1} \rightleftharpoons h^t \cup \{(k(t+1), y)\}, f_p^{t+1} \rightleftharpoons f_p^t, s(p,t+1) \rightleftharpoons s(p,t) \quad \text{for all} \quad p.$$

(b) If (A) holds and, for some p, $B(p)$ holds then define

$$h^{t+1} \leftrightharpoons h^t \cup \{(x_{i_0}, a_1), (y_{i_0}, a_2), (z_0, u_0)\} \cup \{(x, a_1) \mid k(t) < x \leq k(t+1) \ \&\ x \notin \{x_{i_0}, y_{i_0}, z_0\}\}$$

and

$$f_{p_0}^{t+1} \leftrightharpoons f_{p_0}^t \cup \{(u_0, z_0)\}, s(p_0, t+1) \leftrightharpoons s(p_0, t) + 1, f_p^{t+1} \leftrightharpoons f_p^t$$
$$s(p, t+1) \leftrightharpoons s(p, t) \quad \text{for all} \quad p \neq p_0$$

(c) If (A) holds but $B(p)$ fails for all $p \leq t$ then define

$$h^{t+1} \leftrightharpoons h^t \cup \{(x_0, a_1), (y_0, a_2)\} \cup \{(x, a_1) \mid k(t) < x \leq k(t+1) \ \&\ x \notin \{x_0, y_0\}\}$$

and

$$f_p^{t+1} \leftrightharpoons f_p^t, s(p, t+1) \leftrightharpoons s(p, t) \quad \text{for all} \quad p.$$

(d) If (A) fails and $B(p)$ holds for some p then define

$$h^{t+1} \leftrightharpoons h^t \cup \{(z_0, u_0)\}$$

and

$$f_{p_0}^{t+1} \leftrightharpoons f_{p_0}^t \cup \{(u_0, z_0)\}, s(p_0, t+1) \leftrightharpoons s(p_0, t) + 1$$
$$f_p^{t+1} \leftrightharpoons f_p^t, s(p, t+1) \leftrightharpoons s(p, t) \quad \text{for all} \quad p \neq p_0$$

Notice that in this case $z_0 = k(t+1)$. Go to stage $t+2$.

The verification. We show that the following conditions hold:

(1) For every t, $k(t) < k(t+1)$.

(2) The functions h, k, f_p, with $p \in \mathbf{N}$, are $\mathbf{0}'$–computable.

(3) For every x, we have that $\beta(x) = \gamma(h(x))$. Thus β is a Σ^0_{n+2}–computable numbering of some subfamily $\mathcal{A}_0 \subseteq \mathcal{A}$.

(4) For every p, $s(p) \leq s(p+1) \leq s(p) + 1$.

(1)–(4) are obvious.

(5) For every p,

$$W_p \text{ infinite } \Rightarrow \lim_t s(p, t) = \infty.$$

To see this, assume that W_p is infinite and at a stage t_0 we have $s(p, t_0) = m$ for some m. We will show that for some $t_1 \geq t_0$, $s(p, t_1 + 1) > m$.

For all t and $p' > m$ we have $s(p', t) > m$ (recall that $s(p', 0) = p'$). Let $t_1 \geq t_0$ be a stage such that $t_1 \geq p$ and for every $p' \leq m$,

- if $\lim_t s(p', t) < \infty$ than $(\forall t \geq t_1)[s(p', t) = s(p', t_1)]$;
- if $\lim_t s(p', t) = \infty$ than $s(p', t_1) > m$;
- $(k(t_1), k(t_1 + 1)] \cap W_p \neq \emptyset$.

So, either $s(p, t_1) > m$, or $s(p, t_1) = m$. In the latter case, we choose $p = p_0$ at t_1 and increase the counter $s(p, t)$, i.e. $s(p, t_1 + 1) = m + 1$.

(6) For every p, if W_p is infinite then f_p is total, range$(f_p) \subseteq W_p$, and $\gamma = \beta \circ f_p$.

This follows from the way we define f_p and β and the fact that $\lim_t s(p, t) = +\infty$ by (5) and thus every time we increase $s(p, t)$ we also define f_p on the least element u_0 which is not in the domain of f_p^t:

$$\beta(f_p(u_0)) = \beta(z_0) = \gamma(h(z_0)) = \gamma(u_0).$$

(7) For every p, if W_p is infinite then β is a numbering of the whole family and $\beta_{W_p} \nleqslant \alpha$.

This follows immediately from the previous point, and the choice of $\gamma \nleqslant_{0'} \alpha$.

(8) For every p, q if $W_p \setminus W_q$ is infinite then $\beta_{W_p} \nleqslant \beta_{W_p}$.

To see this, assume that $W_p \setminus W_q$ is infinite, and $\beta_{W_p} \leqslant \beta_{W_p}$. By (1c) of Lemma 2.2, let η be a c.e. equivalence relation such that

$$\forall x, y((x, y) \in \eta \;\&\; x \neq y \Rightarrow x, y \in W_p \cup W_q),$$
$$\forall x, y((x, y) \in \eta \Rightarrow \beta(x) = \beta(y)),$$
$$\forall x \in W_p \exists y \in W_q((x, y) \in \eta).$$

Let $\eta = \epsilon_t$ for some $t \in \mathbf{N}$. Since $W_p \setminus W_q$ is infinite it follows that either η has infinitely many nontrivial equivalence classes, or it has an infinite equivalence class. Therefore (A) holds at stage t. By construction, at stage t we proceed through (b) or (c) in Step 3. We then define $h(x) = a_1$, $h(y) = a_2$ for some $x, y > k(t)$ such that $(x, y) \in \eta$. But $(x, y) \in \eta$ implies $\beta(x) = \beta(y)$, and therefore $\gamma(a_1) = \gamma(a_2)$, since $\beta(x) = \gamma(h(x))$ and $\beta(y) = \gamma(h(y))$. This contradicts our choice of $\gamma(a_1) \neq \gamma(a_2)$.

To finish off the proof in the case of an infinite family $\mathcal{A}$, we observe that β satisfies both Lemma 2.3 and Lemma 2.4. (Note that every set of $\mathcal{A}$ has infinitely many β-indices. This is consequence of (6), since there are infinitely many pairwise disjoint infinite c.e. sets.) Hence the interval $[\alpha, \alpha \oplus \beta]$ is isomorphic to ε^* and the ideal $\hat{\beta}$ is isomorphic to $\varepsilon^* \setminus \{\perp\}$.

Finally, if $\mathcal{A}$ is finite, then we may slightly modify stage 0 of our construction as follows. Let $\mathcal{A}$ consist of $m + 1$ sets and let $b_0, b_1, \ldots, b_m$ be γ-indices of these sets. Then at stage 0 we define $k(0) \leftrightharpoons m$, $h(i) \leftrightharpoons b_i$ for all $i \leq m$, and let $f_p^0 \leftrightharpoons \emptyset$, $s(p,0) \leftrightharpoons p$ for every $p \in \mathbf{N}$.

It is easy to check that the modified numbering β satisfies both Lemma 2.3 and Lemma 2.5 and thus in this case the ideal $\hat{\beta}$ is isomorphic to ε_F^* for $F = \{0, 1, \dots, m\}$. The result then follows by observing the trivial fact that the lattices ε_F^* and ε^* are isomorphic. □

Corollary 2.2.1 (S.Podzorov, [12]). *Let $\mathcal{A}$ be any Σ^0_{n+2}–computable family. For every numbering $\alpha \in \mathrm{Com}^0_{n+2}(\mathcal{A})$ there exists a numbering $\beta \in \mathrm{Com}^0_{n+2}(\mathcal{A})$ which is $\mathbf{0}'$–equivalent to α and such that*

(1) $\hat{\beta}$ is isomorphic to $\varepsilon^* \backslash \{\bot\}$ if the family $\mathcal{A}$ is infinite;

(2) $\hat{\beta}$ is isomorphic to ε if the family $\mathcal{A}$ is finite.

Proof. Note that numbering β constructed in Theorem 2.2 is $\mathbf{0}'$–equivalent to γ. The proof is now immediate, since if we do not require that the interval $[\alpha, \alpha \oplus \beta]$ be isomorphic to ε^*, then we do not need the feature that $\gamma \not\leq_{\mathbf{0}'} \alpha$. In other words, we can carry on the construction of Theorem 2.2 starting from the numbering α instead of γ, and thus obtaining the desired conclusions (1) and (2). □

Corollary 2.2.1 and the fact that the elementary theory of ε^* is hereditarily undecidable, [13], immediately yield:

Corollary 2.2.2. *The elementary theory of every non-trivial Rogers semilattice $\mathcal{R}^0_{n+2}(\mathcal{A})$ is hereditarily undecidable.*

Corollary 2.2.1 and Corollary 2.2.2 give us a deep insight into the complexity of Rogers semilattices of Σ^0_{n+2}–computable families. The case of Σ^0_1–computable families is still open:

Question 7. *Is the elementary theory of any non-trivial Rogers semilattice of a Σ^0_1–computable family hereditarily undecidable, or at least undecidable?*

We know that the range of isomorphism types of ε_A^*, for different c.e. sets A, is very wide. For instance, we know that every countable Boolean algebra with a sufficiently effective presentation is isomorphic to some ε_A^* (namely every Σ^0_3–presentable Boolean algebra, see [9]; see also [14]). Under reasonable conditions, we can realize these isomorphism types as intervals of Rogers semilattices of arithmetical numberings.

Theorem 2.3. *Let $\mathcal{A}$ be a Σ^0_{n+3}–computable family, and let A be both an infinite and coinfinite c.e. set. Then for every $\alpha \in \mathrm{Com}^0_{n+3}(\mathcal{A})$ such that α is not $\mathbf{0}^{(2)}$–universal in $\mathrm{Com}^0_{n+3}(\mathcal{A})$, there exists some numbering $\beta \in \mathrm{Com}^0_{n+3}(\mathcal{A})$ such that the subsemilattice $[\alpha, \beta]$ is isomorphic to ε_A^*.*

Proof. Let $\mathcal{A}$, A, and α be given, and let $f : \mathbf{N} \longrightarrow A$ be a computable bijection. We will construct a numbering β, as required. First of all, we define

$$\beta(f(n)) \leftrightharpoons \alpha(n)$$

so that $\alpha \leqslant \beta$ and $\alpha \equiv \beta_A$. It remains to define $\beta(n)$ on each $n \in \overline{A}$: This will be done, stage by stage, by the construction below. Fix some $\gamma \in \mathrm{Com}^0_{n+3}(\mathcal{A})$ such that $\gamma \nleqslant_{\mathbf{0}^{(2)}} \alpha$; define a listing of all c.e. supersets of A by letting, for every i,

$$V_i \leftrightharpoons W_i \cup A;$$

and fix some distinct $C, D \in \mathcal{A}$.

We sketch the strategy for the construction. We define β so that the mapping

$$\psi(V_i) \leftrightharpoons \deg(\beta_{V_i})$$

induces an isomorphism from ε^*_A onto $[\alpha, \beta]$. For this, we first prove:

Lemma 2.6. *ψ (as defined above) induces an isomorphism from ε^*_A onto $[\alpha, \beta]$ if and only if*

$$\forall V_i, V_j \in \varepsilon_A \left(V_i \setminus V_j \textit{ infinite } \Rightarrow \beta_{V_i} \nleqslant \beta_{V_j}\right). \qquad (\dagger)$$

Proof. The condition ($\dagger$) is clearly necessary. Suppose now that ψ satisfies condition ($\dagger$). It follows from (2) and (9) of Lemma 2.2 that

$$V_i \subseteq^* V_j \Rightarrow \beta_{V_i} \leqslant \beta_{V_j}.$$

By (5) of Lemma 2.2 every γ such that $\alpha \leqslant \gamma \leqslant \beta$ satisfies $\gamma \equiv \beta_B$, for some c.e. $B \supseteq A$. So ψ is onto. On the other hand, if $V_i \nsubseteq^* V_j$, then $V_i \setminus V_j$ is infinite, and thus $\beta_{V_i} \nleqslant \beta_{V_j}$. So ψ is an isomorphism. □

The requirements. In view of Lemma 2.6 the construction aims to satisfy the following requirements:

$$R_{p,i,j} : V_i \setminus V_j \text{ infinite and } \varphi_p[V_i] \nsubseteq^* A \Rightarrow \beta_{V_i} \nleqslant \beta_{V_j} \text{ via } \varphi_p$$
$$Q_i : V_i \setminus A \text{ infinite } \Rightarrow \gamma \leqslant_{\mathbf{0}^{(2)}} \beta_{V_i}$$

where we say that $\beta_{V_i} \nleqslant \beta_{V_j}$ *via* φ_p if the following does not happen (see (1b) of Lemma 2.2): $V_i \subseteq \mathrm{dom}(\varphi_p)$, $\varphi_p[V_i] \subseteq V_j$, and, for every $x \in V_i$, $\beta(x) = \beta(\varphi_p(x))$.

The requirement $R_{p,i,j}$ will be satisfied at stage $2t+1$ with $t = \langle p, i, j\rangle$, by looking for some $k \in V_i \setminus A$ such that $\varphi_p(k) \in V_j \setminus A$, and by defining $\beta(k) \neq \beta(\varphi_p(k))$.

The requirement Q_i will be attacked at infinitely many stages $2t+2$, with $\pi_0(t) = i$, each time extending (if $V_i \setminus A$ is infinite), the values of a $\mathbf{0}^{(2)}$–

computable function f_i such that range$(f_i) \subseteq \overline{A}$ and $\gamma = \beta \circ f_i$. This is used to show that $\gamma \leqslant_{0^{(2)}} \alpha$ (as shown in details later), contrary to the assumptions.

Step by step, we will construct an increasing sequence $\{A^t\}_{t \in \mathbf{N}}$ of finite sets such that $\overline{A} = \bigcup_t A^t$, and the (in general, partial) $0^{(2)}$–computable functions $\{f_i\}_{i \in \mathbf{N}}$.

The construction. By stages:
Stage 0) Define $\beta(f(n)) \rightleftharpoons \alpha(n)$ for every n. Let $f_i^0 \rightleftharpoons \emptyset$ for all i, and let $A^0 \rightleftharpoons \emptyset$.

Stage $2t+1$). Assume that $t = \langle p, i, j \rangle$. If $i \neq j$ and the following conditions hold:

(1) $V_i \subseteq \mathrm{dom}(\varphi_p)$ & $\forall y (y \in V_i \Rightarrow \varphi_p(y) \in V_j)$;

(2) $\exists k (k \in V_i \setminus A$ & $\varphi_p(k) \in V_j \setminus A)$ & $k \notin A^{2t}$ & $\varphi_p(k) \notin A^{2t} \cup \{k\})$

then choose the least such number k, let

$$A^{2t+1} \rightleftharpoons A^{2t} \cup \{k, \varphi_p(k)\}$$
$$\beta(k) \rightleftharpoons C$$
$$\beta(\varphi_p(k)) \rightleftharpoons D$$

and move to next stage.

If $i = j$, or either of (1) or (2) does not hold, then do nothing and move to next stage.

Stage $2t+2$). Suppose that $\pi_0(t) = i$, and there exists $x \in V_i \setminus A$ such that $\beta(x)$ has not yet been defined. Fix the least such x, and define

$$A^{2t+2} \rightleftharpoons A^{2t+1} \cup \{x\}$$
$$f_i^{2t+2} \rightleftharpoons f_i^{2t+1} \cup \{(s, x)\}$$

where s is the least number such that $s \notin \mathrm{dom}(f_i^{2t+1})$; finally define

$$\beta(x) \rightleftharpoons \gamma(s).$$

Move to next stage.

Finally, define $f_i = \bigcup_t f_i^t$, for every i.

The verification. The following properties hold:

(1) For every x there exists a stage t such that $\beta(x)$ is defined at stage t.

To see this, notice that if $x \in A$ then $\beta(x)$ is defined at stage 0 of the construction. On the other hand, let i_0 be such that $\mathbf{N} = V_{i_0}$. At every stage $2t+2$, with $\pi_0(t) = i_0$, we define $\beta(x)$ for the least x such that $x \notin A$ and

$\beta(x)$ is undefined by stage $2t+1$. So if $x \notin A$, then there exists some t such that we define $\beta(x)$ at stage $2t+2$.

(2) $\beta \in \mathrm{Com}^0_{n+3}(\mathcal{A})$.

Indeed, range$(\beta) = \mathcal{A}$. This follows from the way β is defined at stage 0. Moreover, the relation "$y \in \beta(x)$" is $\mathbf{0}^{(3)}$–computably enumerable, since $\alpha, \gamma \in \mathrm{Com}^0_{n+3}(\mathcal{A})$ and the conditions (1) and (2) of odd stages, as well as the test relative to each even stage can be effectively tested with oracle $\mathbf{0}^{(2)}$.

(3) For every i, if $V_i \setminus A$ is infinite, then $\beta_{V_i} \nleqslant \alpha$.

Indeed, if $V_i \setminus A$ is infinite then at every stage $2t+2$ such that $\pi_0(t) = i$, we have that $f_i^{2t+2} = f^{2t+1} \cup \{(s, x)\}$, for some s and $x \in V_i$. It is thus easy to see that f_i is $\mathbf{0}^{(2)}$–computable, and $\gamma = \beta \circ f_i$. Therefore $\gamma \leqslant_{\mathbf{0}^{(2)}} \beta_{V_i}$ by (7) of Lemma 2.2. This implies that $\beta_{V_i} \nleqslant \alpha$, since otherwise it would follow that $\gamma \leqslant_{\mathbf{0}^{(2)}} \alpha$.

(4) If $V_i \setminus V_j$ is infinite then $\beta_{V_i} \nleqslant \beta_{V_j}$.

We prove this by contradiction. Assume that $\beta_{V_i} \leqslant \beta_{V_j}$. Then by (1b) of Lemma 2.2 there exists a partial computable function φ_p such that $V_i \subseteq \mathrm{dom}(\varphi_p)$, $\varphi_p[V_i] \subseteq V_j$, and $\beta(x) = \beta(\varphi_p(x))$, for every $x \in V_i$.

Let $V_m = \{x \in V_i \mid \varphi_p(x) \neq x\} \cup A$ and $t = \langle p, m, j\rangle$. It is easy to see that $V_m \setminus A$ is infinite since it contains $V_i \setminus V_j$. We have $\varphi_p[V_m] \nsubseteq^* A$, otherwise $\beta_{V_m} \leqslant \beta_A$ via φ_p, but this contradicts (3). Moreover, $\varphi_p[V_m \setminus (A \cup A^{2t})] \nsubseteq A \cup A^{2t}$. To show this fact suppose that this inclusion holds and consider the partial computable function

$$\varphi_q(x) \leftrightharpoons \begin{cases} \varphi_p(x) & \text{if } \exists y(g(y) = \varphi_p(x) \& \forall z < y(g(z) \notin \{x, \varphi_p(x)\})), \\ x & \text{if } \exists y(g(y) = x \& \forall z < y(g(z) \notin \{x, \varphi_p(x)\})), \\ \uparrow & \text{otherwise,} \end{cases}$$

where g is a computable function from $\mathbf{N}$ onto $A \cup A^{2t}$. It's easy to see that $\mathrm{dom}(\varphi_q) \supseteq V_m$ and $\beta_{V_m} \leqslant \beta_A$ via φ_q which contradicts (3) again.

Now consider the stage $2t+1$. We see that all conditions of this stage hold and we have $\beta(k) = C \neq D = \beta(\varphi_p(k))$ for some $k \in V_m$. But $k \in V_i$ since $V_m \subseteq V_i$ by choice of V_m. □

By requesting more of α, we can show that a result similar to Theorem 2.3 holds also for Σ^0_2–computable families. (We observe that next theorem is a trivial consequence of the previous theorem if $n > 0$; for $n = 0$ it provides a stronger result. We give a proof here for any arbitrary n.)

Theorem 2.4. *Let $\mathcal{A}$ be a Σ^0_{n+2}–computable family and let A be both an infinite and coinfinite c.e. set. If $\alpha \in \mathrm{Com}^0_{n+2}(\mathcal{A})$ and α is not $\mathbf{0}^{(n+1)}$–universal in $Com^0_{n+2}(\mathcal{A})$ then there exists a numbering $\beta \in \mathrm{Com}^0_{n+2}(\mathcal{A})$ such that the interval $[\alpha, \beta]$ is isomorphic to ε^*_A.*

Proof. Let $\mathcal{A}$, A, and α be given as in the statement of the theorem. In view of Lemma 2.6, we construct β so as to satisfy the following requirements, for every p, i, j:

$$R_{p,i,j} : V_i \setminus V_j \text{ infinite } \Rightarrow \beta_{V_i} \not\leqslant \beta_{V_j} \text{ via } \varphi_p$$

where $V_e \leftrightharpoons W_e \cup A$, as in the proof of Theorem 2.3, and "$\beta_{V_i} \not\leqslant \beta_{V_j}$ via φ_p" has the same meaning as in the proof of Theorem 2.3. Thus if $V_i \setminus V_j$ is infinite, satisfaction of $R_{p,i,j}$, for any p, will guarantee that φ_p does not reduce β_{V_i} to β_{V_j} in the sense of (1b) of Lemma 2.2.

For this proof we employ a finite priority argument to computably enumerate, relatively to the oracle $\mathbf{0}^{(n+1)}$, a numbering β so that $\beta \in \mathrm{Com}^0_{n+2}(\mathcal{A})$.

We briefly sketch the strategy to meet $R_{p,i,j}$.

Let $\gamma \in \mathrm{Com}^0_{n+2}(\mathcal{A})$ be such that $\gamma \not\leqslant_{\mathbf{0}^{(n+1)}} \alpha$. Using that $\alpha, \gamma \in \mathrm{Com}^0_{n+2}(\mathcal{A})$, fix $\mathbf{0}^{(n+1)}$–computable sequences of finite sets $\{\alpha^t(x)\}_{t,x\in\mathbf{N}}$ and $\{\gamma^t(x)\}_{t,x\in\mathbf{N}}$ such that, for every t, x,

$$\alpha^t(x) \subseteq \alpha^{t+1}(x), \qquad \alpha(x) = \bigcup_t \alpha^t(x)$$

and

$$\gamma^t(x) \subseteq \gamma^{t+1}(x), \qquad \gamma(x) = \bigcup_t \gamma^t(x).$$

First of all, we make β "look like" α on A, to get $\alpha \leqslant \beta$. Then we monitor (via the notion of t–reducibility, and a suitable counter function) the step by step progress towards obtaining that $V_i \setminus V_j$ is infinite and φ_p reduces β_{V_i} to β_{V_i} (via monitoring, in turn, the progress towards obtaining $V_i \subseteq \mathrm{dom}(\varphi_p)$, $\varphi_p[V_i] \subseteq V_j$ and $\beta(x) = \beta(\varphi_p(x))$, for every $x \in V_i$). If this progress threatens to tend to infinity, then we select (by suitably labelling its elements) a set $G \subseteq V_i \setminus V_j$ (hence $G \subseteq \overline{A}$) on which we define a partial $\mathbf{0}^{(n+1)}$–computable function F_γ such that, for every $x \in G$, $\beta(x) = \gamma(F_\gamma(x))$. On the other hand, if $F_\gamma(x)$ is defined and $\varphi_p(x) \in V_j$, then we define on $\varphi_p(x)$ the value $F_\alpha(\varphi_p(x))$ of a $\mathbf{0}^{(n+1)}$–computable function F_α, so that $\beta(\varphi_p(x)) = \alpha(F_\alpha(\varphi_p(x)))$. As shown in details later, this implies that $\gamma \leqslant_{\mathbf{0}^{(n+1)}} \alpha$, contradiction. Thus φ_p does not reduce β_{V_i} to β_{V_j} (in the sense of (1b) of Lemma 2.2), and eventually the requirement is satisfied, and thus injures only finitely many times the lower priority requirements.

Some of the β-indices will be marked by labels of the form $\boxed{m}$, with $m \in \mathbf{N}$ (where m must be viewed as coding a triple, $m = \langle p, i, j\rangle$). Together with the values of the finite approximation β^t to β at stage t, the construction will define a counter function $s(m, t)$, for every m.

We will say that φ_p *t-reduces β_{V_i} to β_{V_j}* if (where $m = \langle p, i, j\rangle$)

(1) $s(m,t) < \max(V_i \setminus V_j \cap [0,t])$;

(2) for every $x \in V_i \cap [0, s(m,t)]$ we have that $x \in \mathrm{dom}(\varphi_p)$ and $\varphi_p(x) \in V_j$;

(3) for every $x \in V_i \cap [0, s(m,t)]$ we have

$$\beta^t(x) \cap [0, s(m,t)] = \beta^t(\varphi_p(x)) \cap [0, s(m,t)].$$

Fix also a partial computable 1–1 and onto function $\varphi : A \longrightarrow \mathbf{N}$, and a $\mathbf{0}'$–computable function k enumerating $\overline{A}$ in order of magnitude, i.e.

$$\overline{A} = \{k(0) < k(1) < \cdots\}.$$

Finally, at stage t we will define also the approximations F^t_α, F^t_γ of partial $\mathbf{0}^{(n+1)}$– computable functions F_α, F_γ. The arguments of these functions are β– indices, the values of F_α are α– indices, and the values of F_γ are γ– indices. Moreover, we shall have, for every t,

$$\mathrm{dom}(F^t_\alpha) \cap \mathrm{dom}(F^t_\gamma) = \emptyset$$

and

$$\mathrm{dom}(F^t_\alpha) \cup \mathrm{dom}(F^t_\gamma) = \{k(i) \mid i < t\}.$$

The construction. By stages:

Stage 0) Define $\beta^0(x) \leftrightharpoons \alpha^0(\varphi(x))$ for every $x \in A$. Let $\beta^0(x) \leftrightharpoons \emptyset$, for every $x \in \overline{A}$. Let $F^0_\alpha = \emptyset$ and $F^0_\gamma = \emptyset$. Define $s(m,0) = 0$ for all m.

Move to next stage.

Stage $t+1$) For every $m \leq t$ (with, say, $m = \langle p, i, j \rangle$) check whether the following conditions hold.

- Any number, marked by $\boxed{m'}$ with $m' \leq m$, is less then $k(t)$;
- φ_p t– reduces β_{V_i} to β_{V_j};
- $k(t) \in V_i \setminus V_j$;
- $k(t) \in \mathrm{dom}(\varphi_p)$ and $\varphi_p(k(t)) \in V_j$.

If there is no $m \leq t$ with those four properties then go directly to Procedure E below. Otherwise let $m_0 = \langle p_0, i_0, j_0 \rangle$ be the least number satisfying these conditions. Act as follows.

- Remove all labels $\boxed{m}$ with $m > m_0$ from all numbers currently marked by these labels;
- Increase $s(m_0, t)$ (i.e. define $s(m_0, t+1) = s(m_0, t) + 1$);

- Put the label $\boxed{m_0}$ on $k(t)$ (notice that $k(t)$ is not currently marked by any label since we have already removed all markers $\boxed{m}$, with $m > m_0$);

- If $\varphi_{p_0}(k(t)) \notin A$ and $\varphi_{p_0}(k(t))$ is not marked by any markers then put the marker $\boxed{m_0}$ on $\varphi_{p_0}(k(t))$;

- If $\varphi_{p_0}(k(t))$ is marked by $\boxed{m_0}$ (it should be kept in mind that we have already executed the previous item) and $\varphi_{p_0}(k(t)) \in \mathrm{dom}(F_\gamma^t)$ then remove the pair $(\varphi_{p_0}(k(t)), F_\gamma^t(\varphi_{p_0}(k(t))))$from F_γ, find the least pseudopair $\langle y, s\rangle$ such that $\beta^t(\varphi_{p_0}(k(t))) \subseteq \alpha^s(y)$ and put the pair $(\varphi_{p_0}(k(t)), y)$ into F_α.

Go to Procedure E, below.

Procedure E. If $k(t)$ is not marked by any markers then put the pair $(k(t), 0)$ into F_α. If $k(t)$ is marked by $\boxed{m}$ with $m = \langle p, i, j\rangle$, consider the following possibilities:

- $k(t) \in V_i \backslash V_j$. Find the least y such that $y \notin \mathrm{range}(F_\gamma^t)$ or $y \in \mathrm{range}(F_\gamma^t)$ but for every x such that $F_\gamma^t(x) = y$ we have that x is not marked by $\boxed{m}$ and put the pair $(k(t), y)$ into F_γ;

- $k(t) \in V_j$. Put the pair $(k(t), 0)$ into F_α.

Let F_α^{t+1} and F_γ^{t+1} be the sets of all pairs which are currently in F_α and F_γ, respectively, as a consequence of the actions undertaken in the previous stages, and of the action so far undertaken at the current stage. Let $z > t$ be the least number such that for all x, y,

$$(x, y) \in F_\alpha^{t+1} \Rightarrow \beta^t(x) \subseteq \alpha^z(y)$$
$$(x, y) \in F_\gamma^{t+1} \Rightarrow \beta^t(x) \subseteq \gamma^z(y).$$

For every x, let

$$\beta^{t+1}(x) \rightleftharpoons \begin{cases} \alpha^{t+1}(\varphi(x)) & \text{if } x \in A, \\ \alpha^z(y) & \text{if } (x, y) \in F_\alpha^{t+1} \text{ for some } y \\ \gamma^z(u) & \text{if } (x, u) \in F_\gamma^{t+1} \text{ for some } u \\ \emptyset & \text{otherwise.} \end{cases}$$

For any m, if $s(m, t+1)$ has not yet been defined then let $s(m, t+1) = s(m, t)$. Go to the next stage.

The Verification. Let $\beta(x) = \bigcup_{t\in\mathbf{N}} \beta^t(x)$. The construction satisfies the following properties:

(1) All parameters in the construction are $\mathbf{0}^{(n+1)}$– computable and β is Σ^0_{n+2}– computable numbering of the family $\beta[\mathbf{N}]$.

(2) For every t any number can be labelled by at most one marker after stage t. Any marked number must be in $\overline{A}$.

(3) We can remove marker only if we put some smaller marker. If we put a marker $\boxed{m}$ then we increase $s(m,t)$. At every stage we put at most one marker and we mark by it at most two numbers.

(4) For every m, t, $s(m,t) \leq s(m,t+1) \leq s(m,t)+1$.

(5) For every t, $\mathrm{dom}(F^{t+1}_\alpha) \cap \mathrm{dom}(F^{t+1}_\gamma) = \emptyset$, and

$$\mathrm{dom}(F^{t+1}_\alpha) \cup \mathrm{dom}(F^{t+1}_\gamma) = \{k(0), \ldots, k(t)\}.$$

(6) For every $x \notin A$ if, for some t_0, $x \in \mathrm{dom}(F^{t_0}_\alpha)$ then for all $t \geq t_0$ $x \in \mathrm{dom}(F^t_\alpha)$ and $F^t_\alpha(x) = F^{t_0}_\alpha(x)$.

(7) For every $x \notin A$ if, for some t_0, $x \in \mathrm{dom}(F^{t_0}_\gamma)$ then for all $t \geq t_0$ such that $x \in \mathrm{dom}(F^t_\gamma)$ we have $F^t_\gamma(x) = F^{t_0}_\gamma(x)$.

In view of properties (5)–(7) above we can introduce two functions: $F_\alpha = \lim_t F^t_\alpha$ and $F_\gamma = \lim_t F^t_\gamma$.

(8) $\mathrm{dom}(F_\alpha) \cap \mathrm{dom}(F_\gamma) = \emptyset$ and $\mathrm{dom}(F_\alpha) \cup \mathrm{dom}(F_\gamma) = \overline{A}$. F_α is a $\mathbf{0}^{(n+1)}$– computable partial function and for every $\mathbf{0}^{(n+1)}$– computably enumerable G such that $G \subseteq \mathrm{dom}(F_\gamma)$, the partial function $F_\gamma \restriction G$ is $\mathbf{0}^{(n+1)}$– computable.

(9) For every $x \in \mathrm{dom}(F_\alpha)$, $\beta(x) = \alpha(F_\alpha(x))$, and for every $x \in \mathrm{dom}(F_\gamma)$, $\beta(x) = \gamma(F_\gamma(x))$.

(10) β is a Σ^0_{n+2}– computable numbering of $\mathcal{A}$.

We have already noticed in (1) that β is Σ^0_{n+2}– computable. By the definition of β we have that $\beta[A] = \mathcal{A}$. From (8) and (9) we have that $\beta[\overline{A}] \subseteq \mathcal{A}$.

(11) For every m (with, say, $m = \langle p, i, j\rangle$) if $\lim_t s(m,t) = \infty$ then $V_i \setminus V_j$ is infinite and $\beta_{V_i} \leqslant \beta_{V_j}$ via φ_p (in the sense of Lemma 2.2 (1b)).

Suppose that $\lim_t s(m,t) = \infty$, and $V_i \setminus V_j$ is finite or $\beta_{V_i} \nleqslant \beta_{V_j}$ via φ_p. If $\lim_t s(m,t) = \infty$ then let $t_0 \geq m$ be a step such that

- if $V_i \setminus V_j$ is finite than $s(m,t_0) \geq \max(V_i \setminus V_j)$;
- if $V_i \nsubseteq \mathrm{dom}(\varphi_p)$ then there exists $x \in V_i$ such that $x \notin \mathrm{dom}(\varphi_p)$ and $x \leq s(m,t_0)$;
- if $\varphi_p(V_i) \nsubseteq V_j$ then there exists $x \in V_i$ such that $\varphi_p(x) \notin V_j$ and $x \leq s(m,t_0)$;

- if $\beta(x) \neq \beta(\varphi_p(x))$ for some $x \in V_i$ then $x \leq s(m, t_0)$ and for some $y \in (\beta(x) \cup \beta(\varphi_p(x))) \setminus (\beta(x) \cap \beta(\varphi_p(x)))$ we have $y \leq s(m, t_0)$ and $y \in \beta^{t_0}(x) \cup \beta^{t_0}(\varphi_p(x))$.

According to the definition of t-reducibility and in view of property (4) above, we have that for all $t \geq t_0$ φ_p does not t-reduces β_{V_i} to β_{V_j}. So for any such t, m does not satisfy the second condition of stage $t + 1$ and we have $s(m, t + 1) = s(m, t)$. A contradiction.

(12) For any $m = \langle p, i, j\rangle$ if m is the least number such that $V_i \setminus V_j$ is infinite and $\beta_{V_i} \leqslant \beta_{V_j}$ via φ_p then $\lim_t s(m, t) = \infty$.

From (11) we have that $\lim_t s(m', t) < \infty$ for all $m' < m$. Suppose that t_0 is so big that $m \leq t_0$ and $s(m', t_0) = \lim_t s(m', t_0)$ for all $m' < m$.

In order to obtain a contradiction suppose that $\lim_t s(m, t) = s(m, t_1)$ for some $t_1 \geq t_0$. In view of (3) after stage t_1 we do not put markers $\boxed{m'}$ for $m' \leq m$. Let $t \geq t_1$ be a step such that $k(t) \in V_i \setminus V_j$, $k(t)$ is greater that any number marked by $\boxed{m'}$ with $m' \leq m$ at stage t_1, $\max(V_i \setminus V_j \cap [0, t]) > s(m, t_1)$ and for all $x \in V_i \cap [0, s(m, t_1)]$ we have $\beta^t(x) \cap [0, s(m, t_1)] = \beta(x) \cap [0, s(m, t_1)]$ and $\beta^t(\varphi_p(x)) \cap [0, s(m, t_1)] = \beta(\varphi_p(x)) \cap [0, s(m, t_1)]$. Than all four conditions of step $t + 1$ are satisfied for m. They are not satisfied for any $m' < m$ since $s(m', t + 1) = s(m', t)$, by choice of t_0. Then m is the least number such that $m \leq t$ and all conditions of stage $t + 1$ hold for m; therefore $s(m, t + 1) = s(m, t) + 1$ and we have a contradiction with the choice of t_1.

(13) For every m, $\lim_t s(m, t) < \infty$.

Let $m = \langle p, i, j\rangle$ be the least number with $\lim_t s(m, t) = \infty$. By (11) we have that $V_i \setminus V_j$ is infinite and $\beta_{V_i} \leqslant \beta_{V_j}$ via φ_p.

Let t_0 be such that $s(m', t_0) = \lim_t s(m', t)$ for all $m' < m$. In view of (3) we may assume that $k(t_0)$ is greater than any number ever marked by $\boxed{m'}$ with $m' < m$. Let M be the set of all numbers marked by $\boxed{m'}$ with $m' < m$ at the end of stage t_0.

Let $G = \{k(t) : t \geq t_0 \text{ and } s(m, t+1) = s(m, t) + 1\}$. Then G is $\mathbf{0}^{(n+1)}$-computably enumerable. For every $t \geq t_0$ if $k(t) \in G$ then at stage $t + 1$ we put the label $\boxed{m}$ on $k(t)$ and put $k(t)$ into $\text{dom}(F_\gamma^{t+1})$. After this, the marker $\boxed{m}$ can not be removed from $k(t)$ and we have $k(t) \in \text{dom}(F_\gamma^s)$ for all $s > t$. So, $G \subseteq \text{dom}(F_\gamma)$ and $G \subseteq V_i \setminus V_j$.

Notice that G contains almost all numbers in $\text{dom}(F_\gamma)$ that are ever marked by $\boxed{m}$. Indeed if we put $\boxed{m}$ on some $x \notin G$ at some stage $t + 1 > t_0$ then $x \in V_j$ and either $x < k(t)$ and $x \in \text{dom}(F_\alpha^{t+1})$ or $x = k(t')$ for some $t' > t$ and we put x into $\text{dom}(F_\alpha)$ at stage $t' + 1$. Then from Procedure E of the construction we have that $F_\gamma[G]$ is a cofinite subset of $\mathbf{N}$.

On the other hand, for any $x \in G$ $\varphi_p(x) \in V_j$, and either $\varphi_p(x) \in A$ or $\varphi_p(x) \in \mathrm{dom}(F_\alpha)$ or $\varphi_p(x) \in M$. In all three cases we can easily compute an α–index of $\varphi_p(x)$ for any $x \in G$ with oracle $\mathbf{0}^{(n+1)}$ since M is finite, A is $\mathbf{0}^{(n+1)}$– computable and $\beta(x) = \alpha(\varphi^{-1}(x))$ for all $x \in A$.

Putting things together we have that there is a $\mathbf{0}^{(n+1)}$– computable function g, defined on $F_\gamma[G]$, such that $\gamma(x) = \alpha(g(x))$ for all $x \in F_\gamma[G]$. In order to compute $g(x)$ with oracle $\mathbf{0}^{(n+1)}$, in view of (8) and (9) we may enumerate G until we find $y \in G$ such that $F_\gamma(y) = x$ and then define $g(x)$ to be an α–index of $\varphi_p(y)$. Now defining $g(x)$ on $\mathbf{N} \setminus F_\gamma[G]$ in an obvious way we obtain that $\gamma \leqslant_{\mathbf{0}^{(n+1)}} \alpha$ which contradicts the choice of γ.

(14) All requirements $R_{p,i,j}$ are satisfied.

For a contradiction let $m = \langle p, i, j \rangle$ be the least number such that the requirement $R_{p,i,j}$ is not satisfied. Then $\lim_t s(m,t) = \infty$ by (12) and $\lim_t s(m,t) < \infty$ by (13).

Properties (10) and (14) imply that the theorem is true. □

The previous three theorems have been proved under the assumption that the numbering α has special properties of relativized non-universality. We do not know whether in any of these theorems this assumption may be replaced by the weaker assumption of simple non-universality. The following question therefore arises naturally:

Question 8. *Do the claims of the previous three theorems remain true if we just assume that the numbering α is not universal?*

3. Empty Intervals. Minimal Elements and Minimal Covers of Rogers Semilattices

We now turn our attention to empty intervals of Rogers semilattices, i.e. to intervals of the form $(\alpha, \beta) = \emptyset$, with $\alpha < \beta$ (such a pair of numberings is usually described by saying that *β is a minimal cover of α*), and of the form $\hat{\alpha} = \emptyset$, i.e. α is minimal. We say that β is a *strong minimal cover of* α if $\alpha < \beta$ and $\hat{\hat{\beta}} = \hat{\alpha}$.

3.1 Minimal elements of Rogers semilattices

As to the number of minimal elements of Rogers semilattices of Σ^0_{n+1}– computable families, the question has been completely settled for infinite families in the case $n \geq 1$ by Badaev and Goncharov, see Theorem 1.3, [2].

The case $n = 0$ is still open:

Question 9. *Let $\mathcal{A}$ be a Σ^0_1– computable family such that the Rogers semilattice $\mathcal{R}^0_1(\mathcal{A})$ contains at least two minimal elements. Does $\mathcal{R}^0_1(\mathcal{A})$ have infinitely many minimal elements?*

There has been however a considerable progress towards answering this question since it was first posed by Ershov in the late sixties.

We recall the following complete answer to this problem, due to Goncharov, [10], [11], if one just considers Friedberg or positive numberings:

Theorem 3.1 (S.Goncharov). *For every n there exists a Σ^0_1– computable family $\mathcal{A}$ such that $\mathcal{R}^0_1(\mathcal{A})$ has exactly n degrees of Friedberg (positive) numberings.*

Proof. See [10], [11] □

3.2 Covers and strong minimal covers in Rogers semilattices

The following lemma was the starting point of an approach pursued by S. Badaev and S. Goncharov's, [3], towards searching for reasonable conditions under which a Σ^0_{n+2}– computable numbering has minimal covers.

Lemma 3.1. *Let β be any Σ^0_{n+2}– computable numbering of a family $\mathcal{A}$. If for some maximal set M, β_M is a numbering of the whole family $\mathcal{A}$ and $\beta \not\leq \beta_M$ then β is a minimal cover of β_M.*

Proof. Immediate by Lemma 2.2(5), and by maximality of M. □

An evident way to construct a minimal cover β of a given numbering $\alpha \in \mathrm{Com}^0_{n+2}(\mathcal{A})$ is the following. Choose any maximal set M and let f be a computable bijection from $\mathbf{N}$ onto M. For every $x \in M$, define $\beta(x) = \alpha f^{-1}(x)$. It remains to define β on $\overline{M}$ in such a way that $\beta \in \mathrm{Com}^0_{n+2}(\mathcal{A})$ and $\beta \not\leq \alpha$.

The following theorem singles out some conditions under which this can be suitably done.

Theorem 3.2. *Let $\mathcal{A} \subseteq \Sigma^0_{n+2}$ be any Σ^0_{n+2}– computable family, and let α be a Σ^0_{n+2}– computable numbering of $\mathcal{A}$. If either*

(1) there exists a proper subfamily $\mathcal{A}_0$ of $\mathcal{A}$ whose index set $\alpha^{-1}[\mathcal{A}_0]$ is $\mathbf{0}^{(n+1)}$– computable, or

(2) the numbering α is not $\mathbf{0}'$– universal in $\mathrm{Com}^0_{n+2}(\mathcal{A})$, or

(3) there exists a $\mathbf{0}^{(n+1)}$– computable function f such that $\alpha(x) \neq \alpha(f(x))$ for every x,

then α has a minimal cover.

Proof. Let M be a maximal set, with $\overline{M} = \{m_0 < m_1 < m_2 \dots\}$, and let f be a one-to-one computable function with range$(f) = M$. If for every $x \in M$ we define $\beta(x) \leftrightharpoons \alpha(f^{-1}(x))$ then clearly $\alpha \equiv \beta_M$. Let us now define the value $\beta(x)$ for an arbitrary $x \in \overline{M}$ according to which of the assumptions (1), (2) or (3), in the statement of the theorem, holds. Let $x = m_s$ for some $s \in \mathbf{N}$.

(1) Fix two sets A, B such that $A \in \mathcal{A}_0$ and $B \in \mathcal{A} \setminus \mathcal{A}_0$. If $m_s \in \mathrm{dom}(\varphi_s)$ and $\varphi_s(m_s) \in \alpha^{-1}(\mathcal{A}_0)$ then define $\beta(m_s) = B$, otherwise let $\beta(m_s) = A$. Since the sets $\overline{M}, \alpha^{-1}(\mathcal{A}_0)$, and $\{s \mid m_s \in \mathrm{dom}(\varphi_s)\}$ are $\mathbf{0}^{(n+1)}$– computable it follows that β is a Σ^0_{n+2}– computable numbering of $\mathcal{A}$.

To apply Lemma 3.1, it is sufficient to show that $\beta \not\leqslant \alpha$. Suppose $\beta(x) = \alpha(\varphi_s(x))$, $x \in \mathbf{N}$, for some computable function φ_s. Then $m_s \in \mathrm{dom}(\varphi_s)$. If $\varphi_s(m_s) \in \alpha^{-1}(\mathcal{A}_0)$ then $\beta(m_s) = B$. On the other hand, $\alpha(\varphi_s(m_s)) \in \mathcal{A}_0$, a contradiction. Similarly, if $\varphi_s(m_s) \notin \alpha^{-1}[\mathcal{A}_0]$ then $\beta(m_s) = A$, and this is in contradiction with $\alpha(\varphi_s(m_s)) \notin \mathcal{A}_0$.

(2) Since the numbering α is not $\mathbf{0}'$– universal in $\mathrm{Com}^0_{n+2}(\mathcal{A})$, there exists $\gamma \in \mathrm{Com}^0_{n+2}(\mathcal{A})$ such that $\gamma \not\leqslant_{\mathbf{0}'} \alpha$.

Define $\beta(m_s) = \gamma(s)$. By Theorem 1.2, [2], $\beta \in \mathrm{Com}^0_{n+2}(\mathcal{A})$. Since the mapping $\lambda s\, m_s$ is $\mathbf{0}'$– computable, it follows that $\gamma \leqslant_{\mathbf{0}'} \beta$. Therefore, $\beta \not\leqslant \alpha$, otherwise we would obtain $\gamma \leqslant_{\mathbf{0}'} \alpha$.

(3) Let f be a $\mathbf{0}^{(n+1)}$– computable function, such that $\alpha(x) \neq \alpha(f(x))$ for every x. Define

$$\beta(m_s) = \begin{cases} \alpha(0) & \text{if } m_s \notin \mathrm{dom}(\varphi_s), \\ \alpha(f(\varphi_s(m_s))) & \text{if } m_s \in \mathrm{dom}(\varphi_s). \end{cases}$$

Again by Theorem 1.2, [2], $\beta \in \mathrm{Com}^0_{n+2}(\mathcal{A})$. Let us prove that $\beta \not\leqslant \alpha$. By contradiction, assume that β is reducible to α via a computable function φ_s. Then $\beta(x) = \alpha(\varphi_s(x))$ for all x. By definition of β, $\beta(m_s) = \alpha(f(\varphi_s(m_s)))$. Therefore, $\alpha(\varphi_s(m_s)) = \alpha(f(\varphi_s(m_s)))$, i.e. $\varphi_s(m_s)$ is a fixed point of the function f modulo the numbering α, contradicting assumption (3). □

Remark 3.1. The particular case of condition (1) in Theorem 3.2, when $\mathcal{A}_0$ consists of a single set A, was considered by S.Badaev, S.Goncharov in [3]. Conditions (2) and (3) were suggested by S.Badaev and S.Podzorov, [5].

Question 10. *Let $\mathcal{A}$ be any Σ^0_{n+2}– computable family, and assume that $\alpha \in \mathrm{Com}^0_{n+2}(\mathcal{A})$ is not universal in $\mathrm{Com}^0_{n+2}(\mathcal{A})$. How many (up to equivalence) minimal covers can α have?*

The next theorem shows another approach to constructing minimal covers based on the special minimal numberings considered in Theorem 1.3, [2].

Theorem 3.3. *If $\mathcal{A}$ is an infinite Σ^0_{n+2}– computable family and $\alpha \in \mathrm{Com}^0_{n+2}(\mathcal{A})$ is not $\mathbf{0}'$- universal in $\mathrm{Com}^0_{n+2}(\mathcal{A})$, then there exist infinitely many minimal covers (up to equivalence) of α.*

Proof. Let $\mathcal{A}$ be an infinite Σ^0_{n+2}– computable family, and let $\alpha, \beta \in \mathrm{Com}^0_{n+2}(\mathcal{A})$ be numberings such that $\beta \not\leqslant_{\mathbf{0}'} \alpha$.

Let M be a maximal set and let $M' \leftrightharpoons \{2x+1 \mid x \in M\}$. For every $A \in \mathcal{A}$, let us consider the minimal numbering $\beta_{M,A}$ constructed in the proof of Theorem 1.3, [2]. We will now check that the numberings $\alpha \oplus \beta_{M,A}$, $A \in \mathcal{A}$, are minimal covers of β and that $\alpha \oplus \beta_{M,A} \nleqslant \alpha \oplus \beta_{M,B}$ if $A \neq B$.

Since $\beta_{M,A} \equiv_{0'} \beta$, it follows that $\alpha < \alpha \oplus \beta_{M,A}$, otherwise we would obtain a contradiction with $\beta \nleqslant_{0'} \alpha$. Let $\gamma \in \mathrm{Com}^0_{n+2}(\mathcal{A})$ be any numbering such that $\alpha \leqslant \gamma \leqslant \alpha \oplus \beta_{M,A}$. If C_1 stands for the set all even numbers then $\alpha \equiv (\alpha \oplus \beta_{M,A})_{C_1}$. By Lemma 2.2(5), $\gamma \equiv (\alpha \oplus \beta_{M,A})_{C_2}$ for some c.e. set $C_2 \supseteq C_1$.

By construction of the numbering $\beta_{M,A}$, we have $(\alpha \oplus \beta_{M,A})(x) = A$ for every $x \in M'$. Therefore, by applying Lemma 2.2(8) if necessary, we may suppose that $C_2 \supseteq M'$. By Lemma 2.2(6),

$$\gamma \equiv (\alpha \oplus \beta_{M,A})_{C_1} \oplus (\alpha \oplus \beta_{M,A})_{C_2 \setminus C_1}.$$

If $(\alpha \oplus \beta_{M,A})[C_2 \setminus C_1]$ is a finite family then $(C_2 \setminus C_1) \setminus M'$ is a finite set since M is a maximal set. Applying Lemma 2.2(8,9), we have $(\alpha \oplus \beta_{M,A})_{C_1} \equiv (\alpha \oplus \beta_{M,A})_{C_2}$, and, therefore, $\gamma \equiv \alpha$. If $(\alpha \oplus \beta_{M,A})[C_2 \setminus C_1]$ is an infinite family then, by maximality of M, C_2 is a co-finite set and, hence, by Lemma 2.2(9), $\gamma \equiv \alpha \oplus \beta_{M,A}$. Thus, for every $A \in \mathcal{A}$, the numbering $\alpha \oplus \beta_{M,A}$ is a minimal cover of α.

It remains to show that $\alpha \oplus \beta_{M,A} \nleqslant \alpha \oplus \beta_{M,B}$ if $A \neq B$. We prove this by contradiction. Assume that $A \neq B$ and $\alpha \oplus \beta_{M,A} \leqslant \alpha \oplus \beta_{M,B}$. Then $\beta_{M,A} \leqslant \alpha \oplus \beta_{M,B}$. By Lemma 2.2(4), $\beta_{M,A} \equiv (\alpha \oplus \beta_{M,B})_D$ for some c.e. set D. Let D_1 and D_2 consist of the even numbers of D and the odd numbers of D, respectively. By Lemma 2.2(6),

$$\beta_{M,A} \equiv (\alpha \oplus \beta_{M,B})_{D_1} \oplus (\alpha \oplus \beta_{M,B})_{D_2}.$$

Just the same arguments which we have used for the set C_2 above, can be applied to the set D_2 as well. So, we may suppose that $M' \subseteq D_2$. If the family $(\alpha \oplus \beta_{M,B})[D_2 \setminus M']$ is finite then the set $D_2 \setminus M'$ is finite. Then by Lemma 2.2(8,9), $(\alpha \oplus \beta_{M,B})_D \equiv (\alpha \oplus \beta_{M,B})_{D_1}$. Therefore, $\beta_{M,A} \equiv \alpha_{D'}$ with $D' = \{x \mid 2x \in D_1\}$. Hence, $\beta_{M,A} \leqslant \alpha$. Since $\beta_{M,A} \equiv_{0'} \beta$, we obtain a contradiction with $\beta \nleqslant_{0'} \alpha$.

If the family $(\alpha \oplus \beta_{M,B})[D_2 \setminus M']$ is infinite then by maximality of M, the set D_2 is co-finite with respect to the set of all odd numbers. Then Lemma 2.2(6,9) implies that

$$\beta_{M,A} \equiv (\alpha \oplus \beta_{M,B})_D \equiv \alpha_{D'} \oplus \beta_{M,B}.$$

This is in contradiction with $\beta_{M,B} \nleqslant \beta_{M,A}$.

Thus, $\alpha \oplus \beta_{M,A} \nleqslant \alpha \oplus \beta_{M,B}$ if $A \neq B$. □

A useful condition guaranteeing the existence of strong minimal covers is provided by the following theorem.

Theorem 3.4 (S.Badaev, S.Podzorov). *Let $\mathcal{A}$ be any Σ^0_{n+2}–computable family, and let $\alpha \in \mathrm{Com}^0_{n+2}(\mathcal{A})$. If there exist a subfamily $\mathcal{A}_0 \subseteq \mathcal{A}$, a numbering $\gamma \in \mathrm{Com}^0_{n+2}(\mathcal{A}_0)$ with $\gamma \nleq \alpha$, and a computable function f such that $\alpha(f(x)) \subseteq \gamma(x)$, for all x, then α has a strong minimal cover.*

Proof. See [5]. Notice that the proof is based on Lachlan's construction using Chinese boxes, [16]. □

Corollary 3.4.1. *If $\alpha \in \mathrm{Com}^0_{n+2}(\mathcal{A})$ and there exists a subfamily $\mathcal{A}_0 \subseteq \mathcal{A}$ such that $\mathcal{A}_0$ has least element under inclusion and there is a numbering $\gamma \in \mathrm{Com}^0_{n+2}(\mathcal{A}_0)$ with $\gamma \nleq \alpha$, then α has a strong minimal cover.*

Proof. Let $\mathcal{A}, \mathcal{A}_0, \alpha, \gamma$ be as in the statement of the corollary. Let $\bot \in \mathcal{A}_0$ be the least element of $\mathcal{A}_0$ under inclusion, and let a be such that $\alpha(a) = \bot$. Define $f(x) = a$, for every x. Then $\alpha(f(x)) \subseteq \gamma(x)$, for every x. Thus the hypotheses of Theorem 3.4 hold, and α has a strong minimal cover. □

Corollary 3.4.2. *If a Σ^0_{n+2}–family $\mathcal{A}$ has least element with respect to inclusion, then every non–universal numbering $\alpha \in \mathrm{Com}^0_{n+2}(\mathcal{A})$ has a strong minimal cover.*

Proof. Immediate by the previous corollary, taking $\mathcal{A}_0 = \mathcal{A}$. If α is not universal in $\mathrm{Com}^0_{n+2}(\mathcal{A})$, then there exists a numbering $\gamma \in \mathrm{Com}^0_{n+2}(\mathcal{A})$ such that $\gamma \nleq \alpha$. □

Corollary 3.4.3. *For every finite family $\mathcal{A} \subseteq \Sigma^0_{n+2}$, for every numbering $\alpha \in \mathrm{Com}^0_{n+2}(\mathcal{A})$, if α is not universal in $\mathrm{Com}^0_{n+2}(\mathcal{A})$ then α has a minimal cover.*

Proof. If $\mathcal{A}$ contains $\bot$, the least set under inclusion, then α has a minimal cover by the previous corollary.

Suppose that $\mathcal{A}$ has no least element under inclusion. Let $\mathcal{A} = \{A_0, A_1, \ldots$ $\ldots, A_k\}$. It can be easily shown (see [5] for details) that there exists a family $\mathcal{F} = \{F_0, F_1, \ldots, F_k\}$ of finite sets such that for all $i, j \leq k$

$$F_i \subseteq F_j \Leftrightarrow F_i \subseteq A_j \Leftrightarrow A_i \subseteq A_j.$$

Without loss of generality we may assume that $A_0, A_1, \ldots, A_m$, for some $m \leq k$, are all the minimal elements of $\mathcal{A}$ with respect to inclusion and $\alpha(x) = A_x$ for $x \leq m$. Let σ be a permutation of the set $\{0, 1, \ldots, m\}$ with no fixed points. Let $\{\alpha^t(x)\}_{t \in \mathbf{N}}$ be a $\mathbf{0}^{(n+1)}$–computable sequence of finite sets such that for all x, t

$$\alpha^t(x) \subseteq \alpha^{t+1}(x); \qquad \alpha(x) = \bigcup_{t \in \mathbf{N}} \alpha^t(x).$$

It is easy to see that the functions

$$t_0(x) = \mu t(\exists i \leq m(F_i \subseteq \alpha^t(x))),$$
$$f(x) = \mu i \leq m(F_i \subseteq \alpha^{t_0(x)}(x))$$

are $\mathbf{0}^{(n+1)}$–computable and $\alpha(\sigma(f(x))) \neq \alpha(x)$ for all x. Finally we can apply Theorem 3.2(3) for the function $\sigma \circ f$. □

Question 11. *Let* $\alpha \in \mathrm{Com}^0_{n+2}(\mathcal{A})$ *be a numbering which is not universal in* $\mathrm{Com}^0_{n+2}(\mathcal{A})$. *How many strong minimal covers, up to equivalence, can the numbering* α *have?*

Question 12. *Let* $\mathcal{A}$ *be any finite family of* Σ^0_{n+2}*– sets, and let* $\alpha \in \mathrm{Com}^0_{n+2}(\mathcal{A})$ *be not universal in* $\mathrm{Com}^0_{n+2}(\mathcal{A})$. *Does* α *have a strong minimal cover?*

References

[1] S.A. Badaev, *On incomparable enumerations.* Sibirsk. Mat. Zh., 1974, vol. 15, no. 4, pp. 730–738 (Russian); Siberian Math. J. 1974, vol. 15, no. 4, pp. 519–524 (English translation).

[2] S.A. Badaev, S.S. Goncharov, *Theory of numberings: open problems.* In *Computability Theory and its Applications.* P. Cholak, S. Lempp, M. Lerman and R. Shore eds.—Contemporary Mathematics, American Mathematical Society, 2000, vol. 257, Providence, pp. 23-38.

[3] S.A. Badaev, S.S. Goncharov, *On Rogers semilattices of families of arithmetical sets.* Algebra i Logika, 2001, vol. 40, no. 5, pp. 507–522 (Russian); Algebra and Logic, 2001, vol. 40, no. 5, pp. 283–291 (English translation).

[4] S.A. Badaev, S.S. Goncharov, A. Sorbi, *Arithmetical numberings: completeness and universality.* This volume.

[5] S.A. Badaev, S.Yu. Podzorov, *Minimal covers in Rogers semilattices of* Σ^0_n*– computable numberings.* Siberian Math. J., to appear.

[6] Yu.L. Ershov, *Numerations of families of general recursive functions*, Sibirsk. Mat. Zh., 1967, vol. 8, no. 5., pp. 1015-1025 (Russian); Sib. Math. J., 1967, vol. 8, no. 5., pp. 771-778 (English translation).

[7] Yu.L. Ershov, *Hyperhypersimple* m*– degrees.* Algebra i Logika, 1969, vol. 8, pp. 523–552 (Russian); Algebra and Logic, 1969, vol. 8, pp. 298–315 (English translation).

[8] Yu.L. Ershov, I.A. Lavrov, *The upper semilattice* $L(\gamma)$. Algebra i Logika, 1973, vol. 12, no. 2, pp. 167–189 (Russian); Algebra and Logic, 1973, vol. 12, no. 2, pp. 93–106 (English translation).

[9] Yu.L. Ershov, *Theory of Numberings.* Nauka, Moscow, 1977 (Russian).

[10] S.S. Goncharov, *Computable single-valued numerations.* Algebra i Logika, 1980, vol. 19, no. 5, pp. 507–551 (Russian); Algebra and Logic, 1980, vol. 19, no. 5, pp. 325–356 (English translation).

[11] S.S. Goncharov, *Positive computable enumerations.* Dokl. Akad. Nauk, 1993, vol. 332, no. 2, pp. 142–143 (Russian); Russian Acad. Sci. Dokl. Math., 1994, vol. 48, no. 2, pp. 268–270 (English translation).

[12] S.S. Goncharov, A. Sorbi, *Generalized computable numerations and non-trivial Rogers semilattices.* Algebra i Logika, 1997, vol. 36, no. 6, pp. 621–641 (Russian); Algebra and Logic, 1997, vol. 36, no. 6, pp. 359–369 (English translation).

[13] L. Harrington, A. Nies, *Coding in the lattice of enumerable sets.* Adv. Math., to appear

[14] A.B. Khutoretsky, *On the cardinality of the upper semilattice of computable enumerations*, Algebra i Logika, 1971, vol. 10, no. 5, pp. 561–569 (Russian); Algebra and Logic, 1971, vol. 10, no. 5, pp. 348–352 (English translation).

[15] A.H. Lachlan, *On the lattice of recursively enumerable sets.* Trans. Amer. Math. Soc., 1968, vol. 130, pp. 1–37.

[16] A.H. Lachlan, *Two theorems on many–one degrees of recursively enumerable sets.* Algebra i Logika, 1972, vol. 11, no. 2, pp. 216–229 (Russian); Algebra and Logic, 1972, vol. 11, no. 2, pp. 127–132 (English translation).

[17] A.H. Lachlan, *Recursively enumerable many–one degrees.* Algebra i Logika, 1972, vol. 11, no. 3, pp. 326–358 (Russian); Algebra and Logic, 1972, vol. 11, no. 3, pp. 186–202 (English translation).

[18] A.I. Mal'tsev, *Algorithms and Recursive Functions.* Nauka, Moscow, 1965 (Russian) ; Wolters-Noordoff Publishing, Groningen, 1970 (English translation).

[19] S.Yu. Podzorov, *Initial segments in Rogers semilattices of* Σ^0_n *–computable numberings,* Algebra and Logic, to appear.

[20] H. Rogers, *Gödel numberings of partial computable functions.* J. Symb. Logic, 1958, vol. 23, no. 3, pp.49–57.

[21] H. Rogers, Jr. *Theory of Recursive Functions and Effective Computability.* McGraw-Hill, New York, 1967.

[22] V.L. Selivanov, *Two theorems on computable enumerations*, Algebra i Logika, 1976, vol. 15, no. 4, pp. 470–484 (Russian); Algebra and Logic, 1976, vol. 15, no. 4, pp. 297–306 (English translation).

[23] R.I. Soare *Recursively Enumerable Sets and Degrees.* Springer-Verlag, Berlin Heidelberg, 1987.

[24] V.V. V'jugin, *On some examples of upper semilattices of computable enumerations.* Algebra i Logika, 1973, vol. 12, no. 5, pp. 512–529 (Russian); Algebra and Logic, 1973, vol. 12, no. 5, pp. 277–286 (English translation).

[25] V.V. V'jugin, *On upper semilattices of numberings.* Dokl. Akad. Nauk SSSR, 1974, vol. 217, no. 4, pp. 749–751 (Russian); Soviet Math. Dokl., 1974, vol. 15, pp. 1110–1113 (English translation).

ISOMORPHISM TYPES AND THEORIES OF ROGERS SEMILATTICES OF ARITHMETICAL NUMBERINGS

Serikzhan Badaev*
Kazakh National University
Almaty, Kazakhstan
badaev@math.kz

Sergey Goncharov†
Institute of Discrete Mathematics and Informatics
Novosibirsk State University
Novosibirsk, Russia
gonchar@math.nsu.ru

Andrea Sorbi‡
University of Siena
Siena, Italy
sorbi@unisi.it

Abstract We investigate differences in isomorphism types and elementary theories of Rogers semilattices of arithmetical numberings, depending on different levels of the arithmetical hierarchy. It is proved that new types of isomorphism appear as the arithmetical level increases. It is also proved the incompleteness of the theory of the class of all Rogers semilattices of any fixed level. Finally, no Rogers semilattice of any infinite family at arithmetical level $n \geq 2$ is weakly distributive, whereas Rogers semilattices of finite families are always distributive.

* Partial funding provided by grant INTAS *Computability in Hierarchies and Topological Spaces* no. 00-499; and by grant PICS-541-Kazakhstan.
† Partial funding provided by grant INTAS *Computability in Hierarchies and Topological Spaces* no 00-499.
‡ Partial funding provided by grant INTAS *Computability in Hierarchies and Topological Spaces* no 00-499; work done under the auspices of GNSAGA.

Keywords: Numberings; arithmetical hierarchy; computable numberings; Rogers semilattice; elementary theory; isomorphism type; Boolean algebras; hyperhypersimple sets; Feiner's hierarchy; computable algebra.

For the basic notions, notations and methods relative to arithmetical numberings and their Rogers semilattices we will refer to [2] and [3]. In this paper we investigate Rogers semilattices of arithmetical numberings from the point of view of computable algebra. We continue the investigation of the isomorphism types and the elementary types of Rogers semilattices both for finite and infinite arithmetical families. In this direction we are interested in the differences between elementary theories and isomorphism types for different arithmetical levels. For unexplained terminology and notations relative to computability theory, our main references are the textbooks of A.I. Mal'tsev [10], H. Rogers [13] and R. Soare [14]. For the main concepts and notions of the theory of numberings and computable Boolean algebras we refer to the book of Yu.L. Ershov [5] and the book of S.S. Goncharov [7].

1. Distinguishing Rogers Semilattices of Finite Families

Given arithmetical families $\mathcal{A}$ and $\mathcal{B}$, the corresponding Rogers semilattices may look very different from each other. The results illustrated in [3] already enable us to make some comments on this matter. First of all, we already know that if $\mathcal{A}$ and $\mathcal{B}$ are finite families such that $\mathcal{A}$ has least element with respect to inclusion and $\mathcal{B}$ does not, then $\mathcal{R}^0_{n+1}(\mathcal{A})$ and $\mathcal{R}^0_{n+1}(\mathcal{B})$ are not elementarily equivalent, since $\mathcal{R}^0_{n+1}(\mathcal{A})$ possesses a greatest element, whereas $\mathcal{R}^0_{n+1}(\mathcal{B})$ does not: See Theorem 3.2, [2].

2. Distinguishing Rogers Semilattices of Finite Families from Rogers Semilattices of Infinite Families

It is not difficult to point out elementary differences between Rogers semilattices of finite families and Rogers semilattices of infinite families. For instance, we may observe:

Theorem 2.1. *If $\mathcal{A}$ is a finite family of Σ^0_{n+1} sets, then $\mathcal{R}^0_{n+1}(\mathcal{A})$ has least element.*

Proof. Let $\mathcal{A} \leftrightharpoons \{A_0, \dots, A_k\}$. Let $\{X_0, \dots, X_k\}$ be a computable partition of $\mathbf{N}$ into computable sets. For every $x \in X_i$ let $\alpha(x) \leftrightharpoons A_i$. It is easy to see that $\alpha \leqslant \beta$, for all $\beta \in \mathrm{Com}^0_{n+1}(\mathcal{A})$. □

On the other hand, there exist infinite families of c.e. sets $\mathcal{B}$ such that $\mathcal{R}^0_1(\mathcal{B})$ has no minimal elements, [1] and [15]. Moreover, if $\mathcal{B}$ is an infinite Σ^0_{n+2}–computable family of sets, then there exists a numbering $\alpha \in \mathrm{Com}^0_{n+2}(\mathcal{B})$ such that no minimal numbering of $\mathcal{B}$ is reducible to α, [12], [3].

Thus $\mathcal{R}^0_{n+2}(\mathcal{A})$ is always elementarily different from $\mathcal{R}^0_{n+2}(\mathcal{B})$, if $\mathcal{A}$ is finite and $\mathcal{B}$ is infinite.

3. Weak Distributivity

In this section we show another interesting and natural elementary difference between $\mathcal{R}^0_{n+2}(\mathcal{A})$, with $\mathcal{A}$ finite, and $\mathcal{R}^0_{n+2}(\mathcal{B})$, with $\mathcal{B}$ infinite. In the former case we always get a distributive semilattice, whereas in the latter case we always get a nondistributive, not even weakly distributive, semilattice. First we give some definitions.

Definition 3.1. An upper semilattice $\langle L, \vee, \leqslant\rangle$ is called *distributive* if for every $a_1, a_2, b \in L$, if $b \leq a_1 \vee a_2$ then there exist $b_1, b_2 \in L$ such that $b_1 \leq a_1, b_2 \leq a_2$ and $b = b_1 \vee b_2$.

The following proposition states a well known and general fact of the theory of numberings.

Proposition 3.1. *For every numberings $\alpha_0, \alpha_1, \beta$, if $\beta \leqslant \alpha_0 \oplus \alpha_1$ then there exist numberings β_0, β_1 such that $\beta_0 \leqslant \alpha_0, \beta_1 \leqslant \alpha_1$ and $\beta \equiv \beta_0 \oplus \beta_1$.*

Proof. Let f be a computable function such that range(f) contains both even and odd numbers and $\beta(x) = (\alpha_0 \oplus \alpha_1)(f(x))$ for all x. Let $R_0 \leftrightharpoons \{x \mid f(x) \text{ even}\}$ and $R_1 \leftrightharpoons \{x \mid f(x) \text{ odd}\}$. The sets R_0, R_1 are clearly computable and, therefore, there exist computable functions g_i such that $R_i = \text{range}(g_i)$, $i \leq 1$. For $i \leq 1$, define the numbering β_i by $\beta_i(x) \leftrightharpoons \beta(g_i(x))$ for all x. It is now easy to check that $\beta_0 \leqslant \alpha_0, \beta_1 \leqslant \alpha_1$ and $\beta \equiv \beta_0 \oplus \beta_1$. □

Remark 3.1. Note that even if α_0 and α_1 are both numberings of the same family $\mathcal{A}$ then neither β_0 nor β_1 need be mappings from **N** *onto* $\mathcal{A}$.

Theorem 3.1. *For every n and for every finite family $\mathcal{A} \subseteq \Sigma^0_{n+1}$, $\mathcal{R}^0_{n+1}(\mathcal{A})$ is a distributive upper semilattice.*

Proof. Let $\alpha_0, \alpha_1, \beta$, be arbitrary Σ^0_{n+1}–computable numberings of $\mathcal{A}$ such that $\beta \leqslant \alpha_0 \oplus \alpha_1$. By Proposition 3.1, there exist numberings $\beta_0 \leqslant \alpha_0, \beta_1 \leqslant \alpha_1$ such that $\beta \equiv \beta_0 \oplus \beta_1$. Let $\mathcal{A} = \{A_0, A_1, \dots, A_k\}$. For $i \leq k$ and for all x define

$$\beta_i'(x) \leftrightharpoons \begin{cases} A_{\dot{x}} & \text{if } x \leq k, \\ \beta_i(x - k - 1) & \text{if } x > k. \end{cases}$$

It is easy to verify that $\beta_0' \leqslant \alpha_0, \beta_1' \leqslant \alpha_1$ and that $\beta \equiv \beta_0' \oplus \beta_1'$. On the other hand, each β_i' is a numbering of the whole family $\mathcal{A}$. Therefore, $\mathcal{R}^0_{n+1}(\mathcal{A})$ is a distributive upper semilattice. □

The situation is different if we consider infinite families. First of all, we notice:

Remark 3.2. It is easy to see that the three element upper semilattice $L_0 = \{a, b, c\}$, where a and b are incomparable and $c = a \vee b$, is not distributive. There exist many Rogers semilattices which contain L_0 as an ideal, [5], and, therefore, are not distributive. However, if we add $\perp$ to L_0, we do obtain a distributive lattice.

This remark motivates our next definition.

Definition 3.2. An upper semilattice $\mathfrak{L} = \langle L, \leqslant\rangle$ is *weakly distributive* if $\mathfrak{L}_\perp = \langle L \cup \{\perp\}, \leqslant_\perp\rangle$ is distributive, where $\perp \notin \mathfrak{L}$ and

$$\leqslant_\perp \leftrightharpoons \leqslant \cup \{(\perp, a) \mid a \in L \cup \{\perp\}\}.$$

Proposition 3.2. *An upper semilattice $\langle L, \vee, \leqslant\rangle$ is weakly distributive if and only if for every $a_1, a_2, b \in L$, if $b \leq a_1 \vee a_2$ and $b \nleq a_1, b \nleq a_2$ then there exist $b_1, b_2 \in L$ such that $b_1 \leq a_1, b_2 \leq a_2$ and $b = b_1 \vee b_2$.*

Proof. Immediate. □

Lemma 3.1. *Let $\mathcal{A}$ be an infinite Σ^0_{n+2}–computable family. Then there exists a finite subfamily $\mathcal{A}_0$ of $\mathcal{A}$ and numberings $\alpha, \beta \in \mathrm{Com}^0_{n+2}(\mathcal{A})$ and $\gamma \in \mathrm{Com}^0_{n+2}(\mathcal{A}_0)$ such that*

(1) the ideal $\hat{\alpha}$ has no minimal elements in $\mathcal{R}^0_{n+2}(\mathcal{A})$;

(2) β is a minimal numbering;

(3) $\gamma \nleq \beta$.

Proof. Let $\mathcal{A}$ be an infinite Σ^0_{n+2}–computable family. Then by [12] (see also Corollary 1.5.1, [3]) there exists a numbering $\alpha \in \mathrm{Com}^0_{n+2}(\mathcal{A})$ such that no minimal numbering of $\mathcal{A}$ is reducible to α. Given α, numberings β and γ with the desired properties can be constructed as follows. Let M be any maximal set, and let A, B, C be three different sets of $\mathcal{A}$. By the construction of Theorem 1.3, [2], consider the numbering $\beta \leftrightharpoons \alpha_{M,A}$. Now, let $\mathcal{A}_0 \leftrightharpoons \{B, C\}$, and let γ be any Σ^0_{n+2}–computable non-solvable numbering of $\mathcal{A}_0$. The existence of such a γ follows for instance from Theorem 1.3, [3]. Recall that a numbering ν is called *solvable* if the predicate $\nu(x) = \nu(y)$ is computable in x, y.

We claim that $\gamma \nleq \alpha_{M,A}$. Indeed, if this were not the case, then there would be a computable function f such that $\gamma = \alpha_{M,A} \circ f$. From the way $\mathcal{A}_0$ and $\alpha_{M,A}$ are defined, we would have range$(f) \subseteq \overline{M}$; but by maximality of M this implies that range(f) is finite. Therefore γ would be a decidable numbering, a contradiction. □

Theorem 3.2. *For every n, the Rogers semilattice of any infinite Σ^0_{n+2}– computable family is not weakly distributive.*

Proof. Let $\mathcal{A}$ be an infinite Σ^0_{n+2}–computable family. Let the numberings α, β, γ and let a finite subfamily $\mathcal{A}_0 \leftrightharpoons \{A_0, A_1, \dots, A_k\}$ of the family $\mathcal{A}$ satisfy properties (1–3) of Lemma 3.1.

Assume for a contradiction that $\mathcal{R}^0_{n+2}(\mathcal{A})$ is weakly distributive. Let ν_0, ν_1 stand for the numberings $\alpha \oplus \gamma$ and β, respectively. Then, the numbering $\mu = \gamma \oplus \beta$ is reducible to $\nu_0 \oplus \nu_1$, and clearly $\mu \nleqslant \nu_1$. On the other hand we can also argue that $\mu \nleqslant \nu_0$, as follows. Assume for a contradiction that $\gamma \oplus \beta \leqslant \alpha \oplus \gamma$; then $\beta \leqslant \alpha \oplus \gamma$ and therefore by Proposition 3.1 there exist $\beta_0 \leqslant \alpha$ and $\beta_1 \leqslant \gamma$ such that $\beta \equiv \beta_0 \oplus \beta_1$. Note that range$(\beta_0) \supseteq \mathcal{A} \setminus \mathcal{A}_0$. We may in fact assume that range$(\beta_0) = \mathcal{A}$, otherwise define

$$\beta_0'(i)(x) \leftrightharpoons \begin{cases} A_x & \text{if } x \leq k, \\ \beta_0(x-k-1) & \text{otherwise.} \end{cases}$$

It is now easy to see that range$(\beta_0') = \mathcal{A}$, $\beta \equiv \beta_0' \oplus \beta_1$ and $\beta_0' \leqslant \alpha$. But $\beta_0 \equiv \beta$ since β is minimal, contradicting that α does not bound any minimal numbering of $\mathrm{Com}^0_{n+2}(\mathcal{A})$.

By the weak distributivity assumption, there exist numberings μ_0, μ_1 of $\mathcal{A}$ such that $\mu_0 \leqslant \nu_0, \mu_1 \leqslant \nu_1$, and $\mu \equiv \mu_0 \oplus \mu_1$. Since β is minimal, it follows that μ_1 is equivalent to β.

It follows from $\mu_0 \leqslant \alpha \oplus \gamma$ that $\mu_0 \equiv \alpha_0 \oplus \gamma_0$ for some $\alpha_0 \leqslant \alpha$ and $\gamma_0 \leqslant \gamma$. Note that range$(\alpha_0) \supseteq \mathcal{A} \setminus \mathcal{A}_0$. As for β_0, we may suppose that range$(\alpha_0) = \mathcal{A}$.

On the other hand, $\mu_0 \leqslant \gamma \oplus \beta$ and, hence, $\alpha_0 \leqslant \gamma \oplus \beta$. Again, by Proposition 3.1 there exist numberings α_1, γ_1 (where, again, we may suppose that α_1 is a mapping of $\mathbf{N}$ onto $\mathcal{A}$) such that $\gamma_1 \leqslant \gamma, \alpha_1 \leqslant \beta$, and $\alpha_0 \equiv \gamma_1 \oplus \alpha_1$.

So the numbering α_1 of $\mathcal{A}$ is reducible to the minimal numbering β. Therefore, $\alpha_1 \equiv \beta$. It implies $\alpha_0 \equiv \gamma_1 \oplus \beta$ and, hence, $\gamma_1 \oplus \beta \leqslant \alpha$. Therefore, the minimal numbering β is reducible to α, and this again is in contradiction with the choice of α. The theorem is proved. □

Modulo the existence of suitable α, β and γ the argument above holds also for Σ^0_1– computable families.

Corollary 3.2.1. *Let $\mathcal{A}$ be a family of c.e. sets such that the Rogers semilattice $\mathcal{R}^0_1(\mathcal{A})$ has at least one minimal element* deg(β), *contains a principal ideal $\hat{\alpha}$ without minimal elements, and there is a numbering γ of a finite subfamily such that $\gamma \nleqslant \beta$. Then $\mathcal{R}^0_1(\mathcal{A})$ is not weakly distributive. In particular, the Rogers semilattice of the computable numberings of the family of all c.e. sets is not weakly distributive.*

The hypotheses of the previous corollary are not trivial. We recall in fact that for every computable infinite family of computable functions, every ideal of the corresponding Rogers semilattice contains at least one minimal element, [5]. On the other hand, there exist families of c.e. sets $\mathcal{A}$ such that $\mathcal{R}^0_1(\mathcal{A})$ has no minimal elements, [1] and [15] and there exist also infinite families of c.e. sets with trivial Rogers semilattice. In view of these well known facts of the classical theory of numberings, the following question seems to be of some interest.

Question 13. *Does there exist a computable infinite family $\mathcal{A}$ of c.e. sets such that $\mathcal{R}^0_1(\mathcal{A})$ is non-trivial and distributive? Does there exist a computable infinite family $\mathcal{A}$ of c.e. sets such that $\mathcal{R}^0_1(\mathcal{A})$ is non-trivial and weakly distributive?*

4. Distinguishing Rogers Semilattices of Infinite Families

Having seen that the Rogers semilattice $\mathcal{R}^0_{n+2}(\mathcal{A})$ is always elementarily different from $\mathcal{R}^0_{n+2}(\mathcal{B})$ when $\mathcal{A}$ is finite, and $\mathcal{B}$ is infinite, what about elementary differences between Rogers semilattices of distinct infinite families?

Theorem 4.1. *For every n there exist Σ^0_{n+1}–computable infinite families $\mathcal{A}$ and $\mathcal{B}$ such that their Rogers semilattices $\mathcal{R}^0_{n+1}(\mathcal{A})$ and $\mathcal{R}^0_{n+1}(\mathcal{B})$ are not elementary equivalent.*

Proof. It is immediate to see that for every n there are infinite Σ^0_{n+1}–computable families with universal numberings. It suffices to take for every n the family of all Σ^0_{n+1} sets. Then the standard numbering $\rho(e) = W_e^{0^{(n)}}$ is universal in $\mathrm{Com}^0_{n+1}(\Sigma^0_{n+1})$.

On the other hand, there are infinite Σ^0_{n+1}–computable families without universal numberings. This fact follows easily from straightforward relativizations of known result of S.Marchenkov [11] that every non-trivial Rogers semilattice of a family of total computable functions has no greatest element. □

In Theorem 4.1, instead of the quoted result by Marchenkov, one could use also the following fact of the theory of numberings which is an immediate consequence of the Lachlan's paper [8]: The family $\mathcal{F}$ of all finite sets has no computable principal numbering.

This family allows also to exhibit Rogers semilattices which are not elementarily equivalent with respect to $\leqslant_{0^{(i)}}$, for every $i \leq n$.

Theorem 4.2. *For every n there exist Σ^0_{n+1}–computable infinite families $\mathcal{A}$ and $\mathcal{B}$ such that their Rogers semilattices $\mathcal{R}^{0,0^{(i)}}_{n+1}(\mathcal{A})$ and $\mathcal{R}^{0,0^{(i)}}_{n+1}(\mathcal{B})$ are not elementary equivalent for every $i \leq n$.*

Proof. Fix any $i \leq n$. Let $\mathcal{F} \leftrightharpoons \{F \mid F \text{ is a finite set}\}$. It is immediate to see that $\mathcal{F}$ is Σ^0_{n+1}–computable for every n. Assume that $\alpha \in \mathrm{Com}^0_{n+1}(\mathcal{F})$. We will show that there exists $\beta \in \mathrm{Com}^0_{n+1}(\mathcal{F})$ such that $\beta \nleqslant_{0^{(i)}} \alpha$. We begin with building a numbering γ such that range$(\gamma) \subseteq \mathcal{F}$ and $\gamma \not\leqslant_{\mathbf{0}^{(i)}} \alpha$, by stages. For every e, the requirements for building γ are

$$R_e : \varphi_e^{\mathbf{0}^{(i)}} \text{ total } \Rightarrow (\exists x)(\gamma(x) \neq \alpha(\varphi_e^{\mathbf{0}^{(i)}}(x))).$$

We will work with a given $\mathbf{0}^{(n)}$–computable approximation $\{\alpha^s(x)\}_{x,s\in\mathbf{N}}$ to the numbering α.

If at stage $s+1$ we do not explicitly define $\gamma^{s+1}(x)$ for some x then $\gamma^{s+1}(x) = \gamma^s(x)$.

Stage 0) For all e define $\gamma^0(e) \leftrightharpoons \emptyset$.

Stage $s+1$) Consider all $e \leq s$ such that

- $\varphi_{e,s}^{\mathbf{0}^{(i)}}(e)\downarrow$ (say $\varphi_{e,s}^{\mathbf{0}^{(i)}}(e) = y$);
- $\gamma^s(e)) \subseteq \alpha^s(y)$.

For each such e, choose $z \notin \alpha^s(y)$, and define $\gamma^{s+1}(e) \leftrightharpoons \gamma^s(e) \cup \{z\}$.

Finally let $\gamma(x) \leftrightharpoons \bigcup_s \gamma^s(x)$. It is now easy to see that the numbering $\beta \leftrightharpoons \alpha \oplus \gamma$ has the desired properties. First of all β is Σ^0_{n+1}–computable, since γ is Σ^0_{n+1}–computable by construction; range$(\beta) = \mathcal{F}$ since $\alpha \leqslant \beta$; and finally $\beta \not\leqslant_{\mathbf{0}^{(i)}} \alpha$ since $\gamma \not\leqslant_{\mathbf{0}^{(i)}} \alpha$.

Thus, as suitable families $\mathcal{A}$ and $\mathcal{B}$, we could consider the family $\mathcal{F}$ and the family Σ^0_{n+1} of all Σ^0_{n+1}–sets, respectively. □

Remark 4.1. The Rogers semilattice $\mathcal{R}^{0,\mathbf{0}^{(i)}}_{n+1}(\mathcal{A})$ of any finite family $\mathcal{A} \subseteq \Sigma^0_{n+1}$ consists of a single element for every $i > n$ and, therefore, all such semilattices are elementary equivalent.

5. Isomorphism Types

Finally we show that the isomorphism type of Rogers semilattice $\mathcal{R}^0_{n+1}(\mathcal{A})$ of an arbitrary Σ^0_{n+1}–computable family $\mathcal{A}$ may be very different from that of $\mathcal{R}^0_{m+1}(\mathcal{B})$ of some Σ^0_{m+1}–computable family $\mathcal{B}$, for distinct n and m.

The following three lemmas and the notion of an $\mathbf{X}$–computable Boolean algebra play a key role in establishing this claim. Recall (see [7]) that a Boolean algebra $\mathfrak{A}$ is called $\mathbf{X}$–*computable* if its universe, operations, and relation are $\mathbf{X}$–computable.

Lemma 5.1. *Let γ and δ be Σ^0_{n+1}–computable numberings of a family $\mathcal{A}$. If $[\gamma, \delta]$ is a Boolean algebra, then it is $\mathbf{0}^{(n+3)}$–computable.*

Proof. Given n, $\mathcal{A}$, γ, and δ as in the hypothesis of the lemma, we first observe that by (4) and (5) of Lemma 2.2, [3], there exists a c.e. set C such that $\gamma \equiv \delta_C$ and

$$[\gamma, \delta] = \{\deg(\delta_X) \mid X \text{ c.e. and } X \supseteq C\}.$$

For every i, let $U_i \leftrightharpoons C \cup W_i$. This gives an effective listing of all c.e. supersets of C. By Lemma 2.2, [3], (1b), for every i, j, we have $\delta_{U_i} \leqslant \delta_{U_j}$ if and only if

$$\exists p[\forall x(x \in U_i \Rightarrow \exists y(\varphi_p(x) = y \ \&\ y \in U_j)) \\ \&\ \forall x \forall y(x \in U_i \ \&\ \varphi_p(x) = y \Rightarrow \delta(x) = \delta(y))]$$

Since $\delta \in \mathrm{Com}^0_{n+1}(\mathcal{A})$, this implies that $\delta_{U_i} \leqslant \delta_{U_j}$ is a Σ^0_{n+3}–relation in i, j.

Let us consider the equivalence η on $\mathbf{N}$ defined by

$$(i, j) \in \eta \Leftrightarrow \delta_{U_i} \leqslant \delta_{U_j} \ \&\ \delta_{U_j} \leqslant \delta_{U_i}.$$

Let $B \leftrightharpoons \{x \mid \forall y(y < x \Rightarrow (x, y) \notin \eta)\}$. Define a bijection $\psi_1 : B \longrightarrow [\gamma, \delta]$, by letting $\psi_1(i) = \deg(\delta_{U_i})$, for all $i \in B$. It is evident that ψ_1 induces in $\mathcal{R}^0_{n+1}(\mathcal{A})$ a partially ordered set $\mathfrak{B}$ which is a Boolean algebra isomorphic to $[\gamma, \delta]$. The interval $\mathfrak{B}$ is a $\mathbf{0}^{(n+3)}$–computable partial ordered set. It follows from [7, Theorem 3.3.4] and [4], that the Boolean algebra $\mathfrak{B}$ relatively to the Boolean operations is $\mathbf{0}^{(n+3)}$–computable too. □

Lemma 5.2 (L. Feiner). *Let $\mathfrak{F}$ be a computable atomless Boolean algebra. Then for every $\mathbf{X}$ there is an ideal J such that J is $\mathbf{X}$–c.e. and the quotient $\mathfrak{F}/J$ is not isomorphic to any $\mathbf{X}$–computable Boolean algebra.*

Proof. See [6]. □

Below, we will use the following notations. For a given c.e. set A, $\{V_i \mid i \in \mathbf{N}\}$ denotes an effective listing of all c.e. supersets of the set A defined, for instance, by $V_i \leftrightharpoons A \cup W_i$, for all i. We will assume for convenience that $V_0 = A$. As in [3], ε_A stands for the distributive lattice of the c.e. supersets of A. For a given c.e. set $V \supseteq A$, let V^* denote the image of V under the canonical homomorphism of ε_A onto ε^*_A (i.e. ε_A modulo the finite sets), and let $\subseteq^*$ denote the partial ordering relation of ε^*_A. Obviously, if J is an ideal in ε_A then $J^* \leftrightharpoons \{V^* \mid V \in J\}$ is an ideal in ε^*_A.

As is known (see, for instance, [7]), if $\mathfrak{A}$ is a Boolean algebra and J is an ideal of $\mathfrak{A}$, then the universe of the quotient Boolean algebra $\mathfrak{A}/J$ is given by the set of equivalence classes $\{[a]_J \mid a \in \mathfrak{A}\}$ under the equivalence relation $\equiv_J$ given by

$$a \equiv_J b \Leftrightarrow \exists c_1, c_2 \in J(a \vee c_1 = b \vee c_2),$$

and the partial ordering relation is given by

$$[a]_J \leq_J [b]_J \Leftrightarrow a - b \in J.$$

where $a - b$ stands for $a \wedge \neg b$.

Lemma 5.3. *Let $\mathcal{B}$ be a Σ^0_{m+1}–computable family, $\beta \in \mathrm{Com}^0_{m+1}(\mathcal{B})$, and let A be any c.e. set such that ε^*_A is a Boolean algebra. Let $\psi_2 : \varepsilon_A \longrightarrow [\beta_A, \beta]$*

be the mapping given by $\psi_2(V_i) = \deg(\beta_{V_i})$ *for all* i*, and let* I *be any ideal of* ε_A*. Then* ψ_2 *induces an isomorphism of* ε_A^*/I^* *onto* $[\beta_A, \beta]$ *if and only if for every* i, j

(1) $V_i \in I \Rightarrow \beta_{V_i} \leqslant \beta_A$*;*

(2) $V_i - V_j \notin I \Rightarrow \beta_{V_i} \nleqslant \beta_{V_j}$ *(where* $V_i - V_j \leftrightharpoons (V_i \setminus V_j) \cup A$*.)*

Proof. Let $A, \mathcal{B}, \beta, \psi_2$ be given. The "only if" direction is immediate. As to show that the conditions stated in the lemma are also sufficient, we can argue as follows. By (4–5) of Lemma 2.2, [3], we have that every γ with $\beta_A \leqslant \gamma \leqslant \beta$ is of the form $\gamma \equiv \beta_C$ for some c.e. set $C \supseteq A$. Then the mapping induced by ψ_2 is clearly onto.

Suppose now that $[V_i^*]_{I^*} \subseteq_{I^*}^* [V_j^*]_{I^*}$. Then $V_i^* - V_j^* \in I^*$. But $V_i^* - V_j^* = (V_i - V_j)^*$, with $V_i - V_j$ a c.e. superset of A, since ε_A^* is a Boolean algebra. Then $V_i - V_j \in I$. On the other hand,

$$V_i = (V_i - V_j) \cup (V_i \cap V_j).$$

Now, by (1), $\beta_{V_i - V_j} \leqslant \beta_A$, so by (3) of Lemma 2.2, [3], $\beta_{V_i} \equiv \beta_{V_i \cap V_j}$, hence $\beta_{V_i} \leqslant \beta_{V_j}$ by (2) of Lemma 2.2, [3], as $V_i \cap V_j \subseteq V_j$.

Finally, if $[V_i^*]_{I^*} \nsubseteq_{I^*}^* [V_j^*]_{I^*}$ then $V_i - V_j \notin I$, and therefore by (2) $\beta_{V_i} \nleqslant \beta_{V_j}$. □

Theorem 5.1. *For every* n *there exist* $m \geq n$ *and a* Σ_{m+2}^0*–computable family* $\mathcal{B}$ *such that no Rogers semilattice* $\mathcal{R}_{n+1}^0(\mathcal{A})$ *of any* Σ_{n+1}^0*–computable family* $\mathcal{A}$ *is isomorphic to* $\mathcal{R}_{m+2}^0(\mathcal{B})$*.*

Proof. Let n be given. By Lemma 5.1, all Boolean intervals of $\mathcal{R}_{n+1}^0(\mathcal{A})$ are $\mathbf{0}^{(n+3)}$– computable Boolean algebras. Therefore, to show the theorem it is sufficient:

(1) to find a suitable number $m \geq n + 3$;

(2) to consider a computable atomless Boolean algebra $\mathfrak{F}$ and an ideal J of $\mathfrak{F}$ as in Feiner's Lemma such that J is c.e. in $\mathbf{0}^{(m)}$ and $\mathfrak{F}/J$ is not isomorphic to any $\mathbf{0}^{(m)}$– computable Boolean algebra,

(3) to find a Σ_{m+2}^0– computable family $\mathcal{B}$ and Σ_{m+2}^0– computable numberings α and β of $\mathcal{B}$ such that the interval $[\alpha, \beta]$ of $\mathcal{R}_{m+2}^0(\mathcal{B})$ is a Boolean algebra isomorphic to $\mathfrak{F}/J$.

We will determine the number m later. First, we consider item (2) above. Let $\mathfrak{F}$ be a computable atomless Boolean algebra. According to a famous result of Lachlan, [9], there exists a hyperhypersimple set A such that ε_A^* is isomorphic to $\mathfrak{F}$. We fix such a set A.

We refer to the textbook of Soare, [14], for the details of a suitable isomorphism χ of ε_A^* onto $\mathfrak{F}$. We only notice that starting from a computable listing $\{b_0, b_1, \dots\}$ of the elements of $\mathfrak{F}$ one can find a listing $\{B_0, B_1, \dots\}$ of a subfamily of the family ε_A such that $\varepsilon_A^* = \{B_0^*, B_1^*, \dots\}$, $\chi(B_i^*) = b_i$, and the relation "$x \in B_i$" is Σ_3^0.

We will use the techniques for embedding posets into intervals of Rogers semilattices which have been developed in [3]. Let J be any $\mathbf{0}^{(m)}$ - c.e. ideal of $\mathfrak{F}$. Then $I^* \leftrightharpoons \chi^{-1}[J]$ is an ideal of ε_A^* and $\mathfrak{F}/J$ is isomorphic to ε_A^*/I^*. So, instead of the Boolean algebra $\mathfrak{F}/J$ we can just embed ε_A^*/I^*.

Let now $I \leftrightharpoons \{V_i \mid V_i^* \in I^*\}$. Obviously, I is an ideal of ε_A, and

$$I = \{V_j \mid \exists i (V_j =^* B_i \ \&\ b_i \in J)\}.$$

A simple calculation shows that the relation "$V_j =^* B_i$", in i, j, is Σ_5^0.

Now, if we take $m \leftrightharpoons \max\{4, n+3\}$ then I is Σ_{m+1}^0, as J is $\mathbf{0}^{(m)}$ - c.e. Finally, we choose J to be an ideal of $\mathfrak{F}$ satisfying the conclusions of Lemma 5.2, and we let $I^* \leftrightharpoons \chi^{-1}[J]$ as above.

It should be mentioned a very useful property of the Boolean algebra ε_A^* which we will use in our construction of $\mathcal{B}$: if $V_i \notin I$ then $V_i - V \notin I$ for every $V \in I$ and, in particular, $V_i \setminus V$ is an infinite set. This property can be easily verified by contradiction using the following equality:

$$V_i = (V_i - V) \cup (V_i \cap V).$$

We will now construct a Σ_{m+2}^0–computable family $\mathcal{B}$, and a numbering $\beta \in \mathrm{Com}_{m+2}^0(\mathcal{B})$ such that the interval $[\beta_A, \beta]$ is isomorphic to ε_A^*/I^*.

The requirements. First of all, we will construct $\mathcal{B}$ and β so that $\beta[A] = \mathcal{B}$ to guarantee that β_A is a numbering of the whole family $\mathcal{B}$. Then in view of Lemma 5.3 we must satisfy, for every p, i, j, the requirements:

$$P_i : V_i \in I \Rightarrow \beta_{V_i} \leqslant \beta_A$$

$$R_{i,j,p} : V_i - V_j \notin I \Rightarrow \beta_{V_i} \nleqslant \beta_{V_j} \text{ via } \varphi_p$$

where by "$\beta_{V_i} \nleqslant \beta_{V_j}$ via φ_p" we mean that φ_p does not reduce β_{V_i} to β_{V_j} in the sense of Lemma 2.2(1b), [3].

The construction. We use the oracle $\mathbf{0}^{(m+1)}$ in our construction to answer questions such as "$V_i \in I$?" and to verify some properties of c.e. sets and functions. We fix an infinite computable subset R of the set A and a computable partition of R into disjoint infinite computable sets R_i, $i \in \mathbf{N}$.

Initially we define an auxiliary Σ_{m+2}^0– computable numbering β_1 of $\mathcal{B}$ and an auxiliary $\mathbf{0}^{(m+1)}$– computable function $r(i,t)$. If at the end of any stage

$t+1$, $\beta_1^{t+1}(x)$ or $r(x,t+1)$ have not been explicitly modified then they are understood to retain the same value as in the previous stage.

Stage 0) Let $\beta_1^0(x) = \emptyset$, $r(i,0) = 0$ for all x, i.

Stage $t+1$) Let $t = \langle i,j,p\rangle$. Find k such that $V_k = V_i - V_j$ (we can do it with oracle). Check the following conditions:

(i) $V_k \notin I$;

(ii) $V_i \subseteq \mathrm{dom}(\varphi_p)$ and $\varphi_p[V_i] \subseteq V_j$.

If one of (i) or (ii) fails then do nothing. Otherwise let $U_t \leftrightharpoons \bigcup\{V_s : s \leq t \ \&\ V_s \in I\}$. Notice that $A \subseteq U_t$ since we have chosen $V_0 = A$. Choose the least element x in the set

$$\{y \mid V_i \setminus U_t \ \&\ \varphi_p(y) \neq y\} \setminus \{y \mid \beta_1^t(y) \neq \emptyset\}.$$

(See (3) below for the existence of such an x). Take a new number a and define $\beta_1^{t+1}(x) = \beta_1^t(x) \cup \{a\}$. If $\varphi_p(x) \in R_m$ and $m > t$ then define $r(m,t+1) = \max\{r(m,t), \varphi_p(x)\}$. Go to the next stage.

Obviously, β_1 is a Σ^0_{m+2}–computable numbering of the family $\mathcal{B} \leftrightharpoons \{\beta_1(x) \mid x \in \mathbf{N}\}$.

Properties of the construction. The construction satisfies the following properties:

(1) For every $x \in R$, $\beta_1(x) = \emptyset$. Each non-empty set of $\mathcal{B}$ has exactly one index relative to β_1.

(2) For every m, i, if $i \geq m$ then $r(m,i) = r(m, m+1)$.

Properties (1),(2) are evident.

(3) For every i, j, k, p, t, if $V_k = V_i - V_j$ and conditions (i),(ii) above hold then the set $\{x \mid V_i \setminus U_t \ \&\ \varphi_p(x) \neq x\}$ is infinite.

Indeed, $V_i \setminus V_j$ is infinite since $V_k = (V_i \setminus V_j) \cup A$ and $V_k \notin I$ by (i). Condition (ii) implies that $\varphi_p(x) \neq x$ for all $V_i \setminus V_j$.

So, if we assume that the set $\{x \mid V_i \setminus U_t \ \&\ \varphi_p(x) \neq x\}$ is finite then we obtain that $V_i \setminus V_j \subseteq^* U_t$ and, hence, $V_k \subseteq^* U_t$. This is a contradiction with (i) since $U_t \in I$.

Let us now define a numbering $\beta \in \mathrm{Com}^0_{m+2}(\mathcal{B})$. Let $\beta(x) \leftrightharpoons \beta_1(x)$ for all $x \in \overline{R}$. For every $i \in \mathbf{N}$, let ψ_i be a partial computable one-to-one function from $\{x \in R_i : x > r(i,i+1)\}$ onto $V_i \setminus R$. For every i and every $x \in R_i$, let

$$\beta(x) \leftrightharpoons \begin{cases} \beta_1(\psi_i(x)) & \text{if } x > r(i,i+1) \text{ and } V_i \in I\,, \\ \emptyset & \text{otherwise.} \end{cases}$$

For every i if $V_i \in I$ then $\beta_{V_i} \leqslant \beta_R$ via the partial computable function φ such that for every $x \in V_i$

$$\varphi(x) \leftrightharpoons \begin{cases} \psi_i^{-1}(x) & \text{if } x \in V_i \setminus R, \\ x & \text{if } x \in R. \end{cases}$$

We are using again (1b) of Lemma 2.2, [3], so all requirements P_i are satisfied.

We have $\beta[A] = \beta[\mathbb{N}]$ since for every $x \notin A$, one can find i_0 such that $V_{i_0} = A \cup \{x\} \in I$. The requirement P_{i_0} is satisfied, therefore, $\beta_{V_{i_0}} \leqslant \beta_A$, and, in particular, $\beta(x) \in \beta[A]$.

Finally, let us check the requirements $R_{i,j,p}$. Let $i, j \in \mathbb{N}$ and $V_i - V_j \notin I$: we want to show that $\beta_{V_i} \nleqslant \beta_{V_j}$ via φ_p. Let $t \leftrightharpoons \langle i, j, p \rangle$. Consider the number x chosen at stage $t+1$. Notice that $x \notin A$, $\varphi_p(x) \neq x$ and $\beta(x) = \beta_1(x) \neq \emptyset$.

If $\varphi_p(x) \notin R$ then $\beta(x) \neq \beta(\varphi_p(x))$ by property (1) since β and β_1 coincide on $\overline{R}$.

Let now $\varphi_p(x)$ be in R_m for some m. If $V_m \notin I$, or $V_m \in I$ but $\varphi_p(x) \leq r(m, m+1)$, then $\beta(\varphi_p(x)) = \emptyset$, and, hence, $\beta(x) \neq \beta(\varphi_p(x))$.

If $m > t$ then by construction $\varphi_p(x) \leq r(m, m+1)$ and, again we have $\beta(x) \neq \beta(\varphi_p(x))$.

It remains to consider the case when $m \leq t$, $V_m \in I$, and $\varphi_p(x) > r(m, m+1)$. In this case we have $\varphi_p(x) \in \mathrm{dom}(\psi_m)$ and, by construction of β, $\beta(\varphi_p(x)) = \beta_1(\psi_m(\varphi_p(x)))$. Since $\mathrm{range}(\psi_m) \subseteq V_m$, $x \in V_i \setminus U_t$, and $V_m \subseteq U_t$, it follows that $x \neq \psi_m(\varphi_p(x))$. Now (1) implies inequality $\beta(x) \neq \beta(\varphi_p(x))$.

□

References

[1] S.A. Badaev, *On minimal enumerations*, Siberian Adv. Math, 1992, vol. 2, no. 1, pp. 1–30.

[2] S.A. Badaev, S.S. Goncharov, A. Sorbi, *Arithmetical numberings: completeness and universality.* This volume.

[3] S.A. Badaev, S.S. Goncharov, S.Yu. Podzorov, A. Sorbi, *Arithmetical numberings: algebraic properties of Rogers semilattices.* This volume.

[4] V.D. Dzgoev, *Constructive Enumerations of Boolean Lattices.* Algebra i Logika, 1988, vol. 27, no. 6, pp. 641–648 (Russian); Algebra and Logic, 1988, vol. 27, no. 6, pp. 395–400 (English translation).

[5] Yu.L. Ershov, *Theory of Numberings.* Nauka, Moscow, 1977 (Russian).

[6] L. Feiner, *Hierarchies of Boolean algebras.* J. Symb. Logic. 1970, vol. 35, no. 3, pp. 365–374.

[7] S.S. Goncharov *Countable Boolean Algebras and Decidability.*, Plenum, Consultants Bureau, New York, 1997.

[8] A.H. Lachlan, *Standard classes of recursively enumerable sets.* Zeit. Mat. Log. Grund. Math., 1964, vol. 10, pp. 23–42.

[9] A.H. Lachlan, *On the lattice of recursively enumerable sets.* Trans. Amer. Math. Soc., 1968, vol. 130, pp. 1–37.

[10] A.I. Mal'tsev, *Algorithms and Recursive Functions.* Nauka, Moscow, 1965 (Russian) ; Wolters-Noordoff Publishing, Groningen, 1970 (English translation).

[11] S.S. Marchenkov, *On computable enumerations of families of general recursive functions.* Algebra i Logika, 1972, vol. 11, no. 5, pp. 588–607 (Russian); Algebra and Logic, 1972, vol. 11, no. 5, pp. 326–336 (English translation).

[12] S.Yu. Podzorov, *Initial segments in Rogers semilattices Σ^0_n –computable numberings,* Algebra and Logic, to appear.

[13] H. Rogers, Jr. *Theory of Recursive Functions and Effective Computability.* McGraw-Hill, New York, 1967.

[14] R.I. Soare *Recursively Enumerable Sets and Degrees.* Springer-Verlag, Berlin Heidelberg, 1987.

[15] V.V. V'jugin, *On some examples of upper semilattices of computable enumerations.* Algebra i Logika, 1973, vol. 12, no. 5, pp. 512–529 (Russian); Algebra and Logic, 1973, vol. 12, no. 5, pp. 277–286 (English translation).

COMPUTABILITY OVER TOPOLOGICAL STRUCTURES

Vasco Brattka*
Theoretische Informatik I
FernUniversität Hagen
58084 Hagen
Germany
Vasco.Brattka@FernUni-Hagen.de

Abstract Computable analysis is the Turing machine based theory of computability on the real numbers and other topological spaces. Similarly as Eršov's concept of numberings can be used to deal with discrete structures, Kreitz and Weihrauch's concept of representations can be used to handle structures of continuum cardinality. In this context the choice of representations is very sensitively related to the underlying notion of approximation, hence to topology. In this paper we summarize some basic ideas of the representation based approach to computable analysis and we introduce an abstract and purely set theoretic characterization of this theory which can be considered as a generalization of the classical concept of μ-recursive functions. Together with this characterization we introduce the notion of a perfect topological structure. In particular, these structures are effectively categorical, i.e. they characterize their own computability theory. Important examples of perfect structures are provided by metric spaces and additional attention is paid to their effective subsets.

Keywords: Computable analysis, recursion, representations, topological structures.

1. Introduction to Computable Analysis

As one starting point of computable analysis one can surely consider Turing's famous paper [59], which begins with the words:

> The "computable" numbers may be described briefly as the real numbers whose expressions as a decimal are calculable by finite means.

In the same paper Turing introduced his well-known machine model in order to give a precise meaning to the words "calculable by finite means". He already realized that the class of computable real numbers, although countable, is quite

*Work supported by DFG Grant BR 1807/4-1

large and contains all well-known mathematical constants as π and e. As proved by Rice, the class of computable real numbers even forms a real algebraically closed field [46]. Based on Turing's ideas, Banach and Mazur worked out a first theory of computable real number functions [3]. Grzegorczyk and Rasiowa compiled a book on lectures which have been given on this approach by Mazur [40]. The next major step was achieved independently by Grzegorczyk [24, 25] and Lacombe [36] who defined computable real number functions in an equivalent way as it is done today.

While the notion of a computable real number is quite stable with respect to different representations of the real numbers (such as the decimal representation, the binary representation, the continued fraction representation and others), see Robinson [47], it was observed by Mostowski [43] that the notion of a computable function and even the notion of a computable sequence does very sensitively rely on the choice of the representation. Actually, it was already Turing who noticed in a correction [60] of his famous paper that the decimal representation is not suitable to define computable functions and even multiplication by 3 is not computable with respect to this representation. Turing suggested other representations like the interval representation to overcome this difficulty and actually Grzegorczyk's and Lacombe's definitions are based on equivalent representations.

The first systematic study of representations has been presented by Hauck [26, 27], and following this line Kreitz and Weihrauch developed a unified theory of representations which conclusively shows that computational differences of representations are often caused by topological reasons [32, 33, 61, 65, 34, 64]. An alternative but compatible approach to computability on Banach spaces has been presented by Pour-El and Richards [PR89] and also Ko's theory of time complexity on the real numbers [30] is polynomially equivalent to Kreitz and Weihrauch's approach. While all the aforementioned approaches are based on classical logic, there are also several closely related lines of research which are based on different systems of intuitionistic logic. We only mention Bishop school of constructive analysis [6] and Markov's school of Russian constructive analysis [35]. Aberth's work on computable analysis [1] is based on classical logic but closely related to the Russian approach.

A different line of research was devoted to the generalization of classical abstract characterizations of computable functions. Such generalizations have been presented, for instance, by Moschovakis, Friedman, Shepherdson, Fenstad, Tucker and Zucker and Skordev [42, 22, 50, 21, 56, 52]. However, non of these approaches was designed to fit together with computable analysis.

In the following Sections 2 to 6 we will give a brief introduction to certain aspects of the representation based approach to computable analysis. Among other topics we will discuss admissible representations, special representations of the real numbers, the concept of a computable metric space and several notions of effective subsets of such spaces. In Section 7 and 8 we extend the theory to multi-valued operations which are important since even very elementary problems of analysis can only be solved by multi-valued computations. In Sections 9 and 10 we present an abstract characterization of computability which is similar in spirit to the classical characterization by μ–recursive functions. Combined with this characterization we develop a theory of perfect structures which have several nice properties. Especially, they are effectively categorical in the sense that they characterize their own computability theory. In Section 11 we apply this concept to topological structures and we discuss important metric examples. The final Section 12 is devoted to the abstract characterization of effective subsets over structures.[1]

2. Representations and Computable Functions

Inspired by Eršov's theory of numberings [19, 20], Kreitz and Weihrauch have introduced the notion of a representation, see e.g. [32, 33, 61, 62]. A comprehensive and recent account of this theory can be found in [64]. We present some basic concepts of the representation based approach. Roughly speaking, a representation is a (partial) map which assigns at least one "name" to each point of the represented set.

Definition 2.1 (Representation). A *representation* of a set X is a surjective mapping $\delta :\subseteq \mathbb{N}^{\mathbb{N}} \to X$. In this situation (X, δ) is called a *represented space*.

Here and in the following, $\mathbb{N}^{\mathbb{N}}$ denotes the set of sequences $p : \mathbb{N} \to \mathbb{N}$, where $\mathbb{N} := \{0, 1, 2, ...\}$ and the inclusion symbol "$\subseteq$" is used to denote partial maps. A point $x \in X$ is called *δ–computable*, if there exists some computable $p \in \mathbb{N}^{\mathbb{N}}$ such that $\delta(p) = x$. For the definition of computable functions $F :\subseteq \mathbb{N}^{\mathbb{N}} \to \mathbb{N}^{\mathbb{N}}$ we will use the *set of finite sequences* or *words* $\mathbb{N}^*$ of natural numbers. For all $v, w \in \mathbb{N}^*, p \in \mathbb{N}^{\mathbb{N}}$ we will write $v \sqsubseteq w$ and $v \sqsubseteq p$, if v is a *prefix* of w or of p, respectively. Occasionally, we will use the notation wp for the *concatenation* of a word $w \in \mathbb{N}^*$ and a sequence $p \in \mathbb{N}^{\mathbb{N}}$ and we will write $w\mathbb{N}^{\mathbb{N}}$ for the set $\{p \in \mathbb{N}^{\mathbb{N}} : w \sqsubseteq p\}$ of all sequences which extend the word $w \in \mathbb{N}^*$. For any sequence $p \in \mathbb{N}^{\mathbb{N}}$ and $n \in \mathbb{N}$ we denote by $p[n] := p(0)...p(n-1)$ the prefix of p of length n. A function $\varphi : \mathbb{N}^* \to \mathbb{N}^*$ is called *monotone*, if $v \sqsubseteq w \Longrightarrow \varphi(v) \sqsubseteq \varphi(w)$ holds for all $v, w \in \mathbb{N}^*$. Now we can approximate partial functions $F :\subseteq \mathbb{N}^{\mathbb{N}} \to \mathbb{N}^{\mathbb{N}}$ by monotone functions $\varphi : \mathbb{N}^* \to \mathbb{N}^*$.

Definition 2.2 (Computable functions). A function $F :\subseteq \mathbb{N}^{\mathbb{N}} \to \mathbb{N}^{\mathbb{N}}$ is called *computable*, if there exists a computable, total and monotone function

[1] The material contained in Sections 7 to 12 is mainly based on the author's thesis [9] which contains additional results and proofs. An extended abstract on parts of the material has been published in [10] and preliminary versions of some results have been published in [7, 8]. Interesting related considerations of WHILE programs have recently been presented by Tucker and Zucker [57, 58].

$\varphi : \mathbb{N}^* \to \mathbb{N}^*$ such that φ *approximates* F, i.e.

$$F(p) = \begin{cases} \sup\limits_{w \sqsubseteq p} \varphi(w) & \text{if the length of the supremum is not finite} \\ \uparrow & \text{else} \end{cases}$$

for all $p \in \mathbb{N}^{\mathbb{N}}$.

Here we assume that the notion of a computable function $\varphi : \mathbb{N}^* \to \mathbb{N}^*$ is well-known. It could be made precise using some bijective standard numbering $\nu^* : \mathbb{N} \to \mathbb{N}^*$ of the set $\mathbb{N}^*$, defined by $\nu^*(0) := \varepsilon$ and inductively by $\nu^*(\langle\langle n_0, ..., n_k\rangle, k\rangle + 1) := n_0 n_1 ... n_k$ for all $k, n_0, ..., n_k \in \mathbb{N}$ (where ε denotes the *empty word* and $\langle . \rangle$ denotes a tupling function which will be precisely defined below). For any sequence $p \in \mathbb{N}^{\mathbb{N}}$ and $n \in \mathbb{N}$ we denote by $\overline{p}n := (\nu^*)^{-1} p[n]$ the code of the prefix of length n.

The set $\mathbb{N}^{\mathbb{N}}$, considered as *Baire's space*, comes equipped with the product topology of the discrete topology on $\mathbb{N}$. It is well-known that computable functions $F :\subseteq \mathbb{N}^{\mathbb{N}} \to \mathbb{N}^{\mathbb{N}}$ are continuous and their domains are G_δ-sets [62]. Alternatively, one can characterize computable functions by Turing machines which operate on infinite tapes over the infinite[2] alphabet $\mathbb{N}$: a function $F :\subseteq \mathbb{N}^{\mathbb{N}} \to \mathbb{N}^{\mathbb{N}}$ is computable, if and only if there exists a Turing machine with one-way output tape which transfers each input sequence $p \in \text{dom}(F)$ into the corresponding output sequence $F(p)$ by an infinite computation and which diverges on all other sequences. Using computable functions $F :\subseteq \mathbb{N}^{\mathbb{N}} \to \mathbb{N}^{\mathbb{N}}$ one can lift the computability notion to represented spaces.

Definition 2.3 (Computability). Let (X, δ_X), (Y, δ_Y) be represented spaces. Then $f :\subseteq X \to Y$ is called a *(δ_X, δ_Y)–computable function*, if there is a computable function $F :\subseteq \mathbb{N}^{\mathbb{N}} \to \mathbb{N}^{\mathbb{N}}$ such that

$$\delta_Y F(p) = f \delta_X(p)$$

for all $p \in \text{dom}(f\delta_X)$. If, additionally, $p \notin \text{dom}(F)$ for all $p \in \text{dom}(\delta_X) \setminus \text{dom}(f\delta_X)$ holds, then f is called *strongly* (δ_X, δ_Y)*–computable.* We will also say that f is (strongly) (δ_X, δ_Y)–computable *via* F.

If we substitute continuous functions F for computable functions F in the previous definition, then we obtain the corresponding notion of (δ_X, δ_Y)–*continuity*. Figure 1 illustrates the definition.

Using the intuition of Turing machines, a function $f :\subseteq X \to Y$ is (δ_X, δ_Y)–computable, if there exists a Turing machine which transfers each sequence p,

[2] In some references $\mathbb{N}^{\mathbb{N}}$ is used [62], while in other reference, as [64], $\mathbb{N}^{\mathbb{N}}$ is replaced by set of infinite words Σ^ω over some finite alphabet Σ. This has some advantages as soon as complexity is considered but computationally it is essentially equivalent.

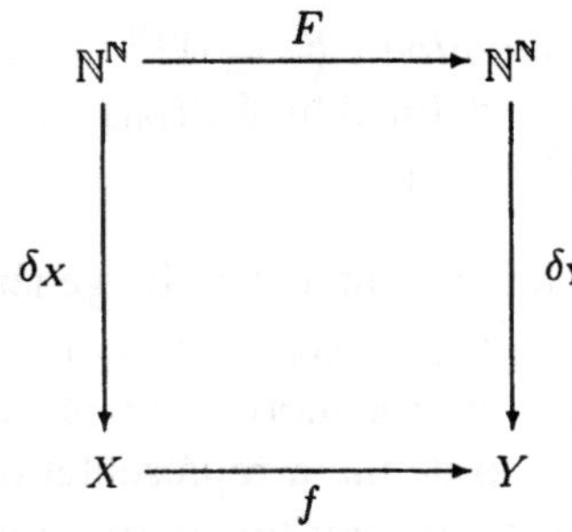

Figure 1. Computability with respect to representations

which represents some $x \in \mathrm{dom}(f)$ with respect to δ_X, into a sequence q, which represents $f(x)$ with respect to δ_Y. The strong notion of computability would additionally require that the machine diverges on sequences p which represent elements outside the domain $\mathrm{dom}(f)$. If we consider the identity $\mathrm{id} : \mathbb{N}^\mathbb{N} \to \mathbb{N}^\mathbb{N}$ as representation of $\mathbb{N}^\mathbb{N}$, then a function $f :\subseteq \mathbb{N}^\mathbb{N} \to \mathbb{N}^\mathbb{N}$ is strongly (id, id)–computable, if and only if it is computable. Moreover, f is (id, id)–computable, if and only if it is the restriction of a computable function $F :\subseteq \mathbb{N}^\mathbb{N} \to \mathbb{N}^\mathbb{N}$. Thus, the "weak" standard notion of computability does not require natural domains (this is similar with the notion of continuity in topology) while the strong notion of computability does require natural domains (which is closer to the classical notion of computability in recursion theory).

From given representations of spaces we can easily construct representations of product and function spaces as we will see in the following. For coding purposes we will use *Cantor's pairing function* $\pi : \mathbb{N}^2 \to \mathbb{N}$, defined by

$$\pi(n,k) := \langle n,k\rangle := \frac{1}{2}(n+k)(n+k+1)+k$$

This definition can be generalized inductively by $\langle n\rangle := n$ and for $i > 1$ by $\langle n_1,...,n_{i+1}\rangle := \langle\langle n_1,...,n_i\rangle, n_{i+1}\rangle$. Sometimes we will also use pairing functions of types $\mathbb{N}^\mathbb{N} \times \mathbb{N}^\mathbb{N} \to \mathbb{N}^\mathbb{N}$, $(\mathbb{N}^\mathbb{N})^\mathbb{N} \to \mathbb{N}^\mathbb{N}$. More precisely, we define

$$\langle p,q\rangle(k) := \begin{cases} p(n) & \text{if } k = 2n \\ q(n) & \text{if } k = 2n+1, \end{cases}$$

and $\langle p_0, p_1, ...\rangle\langle n,k\rangle := p_n(k)$. We will also use the notation $\langle p,n\rangle := np$ for $p \in \mathbb{N}^\mathbb{N}$ and $n \in \mathbb{N}$. It is well-known that all defined pairing functions, as well as the projections of their inverses, are computable. Now we are already prepared to define product and sequence space representations.

Definition 2.4 (Product and sequence spaces). Let (X, δ_X), (Y, δ_Y) be represented spaces.

(1) The *product representation* $[\delta_X, \delta_Y] :\subseteq \mathbb{N}^\mathbb{N} \to X \times Y$ is defined by $[\delta_X, \delta_Y]\langle p,q\rangle := (\delta_X(p), \delta_Y(q))$ for all $p, q \in \mathbb{N}^\mathbb{N}$.

(2) The *sequence representation* $\delta_X^\infty :\subseteq \mathbb{N}^\mathbb{N} \to X^\mathbb{N}$ of the set $X^\mathbb{N}$ of sequences $f : \mathbb{N} \to X$ is defined by $\delta_X^\infty\langle p_0, p_1, ...\rangle(n) := \delta_X(p_n)$, for all $p = \langle p_0, p_1, ...\rangle \in \mathbb{N}^\mathbb{N}, n \in \mathbb{N}$.

The product of representations can easily be generalized to finite products $[\delta_1, ..., \delta_n]$ by $[\delta] := \delta$ and $[\delta_1, ..., \delta_{n+1}] := [[\delta_1, ..., \delta_n], \delta_{n+1}]$. In case of $\delta = \delta_1 = \delta_2 = ... = \delta_n$ we write for short $\delta^n := [\delta_1, ..., \delta_n]$.

In the next step we want to define a representation $[\delta_X \to \delta_Y]$ of the set of (δ_X, δ_Y)–continuous functions. For this purpose we will first define a representation η of certain functions $F :\subseteq \mathbb{N}^\mathbb{N} \to \mathbb{N}^\mathbb{N}$ and then we will use this representation to define $[\delta_X \to \delta_Y]$. For the definition of η we have to encode functions F by sequences $q \in \mathbb{N}^\mathbb{N}$. We introduce such an encoding in two steps. First we define a partial *multiplication* $* :\subseteq \mathbb{N}^\mathbb{N} \times \mathbb{N}^\mathbb{N} \to \mathbb{N}$ as follows[3]:

$$q * p = k :\iff (\exists t)\Big(q(\overline{p}t) = k + 1 \text{ and } (\forall s < t)(q(\overline{p}s) = 0)\Big).$$

In other words $q * p$ is defined, if and only if $t = \mu s[q(\overline{p}s) > 0]$ exists and in this case $q * p = q(\overline{p}t) - 1$. Using this partial function we can define the *application* $(.|.) :\subseteq \mathbb{N}^\mathbb{N} \times \mathbb{N}^\mathbb{N} \to \mathbb{N}^\mathbb{N}$ by $(q|p)(n) := q * \langle p, n\rangle$ for all $p, q \in \mathbb{N}^\mathbb{N}$ and $n \in \mathbb{N}$. Obviously, $(.|.)$ is continuous and thus, any $q \in \mathbb{N}^\mathbb{N}$ encodes a continuous function $\eta_q :\subseteq \mathbb{N}^\mathbb{N} \to \mathbb{N}^\mathbb{N}, p \mapsto (q|p)$. One can prove that in this way a total representation η of the set of continuous functions $F :\subseteq \mathbb{N}^\mathbb{N} \to \mathbb{N}^\mathbb{N}$ with G_δ–domain has been defined [4]. Moreover, a function $F :\subseteq \mathbb{N}^\mathbb{N} \to \mathbb{N}^\mathbb{N}$ is computable, if and only if there is a computable $q \in \mathbb{N}^\mathbb{N}$ such that $F = \eta_q$. Now we are prepared to define the function space representation.

Definition 2.5 (Function spaces). Let $(X, \delta_X), (Y, \delta_Y)$ be represented spaces. We define a representation $[\delta_X \to \delta_Y]$ of the set $\mathcal{C}(\delta_X, \delta_Y)$ of total (δ_X, δ_Y)–continuous functions $f : X \to Y$ by

$$[\delta_X \to \delta_Y](q) = f :\iff (\forall p \in \text{dom}(\delta_X))\ \delta_Y \eta_q(p) = f\delta_X(p)$$

for all $q \in \mathbb{N}^\mathbb{N}$ and $f : X \to Y$.

The representation η has some nice properties: it fulfills a certain smn- and utm-theorem [64]. Or stated differently, the Baire space $\mathbb{N}^\mathbb{N}$ together with the application function $(.|.) :\subseteq \mathbb{N}^\mathbb{N} \times \mathbb{N}^\mathbb{N} \to \mathbb{N}^\mathbb{N}$ forms a partial combinatory algebra [4]. With respect to $[\delta_X \to \delta_Y]$, these properties imply the following result on evaluation and type conversion.

Proposition 2.6. *Let (X, δ_X), (Y, δ_Y), (Z, δ_Z) be represented spaces.*

[3] These definitions have been used for instance by Troelstra [55] and Bauer [4]

(1) **(Evaluation)** *Evaluation* $\mathrm{ev} : \mathcal{C}(\delta_X, \delta_Y) \times X \to Y, (f, x) \mapsto f(x)$ *is* $([[\delta_X \to \delta_Y], \delta_X], \delta_Y)$*–computable,*

(2) **(Type conversion)** $f : Z \times X \to Y$ *is* $([\delta_Z, \delta_X], \delta_Y)$*–computable, if and only if the transposed function* $\check{f} : Z \to \mathcal{C}(\delta_X, \delta_Y)$*, defined by* $\check{f}(z)(x) := f(z, x)$*, is* $(\delta_Z, [\delta_X \to \delta_Y])$*–computable.*

The evaluation property is already hidden in the definition of η while the "only if"-direction of the type conversion property requires a technical but straightforward proof. The "if"-direction follows from the evaluation property. As a consequence one obtains the following result on the category of represented spaces.

Corollary 2.7. *The category of represented spaces is cartesian closed.*

Here the objects are represented spaces, the morphisms are total functions which are continuous (or computable) with respect to the corresponding representations. Products and exponentials are defined with the help of product and function space representations.

3. Topologically Admissible Representations

Up to now we have considered arbitrary representations of sets. But sets like the real numbers $\mathbb{R}$ admit a large variety of different representations. Some of these representations are useful, other rather useless; some representations fit together with an a priori defined topology on the represented set, others do not. In this section we will consider the class of *admissible* representations which are well-behaved in a topological sense. As a very useful tool for the comparison of representations we will use the following concept of reducibility.

Definition 3.1 (Reducibility). Let δ, δ' be representations.

(1) δ is *reducible* to δ', or $\delta \leq \delta'$ for short, if there is a computable function $F :\subseteq \mathbb{N}^{\mathbb{N}} \to \mathbb{N}^{\mathbb{N}}$ with $\delta(p) = \delta' F(p)$ for all $p \in \mathrm{dom}(\delta)$.

(2) δ is *equivalent* to δ', or $\delta \equiv \delta'$ for short, if $\delta \leq \delta'$ and $\delta' \leq \delta$.

In other words, $\delta \leq \delta'$ holds, if and only if the identity id is (δ, δ')–computable. If we replace "computable" by "continuous" in the definition, then we obtain the corresponding concept of *continuous reducibility* which is denoted by "$\leq_t$". It is easy to prove that equivalent representations induce the same (strong) computability on represented sets. Moreover, the product operation on representations is associative up to equivalence, i.e. $[[\delta_1, \delta_2], \delta_3] \equiv [\delta_1, [\delta_2, \delta_3]]$ and equivalent representations have equivalent product, sequence and function space representations.

For the set $\mathbb{N}$ we will use the representation $\delta_{\mathbb{N}} : \mathbb{N}^{\mathbb{N}} \to \mathbb{N}$, which is defined by $\delta_{\mathbb{N}}(p) := p(0)$ for all $p \in \mathbb{N}^{\mathbb{N}}$. We will use the notation $\hat{n}$ for the constant sequence $p \in \mathbb{N}^\mathbb{N}$ with $p(i) = n$ for all $i \in \mathbb{N}$. In particular we obtain $\delta_\mathbb{N}(\hat{n}) = n$. A function $f :\subseteq \mathbb{N} \to \mathbb{N}$ is $(\delta_\mathbb{N}, \delta_\mathbb{N})$–computable, if and only if it is a restriction of a classically recursive function. For any representation δ we obtain $\delta^\infty \equiv [\delta_\mathbb{N} \to \delta]$. Thus, the sequence space representation is compatible with the function space representation.

A representation of a topological space is called *admissible*, if it is continuous and maximal among all continuous representations with respect to continuous reducibility. More precisely:

Definition 3.2 (Admissibility). A representation δ of a topological space X is called *admissible*, if δ is continuous and if $\delta' \leq_t \delta$ holds for all continuous representations δ' of X.

It is straightforward to define examples of admissible representations of second countable T_0–spaces and we will use the following standard representation for such spaces.

Proposition 3.3 (Standard representation). *Let X be a T_0–space with subbase $\{B_n : n \in \mathbb{N}\}$. The representation δ of X, defined by*

$$\delta(p) = x :\iff \text{range}(p) = \{n \in \mathbb{N} : x \in B_n\}$$

for all $p \in \mathbb{N}^\mathbb{N}$ and $x \in X$, is admissible. It is called the standard representation *of the space X.*

Here, we assume that $\bigcup_{n=0}^\infty B_n = X$ for any subbase $\{B_n : n \in \mathbb{N}\}$ of X (such that for any point $x \in X$ there is some n with $x \in B_n$). The following result shows that the admissible representations of second countable T_0–spaces are essentially the continuous representations with surjective and open restrictions. Here, "essentially" means, that this does hold true only for a certain but large class of representations. A representation δ of a topological space X is called *complete*, if the following holds: whenever $\{\delta(p[i]\mathbb{N}^\mathbb{N}) : i \in \mathbb{N}\}$ is a neighbourhood base of some point $x \in X$, then $p \in \text{dom}(\delta)$ and $\delta(p) = x$. For complete representations we obtain the following characterization of admissibility, which is due to Brattka and Hertling [13].

Theorem 3.4. *Let X be a second countable T_0–space and let δ be a complete representation of X. Then δ is admissible, if and only if δ is continuous and has a surjective and open restriction.*

Completeness is not a very strong restriction, since any representation can be completed in a way such that the properties continuity, openness and admissibility are preserved [13]. Admissible representations are very useful because of the Theorem of Kreitz and Weihrauch which shows that for admissible representations relative continuity coincides with ordinary continuity. This theorem has been proved in [33] for second countable T_0–spaces. We present a generalized version for arbitrary topological spaces, which is due to Schröder [49]. Here continuity is replaced by sequential continuity (a function is called *sequentially continuous*, if it maps convergent sequences into convergent sequences).

Theorem 3.5 (Kreitz and Weihrauch). *Let δ_X and δ_Y be admissible representations of topological spaces X and Y, respectively, and let $f :\subseteq X \to Y$ be a function. Then f is sequentially continuous, if and only if f is (δ_X, δ_Y)–continuous.*

Proof. Let $f :\subseteq X \to Y$ be sequentially continuous. Then $f\delta_X$ is sequentially continuous as well, since δ_X is continuous. Since the Baire space $\mathbb{N}^\mathbb{N}$ has a countable base, it is sequential[4] and hence $f\delta_X$ is even continuous. Hence, δ', defined by

$$\delta'(np) := \begin{cases} f\delta_X(p) & \text{if } n = 0 \\ \delta_Y(p) & \text{else} \end{cases}$$

is continuous as well and surjective by definition. Thus, δ' is a continuous representation of Y and we obtain $\delta' \leq_t \delta_Y$, since δ_Y is admissible. Hence, there is a continuous function $G :\subseteq \mathbb{N}^\mathbb{N} \to \mathbb{N}^\mathbb{N}$ such that $\delta' = \delta_Y G$ and $F :\subseteq \mathbb{N}^\mathbb{N} \to \mathbb{N}^\mathbb{N}$, defined by $F(p) := G(0p)$ is continuous as well and we obtain $\delta_Y F(p) = \delta_Y G(0p) = \delta'(0p) = f\delta_X(p)$ for all $p \in \mathrm{dom}(f\delta_X)$. Consequently, f is (δ_X, δ_Y)–continuous.

Now let $f :\subseteq X \to Y$ be (δ_X, δ_Y)–continuous. Then there exists a continuous function $F :\subseteq \mathbb{N}^\mathbb{N} \to \mathbb{N}^\mathbb{N}$ such that $f\delta_X(p) = \delta_Y F(p)$ for all $p \in \mathrm{dom}(f\delta_X)$. Let $x \in \mathrm{dom}(f)$ and let $(x_n)_{n\in\mathbb{N}}$ be a sequence in $\mathrm{dom}(f)$ which converges to x. Now we define a representation δ' of X by

$$\delta'(np) := \begin{cases} x & \text{if } n = 0 \text{ and } p = \hat{0} = 000... \\ x_k & \text{if } n = 0 \text{ and } p = 0^k\hat{1} = 0^k111... \\ \delta_X(p) & \text{if } n \neq 0 \end{cases}.$$

Then δ' is a continuous representation of X and we obtain $\delta' \leq_t \delta_X$ since δ_X is admissible. Hence, there is a continuous function $G :\subseteq \mathbb{N}^\mathbb{N} \to \mathbb{N}^\mathbb{N}$ such that $\delta' = \delta_X G$. Let U be an open neighbourhood of $f(x)$. Since $\delta_Y FG$ is continuous and $\delta_Y FG(\hat{0}) = f\delta_X G(\hat{0}) = f\delta'(\hat{0}) = f(x) \in U$, there is some $k \in \mathbb{N}$ with $\delta_Y FG(0^k\mathbb{N}^\mathbb{N}) \subseteq U$. For all $n \geq k$ we obtain

$$f(x_n) = f\delta'(0^{n+1}\hat{1}) = f\delta_X G(0^{n+1}\hat{1}) = \delta_Y FG(0^{n+1}\hat{1}) \in U.$$

Thus, $(f(x_n))_{n\in\mathbb{N}}$ converges to $f(x)$ and f is sequentially continuous. □

[4] A topological space X is called *sequential*, if all subsets $U \subseteq X$ with the property that any sequence which converges to an element of U is eventually in U, are open.

Second countable T_0–spaces are sequential and in this case ordinary continuity and sequential continuity coincide. Especially, $C(\delta_X, \delta_Y)$ is equal to the set $C(X, Y)$ of continuous functions $f : X \to Y$ if δ_X, δ_Y are admissible representations of sequential spaces. If one tries to generalize Corollary 2.7 to categories of admissibly represented topological spaces, then one has to face the problem that even many categories of topological spaces are not cartesian closed. One has to choose either a subcategory of the category of all topological spaces (as the category of sequential topological spaces, see Schröder [49]) or a larger category (as the category of limit spaces, see Schröder [48]). We formulate the first result.

Theorem 3.6 (Schröder). *The category of admissibly represented sequential T_0–spaces is cartesian closed.*

Again the morphisms are the total continuous functions and products and exponentials of this category are just given by the corresponding product and function space operations on representations, as defined in the previous section.[5]

4. Real Number Representations

In this section we briefly discuss some real number representations. Such representations have implicitly been used for a long time. For instance the identities

$$\pi = 3 + \cfrac{1}{7 + \cfrac{1}{15 + \cfrac{1}{1 + \cfrac{1}{292 + \cfrac{1}{1 + \dots}}}}} = 3.14159265358979323846264338 3...$$

refer to the representations of the number π as *normed continued fraction* and as *decimal expansion*. It is obvious how such ideas can be formalized to define representations $\delta :\subseteq \mathbb{N}^{\mathbb{N}} \to \mathbb{R}$. We define some other representations formally. Therefore, we will use a numbering $\nu_{\mathbb{Q}} : \mathbb{N} \to \mathbb{Q}$ of the rational numbers $\mathbb{Q}$, defined by $\nu_{\mathbb{Q}}\langle i, j, k\rangle := \frac{i-j}{k+1}$. We start with the definition of the Cauchy representation.

Definition 4.1 (Cauchy representation). The *Cauchy representation* ρ of the real numbers $\mathbb{R}$ is defined by

$$\rho(p) := \lim_{n \to \infty} \nu_{\mathbb{Q}} p(n)$$

for all $p \in \mathbb{N}^{\mathbb{N}}$ such that $|\nu_{\mathbb{Q}} p(i) - \nu_{\mathbb{Q}} p(k)| \leq 2^{-k}$ for $i > k$.

[5]The considerations in this section show that the category of admissibly represented sequential T_0–spaces is suitable for semantic purposes. Results of Bauer [4, 5], show that this category is closely related to a subcategory of equilogical spaces, as proposed by Dana Scott.

Intuitively, $\rho(p) = x$, if p is an (encoded) Cauchy sequence of rational numbers which rapidly converges to x. In the following we will call the (ρ^n, ρ)–computable functions $f :\subseteq \mathbb{R}^n \to \mathbb{R}$ just *computable*. Many well-known real number functions, such as the arithmetic operations, the trigonometric functions and many others, are computable [64]. The $(\delta_{\mathbb{N}}, \rho)$–computable sequences and the ρ–computable real numbers will be called *computable* for short. The Cauchy representation characterizes the most important equivalence class of real number representations. For certain purposes other representations are of interest.[6] We define two of them, the representations by enumeration of left and right cuts.

Definition 4.2 (Left and right representations). We define representations $\rho_<, \rho_>$ of the real numbers $\mathbb{R}$ by

(1) $\rho_<(p) = x :\iff \{\nu_{\mathbb{Q}} p(n) : n \in \mathbb{N}\} = \{r \in \mathbb{Q} : r < x\}$,

(2) $\rho_>(p) = x :\iff \{\nu_{\mathbb{Q}} p(n) : n \in \mathbb{N}\} = \{r \in \mathbb{Q} : r > x\}$.

Intuitively, $\rho_<(p) = x$, if and only if p is an (encoded) list of all rational numbers which are less than x. A corresponding statement holds for $\rho_>$. The $\rho_<$– and $\rho_>$–computable numbers $x \in \mathbb{R}$ are called *left computable* and *right computable*, respectively. The $(\rho^n, \rho_<)$– and $(\rho^n, \rho_>)$–computable functions are called *lower semi-computable* and *upper semi–computable*, respectively. It is easy to see that the lower and upper semi–computable functions are lower and upper semi–continuous, respectively. The defined representations $\rho, \rho_<, \rho_>$ are admissible with respect to certain topologies.

Proposition 4.3. *The representations $\rho, \rho_<$ and $\rho_>$ are admissible with respect to the topologies on $\mathbb{R}$, which are generated by the open intervals (r, s), (r, ∞) and $(-\infty, s)$, respectively.*

A proof can be found in [64]. Figure 2 displays reducibility relations of certain real number representations.

Here, any arrow $\delta \to \delta'$ means that δ is computably reducible to δ', i.e. $\delta \le \delta'$, but δ' is not even continuously reducible to δ, i.e. $\delta' \not\le_t \delta$. Up to transitivity the diagram is complete.[7] The classes of computable numbers induced by the representations in Figure 2 which are located below the Cauchy representation

[6] As mentioned in the introduction, the decimal representation is not very suitable: its final topology is the Euclidean topology, but addition is not continuous with respect to the decimal representation [64]. Hence, the decimal representation is not admissible with respect to any topology.

[7] A comprehensive comparison of real number representations can be found in [17] and [64]. It is interesting to mention that the representation with base b expansion is reducible to the representation with base a expansion, if and only if all prime divisors of a are also prime divisors of b.

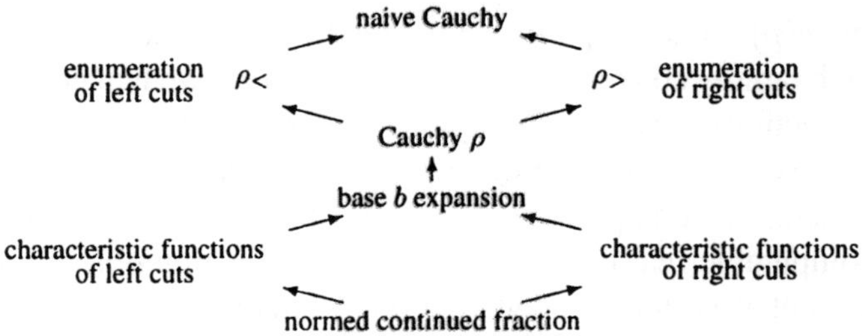

Figure 2. The lattice of some real number representations

ρ, coincide with the computable real numbers. This is not the case for the representations above ρ.[8]

5. Computable Metric Spaces

In this section we present the concept of a *computable metric space*.[9] First, we introduce some notations. By $B(x, \varepsilon) := \{y \in X : d(x, y) < \varepsilon\}$ we denote the *open balls* and by $\overline{B}(x, \varepsilon) := \{y \in X : d(x, y) \leq \varepsilon\}$ the corresponding *closed balls* of a metric space (X, d). Moreover, we use the abbreviation $\overline{n} := \nu_{\mathbb{Q}}(n)$ for the rational numbers given by $n \in \mathbb{N}$. We will say that a sequence $\alpha : \mathbb{N} \to X$ is *dense* in X, if the topological closure of range(α) is equal to X.

Definition 5.1 (Computable metric space). We call (X, d, α) a *computable metric space*, if

(1) $d : X \times X \to \mathbb{R}$ is a metric on X,

(2) $\alpha : \mathbb{N} \to X$ is dense in X,

(3) $d \circ (\alpha \times \alpha) : \mathbb{N}^2 \to \mathbb{R}$ is a computable (double) sequence.

If (X, d, α) just fulfills (1) and (2), then it is sometimes called an *effective metric space* (which is nothing but a separable metric space together with a numbering of a dense subset). For simplicity, we will simply say that (X, d, α) is a *separable metric space* in this case. The space $(\mathbb{R}, d, \nu_{\mathbb{Q}})$ with the *Euclidean metric* $d(x, y) := |x - y|$ is an example of a computable metric space. Similar

[8]The *naive Cauchy representation* is defined as the Cauchy representation but without any restrictions concerning the speed of convergence. It is quite useless, since no finite prefix of such a Cauchy sequence gives any information on the represented number. For further details, see [13]. The class of computable real numbers induced by the naive Cauchy representation is the class of *recursively approximable real numbers*, see [2, 67].

[9]These spaces, sometimes called *recursive metric spaces*, have been used with similar definitions by Lacombe [38], Ceĭtin [16], Moschovakis [41], Weihrauch [63], Spreen [53] and many others. Some of these authors have studied metric spaces restricted to computable points.

as in the real number case, we can introduce the *Cauchy representation* of separable metric spaces in general.

Definition 5.2 (Cauchy representation). Let (X, d, α) be a separable metric space. The *Cauchy representation* δ_X of X is defined by

$$\delta_X(p) := \lim_{n\to\infty} \alpha p(n)$$

for all $p \in \mathbb{N}^{\mathbb{N}}$ such that $d(\alpha p(i), \alpha p(k)) \leq 2^{-k}$ for $i > k$ and $(\alpha p(n))_{n\in\mathbb{N}}$ is convergent.

In case of $(\mathbb{R}, d, \nu_{\mathbb{Q}})$ we obtain $\delta_{\mathbb{R}} = \rho$. The following result characterizes the equivalence class of the Cauchy representation of a computable metric space with the help of the metric $d : X \times X \to \mathbb{R}$ and the *limit operator*

$$\mathrm{Lim} :\subseteq X^{\mathbb{N}} \to X, (x_n)_{n\in\mathbb{N}} \mapsto \lim_{n\to\infty} x_n,$$

which is restricted to rapidly converging sequences, i.e. dom(Lim) contains only convergent sequences $(x_n)_{n\in\mathbb{N}}$ such that $d(x_i, x_k) \leq 2^{-k}$ for all $i > k$.

Proposition 5.3 (Cauchy representation). *Let (X, d, α) be a computable metric space with Cauchy representation δ_X and let δ be a further representation of X. Then $\alpha : \mathbb{N} \to X$ is $(\delta_{\mathbb{N}}, \delta_X)$–computable and*

(1) $\delta \leq \delta_X \iff d : X \times X \to \mathbb{R}$ *is* $([\delta, \delta_X], \delta_{\mathbb{R}})$*–computable,*

(2) $\delta_X \leq \delta \iff \mathrm{Lim} :\subseteq X^{\mathbb{N}} \to X$ *is* $(\delta_X^\infty, \delta)$*–computable.*

Proof. Define $F : \mathbb{N}^{\mathbb{N}} \to \mathbb{N}^{\mathbb{N}}$ by $F(p)(n) := p(0)$ for all $p \in \mathbb{N}^{\mathbb{N}}$ and $n \in \mathbb{N}$. Then F is computable and $\delta_X F(p) = \alpha p(0) = \alpha\delta_{\mathbb{N}}(p)$ for all $p \in \mathbb{N}^{\mathbb{N}}$. Thus, α is $(\delta_{\mathbb{N}}, \delta_X)$–computable via F.
(1) Let $\delta \leq \delta_X$. It suffices to prove that d is $([\delta_X, \delta_X], \delta_{\mathbb{R}})$–computable. Since $d \circ (\alpha \times \alpha) : \mathbb{N} \times \mathbb{N} \to \mathbb{R}$ is computable, we can effectively find some $n_k \in \mathbb{N}$ with $\overline{n_k} < d(\alpha p(k+3), \alpha q(k+3)) < \overline{n_k} + 2^{-k-2}$ for any given $p, q \in \mathrm{dom}(\delta_X)$ and $k \in \mathbb{N}$. Thus, for $x := \delta_X(p)$ and $y := \delta_X(q)$ we obtain

$$\begin{aligned} & |d(x,y) - \overline{n_k}| \\ \leq\ & d(x, \alpha p(k+3)) + d(y, \alpha q(k+3)) + |d(\alpha p(k+3), \alpha q(k+3)) - \overline{n_k}| \\ \leq\ & 2^{-k-3} + 2^{-k-3} + 2^{-k-2} = 2^{-k-1}. \end{aligned}$$

Let $F :\subseteq \mathbb{N}^{\mathbb{N}} \to \mathbb{N}^{\mathbb{N}}$ be the computable function which associates a corresponding value $n_k = F\langle p, q\rangle(k)$ with each $p, q \in \mathrm{dom}(\delta_X)$ and $k \in \mathbb{N}$. Then $\delta_{\mathbb{R}} F\langle p, q\rangle = \lim_{k\to\infty} \overline{n_k} = d(x, y) = d \circ [\delta_X, \delta_X]\langle p, q\rangle$. Thus, d is $([\delta_X, \delta_X], \delta_{\mathbb{R}})$–computable via F.

Now let d be $([\delta, \delta_X], \delta_{\mathbb{R}})$–computable. Since α is $(\delta_{\mathbb{N}}, \delta_X)$–computable, there is a computable function $F :\subseteq \mathbb{N}^{\mathbb{N}} \to \mathbb{N}^{\mathbb{N}}$ which is defined for all $p \in$

$\text{dom}(\delta)$ such that $F(p)(k) = n \implies d(\delta(p), \alpha(n)) < 2^{-k-1}$ for all $p \in \text{dom}(\delta)$, $k, n \in \mathbb{N}$. Thus $\delta(p) = \lim_{k\to\infty} \alpha F(p)(k) = \delta_X F(p)$ for all $p \in \text{dom}(\delta)$, i.e. $\delta \leq \delta_X$.

(2) Let $\delta_X \leq \delta$. It suffices to prove that Lim is $(\delta_X^\infty, \delta_X)$–computable. Define $F : \mathbb{N}^\mathbb{N} \to \mathbb{N}^\mathbb{N}$ by $F\langle p_0, p_1, ...\rangle(k) := p_{k+2}(k+2)$. Then F is computable. Now let $p = \langle p_0, p_1, ...\rangle \in \text{dom}(\text{Lim}\,\delta_X^\infty)$. Then we obtain for all $n > k$

$$\begin{aligned} & d(\alpha p_{n+2}(n+2), \alpha p_{k+2}(k+2)) \\ \leq\ & d(\alpha p_{n+2}(n+2), \alpha p_{n+2}(k+2)) + d(\alpha p_{n+2}(k+2), \delta_X(p_{n+2})) \\ & + d(\delta_X(p_{n+2}), \delta_X(p_{k+2})) + d(\delta_X(p_{k+2}), \alpha p_{k+2}(k+2)) \\ \leq\ & 2^{-k-2} + 2^{-k-2} + 2^{-k-2} + 2^{-k-2} = 2^{-k} \end{aligned}$$

and thus $\delta_X F(p) = \lim_{k\to\infty} \alpha p_{k+2}(k+2) = \lim_{n\to\infty} \lim_{k\to\infty} \alpha p_n(k) = \lim_{n\to\infty} \delta_X(p_n) = \text{Lim} \circ \delta_X^\infty(p)$ for all $p = \langle p_0, p_1, ...\rangle \in \text{dom}(\text{Lim}\,\delta_X^\infty)$. Consequently, Lim is $(\delta_X^\infty, \delta_X)$–computable via F.

Now let Lim be $(\delta_X^\infty, \delta)$–computable. Since α is $(\delta_\mathbb{N}, \delta_X)$–computable and $\delta_X = \text{Lim} \circ \alpha^\mathbb{N}$, it follows that δ_X is $(\delta_\mathbb{N}^\infty, \delta)$–computable. Since $\delta_\mathbb{N}^\infty \equiv \text{id}_{\mathbb{N}^\mathbb{N}}$, we obtain $\delta_X \leq \delta$. □

This result especially shows that the metric d and the limit operator Lim are computable for any computable metric space (X, d, α). Moreover, the result shows that the limit operation can be used to "synthesize" and the metric can be used to "analyze" a metric space.[10]

If X and Y are computable metric spaces with Cauchy representations δ_X and δ_Y, respectively, then we will call a function $f :\subseteq X \to Y$ *computable*, if it is (δ_X, δ_Y)–computable. Similarly as in the real number case, we will call a function $f :\subseteq X \to \mathbb{R}$ *lower semi-computable* and *upper semi-computable*, if it is $(\delta_X, \rho_<)$– and $(\delta_X, \rho_>)$–computable, respectively. A sequence
$s : \mathbb{N} \to X$ is called *computable*, if it is $(\delta_\mathbb{N}, \delta_X)$–computable, and a point $x \in X$ is called *computable*, if the constant sequence with value x is computable. In the real number case it is well-known that the set of computable points cannot be enumerated effectively. The same holds for complete computable metric spaces without isolated points. The result can be derived from an effective version of the Baire Category Theorem [11].

Proposition 5.4. *If X is a computable complete metric space without isolated points, then there exists no computable sequence $(x_n)_{n\in\mathbb{N}}$ in X such that $\{x_n : n \in \mathbb{N}\}$ is the set of computable points of X.*

[10]A corresponding result cannot be obtained for non-separable metric spaces, since one can show that any metric space (X, d) with a representation δ such that d is $([\delta, \delta], \rho_>)$–continuous, necessarily is separable (see Lemma 8.1.1 in [64]).

An alternative approach to computable analysis on Banach spaces has been proposed by Pour-El and Richards [PR89]. In their approach, the notion of a computable sequence has been axiomatized. Yasugi, Mori and Tsujii have presented a generalization of this approach for metric spaces [66]. The approach presented in this section is compatible with these ideas: the set of computable sequences of a computable metric space fulfills the axioms of a computability structure in sense of [66]. We close this section with some examples of computable metric spaces. For the Cauchy representation $\delta_{\mathbb{R}^n}$, associated with the first example, we obtain $\delta_{\mathbb{R}^n} \equiv \rho^n$.

Example 5.5 (Computable metric spaces).

(1) $(\mathbb{R}^n, d_{\mathbb{R}^n}, \alpha_{\mathbb{R}^n})$ with the *Euclidean metric*

$$d_{\mathbb{R}^n}(x, y) := \sqrt{\sum_{i=1}^{n} |x_i - y_i|^2}$$

and some standard enumeration $\alpha_{\mathbb{R}^n}$ of all rational points $\mathbb{Q}^n$ is a computable metric space.

(2) $(\mathcal{C}[0,1], d_{\mathcal{C}}, \alpha_{\mathcal{C}})$ with the set $\mathcal{C}[0,1]$ of continuous real-valued functions $f : [0,1] \to \mathbb{R}$ and the *supremum metric*

$$d_{\mathcal{C}}(f, g) := ||f - g|| := \sup_{x \in [0,1]} |f(x) - g(x)|$$

and some standard numbering $\alpha_{\mathcal{C}}$ of the rational polynomials $\mathbb{Q}[x]$ is a computable metric space. The computable points in this space are exactly the computable functions $f : [0,1] \to \mathbb{R}$.

(3) $(\mathcal{K}(X), d_{\mathcal{K}}, \alpha_{\mathcal{K}})$ with the set $\mathcal{K}(X)$ of non-empty compact subsets of a computable metric space (X, d, α) and the *Hausdorff metric*

$$d_{\mathcal{K}}(A, B) := \max\left\{\sup_{a \in A} \inf_{b \in B} d(a, b), \sup_{b \in B} \inf_{a \in A} d(a, b)\right\}$$

and some standard numbering $\alpha_{\mathcal{K}}$ of the non-empty finite subsets of range(α) is a computable metric space.

6. Computable Subsets of Metric Spaces

In the previous section we have defined computable points, sequences and functions on metric spaces. In this section we want to define computable subsets of metric spaces. Essentially, we want to extend the classical notions of recursively enumerable, co-recursively enumerable and recursive subsets, as they are used in recursion theory [44], to this setting. It turns out that the intuition of decidability is not very fruitful in computable analysis: if (X, d) is a connected metric space, then the *characteristic function*

$$\mathrm{cf}_A : X \to \mathbb{R}, x \mapsto \begin{cases} 0 & \text{if } x \in A \\ 1 & \text{else} \end{cases}$$

of a subset $A \subseteq X$ is continuous (and hence computable) only in the trivial cases $A \in \{\emptyset, X\}$. The solution of this problem is to replace the discontinuous characteristic function by a continuous substitute. Therefore, one can use the *distance function*

$$d_A : X \to \mathbb{R}, x \mapsto \inf_{a \in A} d(x, a),$$

defined for any metric space (X, d) and subsets $A \subseteq X$. Figure 3 displays the characteristic function of the unit interval $A = [0, 1]$ together with its distance function.

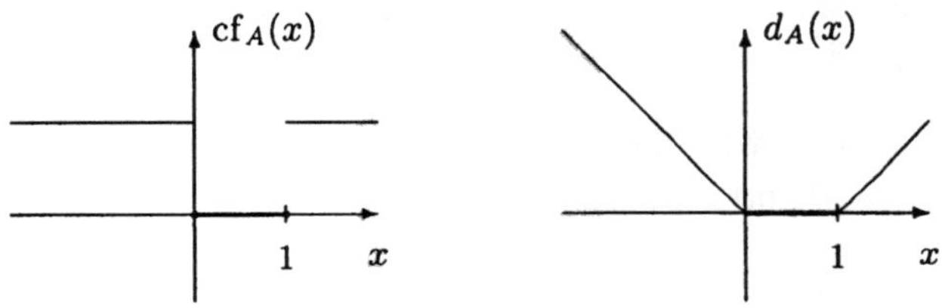

Figure 3. Characteristic function and distance function of the unit interval

In the following definition we use different computability properties of the distance function to define computability notions for subsets.[11]

Definition 6.1 (Located subsets). Let (X, d, α) be a computable metric space and let $A \subseteq X$ be a subset. Then A is called *lower semi-located, upper semi-located* and *located*, if $d_A : X \to \mathbb{R}$ is lower semi-computable, upper semi-computable and computable, respectively.

Since the distance function of a subset $A \subseteq X$ coincides with the distance function of the topological closure $\overline{A}$ of A, we can restrict most considerations about located subsets to closed subsets. For closed subsets (or dually, for open subsets) the following notion of effectivity has already been considered by Lacombe [37].

[11] The idea to use distance functions in order to define *located* subsets goes back to intuitionistic and constructive analysis, see for instance [6]. They have been used in computable analysis by several authors, see for instance [65, 23, 68].

Definition 6.2 (Co-r.e. closed subsets). Let (X, d, α) be a computable metric space and let $A \subseteq X$ be a closed subset. Then A is called *co-r.e. closed*, if there exists an r.e. subset $C \subseteq \mathbb{N}$ such that

$$X \setminus A = \bigcup_{\langle n,k \rangle \in C} B(\alpha(n), \overline{k})$$

It is easy to show that a subset $A \subseteq X$ is co-r.e. closed, if and only if there exists a computable function $f : X \to \mathbb{R}$ such that $f^{-1}\{0\} = A$ and this is the case, if and only if the characteristic function $\mathrm{cf}_A : X \to A$ is lower semi-computable. Another notion which has been considered in some references is the notion of *effective separability.*

Definition 6.3 (Effective separability). Let (X, d, α) be a computable metric space and let $A \subseteq X$ be a subset. Then A is called *effectively separable*, if there exists a computable sequence $f : \mathbb{N} \to X$ such that $\mathrm{range}(f) = \{f(n) : n \in \mathbb{N}\}$ is dense in A.

Finally, we even define four further notions of effectivity.

Definition 6.4 (Recursive subsets). Let (X, d, α) be a computable metric space and let $A \subseteq X$ be a closed subset.

(1) A is called *recursively enumerable closed* (*r.e. closed* for short), if the set $\{\langle n, k \rangle \in \mathbb{N} : A \cap B(\alpha(n), \overline{k}) \neq \emptyset\}$ is r.e.

(2) A is called *recursive closed*, if it is r.e. closed and co-r.e. closed.

(3) A is called *strongly co-recursively enumerable closed* (*strongly co-r.e. closed* for short), if the set $\{\langle n, k \rangle \in \mathbb{N} : A \cap \overline{B}(\alpha(n), \overline{k}) = \emptyset\}$ is r.e.

(4) A is called *strongly recursive closed*, if A is r.e. closed, as well as strongly co-r.e. closed.

Figure 4 shows the logical relations between all introduced notions of effectivity for closed subsets, see [14] for proofs.

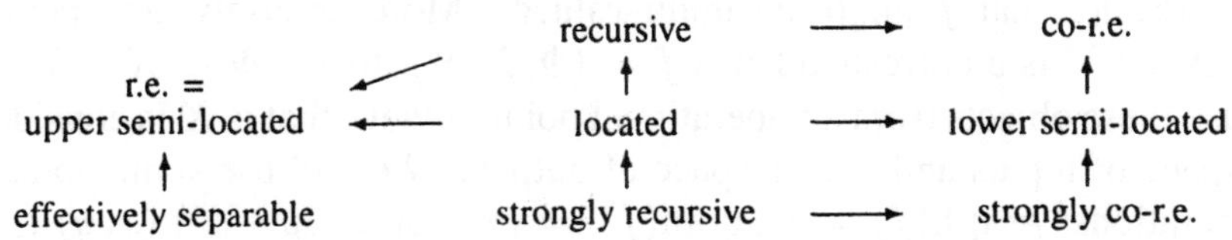

Figure 4. Notions of effectivity for closed subsets of computable metric spaces

Each arrow in the diagram does not only indicate an implication but an effective reducibility: for each notion of effectivity, there exists a corresponding hyperspace representation $\delta :\subseteq \mathbb{N}^{\mathbb{N}} \to \mathcal{A}(X)$ of the set $\mathcal{A}(X)$ of closed subsets

$A \subseteq X$. In this case, the arrows in Figure 4 can be read as reducibilities, similarly as the arrows in Figure 2.

In case of Euclidean space (and other "effectively locally compact" computable metric spaces) the vertical arrows in Figure 4 can be reversed and thus the three horizontal layers of the diagram collapse [15]. Moreover, the recursive compact subsets $K \subseteq \mathbb{R}^n$ are exactly the computable points in $\mathcal{K}(\mathbb{R}^n)$. In contrast to that, the three vertical layers represent essentially different notions of effectivity which generalize the classical notions of effectivity, as the following result shows.

Proposition 6.5. *A subset $A \subseteq \mathbb{N}$ is r.e., co-r.e., or recursive in the classical sense, if and only if A is r.e., co-r.e., or recursive, respectively, considered as a closed subset of Euclidean space $\mathbb{R}$.*

7. Computable Multi-Valued Operations

In this section we want to present an extension of the representation based notion of computability to multi-valued operations. There are at least two motivations for such an extension. On the one hand, multi-valued operations occur in practice and examples like the determination of zeroes of polynomials are important and unavoidable: given the coefficients of a non-constant polynomial, we can effectively find a zero, but in general the corresponding computation is non-extensional and therefore it has to be described by a multi-valued function.[12] Such *indeterministic computations*[13] may lead to different results on the same input, but all possible results have to be valid (in contrast to *non-deterministic* computations in complexity theory, where only some computations have to yield a valid result). On the other hand, one can motivate the introduction of multi-valued operations by formal reasons. If the arrows in the category of admissibly represented spaces are chosen to be multi-valued, then the category gains some nice properties, as we will implicitly see below.

Before we define computability of multi-valued operations, we introduce some notations. By $f :\subseteq X \rightrightarrows Y$ we will denote partial multi-valued functions which we will call for short *operations* in the following. Here the symbol "$\rightrightarrows$" indicates that f might be multi-valued. More precisely, an operation $f :\subseteq X \rightrightarrows Y$ is a correspondence $f = (\Phi, X, Y)$, that is $\Phi \subseteq X \times Y$. We will use these objects from an operational point of view, that is X is considered as a space of inputs and Y as a space of outputs. We will use some notations for operations: $\mathrm{graph}(f) := \Phi$, $\mathrm{dom}(f) := \{x \in X : (\exists y \in Y)\ (x, y) \in \Phi\}$, and $\mathrm{range}(f) := \{y \in Y : (\exists x \in X)\ (x, y) \in \Phi\}$ will be called *graph*,

[12] The idea to compute with multi-valued operations has been considered by several authors in different settings [42, 18, 52, 31].

[13] This notion of indeterminism has already been used by Shepherdson [50].

domain, and *range* of f, respectively. The *image* of $A \subseteq X$ under f will be denoted by $f(A) := \{y \in Y : (\exists x \in A)\ (x, y) \in \Phi\}$, and the *preimage* of $B \subseteq Y$ by $f^{-1}(B) := \{x \in X : (\exists y \in B)\ (x, y) \in \Phi\}$. By $f(x) := f\{x\} = \{y \in Y : (x, y) \in \Phi\}$ we denote the *image* of x under f for each $x \in \text{dom}(f)$. If $f(x)$ is single-valued, i.e. $f(x) = \{y\}$ for some $y \in Y$, then we also write $f(x) = y$, as usual for functions. With each operation $f = (\Phi, X, Y)$ we associate the *inverse operation* $f^{-1} = (\Phi^{-1}, Y, X)$, which is given by $\Phi^{-1} := \{(y, x) : (x, y) \in \Phi\}$.

Computable multi-valued operations $f :\subseteq X \rightrightarrows Y$ are usually defined by substituting "$\delta_Y F(p) \in f\delta_X(p)$" for the condition "$\delta_Y F(p) = f\delta_X(p)$" in Definition 2.3 (see [64]). We will call this kind of computability *weak* (δ_X, δ_Y)–*computability*. Weak computability corresponds to the idea that the realization F on input p with $x = \delta_X(p)$ selects one of the possible results in $f(x)$. For our purposes, however, this definition is too weak, since the realization F does not characterize the whole image $f(x)$. In a second attempt, we could define that f is (δ_X, δ_Y)–computable, if there is a computable function $F :\subseteq \mathbb{N}^\mathbb{N} \to \mathbb{N}^\mathbb{N}$ such that $f(x) = \delta_Y F \delta_X^{-1}\{x\}$ for all $x \in \text{dom}(f)$. But this definition would sensitively rely on the concrete choice of the representation δ_X of X. It is possible that for some representations δ_1 of X the preimage $\delta_1^{-1}\{x\}$ for some specific x is single-valued while for other equivalent representations δ_2 of X the preimage $\delta_2^{-1}\{x\}$ might be very large. Consequently, equivalent representations would induce different kinds of computability of operations. This is one reason why we choose a different definition of computable operations. In our definition the complete image $f(x)$ will be obtained by the help of an additional "oracle" input.

Definition 7.1 (Computable operations). Let $(X, \delta_X), (Y, \delta_Y)$ be represented spaces. Then $f :\subseteq X \rightrightarrows Y$ is called a (δ_X, δ_Y)–*computable operation*, if there is a computable function $F :\subseteq \mathbb{N}^\mathbb{N} \to \mathbb{N}^\mathbb{N}$ such that

$$\delta_Y F\langle p, \mathbb{N}^\mathbb{N}\rangle = f\delta_X(p)$$

and $\langle p, \mathbb{N}^\mathbb{N}\rangle \subseteq \text{dom}(\delta_Y F)$ for all $p \in \text{dom}(f\delta_X)$. Furthermore, f is called a *strongly* (δ_X, δ_Y)–*computable operation*, if, additionally, the condition $\langle p, \mathbb{N}^\mathbb{N}\rangle \not\subseteq \text{dom}(F)$ holds for all $p \in \text{dom}(\delta_X) \setminus \text{dom}(f\delta_X)$.

Here, $\langle p, \mathbb{N}^\mathbb{N}\rangle := \{\langle p, q\rangle : q \in \mathbb{N}^\mathbb{N}\}$. It is easy to see that the previous definition has the property that equivalent representations induce the same kind of (strong) computability of operations. Moreover, a function $f :\subseteq X \to Y$ is (δ_X, δ_Y)–computable, if and only if it is (δ_X, δ_Y)–computable, considered as an operation. And if f is strongly (δ_X, δ_Y)–computable as a function, then it is also strongly (δ_X, δ_Y)–computable as an operation. However, the converse of the latter statement does not hold true in general. Therefore, in case of doubts, strong computability will always mean strong computability in the sense of operations. Our next goal is to show that weak computability and computability of operations are also closely related notions. The following lemma will be helpful for this purpose.

Lemma 7.2 (Oracle Lemma). *There exists a total computable function* $H : \mathbb{N}^{\mathbb{N}} \to \mathbb{N}^{\mathbb{N}}$ *such that*

(1) $\mathrm{range}(H\langle p,q\rangle) = \mathrm{range}(p)$ *for all* $p,q \in \mathbb{N}^{\mathbb{N}}$,

(2) for all $r,p \in \mathbb{N}^{\mathbb{N}}$ *with* $\mathrm{range}(r) = \mathrm{range}(p)$ *there exists a* $q \in \mathbb{N}^{\mathbb{N}}$ *such that* $H\langle p,q\rangle = r$.

Proof. For each $p,q \in \mathbb{N}^{\mathbb{N}}$ we define $r = H\langle p,q\rangle$ by an inductive construction. In step $n = 0,1,...$ let $\langle i,j\rangle := q(n)$.

(a) If there is no k such that $r(k)$ is already defined and equal to $p(n)$, then choose the first $k \geq i$ such that $r(k)$ is still undefined and let $r(k) := p(n)$.

(b) If $r(n)$ is still undefined, then let $r(n) := p(j)$.

The algorithm, described hereby, defines a total computable function $H : \mathbb{N}^{\mathbb{N}} \to \mathbb{N}^{\mathbb{N}}$. After execution of step n, at least $r(0),...,r(n)$ are defined by (b). We have to show that (1) and (2) hold.

(1) The construction obviously guarantees $\mathrm{range}(r) \subseteq \mathrm{range}(p)$. On the other hand, for each n there is some k such that $r(k) = p(n)$ after execution of step n, part (a). Hence, $\mathrm{range}(p) \subseteq \mathrm{range}(r)$.

(2) Now let $r,p \in \mathbb{N}^{\mathbb{N}}$ be such that $\mathrm{range}(r) = \mathrm{range}(p)$. Then define $q \in \mathbb{N}^{\mathbb{N}}$ for each $n \in \mathbb{N}$ by $q(n) := \langle i,j\rangle$, where $i := \mu k[r(k) = p(n)]$ and $j := \mu k[p(k) = r(n)]$. Then $H\langle p,q\rangle = r$. □

It is obvious that any (δ_X,δ_Y)–computable operation $f :\subseteq X \rightrightarrows Y$ is also weakly (δ_X,δ_Y)–computable. The following result shows that a certain converse version of this statement holds true as well.

Proposition 7.3. *Let X be a second countable T_0–space with standard representation δ_X and let (Y,δ_Y) be a represented space. If an operation $f :\subseteq X \rightrightarrows Y$ is weakly (δ_X,δ_Y)–computable, then there is a (δ_X,δ_Y)–computable operation $g :\subseteq X \rightrightarrows Y$ such that* $\mathrm{dom}(g) = \mathrm{dom}(f)$ *and* $\mathrm{graph}(g) \subseteq \mathrm{graph}(f)$.

Proof. Let $f :\subseteq X \rightrightarrows Y$ be weakly (δ_X,δ_Y)–computable. Then there is a computable function $F :\subseteq \mathbb{N}^{\mathbb{N}} \to \mathbb{N}^{\mathbb{N}}$ such that $\delta_Y F(p) \in f\delta_X(p)$ for all $p \in \mathrm{dom}(f\delta_X)$. Let $H : \mathbb{N}^{\mathbb{N}} \to \mathbb{N}^{\mathbb{N}}$ be the computable function which exists by the Oracle Lemma 7.2. Define $G :\subseteq \mathbb{N}^{\mathbb{N}} \to \mathbb{N}^{\mathbb{N}}$ by $G\langle p,q\rangle := FH\langle p,q\rangle$ for all $p,q \in \mathbb{N}^{\mathbb{N}}$. Then G is a computable function. Now let $g :\subseteq X \rightrightarrows Y$

be defined by $\mathrm{dom}(g) := \mathrm{dom}(f)$ and by $g\delta_X(p) := \{\delta_Y G\langle p,q\rangle : q \in \mathbb{N}^{\mathbb{N}}\}$ for all $p \in \delta_X^{-1}(\mathrm{dom}(g))$. By definition of H and δ_X we obtain $\delta_X^{-1}\delta_X(p) = \{H\langle p,q\rangle : q \in \mathbb{N}^{\mathbb{N}}\}$, i.e. g is well-defined and (δ_X, δ_Y)–computable. Now let $x \in \mathrm{dom}(g)$ and $y \in g(x)$. Then there are $p, q \in \mathbb{N}^{\mathbb{N}}$ such that $\delta_X(p) = x$ and $\delta_Y G\langle p,q\rangle = y$. Hence $y = \delta_Y FH\langle p,q\rangle \in f\delta_X H\langle p,q\rangle = f\delta_X(p) = f(x)$. Thus, $\mathrm{graph}(g) \subseteq \mathrm{graph}(f)$. □

We will close this section with a short discussion of continuity. It turns out that computable multi-valued operations are *lower semi-continuous*. As usual, a multi-valued operation $f :\subseteq X \rightrightarrows Y$ on topological spaces X, Y is called *lower semi-continuous*,[14] if $f^{-1}(U)$ is open in $\mathrm{dom}(f)$ for any open subset $U \subseteq Y$. If f is single-valued, then it is lower semi-continuous, if and only if it is continuous in the ordinary sense. Before we formulate our result we introduce another notation. With each representation $\delta_X :\subseteq \mathbb{N}^{\mathbb{N}} \to X$ we can associate its *cylindrification* $\delta^{\mathrm{cyl}} :\subseteq \mathbb{N}^{\mathbb{N}} \to X$, which is defined by $\delta^{\mathrm{cyl}}\langle p,q\rangle := \delta(p)$ for all $p, q \in \mathbb{N}^{\mathbb{N}}$ (cf. [62]). Obviously, each representation is equivalent to its cylindrification.

Theorem 7.4 (Continuity). *Let (X, δ_X), (Y, δ_Y) be admissibly represented second countable T_0–spaces. Then any (δ_X, δ_Y)–computable multi-valued operation $f :\subseteq X \rightrightarrows Y$ is lower semi-continuous.*

Proof. Without loss of generality, we can assume that δ_X, δ_Y are standard representations. Now let $f :\subseteq X \rightrightarrows Y$ be (δ_X, δ_Y)–computable via some computable function $F :\subseteq \mathbb{N}^{\mathbb{N}} \to \mathbb{N}^{\mathbb{N}}$. It is easy to see that δ_Y is continuous and that δ_X is open. The latter fact implies that $\delta_X^{-1} : X \rightrightarrows \mathbb{N}^{\mathbb{N}}$ is lower semi-continuous. Then $g :\subseteq X \rightrightarrows Y$, defined by $g(x) := \delta_Y \circ F \circ \langle \delta_X^{-1}(x), \mathbb{N}^{\mathbb{N}}\rangle$ is an extension of f, that is $g(x) = f(x)$ for all $x \in \mathrm{dom}(f)$. Altogether, it follows that g is lower semi-continuous and hence f, as a restriction of g, is lower semi-continuous as well.[15] □

8. Recursive Closure Schemes

In this section we discuss some set-theoretical closure schemes of computable multi-valued operations. These closure schemes are generalizations of the classical closure schemes *substitution*, *primitive recursion* and *μ–recursion* [44]. In the following we will assume that U, V, X, Y, Z are arbitrary sets. All

[14] A corresponding notion of *upper semi-continuity* is defined with "closed" instead of "open" and in contrast to the single-valued case, upper semi-continuity is different from lower semi-continuity. Both concepts of semi-continuity should not be confused with the corresponding notions of continuity of real number functions.

[15] Implicitly, we have used that lower semi-continuous operations are closed under composition, see Theorem 8.2, where composition is defined as in Definition 8.1.

closure schemes which will be introduced are defined for arbitrary sets, with exception of those places where the set $\mathbb{N}$ of natural numbers is mentioned explicitly.

Definition 8.1 (Recursive closure schemes). The following closure schemes are called *recursive closure schemes*:

(1) **Projection**: If $f :\subseteq X \rightrightarrows Y \times Z$ is an operation, then the *projection* $f_1 :\subseteq X \rightrightarrows Y$ is defined by

$$f_1(x) := \{y : (\exists z)\ (y, z) \in f(x)\}$$

for all $x \in \mathrm{dom}(f_1) := \mathrm{dom}(f)$. The projection $f_2 :\subseteq X \rightrightarrows Z$ on the second component is defined correspondingly.

(2) **Juxtaposition**: If $f :\subseteq X \rightrightarrows Y$ and $g :\subseteq X \rightrightarrows Z$ are operations, then the *juxtaposition* $(f, g) :\subseteq X \rightrightarrows Y \times Z$ is defined by

$$(f, g)(x) := f(x) \times g(x) = \{(y, z) : y \in f(x) \text{ and } z \in g(x)\}$$

for all $x \in \mathrm{dom}(f, g) := \mathrm{dom}(f) \cap \mathrm{dom}(g)$.

(3) **Product**: If $f :\subseteq X \rightrightarrows Y$ and $g :\subseteq U \rightrightarrows V$ are operations, then the *product* $f \times g :\subseteq X \times U \rightrightarrows Y \times V$ is defined by

$$(f \times g)(x, u) := f(x) \times g(u) = \{(y, v) : y \in f(x) \text{ and } v \in g(u)\}$$

for all $(x, u) \in \mathrm{dom}(f \times g) := \mathrm{dom}(f) \times \mathrm{dom}(g)$.

(4) **Composition**: If $f :\subseteq X \rightrightarrows Y$ and $g :\subseteq Y \rightrightarrows Z$ are operations, then the *composition* $g \circ f :\subseteq X \rightrightarrows Z$ is defined by

$$(g \circ f)(x) := g(f(x)) := \{z : (\exists y \in f(x))\ z \in g(y)\}$$

for all $x \in \mathrm{dom}(g \circ f) := \{x : f(x) \subseteq \mathrm{dom}(g)\}$.

(5) **Iteration**: If $f :\subseteq X \rightrightarrows X$ is an operation, then the *iteration* $f^* :\subseteq X \times \mathbb{N} \rightrightarrows X$ is defined by

$$\begin{cases} f^*(x, 0) & := \{x\}, \\ f^*(x, n+1) & := f \circ f^*(x, n) \end{cases}$$

and abbreviated by $f^n(x) := f^*(x, n)$ for all $x \in X$ and $n \in \mathbb{N}$.

(6) **Inversion**: If $f :\subseteq X \times \mathbb{N} \rightrightarrows Y \times \mathbb{N}$ is an operation, then the (*twisted*) *inversion* $f^{\leftrightarrow} :\subseteq X \times \mathbb{N} \rightrightarrows Y \times \mathbb{N}$ is defined by

$$f^{\leftrightarrow}(x,n) := \{(y,k) : (y,n) \in f(x,k)\}$$

for all $(x,n) \in \mathrm{dom}(f^{\leftrightarrow}) := \{(x,n) : (\forall k)\ (x,k) \in \mathrm{dom}(f)$ and $(\exists k)\ n \in f_2(x,k)\}$.

(7) **Evaluation**: If $f :\subseteq X \rightrightarrows Y^{\mathbb{N}}$ is an operation, then the *evaluation* $f_* :\subseteq X \times \mathbb{N} \rightrightarrows Y$ is defined by

$$f_*(x,n) := \{y : (\exists (y_k)_{k\in\mathbb{N}} \in f(x))\ y_n = y\}$$

for all $(x,n) \in \mathrm{dom}(f_*) := \mathrm{dom}(f) \times \mathbb{N}$.

(8) **Transposition**: If $f :\subseteq X \times \mathbb{N} \rightrightarrows Y$ is an operation, then the *transposition* $[f] :\subseteq X \rightrightarrows Y^{\mathbb{N}}$ is defined by

$$[f](x) := \{(y_n)_{n\in\mathbb{N}} : (\forall n)\ y_n \in f(x,n)\}$$

for all $x \in \mathrm{dom}([f]) := \{x : (\forall n)\ (x,n) \in \mathrm{dom}(f)\}$.

(9) **Exponentiation**: If $f :\subseteq X \rightrightarrows Y$ is an operation, then the *exponentiation* $f^{\mathbb{N}} :\subseteq X^{\mathbb{N}} \rightrightarrows Y^{\mathbb{N}}$ is defined by

$$f^{\mathbb{N}}((x_n)_{n\in\mathbb{N}}) := \{(y_n)_{n\in\mathbb{N}} : (\forall n)\ y_n \in f(x_n)\}$$

for all $(x_n)_{n\in\mathbb{N}} \in \mathrm{dom}(f^{\mathbb{N}}) := \{(x_n)_{n\in\mathbb{N}} : (\forall n)\ x_n \in \mathrm{dom}(f)\}$.

(10) **Sequentialization**: If $f :\subseteq X \rightrightarrows \mathbb{N}$ is an operation, then the *sequentialization* $f^{\triangle} :\subseteq X \rightrightarrows \mathbb{N}^{\mathbb{N}}$ is defined by

$$f^{\triangle}(x) := \{(y_n)_{n\in\mathbb{N}} : f(x) = \{y_n : n \in \mathbb{N}\}\}$$

for all $x \in \mathrm{dom}(f^{\triangle}) := \mathrm{dom}(f)$.

The closure schemes projection, juxtaposition and product can be used to handle product spaces and analogously, evaluation, transposition and exponentiation can be used to handle sequence spaces (the latter three schemes are infinite versions of the former three). In presence of projection, juxtaposition and product, the classical schemes of substitution, primitive recursion and minimization can be realized by composition, iteration and inversion. Finally, sequentialization is a scheme which can be used to eliminate indeterminism in certain cases.[16]

[16]The definitions of most of the closure schemes are straightforward, although the details need some care. For instance, a second natural choice for the domain $\mathrm{dom}(g \circ f)$ of the composition of g and f would be the set $\{x : f(x) \cap \mathrm{dom}(g) \neq \emptyset\}$. However, this definition would not satisfy Theorem 8.2.

Theorem 8.2. *On topological spaces, all recursive closure schemes preserve lower semi-continuity of multi-valued operations.*

A proof can be found in [9]. Using the notations of Definition 8.1, we tacitly assume for the previous result that U, V, X, Y, Z are topological spaces and $\mathbb{N}$ is endowed with the discrete topology. Moreover, product and sequence spaces are to be endowed with the corresponding product topologies. Actually, the closure schemes do not only preserve continuity but computability as well. For the following result we assume that $(U, \delta_U), (V, \delta_V), (X, \delta_X), (Y, \delta_Y), (Z, \delta_Z)$ are represented spaces and that $\mathbb{N}$ is represented by $\delta_{\mathbb{N}}$. Product spaces and sequence spaces are represented by the corresponding product and sequence representations.

Theorem 8.3. *On represented spaces, all recursive closure schemes preserve (strong) computability of multi-valued operations.*

Proof. (1) Let $f :\subseteq X \rightrightarrows Y \times Z$ be (strongly) $(\delta_X, [\delta_Y, \delta_Z])$–computable via a computable function $F :\subseteq \mathbb{N}^{\mathbb{N}} \to \mathbb{N}^{\mathbb{N}}$. Define $G :\subseteq \mathbb{N}^{\mathbb{N}} \to \mathbb{N}^{\mathbb{N}}$ by $G\langle p, q\rangle := \pi_1 F\langle p, q\rangle$ for all $p, q \in \mathbb{N}^{\mathbb{N}}$, where $\pi : \mathbb{N}^{\mathbb{N}} \times \mathbb{N}^{\mathbb{N}} \to \mathbb{N}^{\mathbb{N}}$ is the usual pairing function and $\pi_1 := \mathrm{pr}_1 \pi^{-1}$. Then G is computable and $f_1 :\subseteq X \rightrightarrows Y$ is (strongly) (δ_X, δ_Y)–computable via G.

(2) Let $f :\subseteq X \rightrightarrows Y$, $g :\subseteq X \rightrightarrows Z$ be (strongly) (δ_X, δ_Y)–, (δ_X, δ_Z)–computable via computable functions $F, G :\subseteq \mathbb{N}^{\mathbb{N}} \to \mathbb{N}^{\mathbb{N}}$, respectively. Define $H :\subseteq \mathbb{N}^{\mathbb{N}} \to \mathbb{N}^{\mathbb{N}}$ by $H\langle p, \langle q, r\rangle\rangle := \langle F\langle p, q\rangle, G\langle p, r\rangle\rangle$ for all $p, q, r \in \mathbb{N}^{\mathbb{N}}$. Then H is computable and $(f, g) :\subseteq X \rightrightarrows Y \times Z$ is (strongly) $(\delta_X, [\delta_Y, \delta_Z])$–computable via H.

(3) Let $f :\subseteq X \rightrightarrows Y$, $g :\subseteq U \rightrightarrows V$ be (strongly) (δ_X, δ_Y)–, (δ_U, δ_V)–computable via computable functions $F, G :\subseteq \mathbb{N}^{\mathbb{N}} \to \mathbb{N}^{\mathbb{N}}$, respectively. Define $H :\subseteq \mathbb{N}^{\mathbb{N}} \to \mathbb{N}^{\mathbb{N}}$ by $H\langle\langle p, p'\rangle, \langle q, r\rangle\rangle := \langle F\langle p, q\rangle, G\langle p', r\rangle\rangle$ for all $p, p', q, r \in \mathbb{N}^{\mathbb{N}}$. Then H is computable and $f \times g :\subseteq X \times U \rightrightarrows Y \times V$ is (strongly) $([\delta_X, \delta_U], [\delta_Y, \delta_V])$–computable via H.

(4) Let $f :\subseteq X \rightrightarrows Y$, $g :\subseteq Y \rightrightarrows Z$ be (strongly) (δ_X, δ_Y)–, (δ_Y, δ_Z)–computable via computable functions $F, G :\subseteq \mathbb{N}^{\mathbb{N}} \to \mathbb{N}^{\mathbb{N}}$, respectively. Define $H :\subseteq \mathbb{N}^{\mathbb{N}} \to \mathbb{N}^{\mathbb{N}}$ by $H\langle p, \langle q, r\rangle\rangle := G\langle F\langle p, q\rangle, r\rangle$ for all $p, q, r \in \mathbb{N}^{\mathbb{N}}$. Then H is computable and $g \circ f :\subseteq X \rightrightarrows Z$ is (strongly) (δ_X, δ_Z)–computable via H.

(5) Let $f :\subseteq X \rightrightarrows X$ be (strongly) (δ_X, δ_X)–computable via a computable function $F :\subseteq \mathbb{N}^{\mathbb{N}} \to \mathbb{N}^{\mathbb{N}}$. Define $G :\subseteq \mathbb{N}^{\mathbb{N}} \to \mathbb{N}^{\mathbb{N}}$ by

$$\begin{cases} G\langle\langle p, p'\rangle, \langle q_0, q_1, \ldots\rangle\rangle := p & \text{if } p'(0) = 0 \\ G\langle\langle p, p'\rangle, \langle q_0, q_1, \ldots\rangle\rangle := F\langle G\langle\langle p, \hat{n}\rangle, \langle q_0, q_1, \ldots\rangle\rangle, q_n\rangle & \text{if } p'(0) = n + 1 \end{cases}$$

for all $p, p', q_i \in \mathbb{N}^{\mathbb{N}}, i \in \mathbb{N}$. Then G is computable and $f^* :\subseteq X \times \mathbb{N} \rightrightarrows X$ is (strongly) $([\delta_X, \delta_{\mathbb{N}}], \delta_X)$–computable via G.

(6) Let $f :\subseteq X \times \mathbb{N} \rightrightarrows Y \times \mathbb{N}$ be $([\delta_X, \delta_\mathbb{N}], [\delta_Y, \delta_\mathbb{N}])$–computable via a computable function $F :\subseteq \mathbb{N}^\mathbb{N} \to \mathbb{N}^\mathbb{N}$. Thus, there is a computable, total, and monotone function $\varphi : \mathbb{N}^* \to \mathbb{N}^*$ which approximates F. Define a predicate $P \subseteq \mathbb{N}^\mathbb{N} \times \mathbb{N} \times \mathbb{N} \times \mathbb{N}$ by

$$P := \{(p, n, k, i) : \pi_1'\nu^*(i) \sqsubseteq \langle p, \hat{k}\rangle \text{ and } n \sqsubseteq \pi_2'\varphi\nu^*(i)\}.$$

Here, $\pi_1', \pi_2' : \mathbb{N}^* \to \mathbb{N}^*$ denote computable projections such that $w \sqsubseteq \langle p, q\rangle$ is equivalent to $\pi_1'(w) \sqsubseteq p$ and $\pi_2'(w) \sqsubseteq q$. By π_i we will denote the projections of the inverses of the tuple functions $\mathbb{N}^\mathbb{N} \times \mathbb{N}^\mathbb{N} \to \mathbb{N}^\mathbb{N}$, as well as of $\mathbb{N} \times \mathbb{N} \to \mathbb{N}$. Since P is decidable it follows that $h :\subseteq \mathbb{N}^\mathbb{N} \times \mathbb{N} \times \mathbb{N} \to \mathbb{N}$, defined by

$$\begin{cases} h(p, n, 0) & := \quad \mu\langle k, i\rangle[(p, n, k, i) \in P] \\ h(p, n, m+1) & := \quad \mu\langle k, i\rangle[\langle k, i\rangle > h(p, n, m) \text{ and } (p, n, k, i) \in P] \end{cases}$$

for all p, n, m, is computable. Define a function $G :\subseteq \mathbb{N}^\mathbb{N} \to \mathbb{N}^\mathbb{N}$ by $G\langle\langle p, p'\rangle, mq\rangle := \langle \pi_1 F\langle\langle p, \hat{k}\rangle, wq\rangle, \hat{k}\rangle$ for all $p, p', q \in \mathbb{N}^\mathbb{N}$, $m \in \mathbb{N}$, where $\langle k, i\rangle := h(p, p'(0), m)$ and $w := \pi_2'\nu^*(i)$. Then G is computable and one can show that $f^\leftrightarrow$ is computable via G. In case of strong computability one has to restrict G to those inputs $\langle\langle p, p'\rangle, \langle q, q'\rangle\rangle$ such that $\langle\langle p, q\rangle, q'\rangle \in \mathrm{dom}(F)$.

(7) Let $f :\subseteq X \rightrightarrows Y^\mathbb{N}$ be (strongly) $(\delta_X, \delta_Y^\infty)$–computable via a computable function $F :\subseteq \mathbb{N}^\mathbb{N} \to \mathbb{N}^\mathbb{N}$. Define a function $G :\subseteq \mathbb{N}^\mathbb{N} \to \mathbb{N}^\mathbb{N}$ by $G\langle\langle p, p'\rangle, q\rangle := \eta(F\langle p, q\rangle, p'(0))$ for all $p, p', q \in \mathbb{N}^\mathbb{N}$, where the evaluation function $\eta : \mathbb{N}^\mathbb{N} \times \mathbb{N} \to \mathbb{N}^\mathbb{N}$, defined by $\eta(\langle p_0, p_1, ...\rangle, n) := p_n$, is computable. Then G is computable and $f_* :\subseteq X \times \mathbb{N} \rightrightarrows Y$ is (strongly) $([\delta_X, \delta_\mathbb{N}], \delta_Y)$–computable via G.

(8) Let $f :\subseteq X \times \mathbb{N} \rightrightarrows Y$ be (strongly) $([\delta_X, \delta_\mathbb{N}], \delta_Y)$–computable via a computable function $F :\subseteq \mathbb{N}^\mathbb{N} \to \mathbb{N}^\mathbb{N}$. Define $G :\subseteq \mathbb{N}^\mathbb{N} \to \mathbb{N}^\mathbb{N}$ by $G\langle p, \langle q_0, q_1, ...\rangle\rangle := \langle F\langle\langle p, \hat{0}\rangle, q_0\rangle, F\langle\langle p, \hat{1}\rangle, q_1\rangle, ...\rangle$ for all $p, q_i \in \mathbb{N}^\mathbb{N}$, $i \in \mathbb{N}$. Then G is computable and $[f] :\subseteq X \rightrightarrows Y^\mathbb{N}$ is (strongly) $(\delta_X, \delta_Y^\infty)$–computable via G.

(9) Let $f :\subseteq X \rightrightarrows Y$ be (strongly) (δ_X, δ_Y)–computable via a computable function $F :\subseteq \mathbb{N}^\mathbb{N} \to \mathbb{N}^\mathbb{N}$. Define a function $G :\subseteq \mathbb{N}^\mathbb{N} \to \mathbb{N}^\mathbb{N}$ by $G\langle\langle p_0, p_1, ...\rangle, \langle q_0, q_1, ...\rangle\rangle := \langle F\langle p_0, q_0\rangle, F\langle p_1, q_1\rangle, ...\rangle$ for all $p_i, q_i \in \mathbb{N}^\mathbb{N}$, $i \in \mathbb{N}$. Then G is computable and $f^\mathbb{N} :\subseteq X^\mathbb{N} \rightrightarrows Y^\mathbb{N}$ is (strongly) $(\delta_X^\infty, \delta_Y^\infty)$–computable via G.

(10) Let $f :\subseteq X \rightrightarrows \mathbb{N}$ be $(\delta_X, \delta_\mathbb{N})$–computable via a computable function $F :\subseteq \mathbb{N}^\mathbb{N} \to \mathbb{N}^\mathbb{N}$. Define $G' :\subseteq \mathbb{N}^\mathbb{N} \to \mathbb{N}^\mathbb{N}$ by $G'(p)(n) := \delta_\mathbb{N} F\langle p, r_n\rangle = F\langle p, r_n\rangle(0)$ for all $p \in \mathbb{N}^\mathbb{N}$, $n \in \mathbb{N}$, where $r_n := \nu^*(n)000... = \nu^*(n)\hat{0}$. Since $\{r_n : n \in \mathbb{N}\}$ is dense in $\mathbb{N}^\mathbb{N}$ and F is continuous, it follows

$$f\delta_X(p) = \{\delta_\mathbb{N} F\langle p, q\rangle : q \in \mathbb{N}^\mathbb{N}\} = \{G'(p)(n) : n \in \mathbb{N}\}$$

for all $p \in \mathrm{dom}(f\delta_X)$. Now define $G :\subseteq \mathbb{N}^\mathbb{N} \to \mathbb{N}^\mathbb{N}$ by

$$G\langle p, q\rangle\langle n, k\rangle := \begin{cases} H\langle G'(p), q\rangle(n) & \text{if } \langle p, q\rangle \in \mathrm{dom}(F) \\ \uparrow & \text{else} \end{cases}$$

for all $p, q \in \mathbb{N}^\mathbb{N}$ and $n, k \in \mathbb{N}$, where H is the function given by the Oracle Lemma 7.2. Then G is computable. By definition of H it follows

$$\begin{aligned} f^\triangle\delta_X(p) &= \{r \in \mathbb{N}^\mathbb{N} : f\delta_X(p) = \mathrm{range}(r)\} \\ &= \{H\langle G'(p), q\rangle : q \in \mathbb{N}^\mathbb{N}\} \\ &= \{\delta_\mathbb{N}^\infty G\langle p, q\rangle : q \in \mathbb{N}^\mathbb{N}\} \end{aligned}$$

for all $p \in \mathrm{dom}(f^\triangle\delta_X)$, i.e. $f^\triangle :\subseteq X \rightrightarrows \mathbb{N}^\mathbb{N}$ is $(\delta_X, \delta_\mathbb{N}^\infty)$–computable via G. Correspondingly, the case of strong computability can be treated. □

9. Recursive Operations over Structures

In this section we will use the recursive closure schemes to introduce the notion of a recursive operation over a structure. We will consider many-sorted structures $(X_1, ..., X_n; f_1, ..., f_k)$ with a finite number of pairwise disjoint sets $X_1, ..., X_n$ and a finite number of operations $f_1, ..., f_k$. We start with the definition of a *prestructure* (prestructures have the property that the operations are not necessarily related to the universes).

Definition 9.1 (Prestructures). $S = (X_1, ..., X_n; f_1, ..., f_k)$ is called a *many-sorted prestructure* with *universe* $X_1 \times ... \times X_n$, if $X_1, ..., X_n$ are pairwise disjoint sets and $f_1, ..., f_k$ are arbitrary operations, called the *initial operations* of S.

Usually, we will say for short *prestructure* instead of many-sorted prestructure. In the next step we define the notion of a set over a prestructure $S = (X_1, ..., X_n; f_1, ..., f_k)$. Roughly speaking, a set over a prestructure S is a set that can be constructed as product or sequence set of the sets $X_1, ..., X_n$. This definition takes into account that we want to handle product spaces as well as sequential spaces.

Definition 9.2 (Sets and operations over prestructures). Let $S = (X_1, ...$ $..., X_n; f_1, ..., f_k)$ be a prestructure. Then the *class of sets over* S is the smallest class of sets such that:

(1) $X_1, ..., X_n$ are sets over S,

(2) $X \times Y$ and $X^\mathbb{N}$ are sets over S, if X, Y are sets over S.

An operation $f :\subseteq X \rightrightarrows Y$ is called an *operation over* S, if X, Y are sets over S.

Now we will single out those prestructures S whose initial operations are operations over S and we will call them *structures*.

Definition 9.3 (Structures). A prestructure S is called a *structure*, if all initial operations of S are operations over S.

The reason why the universe of a structure S is defined as product is a technical one: it ensures that the universe itself is a set over S. If $S = (X_1, ..., X_n; f_1, ..., f_k)$ and $T = (Y_1, ..., Y_m; g_1, ..., g_l)$ are structures such that $Y_1, ..., Y_m \in \{X_1, ..., X_n\}$ and $g_1, ..., g_l \in \{f_1, ..., f_k\}$, then T is called a *substructure* of S and we write $T \sqsubseteq S$. Now we proceed to define recursive operations over structures.

Definition 9.4 (Recursive operations over structures). The class of *recursive operations over a structure* S is the smallest class of operations which contains all initial operations of S and which is closed under the recursive closure schemes.

From the definition of the closure schemes it is obvious that each recursive operation over a structure is an operation over that structure.

A constant $x \in X$ is called a *recursive constant over a structure* S, if the operation $\{()\} \to X, () \mapsto x$ is recursive over S (we assume $X^0 = \{()\}$ for each set X). If there is a constant $c \in X$ among the initial operations of a structure, then we will consider it as zero-ary constant function $X^0 \to X, () \mapsto c$ with value c. For completeness, we define a representation $\delta : \mathbb{N}^{\mathbb{N}} \to \{()\}$ by $\delta(p) := ()$ for all $p \in \mathbb{N}^{\mathbb{N}}$, but usually we will not use this representation explicitly.

A very important structure is the *structure of natural numbers* which is given by

$$\mathbf{N} := (\mathbb{N}; 0, n, n + 1).$$

By "0" we denote the *zero-ary constant function* $\mathbb{N}^0 \to \mathbb{N}, () \mapsto 0$ with value 0, by "n" the *identity* $\mathrm{id}_{\mathbb{N}} : \mathbb{N} \to \mathbb{N}, n \mapsto n$, and by "$n + 1$" the *successor function* $S : \mathbb{N} \to \mathbb{N}, n \mapsto n + 1$. Here and in the following we will often use bold letters, as "**N**", to distinguish a structure from its universe. It is routine to prove the following result which shows that the class of recursive operations over **N** is quite rich.

Proposition 9.5. *Each computable function* $F :\subseteq \mathbb{N}^{\mathbb{N}} \to \mathbb{N}^{\mathbb{N}}$ *is recursive over* **N**.

One can even prove this result without using the closure scheme of sequentialization, see [9]. In the following we will often need structures which include the natural numbers and some essential operations. These structures will be called *natural*. By $\mathrm{id}_X : X \to X, x \mapsto x$ we denote the *identity* of X.

Definition 9.6 (Natural structures). A structure S with universe X is called *natural*, if $\mathbf{N}$ is a substructure of S, id_X is recursive over S and there is a recursive constant $c \in X$ over S.

It is easy to see that for natural structures $S = (X_1, ..., X_n; f_1, ..., f_k)$ the identity of each set over S is recursive. Especially, the identities $\mathrm{id}_{X_1}, ..., \mathrm{id}_{X_n}$ are recursive, and each set $X_1, ..., X_n$ contains at least one recursive constant.

Sometimes it is also suitable to consider structures where a special constant operation is available. For each set $A \subseteq X$ we define the *omnipotent operation* $\Omega_A : \{()\} \rightrightarrows X, () \mapsto A$ of A, i.e. Ω_A is the constant operation which yields as result $\Omega_A() = A$.

Definition 9.7 (Complete structure). A structure S with universe X is called a *complete structure*, if Ω_X is recursive over S.

Since $\Omega_{\mathbb{N}} = z^{\leftrightarrow} \circ 0$, where $z = (0 \times \mathrm{id}_{\mathbb{N}})_1$ denotes the constant zero function $z : \mathbb{N} \to \mathbb{N}, n \mapsto 0$ and $\Omega_{\mathbb{N}^{\mathbb{N}}} = [(\Omega_{\mathbb{N}} \times \mathrm{id}_{\mathbb{N}})_1]$, we obtain that $\Omega_{\mathbb{N}}$ and $\Omega_{\mathbb{N}^{\mathbb{N}}}$ are recursive over each natural structure. Especially, the structure $\mathbf{N}$ is complete.

Our next goal is to prove that recursive operations are also computable. Therefore, we will first define representations of structures and computability on structures.

Definition 9.8 (Representations of structures). Consider a structure $S = (X_1, ..., X_n; f_1, ..., f_k)$ and let $\delta_1, ..., \delta_n$ be representations of $X_1, ..., X_n$, respectively. Then $\delta := [\delta_1, ..., \delta_n]$ is called a *representation* of S and (S, δ) is called a *represented structure*.

Using this definition we can extend the notion of computability to structures.

Definition 9.9 (Computability on structures). Consider a structure $S = (X_1, ..., X_n; f_1, ..., f_k)$ with representation $\delta = [\delta_1, ..., \delta_n]$. Now, let $f :\subseteq Y \rightrightarrows Z$ be an operation over S and let δ_Y, δ_Z be representations of Y, Z, respectively, which are finitely generated from $\delta_1, ..., \delta_n$, correspondingly as Y, Z are finitely generated from $X_1, ..., X_n$. Then f is called *(strongly) computable* with respect to δ, if it is (strongly) (δ_Y, δ_Z)–computable.

It should be clear how the representations δ_Y, δ_Z have to be constructed correspondingly to Y, Z. If, for instance, $Y = (X_1 \times X_2)^{\mathbb{N}}$, then the corresponding representation is $\delta_Y = [\delta_1, \delta_2]^\infty$. In the next step we define the notion of an effective structure.

Definition 9.10 (Effective structures). A structure S is called a *(strongly) effective structure*, if there is a representation δ of S such that all initial operations of S are (strongly) computable with respect to δ. In this situation S is also called *(strongly) effective via* δ.

It is easy to see that the structure **N** is strongly effective via $\delta_{\mathbb{N}}$. In the following we will use effective structures where the effectivity on natural numbers is fixed. We will say that $\delta = [\delta_1, ..., \delta_n]$ is a *natural representation* of a natural structure $S = (X_1, ..., X_n; f_1, ..., f_k)$, if $X_i = \mathbb{N}$ implies $\delta_i \equiv \delta_{\mathbb{N}}$. By structural induction we obtain the following corollary of Theorem 8.3.

Corollary 9.11 (Recursive operations over effective structures). *If S is a natural structure which is (strongly) effective via a natural representation δ, then each operation which is recursive over S is also (strongly) computable with respect to δ.*

10. A Stability Theorem for Perfect Structures

In this section we introduce the class of recursive structures and we will prove that over recursive structures each strongly computable operation is also recursive. This leads us to the definition of perfect structures which are both: effective and recursive. Perfect structures have several nice properties, especially they are categorical in the sense that they characterize their own computability theory. In order to define recursive structures, we will use the notion of a recursive retraction.

Definition 10.1 (Recursive retraction). An operation $f :\subseteq X \rightrightarrows Y$ over a structure S is called a *recursive retraction* over S, if it is recursive and if it admits a recursive right inverse operation $f^- : Y \rightrightarrows X$ over S, i.e. $f \circ f^- = \mathrm{id}_Y$.

If f is a recursive retraction, then the restriction of f to range(f^-) is a surjective function. A structure will be called strongly recursive, if it admits a representation which is a recursive retraction.

Definition 10.2 (Recursive structure). Let S be a structure. Then

(1) S is called a *recursive structure*, if there is a representation δ of S, which admits a recursive extension, as well as a recursive right inverse operation over S,

(2) S is called a *strongly recursive structure*, if there is a representation δ of S which is a recursive retraction over S.

In these cases S is also called *(strongly) recursive via δ.*

Intuitively, a strongly recursive structure is a structure with a representation which can be "synthesized" as well as "analyzed" within the structure. By the following lemma $\delta = [\delta_1, ..., \delta_n]$ is a recursive retraction, if $\delta_1, ..., \delta_n$ are.

Lemma 10.3. *Let S be a natural structure and let $\delta_i :\subseteq \mathbb{N}^{\mathbb{N}} \to X_i$, $i = 1, ..., n$, $\delta :\subseteq \mathbb{N}^{\mathbb{N}} \to X$ be representations. If $\delta_1, ..., \delta_n, \delta$ are recursive retractions over S, then so are $[\delta_1, ..., \delta_n]$ and δ^∞.*

Vice versa, it is easy to see that $\delta_1, ..., \delta_n$ are recursive retractions over S, if $\delta = [\delta_1, ..., \delta_n]$ is a recursive retraction over S. Now we prove that the structure of natural numbers is recursive.

Proposition 10.4. *The structure* $\mathbf{N}$ *is strongly recursive via* $\delta_{\mathbf{N}}$.

Proof. The representation $\delta_{\mathbf{N}} = (\mathrm{id}_{\mathbb{N}^{\mathbb{N}}})_* \circ (\mathrm{id}_{\mathbb{N}^{\mathbb{N}}} \times 0)$ is recursive over $\mathbf{N}$. Define $f : \mathbb{N} \times \mathbb{N} \rightrightarrows \mathbb{N}$ by $f(k, 0) := k$ and $f(k, n+1) := \Omega_{\mathbf{N}}()$. Then f is recursive over $\mathbf{N}$ and so is $\delta_{\mathbf{N}}^{-1} = [f]$. □

Since the structure of natural numbers is recursive it makes sense to apply the notion of recursiveness to natural structures. Now we can prove the announced converse version of Corollary 9.11.

Theorem 10.5 (Comp. operations over recursive structures). *If* S *is a natural structure which is strongly recursive via a representation* δ*, then each operation over* S *which is strongly computable with respect to* δ *is also recursive over* S.

Proof. Let $S = (X_1, ..., X_n; f_1, ..., f_k)$, let $\delta = [\delta_1, ..., \delta_n]$, and let $f :\subseteq X \rightrightarrows Y$ be an operation over S. Then X, Y are finitely generated by product and sequence constructions from the sets $X_1, ..., X_n$. Let δ_X, δ_Y be the corresponding representations, constructed from $\delta_1, ..., \delta_n$. By structural induction one can deduce from Lemma 10.3 that δ_X, as well as δ_Y, are recursive retractions. Now let $\delta_X^- : X \rightrightarrows \mathbb{N}^{\mathbb{N}}$ be a right inverse of δ_X, such that δ_X^- is recursive over S. Let $f :\subseteq X \rightrightarrows Y$ be strongly (δ_X, δ_Y)–computable via a computable function $F :\subseteq \mathbb{N}^{\mathbb{N}} \to \mathbb{N}^{\mathbb{N}}$. Then one obtains $f = \delta_Y \circ F \circ \langle \delta_X^- \times \Omega_{\mathbb{N}^{\mathbb{N}}} \rangle$ and it follows that f is recursive over S: since F is computable, it is recursive over $\mathbf{N}$ by Proposition 9.5 and $\langle \delta_X^- \times \Omega_{\mathbb{N}^{\mathbb{N}}} \rangle$ is recursive over S, since the tupling function $\langle . \rangle : \mathbb{N}^{\mathbb{N}} \times \mathbb{N}^{\mathbb{N}} \to \mathbb{N}^{\mathbb{N}}$, δ_X^- and $\Omega_{\mathbb{N}^{\mathbb{N}}}$ are recursive over S. Altogether, f is recursive over S. □

From the proof we can also conclude the following property of computable operations over recursive structures.

Corollary 10.6 (Comp. operations over recursive structures). *If* S *is a natural structure which is recursive via a representation* δ*, then each operation over* S *which is computable with respect to* δ *also admits a recursive extension over* S.

We have seen that over effective structures recursive operations are computable and over recursive structures computable operations are recursive. Thus, it makes sense to consider structures which are both, recursive and effective. We will see that such structures have very nice properties. The first result states that for such structures there is a unique effectivity which is characterized by the structure itself.

Theorem 10.7 (Stability theorem). *If S is a natural structure which is recursive via a representation δ and which is effective via a natural representation δ', then $\delta \equiv \delta'$.*

Proof. Let S be a natural structure which is recursive via a representation δ and which is effective via a natural representation δ'. Without loss of generality, let $X = \mathbb{N} \times X_2 \times ... \times X_n$ be the universe of S and let $\delta' = [\delta'_1, ..., \delta'_n]$. Since δ' is a natural representation $\delta_1'^{\infty} \equiv \delta_{\mathbb{N}}^{\infty} \equiv \mathrm{id}_{\mathbb{N}^{\mathbb{N}}}$. Let δ_+ be an extension of δ and let δ^- be a right inverse operation of δ which are recursive over S. Then $\delta_- := (\delta^-)^{-1}$ is a representation of S too. We prove $\delta \equiv \delta_+ \equiv \delta' \equiv \delta_-$.

"$\delta \leq \delta_+$" Since δ_+ is an extension of δ, we obtain $\delta \leq \delta_+$.

"$\delta_+ \leq \delta'$" By Corollary 9.11 it follows that δ_+ is computable with respect to δ', i.e. δ_+ is $(\delta_1'^{\infty}, \delta')$–computable. Thus, δ_+ is $(\mathrm{id}_{\mathbb{N}^{\mathbb{N}}}, \delta')$–computable. But this means $\delta_+ \leq \delta'$.

"$\delta' \leq \delta_-$" Again we can deduce by Corollary 9.11 that δ^- is $(\delta', \delta_1'^{\infty})$–computable and thus $(\delta', \mathrm{id}_{\mathbb{N}^{\mathbb{N}}})$–computable, i.e. there is a computable function $F :\subseteq \mathbb{N}^{\mathbb{N}} \to \mathbb{N}^{\mathbb{N}}$ such that $\delta^-\delta'(p) = \{F\langle p, q\rangle : q \in \mathbb{N}^{\mathbb{N}}\}$ for all $p \in \mathrm{dom}(\delta^-\delta') = \mathrm{dom}(\delta')$. Then $G :\subseteq \mathbb{N}^{\mathbb{N}} \to \mathbb{N}^{\mathbb{N}}$, defined by $G(p) := F\langle p, p\rangle$ for all $p \in \mathbb{N}^{\mathbb{N}}$ is computable and $\delta'(p) = \delta_- G(p)$ for all $p \in \mathrm{dom}(\delta')$, i.e. $\delta' \leq \delta_-$.

"$\delta_- \leq \delta$" Since δ^- is a right inverse operation of δ, it follows that δ_- is a restriction of δ and thus $\delta_- \leq \delta$. □

We can deduce that, if a structure S is recursive, then all natural representations δ such that S is effective via δ are equivalent. Thus, if S is recursive and effective, then it is *effectively categorical*, a notion that has been used in a similar setting by Hertling [29]. But even more, if S is effective via a natural representation, then all representations δ which are recursive retractions over S are also equivalent. Thus, if S is recursive and effective, then it has a further property which could be called *recursively categorical.*

It should be mentioned that there is no hope for a corresponding result without the restriction to natural representations. As long as we do not demand any evaluation property for the output, we can effectivize a structure just by using terms and their evaluations. The situation changes if we do fix effectivity for at least one "output sort". For instance, in the classical setting this can be done by fixing effectivity for the boolean sort, i.e. the "output sort" of predicates.[17] In our setting, we have fixed effectivity on the natural numbers. The Stability Theorem gives rise to the following definition.

[17]This is the case in Mal'cev's Stability Theorem for finitely generated algebras [39, 54], which can be considered as a special version of the Stability Theorem 10.7, see [10].

Definition 10.8 (Perfect structures). A natural structure S is called (*strongly*) *perfect*, if it is (strongly) recursive and (strongly) effective via a natural representation. In this situation each natural representation δ such that S is (strongly) recursive via δ is called a (*strong*) *standard representation* of S.

It is easy to see that $\mathbf{N}$ is a strongly perfect structure with strong standard representation $\delta_{\mathbf{N}}$. The Stability Theorem states that perfect structures uniquely characterize their effectivity. Especially, all standard representations of a perfect structure and all natural representations which make this structure effective belong to the same equivalence class. Therefore, we can define computability over perfect structures without mentioning any special representation.

Definition 10.9 (Comp. operations over perfect structures). An operation f over a perfect structure S is called (*strongly*) *computable over* S, if it is (strongly) computable with respect to a standard representation δ of S.

Now we can formulate a corollary of Theorem 10.5 and Corollary 9.11 which characterizes recursive operations over strongly perfect structures. The corresponding statement for perfect structures follows from Corollary 10.6 and Corollary 9.11

Theorem 10.10 (Operations over perfect structures). *An operation over a strongly perfect structure is strongly computable, if and only if it is recursive. An operation over a perfect structure is computable, if and only if it admits a recursive extension.*

Finally, we obtain an Extension Theorem for operations over strongly perfect structures as a combination of both results.

Corollary 10.11 (Extension Theorem). *Each computable operation over a strongly perfect structure admits a strongly computable extension.*

Another nice property of strongly perfect structures is the property of *conservative extension*. Consider a natural structure S with an *extension* S', i.e. a structure S' such that $S \sqsubseteq S'$. In general, the additional initial operations of S' could increase the class of recursive operations even over the universe of S. The following theorem states that this cannot happen over perfect structures.

Theorem 10.12 (Conservation Theorem). *Let $S \sqsubseteq S'$ be strongly perfect structures and let f be an operation over S. Then f is recursive over S, if and only if f is recursive over S'.*

Proof. Without loss of generality, we assume $S = (X_1, ..., X_n; f_1, ..., f_k)$ and $S' = (X_1, ..., X_{n+1}; f_1, ..., f_m)$ with $m \geq k$. The general case can be deduced by induction. Let $\delta = [\delta_1, ..., \delta_n]$ and $\delta' = [\delta'_1, ..., \delta'_{n+1}]$ be standard representations of S, S', respectively. Then S' is effective via δ' and hence S is effective via $\delta'' := [\delta'_1, ..., \delta'_n]$. On the other hand, S is effective via δ and since S is perfect we can conclude $\delta \equiv \delta''$ by the Stability Theorem 10.7. If f is recursive over S, then f obviously is recursive over S' too. If f is an operation over S which is recursive over S', then f is strongly computable over S' by Theorem 10.10, since S' is strongly perfect. Thus, f is strongly computable with respect to δ' and thus with respect to δ''. Consequently, f is strongly computable with respect to δ too and thus it is recursive over S by Theorem 10.10, since S is strongly perfect. □

If we revisit the proof, then we obtain the following version of the Conservation Theorem for perfect structures.

Corollary 10.13. *Let $S \sqsubseteq S'$ be perfect structures and let f be an operation over S. Then f admits a recursive extension over S, if and only if it admits a recursive extension over S'.*

We close this section with a short meta-analysis of our results. We have already mentioned that for the proof of Proposition 9.5 it suffices to use all recursive closure schemes besides sequentialization. Proposition 9.5 has been used for the proof of Theorem 10.5. But neither for this theorem nor for the other results of this section we have used the sequentialization operator. Indeed, all results on perfect structures remain true, if we exclude sequentialization from our recursive closure schemes. However, the class of perfect structures could be smaller in this case and this is the reason why we have included sequentialization.

11. Topological and Metric Structures

In the previous section we have seen that perfect structures have some nice properties but we have not proved that there are any interesting perfect structures. In this section we will show that T_0–spaces with countable bases admit perfect topological structures and we will define a perfect standard structure for separable metric spaces. We start with the definition of the notion of a topological structure. It is natural to assume that the universes of such structures are topological spaces and that the initial operations are continuous.

Definition 11.1 (Topological structure). A *topological structure* is a structure $S = (X_1, ..., X_n; f_1, ..., f_k)$ where $X_1, ..., X_n$ are endowed with topologies and all initial operations $f_1, ..., f_k$ are lower semi-continuous with respect to the corresponding product topologies. Moreover, S is called *natural topological structure*, if S is a natural structure and the natural numbers come equipped with the discrete topology.

If Y is a set over a structure S, then it inherits a product topology which is generated from the topologies of $X_1, ..., X_n$, correspondingly as Y is generated from the sets $X_1, ..., X_n$. In this situation we will say that Y is a *topological space over* S. In the following we will sometimes use short names for topological structures with additional properties. For instance, a "metric structure" will be a topological structure where the universe is endowed with a metric. As we have seen in Theorem 8.2, all recursive closure schemes preserve lower semi-continuity. Therefore, we obtain immediately the following corollary.

Corollary 11.2 (Continuity of recursive operations). *All recursive operations over natural topological structures are lower semi-continuous.*

In the following, by $S \oplus T := (X_1, ..., X_n, Y_1, ..., Y_m; f_1, ..., f_k, g_1, ..., g_l)$ we will denote the *union* of a prestructure $S = (X_1, ..., X_n; f_1, ..., f_k)$ with a prestructure $T = (Y_1, ..., Y_m; g_1, ..., g_l)$. If X is a second countable T_0–space with standard representation δ, then δ is open and continuous, hence δ, as well as $\delta^{-1} : X \rightrightarrows \mathbb{N}^{\mathbb{N}}$, are lower semi-continuous. If there is some computable point $p \in \mathrm{dom}(\delta)$, then $\mathbf{X} = \mathbf{N} \oplus (X; \delta, \delta^{-1})$ is a natural topological structure. Obviously, $\delta' := [\delta_{\mathrm{N}}, \delta]$ is a representation of $\mathbf{X}$ and $\mathbf{X}$ is recursive via δ'. Moreover, $\mathbf{X}$ is also effective via δ' since δ is $(\mathrm{id}_{\mathbb{N}^{\mathbb{N}}}, \delta)$–computable and δ^{-1} is $(\delta, \mathrm{id}_{\mathbb{N}^{\mathbb{N}}})$–computable by the Oracle Lemma 7.2. Altogether we obtain the following result.

Theorem 11.3 (Perfect topological structures). *If X is a second countable T_0–space with standard representation δ such that there is some computable point in* $\mathrm{dom}(\delta)$, *then $\mathbf{X} = \mathbf{N} \oplus (X; \delta, \delta^{-1})$ is a perfect topological structure with standard representation* $[\delta_{\mathrm{N}}, \delta]$.

Here, second countability is not only a sufficient but even a necessary condition. If $\mathbf{X}$ is a natural topological structure which is recursive via some representation δ, then there exists some right inverse δ^- of δ which is recursive over $\mathbf{X}$. Then $\{(\delta^-)^{-1}(w\mathbb{N}^{\mathbb{N}}) : w \in \mathbb{N}^*\}$ is a countable base of X. Thus, we have proved the following result.

Proposition 11.4. *If S is a recursive topological structure with universe X, then the topology of X has a countable base.*

By Theorem 11.3 we know that any second countable T_0–space admits a perfect topological structure, but the structure given in Theorem 11.3 is quite artificial because it contains a representation and its inverse as initial operations. In the following we will see that separable metric spaces admit far more natural structures. As a special case we first investigate a structure for the real numbers. As usual, 0 and 1 denote the corresponding real constants, $+, -, \cdot, /$ the usual arithmetic operations on the real numbers with their natural domains. By $<$ we denote the *semi-characteristic operation* of the usual order relation:

$$c_< : \mathbb{R} \times \mathbb{R} \rightrightarrows \mathbb{N}, (x, y) \mapsto \begin{cases} \{0, 1\} & \text{if } x < y \\ \{1\} & \text{else} \end{cases}$$

This definition will be generalized and discussed below. As before, $\text{Lim} :\subseteq \mathbb{R}^{\mathbb{N}} \to \mathbb{R}$ denotes the *limit operator* (restricted to rapidly converging sequences) with respect to the Euclidean distance.

Theorem 11.5 (The structure of the real numbers). *The structure*

$$\mathbf{R} := \mathbf{N} \oplus (\mathbb{R}; 0, 1, x + y, -x, x \cdot y, 1/x, \text{Lim}, x < y)$$

is a strongly perfect complete topological structure and the Cauchy representation yields a standard representation of this structure.

Proof. Obviously, all arithmetic operations are continuous with respect to the Euclidean topology. The limit operation is continuous (with respect to the product topology on $\mathbb{R}^{\mathbb{N}}$), since it is restricted to rapidly converging sequences and $c_<$ is lower semi-continuous since $<$ is open. Altogether, $\mathbf{R}$ is a natural topological structure. It is easy to see that the notation of rational numbers $\nu_{\mathbb{Q}}$ is recursive over $\mathbf{R}$ and thus the Cauchy representation $\delta_{\mathbb{R}} = \text{Lim} \circ \nu_{\mathbb{Q}}^{\mathbb{N}}$ is recursive over $\mathbf{R}$ as well. By the Heron algorithm one can prove that the square root function $x \mapsto \sqrt{x}$ is recursive over $\mathbf{R}$ and it follows that the absolute value function and hence the Euclidean metric $d_{\mathbb{R}}(x, y) = |x - y|$ are recursive over $\mathbf{R}$. Using these facts, one can show that there is a recursive function $f : \mathbb{R} \times \mathbb{N} \rightrightarrows \mathbb{N}$ over $\mathbf{R}$ such that $\text{graph}(f) = \{(x, k, n) \in \mathbb{R} \times \mathbb{N} \times \mathbb{N} : d_{\mathbb{R}}(\nu_{\mathbb{Q}}(n), x) < 2^{-k-1}\}$. Hence, $\delta^- := [f] : \mathbb{R} \rightrightarrows \mathbb{N}^{\mathbb{N}}$ is recursive over $\mathbf{R}$ and it is a right inverse of $\delta_{\mathbb{R}}$. Altogether, it follows that $\mathbf{R}$ is strongly recursive via $\delta_{\mathbb{R}}$. It remains to prove that $\mathbf{R}$ is strongly effective via $\delta_{\mathbb{R}}$ as well. It is routine to prove that the arithmetic operations are strongly computable with respect to the Cauchy representation. By Proposition 5.3 it follows that the limit operation is computable as well and using completeness one can show that Lim is even strongly computable. Finally, the semi-characteristic operation $c_< : \mathbb{R}\times\mathbb{R} \rightrightarrows \mathbb{N}$ is strongly computable via the function $F : \mathbb{N}^{\mathbb{N}} \to \mathbb{N}^{\mathbb{N}}$, defined by

$$F\langle\langle p, p'\rangle, q\rangle := \begin{cases} \hat{0} & \text{if} \quad \nu_{\mathbb{Q}} pq(0) < \nu_{\mathbb{Q}} p'q(0) - 2^{-q(0)+1} \text{ and } q(1) = 0 \\ \hat{1} & \text{else} \end{cases}$$

for all $p, p', q \in \mathbb{N}^{\mathbb{N}}$. □

More precisely, $[\delta_{\mathbb{N}}, \delta_{\mathbb{R}}]$ is a standard representation of $\mathbf{R}$. As a direct consequence of this theorem and Theorem 10.10, we can deduce that the recursive functions over $\mathbf{R}$ are exactly the strongly computable ones. As a special case we obtain that the total recursive functions $f : \mathbb{R}^n \to \mathbb{R}$ over $\mathbf{R}$ are exactly the classically computable functions (according to Grzegorczyk's and Lacombe's definitions [24, 36]).

Corollary 11.6. *The recursive functions $f : \mathbb{R}^n \to \mathbb{R}$ over $\mathbf{R}$ are exactly the classically computable real functions.*

As long as we are only interested in total functions, we can obtain the same class of recursive functions even without using the order at all. More precisely, we will consider the order-free structure of the real numbers $\mathbf{R}^\star := \mathbf{N} \oplus (\mathbb{R}; 0, 1, x + y, -x, x \cdot y, 1/x, \mathrm{Lim})$. Shepherdson observed that the effective version of the Weierstraß Approximation Theorem implies that computable functions $f : [0,1] \to \mathbb{R}$ have straight-line programs [51]. Generalizing this idea one can prove the following theorem on order-free recursion. The proof uses special properties of $\mathbb{R}$, such as local compactness (see [8, 9] for details).

Theorem 11.7 (Order-free recursion). *A function $f : \mathbb{R}^n \to \mathbb{R}$ is recursive over $\mathbf{R}^\star$, if and only if it is recursive over $\mathbf{R}$.*

However, one cannot expect that the structure $\mathbf{R}^\star$ is perfect since otherwise the semi-characteristic operation $c_<$ of the order relation would have to be recursive over $\mathbf{R}^\star$ by the Conservation Theorem, i.e. Corollary 10.13. As we will see in the following, the perfectness result on $\mathbf{R}$ can be generalized to recursive metric spaces. As usual, we denote by $\mathrm{Lim} :\subseteq X^{\mathbb{N}} \to X$ the limit operation (restricted to rapidly converging sequences).

Theorem 11.8 (Metric structures). *If (X, d, α) is a (complete) recursive metric space, then*

$$\mathbf{X} := \mathbf{R} \oplus (X; \alpha, \mathrm{id}, d, \mathrm{Lim})$$

is a (strongly) perfect (complete) topological structure. The Cauchy representation yields a standard representation of this structure.

More precisely, if δ_X is the Cauchy representation of the space X, then $[\delta_{\mathbb{N}}, \delta_{\mathbb{R}}, \delta_X]$ is a standard representation of $\mathbf{X}$. The proof is based on Proposition 5.3 and a generalization of the proof of Theorem 11.5. The structure $\mathbf{X}$ is already a bit more natural than the structure given in Theorem 11.3. However, in many concrete cases it is even possible to "decompose" α with the help of the algebraic structure of the given space. We have seen this already in case of the real number structure $\mathbf{R}$. We give two further examples. Again, we will use a λ-like notation for operations as injection $X \to \mathcal{K}(X), x \mapsto \{x\}$, union of compact subsets $\mathcal{K}(X) \times \mathcal{K}(X) \mapsto \mathcal{K}(X), (A, B) \mapsto A \cup B$, and the scalar product of the function space $\mathbb{R} \times \mathcal{C}[0,1] \to \mathcal{C}[0,1], (y, f) \mapsto y \cdot f$ etc.

Example 11.9 (Perfect metric structures).

(1) If (X, d) is a (complete) recursive metric space, then the structure

$$\mathcal{K}(\mathbf{X}) := \mathbf{X} \oplus (\mathcal{K}(X); \{x\}, A, A \cup B, d_{\mathcal{K}}, \mathrm{Lim})$$

is a (strongly) perfect topological structure with respect to the topology generated by the Hausdorff metric $d_{\mathcal{K}}$.

(2) The structure

$$\mathcal{C}[\mathbf{0},\mathbf{1}] := \mathbf{R} \oplus (\mathcal{C}[0,1]; 1, f, y \cdot f, f + g, f \cdot g, ||f||, \mathrm{Lim})$$

is a strongly perfect topological structure with respect to the topology of uniform convergence generated by the supremum norm $||\ ||$.

Here "1" denotes the constant function in $\mathcal{C}[0,1]$ with value 1 and Lim denotes the limit operation on the corresponding space. Proofs and a detailed discussion of further examples can be found in [9]. By the Stability Theorem 10.7 the results of this section especially show that the structures **R** and **X** are effectively categorical. This has already been proved in a slightly different setting by Hertling [29]. By results of Hemmerling [28] it follows that effectively categorical structures for recursive metric spaces necessarily include infinitary operations (as Lim). Since second countable T_0–spaces are quasi-metrizable, it is promising to transfer the considerations of this section to computable quasi-metric spaces [12].

12. Recursive Sets over Structures

In Section 6 we have discussed effective subsets of recursive metric spaces. In this section we want to characterize these notions by some canonical abstract definitions for sets over structures which are presented in the following definition. If $A \subseteq X$, then we sometimes write $A^c := X \setminus A$ for the *complement* of A.

Definition 12.1 (Recursive sets over structures). Let S be a natural structure, let X be a set over S and let $A \subseteq X$.

(1) A is *initially semi-recursive* over S, if there exists a total recursive operation $f : X \rightrightarrows \mathbb{N}$ over S such that $A = f^{-1}\{0\}$.

(2) A is *finally semi-recursive* over S, if there exists a partial recursive operation $f :\subseteq \mathbb{N} \rightrightarrows X$ over S such that $A = f\{0\}$.

(3) A is *recursive* over S, if A is finally semi-recursive and $X \setminus A$ is initially semi-recursive over S.

(4) A is *decidable* over S, if there exists a total recursive function $f : X \to \mathbb{N}$ over S such that $A = f^{-1}\{0\}$.

In the following we will say for short *semi-recursive* instead of initially semi-recursive. Obviously, the empty set $\emptyset$, considered as a subset of a set X over a natural structure S, is always recursive and decidable. Furthermore, the set X, considered as a subset of itself, is always initially semi-recursive and decidable over S and X is also finally semi-recursive and hence recursive over S, if and only if Ω_X is recursive over S. In general, it is easy to see that over a natural structure S a non-empty subset A is finally semi-recursive, if and only if Ω_A is recursive over S. Especially, all sets over complete structures are finally semi-recursive and a point $x \in X$ is recursive over S, if and only if $\{x\}$ is finally semi-recursive over X. Over natural structures S a subset A is decidable, if and only if its *characteristic function* cf_A is recursive over S. We introduce the *semi-characteristic operation* to characterize semi-recursive sets A in a corresponding way.

Definition 12.2 (Semi-characteristic operation). For each set X and each subset $A \subseteq X$ we define the *semi-characteristic operation* $c_A : X \rightrightarrows \mathbb{N}$ of A by

$$c_A(x) := \begin{cases} \{0,1\} & \text{if } x \in A \\ \{1\} & \text{else} \end{cases}$$

With this notation a subset A over a natural structure S is semi-recursive, if and only if c_A is recursive over S. The intuition behind a semi-characteristic operation c_A is that it operates as follows: for any input $x \in X$ the operation can answer "no" (indicated by the value 1), but exactly for all inputs $x \in A$ there has to be a computation such that the operation yields the answer "yes" (indicated by the value 0).

It is easy to see that semi-recursive and decidable sets are closed under intersection and semi-recursive, finally semi-recursive, recursive and decidable sets are closed under union. Decidable subsets are obviously closed under complement as well and, as in the classical case we obtain that a subset $A \subseteq X$ of a set X over a natural structure S is decidable over S, if and only if A and A^c are semi-recursive over S. Moreover, semi-recursive sets are closed under preimage of recursive operations and finally semi-recursive sets under the image (provided they are contained in the domain). Over complete natural structures, semi-recursiveness implies final semi-recursiveness.

Proposition 12.3. *Over complete natural structures each semi-recursive set is also finally semi-recursive and each decidable subset is recursive.*

For any operation $f :\subseteq X \rightrightarrows Y \times \mathbb{N}$ we call $f_0 :\subseteq X \rightrightarrows Y$, defined by $f_0 := ((f \circ \mathrm{pr}_1)^{\leftrightarrow})_1 \circ (\mathrm{id}_X \times 0)$, the *section* of f. This name is justified since one obtains $f_0(x) = \{y \in Y : (y,0) \in f(x)\}$. The proof of the proposition follows immediately from $\Omega_A = ((\mathrm{id}_X, c_A) \circ \Omega_X)_0$. The next important property is an "uniformization property" which states that semi-recursive sets are graphs of recursive operations.

Theorem 12.4 (Uniformization Theorem). *Let X, Y be sets over a natural structure S and let $A \subseteq X \times Y$. If Y is finally semi-recursive and A semi-recursive over S, then there is a recursive operation $f :\subseteq X \rightrightarrows Y$ such that* $\mathrm{graph}(f) = A$.

Proof. Define $g : X \times Y \rightrightarrows Y \times \mathbb{N}$ by $g(x, y) := (y, c_A(x, y))$ for all $(x, y) \in X \times Y$. Then g is recursive over S and so is $f :\subseteq X \rightrightarrows Y$, defined by $f := (g \circ (\mathrm{id}_X \times \Omega_Y))_0$. Finally, $\mathrm{graph}(f) = A$. □

Implicitly, the Uniformization Theorem has already been applied in the proof of Theorem 11.5. For a special class of structures we can formulate further implications among the introduced notions of effectivity. If X is a set over a structure S, then the *equality on* X is the predicate $=_X$, that is the set $\{(x, x) : x = x\} \subseteq X \times X$. Over natural structures with semi-recursive equality each finally semi-recursive subset is also semi-recursive. The proof immediately follows from the formula $c_A = c_{=_X} \circ (\Omega_A \times \mathrm{id}_X)$ for all $A \subseteq X$. Together with Proposition 12.3 we get the following corollary.

Corollary 12.5. *Over complete natural structures with semi-recursive equality a subset is semi-recursive, if and only if it is finally semi-recursive and a subset is decidable, if and only if it is recursive.*

This situation is similar to classical recursion theory. Analogously, we can prove the following result.

Theorem 12.6 (Graph Theorem with equality). *Let S be a complete natural structure, $f :\subseteq X \rightrightarrows Y$ an operation over S, and let the equality $=_Y$ be semi-recursive over S. Then f is recursive and* $\mathrm{dom}(f)$ *is semi-recursive over S, if and only if* $\mathrm{graph}(f)$ *is semi-recursive.*

Proof. Let f be recursive and $\mathrm{dom}(f)$ semi-recursive. Since we obtain $\Omega_{\mathrm{graph}(f)} = (\mathrm{id}_X, f) \circ \Omega_{\mathrm{dom}(f)}$ it follows by Corollary 12.5 that $\mathrm{graph}(f)$ is semi-recursive. Now let $\mathrm{graph}(f)$ be semi-recursive. Then f is recursive over S by the Uniformization Theorem 12.4. Moreover, $c_{\mathrm{dom}(f)} = c_{\mathrm{graph}(f)} \circ (\mathrm{id}_X \times \Omega_Y)$, i.e. $\mathrm{dom}(f)$ is semi-recursive. □

However, perfect structures with semi-recursive equality have necessarily a countable universe (which follows from an easy cardinality argument, see [9]). Thus, structures with semi-recursive equality are not very relevant for spaces which occur in analysis. The next best property which one could expect is semi-recursiveness of inequality. Corresponding to equality, *inequality on* X is the predicate $\neq_X$, that is the set $\{(x, y) : x \neq y\} \subseteq X \times X$. Over complete natural structures semi-recursiveness of the inequality already implies recursiveness of equality. This follows since $=_X$ is equal to $\mathrm{range}((\mathrm{id}_X, \mathrm{id}_X) \circ \Omega_X)$. Next, we will prove that over structures with inequality a point is recursive, if and only if the corresponding single-valued set is recursive.

Proposition 12.7. *Let S be a natural structure, X a set over S such that inequality $\neq_X$ is semi-recursive over S, and let $x \in X$. Then x is recursive over S, if and only if $\{x\}$ is recursive over S.*

Proof. Obviously, x is recursive over S, if and only if $\{x\}$ is finally semi-recursive over S. Thus we only have to prove that the complement $\{x\}^c$ is semi-recursive, if $\{x\}$ is finally semi-recursive over S. This follows immediately from $c_{\neq_X}(\Omega_{\{x\}}, y) = c_{\{x\}^c}(y)$ for all $y \in X$. □

Finally, we mention that for perfect structures with inequality we can at least obtain one direction of the Graph Theorem.

Theorem 12.8 (Graph Theorem with inequality). *Let S be a perfect structure, $f :\subseteq X \to Y$ a function over S and let $\neq_Y$ be semi-recursive over S. If* $\mathrm{dom}(f)$ *and f are recursive over S, then* $\mathrm{graph}(f)$ *is recursive over S.*

The proof can be found in [9]. Next we want to consider recursive, semi-recursive and decidable sets over topological structures. Let X be a topological space and $A \subseteq X$. Since $\{0\}$ is an open subset of $\mathbb{N}$ and $c_A^{-1}\{0\} = A$, the characteristic operation $c_A : X \rightrightarrows \mathbb{N}$ is lower semi-continuous, if and only if A is open. Thus we can immediately deduce some facts about (semi-)recursive sets over topological structures.

Proposition 12.9. *Over natural topological structures semi-recursive subsets are open, recursive subsets are closed and decidable subsets are open and closed.*

We can directly conclude that there are no non-trivial decidable subsets over connected topological structures.

Corollary 12.10. *Let S be a topological structure, X a connected topological space over S. If $A \subseteq X$ is a decidable subset over S then $A = \emptyset$ or $A = X$.*

As a further corollary we can derive necessary topological properties of topological structures with equality and inequality.

Corollary 12.11. *Let S be a natural topological structure and let X be a topological space over S.*

(1) If $\neq_X$ is semi-recursive over S, then X is a Hausdorff space.

(2) If $=_X$ is semi-recursive over S, then X is a discrete space.

The proof follows directly, since in the first case, $=_X$ is a closed subset of $X \times X$ and in the second case an open subset. If, in the second case, the structure S, additionally, is effective via a natural representation, then we know that X is even a countable discrete space. Our metric standard structures actually have semi-recursive inequalities which is easy to see.

Proposition 12.12. *If X is a recursive metric space, then $\neq_X$ is semi-recursive over* **X** *and if X, additionally, is complete, then $\neq_X$ is even recursive over* **X**.

Here, the second part of the statement follows directly from Theorem 11.3 and Proposition 12.3. We close this section with a characterization of effective subsets over metric structures.

Theorem 12.13 (Recursive sets over metric structures). *Let X be a recursive metric space and $A \subseteq X$. Then A is semi-recursive over $\mathbf{X}$, if and only if A^c is co-r.e. closed. If, additionally, X is complete and A closed, then A is finally semi-recursive over $\mathbf{X}$, if and only if A is r.e. closed and A is recursive over $\mathbf{X}$, if and only if A is recursive.*

This result shows that the notions which we have introduced in Definition 12.1 are compatible with those of Section 6. The details are left to the reader and can be found in [9].

Acknowledgement

The author would like to thank Klaus Weihrauch, Matthias Schröder and the anonymous referee for several helpful comments, suggestions and corrections.

References

[1] O. Aberth, *Computable Analysis*. McGraw-Hill, New York 1980.

[2] K. Ambos-Spies, K. Weihrauch, and X. Zheng, Weakly computable real numbers, *J. Complexity* **16** (2000) 676–690.

[3] S. Banach and S. Mazur, Sur les fonctions calculables, *Ann. Soc. Pol. de Math.* **16** (1937) 223.

[4] A. Bauer, *The Realizability Approach to Computable Analysis and Topology*, PhD thesis, School of Computer Science, Carnegie Mellon University, Pittsburgh 2000.

[5] A. Bauer, A relationship between equilogical spaces and type two effectivity, in: S. Brooks and M. Mislove (eds.), *Seventeenth Conference on the Mathematical Foundations of Programming Semantics*, vol. 45 of *Elect. Not. Theoret. Comp. Sci.*, Elsevier, Amsterdam 2001.

[6] E. Bishop and D. S. Bridges, *Constructive Analysis*. Springer, Berlin 1985.

[7] V. Brattka, Recursive characterization of computable real-valued functions and relations, *Theoret. Comp. Sci.* **162** (1996) 45–77.

[8] V. Brattka, Order-free recursion on the real numbers, *Math. Log. Quart.* **43** (1997) 216–234.

[9] V. Brattka, Recursive and computable operations over topological structures, Informatik Berichte 255, FernUniversität Hagen, 1999.

[10] V. Brattka, A stability theorem for recursive analysis, in: C. S. Calude and M. J. Dinneen (eds.), *Combinatorics, Computation & Logic*, Springer, Singapore 1999, 144–158.

[11] V. Brattka, Computable versions of Baire's category theorem, in: J. Sgall, A. Pultr, and P. Kolman (eds.), *Mathematical Foundations of Computer Science 2001*, vol. 2136 of *Lect. Not. Comp. Sci.*, Springer, Berlin 2001, 224–235.

[12] V. Brattka, Recursive quasi-metric spaces, *Theoret. Comp. Sci.* (accepted for publication).

[13] V. Brattka and P. Hertling, Topological properties of real number representations, *Theoret. Comp. Sci.* (to appear).

[14] V. Brattka and G. Presser, Computability on subsets of metric spaces, *Theoret. Comp. Sci.* (accepted for publication).

[15] V. Brattka and K. Weihrauch, Computability on subsets of Euclidean space I: Closed and compact subsets, *Theoret. Comp. Sci.* **219** (1999) 65–93.

[16] G. Ceĭtin, Algorithmic operators in constructive metric spaces, *Tr. Mat. Inst. Steklov* **67** (1962) 295–361.

[17] T. Deil, Darstellungen und Berechenbarkeit reeller Zahlen, Informatik Berichte 51, FernUniversität Hagen, 1984.

[18] S. Eilenberg and C. C. Elgot, *Recursiveness*. Academic Press, New York 1970.

[19] J. L. Eršov, Theorie der Numerierungen I–III, *Zeitschr. Math. Logik Grundl. Math.* **19,21,23** (1973,1975,1977) 289–388,473–584,289–371.

[20] J. L. Eršov, Theory of numberings, in: E. R. Griffor (ed.), *Handbook of Computability Theory*, Elsevier, Amsterdam 1999, 473–503.

[21] J. E. Fenstad, *General Recursion Theory*. Springer, Berlin 1980.

[22] H. Friedman, Algorithmic procedures, generalized Turing algorithms, and elementary recursion theory, in: R. Gandy and C. Yates (eds.), *Logic Colloquium'69*, vol. 61 of *Stud. Logic Found. Math.*, North-Holland, Amsterdam 1971, 361–389.

[23] X. Ge and A. Nerode, On extreme points of convex compact Turing located sets, in: A. Nerode and Y. V. Matiyasevich (eds.), *Logical Foundations of Computer Science*, vol. 813 of *Lect. Not. Comp. Sci.*, Springer, Berlin 1994, 114–128.

[24] A. Grzegorczyk, Computable functionals, *Fund. Math.* **42** (1955) 168–202.

[25] A. Grzegorczyk, On the definitions of computable real continuous functions, *Fund. Math.* **44** (1957) 61–71.

[26] J. Hauck, Berechenbare reelle Funktionen, *Zeitschr. Math. Logik Grundl. Math.* **19** (1973) 121–140.

[27] J. Hauck, Konstruktive Darstellungen reeller Zahlen und Folgen, *Zeitschr. Math. Logik Grundl. Math.* **24** (1978) 365–374.

[28] A. Hemmerling, Standard representations of effective metric spaces, in: J. Blanck, V. Brattka, and P. Hertling (eds.), *Computability and Complexity in Analysis*, vol. 2064 of *Lect. Not. Comp. Sci.*, Springer, Berlin 2001, 48–68.

[29] P. Hertling, A real number structure that is effectively categorical, *Math. Log. Quart.* **45** (1999) 147–182.

[30] K.-I. Ko, *Complexity Theory of Real Functions*. Birkhäuser, Boston 1991.

[31] M. Konečný, *Many-Valued Real Functions Computable by Finite Transducers using IFS-Representations*, PhD thesis, School of Computer Science, University of Birmingham, 2000.

[32] C. Kreitz and K. Weihrauch, A unified approach to constructive and recursive analysis, in: M. Richter, E. Börger, W. Oberschelp, B. Schinzel, and W. Thomas (eds.), *Computation and Proof Theory*, vol. 1104 of *Lect. Not. Math.*, Springer, Berlin 1984, 259–278.

[33] C. Kreitz and K. Weihrauch, Theory of representations, *Theoret. Comp. Sci.* **38** (1985) 35–53.

[34] C. Kreitz and K. Weihrauch, Compactness in constructive analysis revisited, *Ann. Pure Appl. Logic* **36** (1987) 29–38.

[35] B. A. Kušner, *Lectures on Constructive Mathematical Analysis*. American Mathematical Society, Providence 1984.

[36] D. Lacombe, Extension de la notion de fonction récursive aux fonctions d'une ou plusieurs variables réelles I-III, *Comptes Rendus* **240,241** (1955) 2478–2480,13–14,151–153.

[37] D. Lacombe, Les ensembles récursivement ouverts ou fermés, et leurs applications à l'Analyse récursive, *Comptes Rendus* **246** (1958) 28–31.

[38] D. Lacombe, Quelques procédés de définition en topologie récursive, in: A. Heyting (ed.), *Constructivity in mathematics*, North-Holland, Amsterdam 1959, 129–158.

[39] A. I. Mal'cev, Constructive algebras. I, in: *The Metamathematics of Algebraic Systems*, North-Holland, Amsterdam 1971, 148–214.

[40] S. Mazur, *Computable Analysis*. Razprawy Matematyczne, Warsaw 1963.

[41] Y. N. Moschovakis, Recursive metric spaces, *Fund. Math.* **55** (1964) 215–238.

[42] Y. N. Moschovakis, Abstract first order computability I, II, *Transactions of the American Mathematical Society* **138** (1969) 427–464,465–504.

[43] A. Mostowski, On computable sequences, *Fund. Math.* **44** (1957) 37–51.

[44] P. Odifreddi, *Classical Recursion Theory*. North-Holland, Amsterdam 1989.

[45] M. B. Pour-El and J. I. Richards, *Computability in Analysis and Physics*. Springer, Berlin 1989.

[46] H. Rice, Recursive real numbers, *Proc. Amer. Math. Soc.* **5** (1954) 784–791.

[47] R. Robinson, Review of "Peter, R., Rekursive Funktionen", *J. Symbolic Logic* **16** (1951) 280–282.

[48] M. Schröder, Admissible representations of limit spaces, in: J. Blanck, V. Brattka, and P. Hertling (eds.), *Computability and Complexity in Analysis*, vol. 2064 of *Lect. Not. Comp. Sci.*, Springer, Berlin 2001, 273–295.

[49] M. Schröder, Extended admissibility, *Theoret. Comp. Sci.* (to appear).

[50] J. Shepherdson, Computation over abstract structures: serial and parallel procedures and Friedman's effective definitional schemes, in: H. Rose and J. Shepherdson (eds.), *Logic Colloquium'73*, vol. 80 of *Stud. Logic Found. Math.*, North-Holland, Amsterdam 1975, 445–513.

[51] J. Shepherdson, On the definition of computable functions of a real variable, *Zeitschr. Math. Logik Grundl. Math.* **22** (1976) 391–402.

[52] D. G. Skordev, *Computability in Combinatory Spaces*. Kluwer Academic Publishers, Dordrecht 1992.

[53] D. Spreen, On effective topological spaces, *J. Symbolic Logic* **63** (1998) 185–221.

[54] V. Stoltenberg-Hansen and J. V. Tucker, Effective algebras, in: S. Abramsky, D. Gabbay, and T. Maibaum (eds.), *Handbook of Logic in Computer Science, Volume 4*, Clarendon Press, Oxford 1995, 357–527.

[55] A. Troelstra, Comparing the theory of representations and constructive mathematics, in: E. Börger, G. Jäger, H. Kleine Büning, and M. Richter (eds.), *Computer Science Logic*, vol. 626 of *Lect. Not. Comp. Sci.*, Springer, Berlin 1992, 382–395.

[56] J. Tucker and J. Zucker, *Program Correctness over Abstract Data Types, with Error-State Semantics*. North-Holland, Amsterdam 1988.

[57] J. Tucker and J. Zucker, Computation by 'While' programs on topological partial algebras, *Theoret. Comp. Sci.* **219** (1999) 379–420.

[58] J. Tucker and J. Zucker, Abstract versus concrete computation on metric partial algebras, Technical Report CAS-01-01-JZ, McMaster University, Hamilton 2001.

[59] A. M. Turing, On computable numbers, with an application to the "Entscheidungsproblem", *Proc. London Math. Soc.* **42** (1936) 230–265.

[60] A. M. Turing, On computable numbers, with an application to the "Entscheidungsproblem". A correction., *Proc. London Math. Soc.* **43** (1937) 544–546.

[61] K. Weihrauch, Type 2 recursion theory, *Theoret. Comp. Sci.* **38** (1985) 17–33.

[62] K. Weihrauch, *Computability*. Springer, Berlin 1987.

[63] K. Weihrauch, Computability on computable metric spaces, *Theoret. Comp. Sci.* **113** (1993) 191–210.

[64] K. Weihrauch, *Computable Analysis*. Springer, Berlin 2000.

[65] K. Weihrauch and C. Kreitz, Representations of the real numbers and of the open subsets of the set of real numbers, *Ann. Pure Appl. Logic* **35** (1987) 247–260.

[66] M. Yasugi, T. Mori, and Y. Tsujii, Effective properties of sets and functions in metric spaces with computability structure, *Theoret. Comp. Sci.* **219** (1999) 467–486.

[67] X. Zheng and K. Weihrauch, The arithmetical hierarchy of real numbers, *Math. Log. Quart.* **47** (2001) 51–65.

[68] Q. Zhou, Computable real-valued functions on recursive open and closed subsets of Euclidean space, *Math. Log. Quart.* **42** (1996) 379–409.

INCOMPUTABILITY IN NATURE

S. Barry Cooper*
Department of Pure Mathematics
University of Leeds
Leeds LS2 9JT
U.K.
pmt6sbc@leeds.ac.uk

Piergiorgio Odifreddi†
Dipartimento di Matematica
Università degli Studi di Torino
Via Carlo Alberto 10
10123 Torino
Italy
odifreddi@dm.unito.it

Abstract To what extent is incomputability relevant to the material Universe? We look at ways in which this question might be answered, and the extent to which the theory of computability, which grew out of the work of Gödel, Church, Kleene and Turing, can contribute to a clear resolution of the current confusion. It is hoped that the presentation will be accessible to the non-specialist reader.

Keywords: computability, incomputability

1. An Historical Parallel

To the average scientist, incomputability in nature must appear as likely as 'action at a distance' must have seemed before the appearance of Newton's *Principia*. One might expect expertise in the theory of incomputability — paralleling that of alchemy in the seventeenth century — to predispose one to an acceptance of such radical new ideas. But specialist recursion theorists as

*The first author would like to acknowledge the support of EPSRC Research Grant no. GR/M 91419 "Turing Definability" during the preparation of this paper, and also the encouragement and clarification derived from a number of useful conversations with George Barmpalias while preparing the final version for publication.
†Both authors' research was partially supported by INTAS-RFBR Research Grant no. 97-0139, "Computability and Models".

a group have shown remarkably little interest in the more general implications of their work.[1] Part of the reason for this is tricky technical obstacles to even formulating appropriate notions and questions concerning computability in the real world.

However, at a purely intuitive level, one has a seemingly unproblematic model of a deterministic, even mechanical, Universe. Its scientific foundations were Newtonian mechanics, and its philosophical basis most clearly traced back to Laplace's predictive demon.[2] The model lives on in contemporary science surviving, at least in the recesses of the scientists mind, the proliferating challenges to old certainties. In particular, it has achieved an uneasy coexistence with notions of randomness (a very special kind of incomputability) forced on us by greater knowledge of subatomic phenomena.

And so, despite the development of the mathematical theory of computability initiated by Gödel, Turing, Church, Post and Kleene in the 1930s, one is still left with uncertainty as to how it applies to the real world. One can get an idea of the scope for quite contrasting, but arguably valid, approaches to quantifying incomputability in nature from the second author's discussion of Church's thesis and its extensions (in the first volume of his book on "Classical Recursion Theory"). So what is missing from the picture we have? How could a belief (or otherwise) in incomputability in nature be substantiated? And how can one enrich, improve, and make more explicit a mathematical model for what is happening?

One approach — the search for overt incomputability — is to take mathematical equations known to accurately describe some natural phenomenon, and to extract solutions exhibiting incomputability in some generally convincing form. Or, on the other hand, to argue that the persistent failure of this approach increasingly confirms there to be no such incomputability to identify.

What weakens this latter position is the obvious observation that efforts to consolidate the Laplacian model, on a foundation of new science and an appropriate model of the underlying mechanisms, seem even more hopeless. While one can point to quite plausible obstacles to overtness. There may, for instance, be *mathematical* constraints on what incomputabilities can be described, which make it hard to get to grips with the computability, or otherwise, of those natural phenomena for which we do have correlative mathematics.

---

[1] Some have seen the developing terminology of the subject itself — the replacing of Turing's 'computable' by Kleene's 'recursive' — as being intimately related to a turning away from its roots in real problems. It may be no coincidence that the recent reinstatement of Turing's terminology has come at the same time as a growing concern, both within and without the specialism, about whether classical recursion theory *has* any significant general implications.

[2] Of course, Laplace himself did not set out to provide any such model, and it is only twentieth century science which allows us to view the conceptual framework he helped establish in such terms.

For instance, we have in mind the failure to discover 'natural' examples of incomputable sets of integers from amongst the rich variety of incomputable 'almost computable' (i.e. *computably enumerable*) sets. This current lack of mathematically natural examples of incomputable sets has been used as an argument against being interested in such things. But naturalness in the rarified atmosphere of the university pure mathematics department is a very different thing to that of everyday usage. That very mathematical unnaturalness of most sets of numbers may be what presents an obstacle to overt incomputability of objects 'existing in or caused by nature'.[3] It is well-known that one needs no more than high school mathematics to describe incomputable sets (all computably enumerable sets are *diophantine*, got by looking for solutions to appropriate equations from basic school arithmetic); but that those *known* to be incomputable come with descriptions which can be tied in closely to diagonalising techniques in computability theory (definitely not part of the normal school syllabus!).

One can relax the overt approach by looking for incomputability which emerges from mathematics which *looks somewhat like* the mathematics one sees applied scientifically. But despite various proposals[4] in this direction, sceptics tend not to be impressed. They find plenty of scope for arguing that the mathematics involved is *not* typical, or for throwing doubt on the role of incomputability in it.

Finally one must admit that to those not involved, such detailed discussions seem fairly academic. Why should those without a direct career interest care whether actual incomputability (suitably formalised) occurs in Nature? Even if it did occur, for all practical purposes, how would it be distinguishable from theoretically computable but very *complex* phenomena? Whether chaotic phenomena – such as turbulence – involve complexity or incomputability is interesting, but does it really *matter*?

Fortunately, there is another approach — let's call it the *mathematical* approach — which renews the link with Newton. This is a direction rooted in the old debate about whether computability theory has any useful consequences for mathematics other than those whose statements depend on recursion theoretic terminology. Until recently the evidence was even less promising. Recently Soare has sought to present some interesting mathematics originating with researchers from outside computability theory, which both depends on the theory of incomputable sets, and at least looks like mathematics with real-world ex-

---

[3]From the definition of 'natural' given in *The New Oxford Dictionary of English*, 1998 edition.

[4]Probably the best-known and most convincing of these is that of Pour-El and Richards for non computable solutions to the wave equation. In fact, some claim this does rather more than "look .. somewhat like the mathematics one sees applied scientifically".

planatory power. This gives an, admittedly very weak, parallel with Newton and action at a distance.

It is worth noting that action at a distance was not Newton's direct concern, but was an incidental, if remarkable, ingredient in a comprehensive mathematical theory with what was rapidly acknowledged to be invaluable explanatory power. Not even the inverse square law originated in isolation with Newton, but was familiar to Robert Hooke and others. What Newton had done was to go far beyond Hooke's observation, and to develop a body of mathematics which, with the sort of difficulty one is familiar with in correlating mathematics with nature, was capable of explaining observations which had no other known explanation. He proved a striking parallel between mathematical theory and observed physical phenomena, to the point where *predictions* could be made and confirmed. And acceptance of action at a distance became, at that time, an unavoidable by-product of the theory.

Of course there is no corresponding 'big picture' obviously emerging from what Soare describes. But it does seem important as an indicator of how certain aspects of the complicated mathematics needed to describe the material universe may well be implicit in the richness of the theory of computability. This would not be surprising given that algorithms seem to underlie the generation of structure in the classical universe. But before speculating on how material phenomena can be located within a theoretical framework based on this underlying algorithmic content, we should first ask: Is there anything very important left for computability theory to explain? Instead of looking for a role for computability theory, let's very briefly examine the nature of gaps in the scientific 'big picture', and the ways in which the content of proposed remedies points to an added role for mathematical theory.

2. Matter and Mathematical Definability

There is no shortage of writings, both popular and specialist, looking at the many fundamental questions still facing science, and most of us have a fairly good idea of what remains to be explained. Let us step back, and get an overview of some of the more intractable questions puzzling workers in the sciences and humanities, and ask: What is the missing mathematics (if any), and how can it help?

A key element in very many controversies appears to be the interaction between the local and the global, and what appear to be breakdowns in the reductive structures commonly relied on in science and epistemology. The deterministic structures we rely on appear to be punctuated by what one can best describe as phase transitions between different levels of familiar relationships. Apart from the obvious puzzle of quantum wave reductions and associated and its relevance to social and psychological phenomena; about the origins and exact nature of consciousness, and of concepts and creative ideas; concerning the origins of life on earth, and of the laws of nature themselves; and even concerning the nature of truth in mathematics. At the same time, new science is often based on situations where the traditional reductions are no longer adequate (chaos theory being particularly relevant here). As one observes a rushing stream, one is aware that the dynamics of the individual units of flow are well understood. But the relationship between this and the continually evolving forms manifest in the streams surface is not just too *complex* to analyse — it seems to depend on globally emerging relationships not derivable from the local analysis. The form of the changing surface of the stream appears to constrain the movements of the molecules of water, while at the same time being traceable back to those same movements. The mathematical counterpart is the relationship between familiar operations and relations on structures, and globally arising new properties based on those locally encountered ones. For example (one relevant to new theories of the Universe), consistency is a property of theories based on the atomic proof theoretic features of the theory — namely, the assumptions and permitted deductions. Consistency in turn can be perceived as a constraint on those same atomic elements, and on what the theory describes.

The drive for a completely satisfactory interpretation of quantum theory has led to theories, such as that of decoherence, which routinely have recourse to such global notions, notions which were until relatively recently of interest only to those with a special interest in logic. Now even Gödel's Incompleteness Theorem is cited (usually inappropriately!) in relation to quantum and epistemological uncertainty. Unfortunately, the essential service logic has given to mathematics and other areas, such as computing, does not earn it remission from its traditional marginalisation! New notions coming out of current research are absorbed as slowly as ever. In this context, the little understood — even by mathematicians — notion of mathematical *definability* is the key mathematical concept, with the potential to clarify a broad range of fundamental problems. Many questions in quantum theory, proof theory, and epistemology, can be best understood as a breakdown of definability in an appropriate underlying mathematical model. At the same time, other mysteries, such as how the classical universe escapes the underlying quantum ambiguity, and how natural laws arise, can be traced back to the right notion of definability in the right mathematical structure.

Without going into mathematical detail, it is easy to give an intuitive idea of what definability is, and how it relates to another useful notion, that of *invariance*. This not necessarily because the notions are very simple ones (they are not), but because they do correspond to phenomena in the real world which we already, at some level, are very familiar with. We have already had the example of the stream, with its micro- and macro-dynamics. The emergence (a vogue word) of form from chaos, of global relations within turbulent environments, is a particularly vivid metaphor for the assertion of definability, or invariance. Let us take a simple mathematical example from arithmetic.

Given the usual operation $+$ of addition on the set $\mathbb{Z}$ of integers, it is easy to see that the set Ev of even integers is *definable* from $+$ within $\mathbb{Z}$ via the formula

$$x \in Ev \iff (\exists y)(y + y = x).$$

So all we mean by a relation being *definable* from some other relations and/or functions on a given domain is that it can be *described* in terms of those relations and/or functions in some agreed standard language. Of course, there are languages of varying power we can decide on, in which say one can allow say different levels of quantification. In the above example, we have used very basic *first order* language, with finitary quantification over individual elements of the intended domain — we say that Ev is *first order definable* from $+$ over $\mathbb{Z}$. What has happened is that we started off with just an arithmetical operation on $\mathbb{Z}$, but have found it distinguishes certain subsets of $\mathbb{Z}$ from all its other subsets. Intuitively, we first focused on a dynamic flow within the structure given locally by applications of the form $n+m$ to arbitrary integers m, n. But then, standing back from the structure, we observed something global — $\mathbb{Z}$ seemed to fall into two distinct parts, with flow relative to even integers constrained entirely within Ev, and flow from outside Ev being directed into Ev — with Ev being a maximal such subset of $\mathbb{Z}$. From within the structure, $+$ is observable and can be algorithmically captured. Further than that, we are dealing with 'laws' which cannot be related to the local without some higher analysis. This feature of the integers is not of course a deep one, but it does act as a basic metaphor for other ways in which more or less unexpected global characteristics of structures emerge quite deterministically from local infrastructure.

The notion of *invariance* gives a useful, if slightly more abstract, way of looking at such phenomena. Being able to uniquely *describe* a feature of a structure is a measure of its uniqueness. But some feature of a structure may be quite unique, without one being able to describe that uniqueness in everyday language. Mathematically, we use the notion of *automorphism* to capture the idea of a *reorganisation* of a structure which does not change any of its properties. A feature of that structure is *invariant* if it is left fixed by any automorphism of the structure. Obviously if one can uniquely describe such a feature, it must be invariant, but not necessarily conversely. On the other hand, any relation invariant in a structure can be defined in that structure, if one allows a suitable strong (but no longer very natural maybe) language, so definability can be thought of as providing a hierarchy of invariant relations.

The main remaining question is: How can one turn a metaphor into a model? In what sort of mathematical structure can we find definable relations, and corresponding failures of definability, with real world explanatory power?

3. Finitism in a Universe with Algorithmic Content

One often hears the view put that considerations of computability must have limited relevance to a clearly finite Universe. In outline, the argument is that one can analyse the Universe in terms of its quantum structure. This entails a discrete — and according to all the evidence — finite model. Since the model is large, computability is relevant to the scientific project. But incomputability has about as much significance for a complete description of the Universe as it does for any other finite relational structure, such as a graph — that is, none. In fact (see the discussion of Church's thesis in volume I of *Classical Recursion Theory*) no discrete model — finite or otherwise — presents a likely host for incomputable phenomena.[5]

Quantum indeterminacy presents little problem for such an outlook. One either expects an improved scientific description of the Universe in more classical terms, or, more commonly, one takes quantum *randomness* as a *given*, and superimposes more traditional certainties on top of that.

The latter perspective is also common to world views that make no assumptions about discreteness. It has the advantage (for the Laplacian in quantum clothing) of incorporating incomputability in the particular form of randomness, without any need for any *theory* of incomputability. The origins of incomputability in mathematics may be theoretical, but not in the real world, the view is.

Let us try to develop a more coherent alternative world view.

We first look at how science is done, and what relevance this has to what scientists are trying to describe. As Richard Feynman once pointed out[6] "It is really true, somehow, that the physical world is representable in a discretized way, and ... we are going to have to change the laws of physics." But it does not follow that the mathematics needed is correspondingly discrete. There do exist certain arcane proposals in logic for reconstructing fragments of mathematics, but nothing of practical use to the working scientist can be claimed. Much of applied mathematics seems irrevocably dependent on limiting processes and descriptions in terms of real numbers. Moreover, work on nonlinear phenomena have made us very aware of the fact that one cannot accurately describe complex environments within a fixed level of approximation — one cannot avoid working within the full system of real numbers. In the broader context, there is widespread awareness of the inadequacy of finitary language — everyday sentences — for capturing truth of a more abstract nature than scientific facts. Philosophy often seems an unending process of translation of intuition

---

[5]On the other hand, if one limits oneself to the usual computability models, the notion of randomness of *finite* strings seems to provide a first step toward a much needed theory of incomputability of finite objects.
[6]see [Fe82]

into words, while the reading of a philosophical text usually entails a creative, rather than formal, recreation of intuitive content. The epistemological role of sentences in everyday language seems not unlike that of the rational numbers in science.

We have to ask: Is it plausible that a Universe that compels us to use the mathematics of the reals to describe it, actually derives its properties from a mathematics qualitatively no different to that of a very big graph? One surely finds such a discontinuity between reality and description even less satisfactory than that between the mathematics of incomputability and an unexplained quantum uncertainty. Better to put to one side for the moment the possibility of a Universe based entirely on the mathematics of the discrete. It seems more likely that the observed discreteness is something *imposed* by natural laws on an underlying indiscrete mathematical model. And that these natural laws emerge from the model itself much as does the surface form of our rushing stream envisaged previously.

There are also more mathematical reasons for looking for an indiscrete model.

There may be little science can say about why things exist, but it is with mathematics that the scientific project begins. The way in which form develops in that which comes to exist is based on mathematical structure. Mathematics seems to prefigure. But — like joining a club, or enlisting into a culture — the opportunities for development provided come with accompanying rules, pressures and responsibilities. And these may not be predictable — the ordinary citizen of nineteen-thirties Germany had little idea that they were part of a social formation that would build so monstrously on their individual aspirations and ideals. From its origins (whatever that means!) the Universe relies on a mathematical blueprint to narrow down its various potential incarnations. It 'signs up' to an underlying mathematical model, not by choice, but just by the act of coming into existence. Can we say anything about the mathematics prefiguring the Universe?

What transforms inchoate development and gives form to creation, is the way in which change become process. Whatever takes place in the nascent Universe appears to be comprised of basic kinds of atomic acts of development, with repetition forming the basis of further development. Processes are fundamental to matter and the laws which govern its formations and development. They do not seem to be finitarily determined, for example by piecemeal dependency on the members of their real world domain, but can be envisaged relative to a wide range of extrapolations from what we actually observe. These processes are what we broadly describe as 'the algorithmic content' of the Universe. Most of these fundamental processes identified by science are characterised by enough predictability of immediate effect to be covered by the connotations of such terminology. Science since the time of Newton, at least, has been largely based on the identification and mathematical description of algorithmic content in the Universe. We will look at phenomena — primarily subatomic phenomena — which appear to defy such description, in section 6 below. It turns out that these exceptions can be understood as arising at the borderline between invariance and noninvariance in an appropriate model. In this context they actually consolidate our view of the algorithmic nature of the fundamental processes underlying material development.

Anyway, the main observation here is that a finite structure with the sort of algorithmic content we find in the Universe is very different to, say, a graph, in that it entails 'uncompleted infinities'. A fragment of the natural numbers over which one allows the usual operations of arithmetic, is best understood in the context of the theory of the whole structure. In the same way, the algorithmic content of the Universe makes it subject to the principles governing the mathematical whole of which it is a part. Finiteness of the collection of all atomic particles (even assuming one can make sense of such a conception) cannot remove the general algorithmic content, the accompanying limiting processes, or obviate the higher mathematical relations these entail.

The admission of algorithmic content is what makes a nonsense of a strictly finitist view of the universe, and, as we will see in section 6, offers yet another example of how particular explanations emerge from larger mathematical contexts than have immediate counterparts in nature — such as with imaginary numbers, or at a mathematical level, results in number theory coming from nondiscrete mathematics.

Moreover, the mathematical model emerging not only transcends finitism, but also discretism. One can envisage how real numbers in the form of uncompleted infinities feed into physical reality, and determine a mathematical model which is not discrete. For instance, people routinely attempt to mentally simulate events in the interests of reconstructing history and predicting future circumstances, and modify their activity accordingly. In that this simulation can involve sequences with no specific bound, as when one say tries to act towards a world without war and famine, one can say that there is an uncompleted infinity which has a direct real world impact. One hesitates to admit cerebral phenomena which qualitatively transcend what goes on in other parts of the real universe. In fact one can argue that such mental simulations derive from the brain's capacity to physically simulate other complex material phenomena. And that the Universe is quite capable of accelerating uncompleted infinities into the here and now via its own nonlinear, often turbulent activity, involving a globally extending network of interactions relevant to a particular local description. The limiting processes underlying local descriptions in terms of global phenomena, intrinsic to turbulent environments, can be better understood via the standard hierarchies of real numbers found in computability theory, and we will return to this in the next section.

But to summarise, if the ingredients of a simple mathematical construction — such as that of the reals from the natural numbers — are materially manifest, then one should not expect the Universe to be excluded from that construction. Specifically, this leads to a convergence between the mathematics, and the scientific priority given to descriptions in terms of reals and relations between them. We now take this a little further.

4. The Inseparability of Algorithmic Content, Complexity and Incomputability

Mathematically, algorithmic content and incomputability are inseparably linked in sufficiently complex environments. An elementary construction (involving *diagonalising* through a list of all machine computable reals) gives an incomputable real. But for our purposes, we need to limit our attention to mathematical constructions for which the basic ingredients have an obvious physical counterpart, and diagonalisation at first sight appears to be limited to what mathematicians do. More relevantly, *all* real numbers are derivable as the limit of a sequence of *computable* — in fact rational — numbers. Even if one limits oneself to computable *sequences* of computable numbers — that is *uniformly* computable sequences of reals — one still gets limiting reals which are not computable.

However, one needs to look more closely at how incomputability arises from the mathematics of algorithms to be sure that all the basic ingredients are apparent in, say, the turbulent environments we have focused on previously. What is it makes it impossible to compute a given incomputable, but computably enumerable set of natural numbers? The members of the set are of course enumerated by some computable function from numbers to numbers — but not in order of magnitude. The relation between the magnitude of input and that of the output is broken. And since we are only interested in the output, we would need a possibly uncompletable search to computably discover if a particular number was in the set or not. In the case of an incomputable real appearing as the limit of a uniformly computable sequence of reals, this appears as an absence of a computable *modulus of convergence* of the sequence. The breaking of the link between input and output happens in the sense that one cannot be clear at any given point in the sequence how close an approximation to the limiting real is being currently provided. What is there in the case of chaotic situations which corresponds?

A well-known example of chaotic behaviour is that of a pile of sand being accumulated grain by grain. There is an unpredictable link between the order in which the grains of sand are deposited on the heap, and the consequent impact on the pile in terms of settlement of existing grains. This is just the sort of unpredictability of association between input and output that we seem to need. In fact, it has been said that the "Butterfly Effect", or more technically the "sensitive dependence on initial conditions", is the essence of chaos. So a sufficiently sensitive mathematical model for turbulent environments not only cannot be discrete, but will involve limiting processes which have the sort of ingredients needed to generate incomputable elements. Mathematically, the need for an infinitary sequence of approximations does not depend on what is conventionally regarded as a chaotic — or turbulent — environment. Even apparently simple phenomena — such as a dripping tap — exhibit chaotic behaviour. The non-linearity of the mathematics can be approached via a superposition of infinitely many linear factors, closely related to how one would naturally approximate the event in question via individual computable interactions. The association of incomputability with simple chaotic situations is not new. For instance, Georg Kreisel sketched in [Kr70] a collision problem related to the 3-body problem as a possible source of incomputability.

At this point the reader might object that we are doing exactly what we earlier suggested was a pointless exercise, namely trying to provide direct argumentation in favour of incomputability in Nature. No, we are under no illusion that what we have said so far provides any more than a preliminary basis for a belief that the universe is incomputable. What we are trying to do is convince the reader that he or she should seriously consider taking the mathematical model suggested by the above discussion 'on approval', as it were. Any good salesperson (and we hope to use the term without its usual pejorative associations) would admit the prospective purchaser their doubts. The aim would be to provide sufficient technical background and informative argumentation to at least justify the time and effort involved in getting a first-hand experience of what the product can actually deliver. The salesperson may aver that there is no substitute for having the product in your hands, and trying it out in the privacy of your own home. But one must be persuaded that there is actually some point in doing this, and for this reason we have hopefully presented the scenario in which one might want to at least consider what the model — the Turing model — potentially provides in the way of explanatory power.

It is now time to must describe the model in more detail, and discuss certain technical questions concerning it which as yet remain unresolved.

5. A Closer Look at the Turing Model

In 1939 Alan Turing, puzzled by the role of incomputability in mathematics and logic, introduced Turing machines with oracles, as a way of comparing the (in)computability of different reals. Essentially, oracle Turing machines were machines capable of working in the real world of information, in the form of real numbers. The result was a natural model for structures describable in terms of computable relations over real numbers. Our Universe would be an obvious candidate for such a structure, were it not for the fact that certain basic natural processes are not known to give rise to *computable* relations on reals. What we have promised though is that such phenomena *are* modelled via an analysis of invariance and definability in the basic Turing model.

Historically, the investigation of the properties of the Turing model were initiated by Emil Post in the early nineteen forties, and a seminal paper written largely by Stephen Kleene, but co-authored with Post, appeared in 1954. By that time, the interest had become almost entirely mathematical, as reflected in the name 'recursive function theory' given by Kleene to the new subject. The original motivations of Turing had been — necessarily it can be argued — sidelined in favour of an intensive technical development, just part of which involved efforts to describe the Turing universe in more detail. It was not surprising that real world applications did not figure throughout the recursion theoretic period (roughly the sixty years starting with the discovery of incomputable objects by Church and Turing in 1936). On the one hand one had the confusing way in which real world incomputability manifested itself, and on the other one had an as yet technically inadequate subject, quite insufficient to develop and explain any modelling process it might be part of.

By the 1990s few could ignore the signs that not everything was going well with the subject. The long-term survival of a mathematical field with no real-world applications depends on what it delivers, year in, year out, to the wider mathematical community. And in particular, what it offers to new students and professionals, to mathematicians working in other areas, and to the institutions within which they work. To outsiders classical computability had become a deep subject which was maybe *too* deep; in which core research had become hazardous and, even by the standards of fundamental scientific research, lacking predictability of outcome; in which mathematical applications depended on recursion theoretic terminology; and in which the undoubted contribution to theoretical computer science and constructive mathematics did not depend on the sort of things that recursion theorists currently occupied themselves with. Even insiders who instinctively reacted against any consciously political response found problems in pursuing a purely mathematical agenda, and were themselves often caught up in the drift away from core research.

Things started to change in earnest around 1995–96. These changes were rooted in two seemingly unrelated developments, one philosophical and political in content, and the other technical. The first involved a deliberate attempt to reinstate Turing's terminology in keeping with the subject's origins in real world questions — 'computable' in place of 'recursive' etc. — a project outlined in Robert Soare's 1996 paper on 'Computability and recursion'. The other originated in 1995 with the first serious challenge to the mathematical perspective which had dictated the direction of core research for more than ten years, and involved the discarding of mathematical certainties (in more ways than one) in favour of complexities which startlingly seemed to parallel some of those apparent in the real world. We will say more about the specifics of this below. It is not clear to many what will be the eventual outcome to this transitionary period, and so far the impact of the accompanying confusion on the standing of computability theory has been largely negative.

What lost its revered status around 1995 was the so-called 'bi interpretability conjecture', which it was hoped would lead to an extremal characterisation of the Turing definable relations, to the point where the Turing universe, as a mathematical structure, would be rigidly pinned down by those relations. The idea had been that one could avoid direct involvement with the multiplying complexities of Turing's structure, beyond what was needed to translate it into the more familiar theory of (second order) arithmetic. This relationship would deprive the Turing universe of much of its potential to surprise, and make detailed investigation of Turing definability of little more mathematical interest than the details of arithmetical calculations. And for a number of years the bi interpretability conjecture held a gorgon-like fascination for recursion theorists,[7] while offering a mathematically elegant but no less stony end to the subject.

Let us say a bit more about rigidity for the nonspecialist. When one says a structure is 'rigid' it means intuitively, that travellers within the structure can uniquely pin down their location by looking around them. Or more precisely, cartographically the traveller cannot mistake his or her position — for example it is clear if the map is being held the wrong way round! Every feature of the structure has a uniquely determined global context. It is like a team in which every player has a uniquely fulfillable role. If the players exchange jerseys amongst themselves, the team is no longer what it was — the structure has been changed by this relabelling. A relabelling of a structure under which it is *not* changed is called an *automorphism*. A rigid structure is one which has no nontrivial automorphisms. In computability theory a *reducibility* is an attempted ordering of mathematical objects according to a given level of *perception* of information content. It consists of the collection of those specific *reductions* conforming to the permitted level of perception. Turing reducibility is a particularly important reducibility, comprised of reductions based on everyday algorithmic practice, whereby one computes relative to finite collections of data extracted from the real world. Reducibilities can be limited as to the size of the collections of data used, or as to the way they are presented, or as to the manner in which one computes on them. These are called *strong* reducibilities. There is an histori-

[7] At an international conference in Helsinki in 1990, the first author's suggestion, in answer to a question from the audience, that the Turing universe might *not* be rigid, attracted indulgent smiles. The next day, a distinguished special sessions speaker found Turing rigidity 'almost certain' in the light of so many known Turing definable relations.

cally important strong reducibility called *many-one reducibility* which can be regarded as an abstraction of translations between axiomatic theories, and this gives rise to a structure which has few definable relations, and is very un-rigid. On the other hand there are reducibilities which are allowed a level of *infinitary* interrogation of a given real, or even of objects more complicated than reals (say infinite sets of reals, or sets of transfinite ordinals). Such reducibilities are the domain of *generalised* recursion (or computability) theory, and although so far lacking obvious real-world counterparts, often have very strong intuitive content. An example of a structure arising from a more general reducibility is that of the *hyperdegrees*, where algorithms relative to reals are replaced by very general language-based reductions — and this structure does turn out to be rigid.

The intuition was that Turing reductions were mathematically the weakest adequate to extract enough information from a given real to uniquely fix its relationship to the universe of the reducibility. Of course, this is modulo the intercomputability of certain pairs of reals. To take account of this, it is usual to gather together into *degrees* collections of objects of the domain of the reducibility which are mathematically equivalent in this sense. So a Turing *degree* is a maximal set of reals any two of which can be Turing computed from each other. These reals can be thought of as presentations of the same Turing accessible information content. It was the frustration of this intuition that gave rise to what may seem in hindsight a bizarre response to Turing nonrigidity, that is a disappointment that the reducibility was, perhaps, not the 'right' one mathematically. One would now look for some recognition of the parallel with the encountered inadequacies of scientific observation, and more generally, of the power of the Universe to uniquely determine its own structure.

There are two other observations of a technical nature to be made. The first is that Turing nonrigidity is not yet finally established according to the usual criteria of the academic community. What has happened is that progress in the other direction has all but halted, despite the most determined efforts of leading researchers in the field. And no significant technical challenge, either in the form of counterexamples or technical queries, has been made to the increasingly detailed proposal for a nontrivial automorphism. The latter has been presented now at a number of international meetings, most convincingly during a six-hour invited talk at the AMS-IMS-SIAM Joint Summer Research Conference in Computability Theory and Applications at the University of Colorado, Boulder, in June 1999. The fact that the issue remains unresolved is partly a measure of the newness and initial complexity of the techniques, partly a manifestation of a severely under-resourced research area signally failing to keep pace with its own technical development, and (perhaps most significantly!) due to the mathematical and presentational limitations of those few individuals directly responsible. Also — and this is the main reason for writing about the consequences of work in progress — what is happening has every appearance of a paradigm shift in progress. And not just a minor technical shift in an obscure area of pure mathematics, but one of very general importance. A change as global in nature as the phenomena it concerns, which transcends the incremental processes of normal science, and in which an appreciation of context may feed back into a deeper understanding of narrowly technical issues. The development of string theory is an example of the way in which the sheer explanatory power of a theory can actually contribute to the mathematics.

Secondly, one has to mention that just as important as nonrigidity is the counterbalancing body of results establishing a high level of invariance in the Turing universe. One of these says that there are relatively few — at most *countably many* — Turing automorphisms. To understand some of the most relevant of these results one must have some appreciation of the close relationship between the way one might describe a real in standard mathematical language, and its location within the Turing universe. It turns out that if one has such a description of a real, which one knows to be the simplest possible, then one can roughly locate its position in the Turing universe in a very natural way. Important landmarks in this location are provided by naturally arising reals whose descriptions involve a relatively small number of quantifiers. Most importantly, one has the Turing equivalent group of incomputable reals discovered by Church and Turing in the 1930s, degree theoretically denoted by $\mathbf{0}'$. Occupying a position above $\mathbf{0}'$ similar to that of $\mathbf{0}'$ above $\mathbf{0}$ (the set of all computable reals) is a degree $\mathbf{0}''$. What is remarkable is that above this level a Turing automorphism can do no more than move a given real to one which is Turing equivalent to it. That is, the Turing universe above $\mathbf{0}''$ is rigid *within* the whole structure.

6. Scientifically Presenting the Universe

We need to say a little more about what we mean by a mathematical model for the Universe. This has been left deliberately vague up until now, and as we shall see, a degree of vagueness is inherent in the nature of scientific knowledge itself. Flexibility is also a feature, and a strength, of scientific practice as regards the details of mathematical modelling via assignments of reals to parameters. What we are talking about is a *scientific presentation* of the Universe via some informative mathematical structure. The mathematics needs to reflect the ingredients of the current scientific picture we have of the Universe. As we have tried to argue above, scientific descriptions are based on the real numbers. This is not because we necessarily believe that the real numbers actually enable us to capture reality, but because they are what we have found by experience to provide a practical scientific framework. The sort of scientific presentation we have in mind is not one currently used in practice to deal with the sort of fragments of the Universe which science needs to focus on in isolation. We stayed close to such local presentations in our earlier discussion of turbulence and the generation of incomputability. A feature of that discussion was the way in which the new incomputable reals derived their specific properties from the algorithmic content of the chaotic environment. This is what is required in a comprehensive mathematical model which describes the Universe in terms of 'atomic' information content, described via reals. That is, the algorithmic content of the Universe, as revealed to us by the scientific project, needs to be reflected by the comprehensive presentation in terms of reals. This will enable one to correlate the mathematical description with the physical processes in an informative way.

In summary, the model must reflect the picture which science *in principle* provides of the way we experience the Universe, with a comprehensive structuring of the information content of that description in accordance with the observed processes at work. Of course, without making any assumptions regarding underlying reality — even to assert that a given object exists, using some agreed language, is to translate our experience into a scientific statement — all that is being sought is something which is in accordance with what we observe. That is, we look for a mathematical structure within which we may informatively interpret the current state of the scientific enterprise. This presentation may be done in different ways, one must assume, but if differing modes of presentation yield results which build a cohesive description of the Universe, then we have an appropriate modelling strategy.

Part of our experience of science is what seems an intrinsic fragmentation, in which the reductive structure of the Universe is not a good guide to what happens at different levels of scientific application. In section 7 we will look more closely at how the proposed model explains this in terms of internal invariance. All that needs to be observed for the moment is that just as different fragments of the scientific enterprise are based on their own basic relations, whose complex reductive relationships to more basic fragments are suppressed, so it may turn out that our current method of presentation ignores other yet to be discovered basic levels. What is important is that the scientific picture is now sufficiently filled out for the Turing model to have something to tell us about the way the world we live in is as we currently experience it. One may have to make certain assumptions about *how much* of the Turing model is needed to describe our Universe, but the mathematical indications are that a large amount of information content is needed for the emergence of anything like the classical universe of well-defined individual objects we daily encounter.

Anyway, a level of flexibility in the way one envisages using the Turing universe as a model is an advantage, allowing us to apply it with differing perspectives. In particular, it means we can envisage a modelling of the Universe as part of an extended epistemological structure, consisting of statements we can make (in some language) about the Universe and its mathematics.

7. What the Turing Model Delivers

The claim is that many seeming anomalies in science, the humanities, and human affairs in general, stem from failures of reductive analyses which have served us too well in the past to be easily bypassed. Further, that what is lacking is not an attack on the validity of such widely applicable analyses, but a model in which relevant global factors can be clarified. And crucially, that a key global notion is that of definability or more generally, mathematical invariance. That there are sufficient indicators, of both a practical and theoretical nature, for us to look for a model for the Universe based on presentations in terms of reals. And that the appropriate abstraction of such a Universe and its processes is that arising from Alan Turing's work in the late 1930s.

Let us assume for the moment that we are prepared to entertain such a perspective. It is then time to deliver — So what *does* computability theory have to offer? We revisit some of the themes touched on in the first author's 1999 paper. We have tried to make the comments independent of outside references, which might be thought of little help to the majority of readers, who will have read many books and articles of course, but not necessarily those preferred by the authors! We will also try to be concise, at the risk of oversimplification. What we hope will be striking is not the extent to which the model surprises, but the way in which it supports and theoretically substantiates existing perceptions and trends in science and beyond.

Our model says nothing about the mystery of material existence. But it does offer a framework in which a breakdown in reductionism is a commonplace, certainly not inconsistent with the picture given of levels we do have some hope of understanding. It can tell us, in a characteristically schematic way, how *things* come to exist. Our basic premise, nothing new philosophically, is that existence takes the most general form allowed by considerations of internal consistency. Where that consistency is governed by the mathematics of the universe within which that existence has a meaning. Of course, within that universe, not even the terms in which we seek to discuss it can be assumed to have any meaningful existence. But proceeding regardless, assuming our intuitive concept of 'creation' to have some content, we assume, as before, our Universe to embody some process of development, in which processes are basic. We assume the appropriateness of presentations of the Universe in terms of information content and algorithmic relationships on that information content. We can then appeal to the mathematics of the Turing model to tell us something about what we might expect concerning 'thingness' in our Universe. This will be a first step in filling in a picture of the Universe characterised by content. We can then appeal to the mathematics of the Turing model to tell us something about what we might expect concerning 'thingness' in our Universe. This will be a first step in filling in a picture of the Universe characterised by more determinism, more immanence, and less dependence on metaphysical speculation.

What would the model lead us to expect regarding the basic structure of matter and the laws governing it? The mathematical indications are that a low level of information content goes with nonrigidity and a lack of Turing invariant individuals. In other words, the structure of the Turing universe appears to be consistent with the possibility that a given individuals information content may be insufficient to guarantee it a unique relationship to the global structure. The corresponding prediction for a Universe in tune with that model would be that its most basic components may materialise ambiguously[8] — a prediction confirmed by a number of classic experiments on subatomic particles. (The assumption has been since at least the time of Leibniz, that reality takes all forms consistent the underlying laws of existence, which from a contemporary point of view tends to mean consistency with the *mathematics* pertaining.) On the other hand, mathematically entangling such low level information content, perhaps with content at levels of the Turing universe at which rigidity sets in, will inevitably produce new content corresponding to a Turing invariant real. The prediction is that there is a level of material existence which does not display such ambiguity as seen at the quantum level, and whose interactions with the quantum level have the effect of removing such ambiguity — confirmed by our everyday experience of a classical level of reality, and by the familiar 'collapse of the wave function' associated with observation of quantum phenomena. Since there is no obvious mathematical reason why quantum ambiguity should remain locally constrained, there may be an apparent non-locality attached to the collapse. Such a non-locality was first suggested by the well-known Einstein-Podolsky-Rosen thought experiment, and, again, has been confirmed by observation. The way in which definability asserts itself in the Turing universe is not known to be computable, which would explain the difficulties in predicting exactly how such a collapse might materialise in practice, and the apparent randomness involved.

One can only speculate about the origins of subatomic structure, particularly since we do not seem to have a complete description of it as yet. One guess is that when one observes atomic structure, one is looking at *relations* defined on some lower level level of matter lacking any sort of observable form. This could be envisaged as a kind of formless soup of information content out of which arise peaks of definability observed by us as subatomic particles. Such

[8] But in keeping with the limitation on possibilities presented by the existence of only countably many Turing automorphisms.

an implementation of the model might be used to explain why the hypothesised 'dark matter' in the Universe has not been observed. Until it is organised into relations, of which particles are the instantiations, we have no structure capable of being interacted with. It would be as alien to the world of particle physics as that world is to our classical level of human existence.[9]

The mathematics leads to other scientifically appropriate predictions. In particular, there is the question of how the laws of nature immanently arise, how they collapse near the big bang 'singularity', and what the model says about the occurrence or otherwise of such a singularity.

What we have in the Turing universe are not just invariant individuals, but a rich infrastructure of more general Turing definable relations. These relations grow out of the structure, and constrain it, in much the same sort of organic way that the forms observable in our rushing stream appear to. These relations operate at a universal level. The prediction is that a Universe *with sufficiently developed information content* to replicate the defining content of the Turing universe will manifest corresponding material relations. The existence of such relations one would expect to be susceptible to observation, these observations in turn suggesting regularities capable of mathematical description. But then this is a prediction which adumbrates the familiar historical process of the scientific mapping of natural laws. The conjecture is that there is a corresponding parallel between natural laws and relations which are definable in an appropriate fragment of the Turing universe.

The early Universe one would not expect to replicate such a fragment. The homogenisation and randomisation of information content consequent on the extreme interconnectivity of matter would militate against higher order structure. The manifest fragment of the Turing universe, based on random reals, might still contain high information content, but content dispersed and made largely inaccessible to the sort of Turing definitions predicted by the theory. Projected singularities, such as within black holes or associated with boundary states of the Universe, depend on a constancy of the known laws of physics. But immanently originating laws must be of global extraction. This means that their detailed manifestations may vary with global change, and disappear even. Scientific evidence for time-related variations in natural laws is at best inconclusive, as is evidence for the collapse of the laws of physics near the

---

[9]On a technical note, one should mention that it is not known if the Turing universe is rigid above $\mathbf{0}'$ — only that $\mathbf{0}'$ it is definable, and so invariant. A material manifestation of a Turing defined relation will still be scientifically perceived as an entity describable by a real, despite the underlying mathematics by which it is constrained. And the information content of that defined relation will correspond to an infinitary join adumbrated by local information information content organised according to some corresponding description. In which case one can speculate that the level of elevation of information content implicit in a Turing definition is due to some canonical form of description needed for persistence in time, or arises from the conjectured subsubatomic level of information content.

continuity between the current Universe and one bereft of the familiar (and not so familiar!) laws of physics.

The reader will already be aware of the many attempts to stretch the standard model to deal with unsatisfactory gaps and anomalies in the theory. The area abounds with theoretical constructs — such as inflation, many worlds, decoherence, the pilot wave, gauge theory, superstring theory, M-theory — all intended to extend the standard model in ways not yet justified by observation. Some of these are proposals of a limited technical nature designed to address particular problems with the standard model in a relatively piecemeal way. Others are conceptually more radical, and can be located within the present schema. This applies particularly to theories of decoherence, which in trying to reconcile quantum ambiguity with classical reality do seem to deal with, in a very basic way, the underlying mechanics of mathematical definability. However, without locating this within the mathematical theory of definability, one is still left with having to resort to the metaphysics of parallel classical realities, with its unsatisfactory lack of economy, and accompanying problems with explaining our own special relationship to the reality we see around us. Still other attempts to stretch the standard model — such as superstrings and M-theory — present the standard model as a very small fragment of a mathematical theory which goes far beyond that arising from empirical science. These represent an approach which, while not actually inconsistent with that described here, appears to share very little in the way of new thinking.

Logically, it is well known that one can extend a given incomplete theory in many different ways. And that, in general, such an incomplete theory will have infinitely many completions. In the real world one can justify any addition to the standard model via suitably elaborated auxiliary hypotheses.[10] So it is not very surprising that a variety of such proposed extensions to the standard model of the Universe have appeared. Or that more than one of these has achieved the status of a fully-fledged paradigm with its own committed group of supporters. Occam's Razor, and the Principle of Parsimony, are what have been provided as a formal counterpart to our natural scepticism. These say, roughly speaking, that one should not increase, beyond what is necessary, the number of extra factors required to explain something. If one takes into account not just the *number* of added factors, but also the likelihood of those factors being substantiated by observation (either in a positive way or by expected negative data not being found), then the interpretation we have attempted based on physical entities as information is a clear winner. How can we put a computable Universe amongst many, top of our list? Especially since other assumptions must be added to that of many worlds to deal with the still remaining anomalies and incompleteness

[10]This is roughly what the Duhem-Quine thesis says.

of the theory. Superstring and M-theory make very different assumptions, which are however no more accessible to experimental verification. And there is even less clarity about what the theory actually is. There are some very detailed implications for the theoretical basis of the standard model, which the Turing model has as yet nothing to say about. In fact, superstring theory seems capable of sitting quite comfortably within the Turing framework. The aims and consequences of the two approaches are rather different, and potentially complementary.

We will look very briefly at the Turing interpretation of other areas. Leaving quantum theory, one can still find abundant evidence for both nonrigidity, and what appear to be new laws corresponding to invariant relations emerging at higher levels, and relating to new entities reducible to ones at previous levels. However, this evidence may be less scientific, and more subject to controversy. Firstly, the Turing model does tell us why there is this fragmentation of science, and human knowledge in general, and why we do not have computable reductions of one to another. As we have said previously, in general a Turing definition of a given relation does not necessarily yield a computable relationship with the defining factors. But working within the relations at a given level, there may well be computable relationships emerging, which may become the basis for a new area of scientific investigation. For instance research concerning the cells of a living organism may not be usefully reduced to atomic physics, but deals with a higher level of directly observed regularities. Sociologically, one studies the interactions governing groups of people with only an indirect reference to psychological or biological factors. Entire relations upon cells (humans) defined in some imperfectly understood way by the evolutionary process provide the raw material underlying the new discipline, which seeks to identify a further level of algorithmic content. This algorithmic content may not be directly expressed in terms of numbers. But inasmuch as the area in question does have basic notions, corresponding to the new emergent relations, shared by workers in the field, and descriptions of entities and regularities are formulated in a shared language, the algorithmic content is not dissimilar in kind to that at lower levels.

The first author's 1999 paper mentioned a number of areas in which one can observe qualitatively similar problems, all connected with parallel issues of definability and nonrigidity. We briefly review two of these, and leave the reader to look there for further comments.

A particularly puzzling problem is that of the early origin of life on Earth. Wherever one encounters explanations in terms of 'emergence', one can expect an understanding of the notion of definability, relative to the basic relations operating, to reinforce the necessarily vague intuitions underlying such explanations. And at any level, there are initially no more basic relations than those presented by the fundamental processes pertaining. There are more specific characteristics established on this basis, and there is no intention to deny the importance of the science based on such specificities. In any case, in this instance there is an undeniably close parallel between the intuitions underlying definability and those feeding into more recent explanations of how life appeared on earth in the time available. And some theoretical breakthrough is needed, given the confusion betrayed by such speculative explanations as those enlisting extraterrestrial intervention.

An even more high profile debate concerns the exact nature of evolution, and again one can find on the side of those dissatisfied with mechanistic determinism appeals to notions of emergence and chaos theory. And again one can consolidate these more global perspectives via the mathematics of definability. Determinism is defended, but not reduced to a clearly inadequate mechanism.

And there are many other ways in which the perspective described here can shed new light on thorny problems. For instance, there is the mysterious emergence of large scale structure in the Universe, and on the other hand the remarkable fact that the Universe appears to present a very similar view from whatever observation point one chooses. Once one accepts the basic underlying nature of process and its mathematical counterpart of algorithm, and admits the possibility that the Universe is deeply imbued with incomputability and its mathematics, then many troublesome problems can be placed within a helpful explanatory context.

In the 1999 paper mentioned above, there is a section on epistemological relativism. There is a basic intuition that an analysis of the epistemology derived from our Universe is potentially just as complex as that of the Universe itself. So it should not be surprising that the mathematics of definability should be relevant here. And it is not surprising that without such a conceptual framework as proposed here, different epistemological approaches should give rise to such differing views of the world. And that such controversies as currently characterise discussions between scientists and those engaged in the humanities should thrive in the unfilled vacuum. For reasons of space, we limit ourselves to a few comments extending the perspective.

We do this via what at first sight seems a less promising example. Mathematics, as we know it, is expressed via finitary statements, and has at best a problematical relationship with the material world. The exact character of this relationship is still argued over, as is the extent to which mathematics owes its existence to mathematicians. A puzzling anomaly arises from the dominance of the axiomatic model for mathematical activity, and its incompatibility with experience of creative mathematicians. It is quite true, one observes, that having discovered some new and interesting mathematical theorem, one can usually translate its proof into one in some standard axiomatic theory. The problem is that the discovery of the theorem was not obviously axiomatic, and the formal proof when assembled may do much to obscure the mathematical intuition telling us why the theorem is true. If we look more closely at how we derive the theory itself, we may better understand how it is — in the context of our model — that our discovery of the theorem comes via a route which apparently bypasses the rules of that theory.

One first makes the observation that if the formal theory does not explain the emergence of the axioms, then what it contributes to our understanding of theorems derived from these axioms must be limited. The unconscious assumption is that since an axiom seems simple, so must be the process of its derivation. And that the formal connection between the axiom and the theorem must be a good guide to why the theorem is 'true'. But even the natural numbers, before any statements are made about them, are complex in origin. Historically their emergence is linked to that of language, and its usage in relation to recognition of patterns, or relations on matter. Numbers as such relations are already present in the material world, and on the other hand seem to emerge in ones consciousness independently of any such exterior source. But what one can say is that there exists some process of definition of these basic relations, which is manifest both in nature and within the human mind. And that this does not depend on the exact balance of defining factors in a particular context. It may well be that recognition of such relations on the real world depends on some sort of parallel definition within our consciousness, which one can regard as a simulation within the mind of phenomena within the exterior world. Mental processes, while being a microcosm of the greater universe of which they are a part, do appear to mirror some of its complexity and hence its capacity for global definition. It is possible to regard the mind as a functional Universe in miniature. Anyway, what is important is that the emergence of number does show the familiar irreducibility of phenomenon so characteristic of a phase transition, to which modelling in terms of definability is relevant. The natural conjecture is that the forgotten but complex process underlying the emergence of the axioms is what is being revisited via the discovery of theorems, and that in both cases the basis is definability in terms of the atomic ingredients that constitute our intuitive grasp of what happens in the real world. Horizontal connections between different mathematical statements may play a role, particularly in the translation of personal intuitions into a shared framework of proof, but that is not always the most natural route to new mathematics.

What we have argued above is that it may be possible to reconstitute the mathematical framework governing the scientific approach, in such a way as to revive the sense of universal relevance once associated with 'natural science'. The state of human knowledge lends an inevitably speculative quality to what we have written, but we hope this article will be accepted with the openness and generosity of spirit in which it is offered. How appropriate that the concluding remarks above have the effect of giving theoretical support to those who view contemporary mathematical research as being very much a part of the scientific enterprise.

References

[Co99] S. B. Cooper, Clockwork or Turing U/universe? – remarks on causal determinism and computability, in *Models and Computability* (S. B. Cooper and J. K. Truss, eds.), London Mathematical Society Lecture Note Series 259, Cambridge University Press, Cambridge, New York, Melbourne, 1999, pp. 63–116.

[Fe82] R. P. Feynman, Simulating physics with computers, Int. J. Theor. Phys., **21** (1982), 467–488.

[Kr70] G. Kreisel, Church's Thesis: a kind of reducibility axiom for constructive mathematics, in *Intuitionism and proof theory: Proceedings of the Summer Conference at Buffalo N.Y. 1968* (A. Kino, J. Myhill and R. EVesley, eds.) North-Holland, Amsterdam, London, 1970, pp. 121–150.

[Od89] P. Odifreddi, *Classical Recursion Theory*, North-Holland, Amsterdam, New York, Oxford, 1989.

[PR83] M. B. Pour-El and J. I. Richards, Noncomputability in analysis and physics, Adv. Math., **48** (1983), 44–74.

[PR89] M. B. Pour-El and J. I. Richards, *Computability in Analysis and Physics*, Springer-Verlag, Berlin, Heidelberg, New York, London, Paris, Tokyo, 1989.

[So96] R. I. Soare, Computability and recursion, Bull. Symbolic Logic, **2** (1996), 284–321.

GEMS IN THE FIELD OF BOUNDED QUERIES

William Gasarch
Department of Computer Science
University of Maryland at College Park
MD 20742, USA
gasarch@cs.umd.edu

Abstract Let A be a set. Given $\{x_1, \ldots, x_n\}$, I may want to know (1) which elements of $\{x_1, \ldots, x_n\}$ are in A, (2) how many elements of $\{x_1, \ldots, x_n\}$ are in A, or (3) is $|\{x_1, \ldots, x_n\} \cap A|$ even. All of these can be determined with n queries to A. For which A, n can we get by with fewer queries? Other questions involving 'how many queries do you need to ...' have been posed and (some) answered. This article is a survey of the gems in the field—the results that both answer an interesting question and have a nice proof.

Keywords: Queries, Computability

Introduction

Let the halting problem K be the set of all programs which halt on 0. Assume that I give you 1000 programs and ask you which of them halt, that is, which of them are in K. You cannot answer since K is undecidable. What if I allow you to ask 999 questions to K? Now can you determine which of the 1000 are in K? You can! First build, for each i $0 \leq i \leq 1000$, a program P_i which halts if at least i of the given programs halt. By asking whether $P_i \in K$ you can find out the answer to the query "Do at least i of the given programs halt?" Now binary search allows you to find with 10 queries the number of programs which halt. Say you find out that exactly 783. of the 1000 programs halt. You can run all 1000 of them until you see 783 of them halting, and then you know that the rest do not halt. More generally, if you have $2^n - 1$ programs, you can find out which ones halt by asking n questions. This is the first theorem in the field of Bounded Queries. It was discovered independently by Beigel, Hay, and Owings in the early 1980's.

This observation leads to many other questions of interest. The field of Bounded Queries, founded independently by Beigel [Be87] and Gasarch [Ga85], raises the following types of questions:

(1) Given a function f and a set X, how many queries to X are needed to compute f?

(2) Given a set X and an $n \geq 1$, are there functions that can be computed with n queries to X but not with $n-1$?

This paper is a survey of the nicest results in the field of bounded queries. For a more complete exposition of the field, see [GM99].

1. Definitions

We use notation from [So87], with the notable exception that we use "computable" instead of "recursive" and "c.e." instead of "r.e." This change of terminology was proposed by Soare [So96] for reasons that we agree with; hence we use it. In addition it is being accepted by the community. We remind the reader of some standard notations.

Notation 1.1.

(1) $M_0, M_1, \ldots$ is a list of all Turing machines.

(2) $\varphi_0, \varphi_1, \varphi_2, \ldots$ is a list of all computable partial functions. We obtain this by letting φ_e be the partial function computed by M_e.

(3) $W_0, W_1, \ldots$ is a list of all c.e. sets. We obtain this by letting W_e be the domain of M_e.

(4) $D_0, D_1, \ldots$ is a list of all finite sets indexed in a way that you can effectively recover the elements of the set D_i from the index i. One can view i as a bit vector, so $D_{10111011}$ would represent $\{0, 1, 3, 4, 5, 7\}$. Note that $D_0 = \emptyset$.

(5) $M_0^{()}, M_1^{()}, \ldots$ is a list of all oracle Turing machines.

1.1 Functions of Interest

This paper examines the complexity of the following functions and sets.

Definition 1.2. Let A be a set, and let $n \geq 1$.

(1) $\mathrm{C}_n^A : \mathsf{N}^n \rightarrow \{0,1\}^n$ is defined by

$$\mathrm{C}_n^A(x_1, \ldots, x_n) = A(x_1)A(x_2)\cdots A(x_n).$$

(2) $\#_n^A : \mathsf{N}^n \rightarrow \{0, \ldots, n\}$ is defined by

$$\#_n^A(x_1, \ldots, x_n) = |\{i : x_i \in A\}|.$$

(3) $\mathrm{ODD}_n^A = \{(x_1, \dots, x_n) \in \mathsf{N}^n : \#_n^A(x_1, \dots, x_n) \text{ is odd}\}$.

These functions are interesting because they can all be computed with n parallel queries easily; hence the question of whether they can be computed in less than n (perhaps sequential) is intriguing. Also note that C_n^A gives more information then $\#_n^A$ which gives more information then ODD_n^A.

1.2 Bounded Query Classes

Definition 1.3. Let f be a function, $A \subseteq \mathsf{N}$, and $n \in \mathsf{N}$. $f \in \mathrm{FQ}(n, A)$ if $f \leq_{\mathrm{T}} A$ via an algorithm that makes at most n queries to A.

In the introduction we proved that $\mathrm{C}_{2^n-1}^K \in \mathrm{FQ}(n, K)$. Two aspects of that result motivate the next definition.

Aspect 1: When trying to determine $\mathrm{C}_7^K(x_1, \dots, x_7)$, our first question is "Do at least 4 of the programs halt?" If the answer is YES, we then ask "Do at least 6 of the programs halt?" If the answer to the first question is NO, however, the second question is "Do at least 2 of the programs halt?" Note that the second question asked depends on the answer to the first. Thus the queries are *sequential.* We may want to determine the smallest m such that we can compute $\mathrm{C}_{2^n-1}^K$ with m *parallel* queries to K.

Aspect 2: Let's say that of 7 programs, exactly 3 halt. But suppose that when the queries are made, the answers given are incorrect, so you think there are 4 that halt. If you run them looking for 4 to halt, you will wait forever, so your computation will not terminate. Is there a way to compute $\mathrm{C}_{2^n-1}^K$ with n queries such that even the use of incorrect answers leads to convergence (though perhaps to the wrong bit string)?

Definition 1.4. Let f be a function, $A \subseteq \mathsf{N}$, and $n \in \mathsf{N}$.

(1) $f \in \mathrm{FQ}_{||}(n, A)$ if $f \leq_{\mathrm{T}} A$ via an algorithm that makes at most n queries to A, with the restriction that the queries must be made in *parallel* (i.e., they are nonadaptive). (The symbol || stands for *parallel.*)

(2) $f \in \mathrm{FQC}(n, A)$ if $f \leq_{\mathrm{T}} A$ via an algorithm that makes at most n queries to A, with the restriction that the algorithm must converge (perhaps to the wrong answer) regardless of the choice of oracle (The C stands for *converge.*)

(3) $f \in \mathrm{FQC}_{||}(n, A)$ if $f \leq_{\mathrm{T}} A$ via an algorithm that makes at most n queries to A, with the restrictions that the queries must be made in parallel *and* the algorithm must converge regardless of the choice of oracle.

We now define several bounded-query classes consisting of *sets* that can be decided by making queries to an oracle.

Definition 1.5. Let A, B be sets, and let $n \in \mathsf{N}$.

(1) $B \in \mathrm{Q}(n, A)$ if $\chi_B \in \mathrm{FQ}(n, A)$.

(2) $B \in \mathrm{Q}_{||}(n, A)$ if $\chi_B \in \mathrm{FQ}_{||}(n, A)$.

(3) $B \in \mathrm{QC}(n, A)$ if $\chi_B \in \mathrm{FQC}(n, A)$.

(4) $B \in \mathrm{QC}_{||}(n, A)$ if $\chi_B \in \mathrm{FQC}_{||}(n, A)$.

Definition 1.6. Let f, g be functions. The notions of $f \in \mathrm{FQ}(n, g)$, $f \in \mathrm{FQ}_{||}(n, g)$, etc. can be easily defined.

Definition 1.7. For all the notions in this subsection we can define relativized versions. For example, $\mathrm{FQ}^X(n, A)$ is the set of functions that can be computed with n queries to A and an unlimited number of queries to X.

1.3 Enumerability Classes

The notion of enumerability is very useful in the study of bounded queries. The concept within computability theory is due to Beigel [Be87]. The concept within complexity theory is due independently to Beigel [Be87] and Cai & Hemachandra [CH89]. The term "enumerability" is due to Cai & Hemachandra.

Definition 1.8. Let f be a function, and let $m \geq 1$. We define $f \in \mathrm{EN}(m)$ (f is m-enumerable) in two different ways. We leave it to the reader to show that they are equivalent.

(1) $f \in \mathrm{EN}(m)$ if there exist computable partial functions $g_1, \ldots, g_m$ such that $(\forall x)[f(x) \in \{g_1(x), \ldots, g_m(x)\}]$.

(2) $f \in \mathrm{EN}(m)$ if there is a computable function h such that, for every x, $f(x) \in W_{h(x)}$ and $|W_{h(x)}| \leq m$.

Definition 1.9. Let f be a function, and let $m \geq 1$. We define $f \in \mathrm{SEN}(m)$ (f is strongly m-enumerable) in two different ways. We leave it to the reader to show that they are equivalent.

(1) $f \in \mathrm{SEN}(m)$ if there exist computable functions $g_1, \ldots, g_m$ such that $(\forall x)[f(x) \in \{g_1(x), \ldots, g_m(x)\}]$.

(2) $f \in \mathrm{SEN}(m)$ if there is a computable function h such that, for every x, $f(x) \in D_{h(x)}$ and $|D_{h(x)}| = m$. (One can easily show that the requirements $|D_{h(x)}| \leq m$ and $|D_{h(x)}| = m$ define the same set of functions.)

The following theorem establishes the relationship between query complexity and enumeration complexity.

Theorem 1.10. *Let f be a function, and let $n \in \mathsf{N}$. Then the following statements are equivalent.*

(1) $(\exists X)[f \in \mathrm{FQ}(n, X)]$.

(2) $f \in \mathrm{EN}(2^n)$.

(3) $(\exists Y \equiv_{\mathrm{T}} f)[f \in \mathrm{FQ}_{||}(n, Y) \wedge Y \in \mathrm{Q}(1, f)]$.

Proof: We leave the (easy) proof that (1)⇒(2) to the reader. The fact that (3)⇒(1) is obvious. We present the proof that (2)⇒(3). If $n = 0$, then f is computable, so (2) ⇒(3) is obvious. Hence we assume that $n > 0$.

Suppose that $f \in \mathrm{EN}(2^n)$. We define a set $Y \equiv_{\mathrm{T}} f$ that codes information about f into it so that $f \in \mathrm{FQ}_{||}(n, Y)$; however, Y uses a small amount of information about f so we will have $Y \in \mathrm{Q}(1, f)$.

Assume f is 2^n-enumerable via $g_0, \ldots, g_{2^n-1}$. Let $g_{i,s}(x)$ denote what happens when you run the computation for g_i on input x for s steps. Let t, i be the following functions.

$$t(x) = \mu s[(\exists j)[g_{j,s}(x) \downarrow= f(x)]],$$

$$i(x) = \mu j[g_{j,t(x)}(x) \downarrow= f(x)].$$

We represent $i(x)$ in base 2. We refer to the rightmost bit as the '0th bit', the next bit as the '1st bit', etc. For every k with $0 \leq k \leq n-1$, we define

$$i_k(x) = \text{the } k\text{th bit of } i(x)\ .$$

Let

$$Y = \{(x, k) : i_k(x) = 1\}.$$

It is easy to see that $Y \in \mathrm{Q}(1, f)$ and $f \in \mathrm{FQ}_{||}(n, Y)$ ▮

The next theorem is an analogue of Theorem 1.10 for strong enumeration. We leave the proof to the reader.

Theorem 1.11. *Let f be a function, and let $n \in \mathsf{N}$. Then the following statements are equivalent.*

(1) $(\exists X)[f \in \mathrm{FQC}(n, X)]$.

(2) $f \in \mathrm{SEN}(2^n)$.

(3) $(\exists Y \equiv_{\mathrm{T}} f)[f \in \mathrm{FQC}_{||}(n, Y) \wedge Y \in \mathrm{QC}(1, f)]$.

Note 1.12. All the notions in this subsection can be relativized. For example, $f \in \mathrm{SEN}^X(m)$ if there exist computable-in-X functions $g_1, \ldots, g_m$ such that $(\forall x)[f(x) \in \{g_1(x), \ldots, g_m(x)\}]$.

By Theorems 1.10 and 1.11, any result we obtain about enumerability implies a result about bounded queries. In practice, the results about enumerability are sharper. We state results of both types.

1.4 Definitions from Computability Theory

1.4.1 Selective Sets.

We will use selective sets as defined by Jockusch [Jo68]. (He called them "semirecursive" but he now agrees that "selective" would have been a better name.) These sets have nice properties in terms of bounded queries. In particular, if A is selective, then $(\forall k \geq 1)[\mathrm{C}_k^A \in \mathrm{SEN}(k+1)]$. In Section 4 we will use these sets to help us prove theorems about c.e. sets.

We define selective sets in two ways that are provably equivalent. The equivalence is due to McLaughlin and Appel but was presented in [Jo68].

Definition 1.13. A set A is selective if one of the following two equivalent conditions holds.

(1) There exists a computable function $f\colon \mathsf{N}^2 \to \mathsf{N}$ such that, for all x, y,

- $f(x,y) \in \{x,y\}$, and
- $A \cap \{x,y\} \neq \emptyset \Rightarrow f(x,y) \in A$.

The terminology 'selective set' comes from the fact that f selects which of x, y is more likely to be in A.

(2) There exists a computable linear ordering $\sqsubseteq$ such that A is closed downward under $\sqsubseteq$, i.e., $(\forall x,y)[(x \in A \wedge y \sqsubseteq x) \Rightarrow y \in A]$.

Note 1.14. Let X be a set. We define "selective in X" by making $\sqsubseteq$ and f computable in X in definition 1.13.

Lemma 1.15. *If A is selective, then* $(\forall k \geq 1)[\mathrm{C}_k^A \in \mathrm{SEN}(k+1)]$.

Proof: Let A be selective via ordering $\sqsubseteq$, and let $k \geq 1$. The following algorithm shows that $\mathrm{C}_k^A \in \mathrm{SEN}(k+1)$.

(1) Input $(x_1, \ldots, x_k)$. Renumber so that $x_1 \sqsubseteq \cdots \sqsubseteq x_k$.

(2) Output the set of possibilities $\{1^i 0^{k-i} \mid 0 \leq i \leq k\}$.

∎

1.4.2 Extensive Sets.

We will use extensive sets. These are nice since they are 'almost computable.' In Section 3 we will use them to obtain a lower bound on the enumerability of a function from a lower bound on its strong enumerability. In Section 4 we will use these sets (along with selective sets) to help us prove theorems about c.e. sets.

Definition 1.16. A set X is *extensive* if, for every computable partial function g with finite range, there is a total function $h \leq_T X$ such that h extends g. (The Turing degrees of the extensive sets are the same as the Turing degrees of the consistent extensions of Peano arithmetic [Sc62], [Od89, pages 510–515], but this is not important for our purposes. For this reason they are sometimes called PA sets. They have also been referred to as DNR_2 sets, which stands for Diagonally Non Recursive; see [Jo89].)

Note 1.17. Clearly K is extensive. It is easy to show that all extensive sets are not computable. What is of more interest, although we will not use it, is that there are low extensive sets [JS72].

Jockusch and Soare [JS72] proved the following theorem. We will use it in Theorems 3.4 and 4.4 to show that a set A is computable by showing that A is computable in *every* extensive set.

Theorem 1.18. *There exists a minimal pair of extensive sets. That is, there exist extensive sets X_1 and X_2 such that, for all A, if $A \leq_T X_1$ and $A \leq_T X_2$ then A is computable.*

1.4.3 Computably Bounded Sets.

We will use computably bounded sets. These are nice in terms of convergence (see Lemma 1.20 below) and hence will be used when we study the difference between $\mathrm{FQ}(n, A)$ and $\mathrm{FQC}(n, A)$ (also between $\mathrm{Q}(n, A)$ and $\mathrm{QC}(n, A)$).

Definition 1.19. A set X is *computably bounded* (abbreviated c.b.) if, for every (total) function $f : \mathsf{N} \rightarrow \mathsf{N}$ such that $f \leq_T X$, there exists a computable function g such that $(\forall x)[f(x) < g(x)]$. (In the literature, the Turing degrees of c.b. sets are said to be *hyperimmune free*; this term comes from an equivalent definition that we are not using.)

The following theorem is due to Miller and Martin [MM68]; the proof can also be found in [Od89]. In Section 5.2 we will use this theorem to explore questions about more queries being more powerful.

Lemma 1.20.

(1) There exist c.b. sets $B \leq_T \emptyset''$.

(2) If B is a c.b. set and $A \leq_T B$, then $A \leq_{tt} B$.

2. The Complexity of C_n^A

The function $C_n^A(x_1, \ldots, x_n) = A(x_1)\cdots A(x_n)$ can clearly be computed with n queries to A and is clearly 2^n-enumerable. We have already seen that neither result is tight, as C_n^K can be computed with $O(\log n)$ queries to K and C_n^K is $(n+1)$-enumerable.

Is C_n^K n-enumerable? What about sets other than K? This section will address these and other questions.

2.1 The Complexity of C_n^A for General Sets

In this section we show that, for every set A, if $(\exists n)(\exists X)[C_{2^n}^A \in \mathrm{FQ}(n, X)]$ then A is computable. By Theorem 1.10, it suffices to show that if $(\exists n)[C_{2^n}^A \in \mathrm{EN}(2^n)]$ then A is computable. Note that 2^n appears in two places. It is cleaner to prove the stronger statement that if $(\exists n)[C_n^A \in \mathrm{EN}(n)]$ then A is computable.

Recall that in the introduction we answered 1000 instances of K by asking 10 *sequential* queries to K. Corollary 2.4 below will show that the questions *need* to be sequential; that is, 999 *parallel* queries would not suffice.

The following theorem was first proved by Beigel in his thesis [Be87]. It also appears in [BGGO93].

Definition 2.1. If $1 \le i \le j \le n$ and $b = b_1b_2\cdots b_n \in \{0,1\}^n$, then $b[i:j]$ is defined as $b_ib_{i+1}\cdots b_j$.

Theorem 2.2. *Let* $m \ge 1$, *and let* A *be a set such that* $C_m^A \in \mathrm{EN}(m)$. *Then* A *is computable.*

Proof: The proof is by induction on m. The conclusion of the theorem is obvious if $m = 1$; hence the base case is established. So assume that $m > 1$, and that the $m-1$ case is true. We show that A is computable.

Assume $C_m^A \in \mathrm{EN}(m)$ via $g_1, \ldots, g_m$. We would *like* to show that $C_{m-1}^A \in \mathrm{EN}(m-1)$; by the induction hypothesis, this would prove that A is computable. So we attempt to show the existence of computable partial functions $h_1, \ldots, h_{m-1}$ such that $C_{m-1}^A \in \mathrm{EN}(m-1)$ via $h_1, \ldots, h_{m-1}$. Either our algorithm, A1, works, or its very failure to do so leads to an algorithm, A2, that decides A outright.

We have $C_m^A \in \mathrm{EN}(m)$. We want $C_{m-1}^A \in \mathrm{EN}(m-1)$. Given $(x_1, \ldots$ $\ldots, x_{m-1})$ we want to somehow use the $C_m^A \in \mathrm{EN}(m)$ algorithm. We will do this by adding to $(x_1, \ldots, x_{m-1})$ a variety of y's to form m elements and then running the $C_m^A \in \mathrm{EN}(m)$ algorithm. The hope is to find a y such that we know

$$|\{g_1(x_1, \ldots, x_{m-1}, y)[1:m-1], \ldots$$
$$\ldots, g_m(x_1, \ldots, x_{m-1}, y)[1:m-1]\}| \le m-1.$$

ALGORITHM A1

(1) Input $(x_1, \ldots, x_{m-1})$.

(2) Search for i, j, y such that $1 \le i < j \le m$, $g_i(x_1, \ldots, x_{m-1}, y) \downarrow$, $g_j(x_1, \ldots, x_{m-1}, y) \downarrow$, and

$$g_i(x_1, \ldots, x_{m-1}, y)[1 : m-1] = g_j(x_1, \ldots, x_{m-1}, y)[1 : m-1].$$

(3) (If this step is reached, then i, j, and y were found in step 2.) For every l with $1 \le l \le m-1$, let

$$h_l(x_1, \ldots, x_{m-1}) = \begin{cases} g_l(x_1, \ldots, x_{m-1}, y)[1 : m-1], & \text{if } l < j; \\ g_{l+1}(x_1, \ldots, x_{m-1}, y)[1 : m-1], & \text{if } l \ge j. \end{cases}$$

Let $x_1, \ldots, x_{m-1} \in \mathrm{N}$. We show that *if* $\mathrm{A1}(x_1, \ldots, x_{m-1}) \downarrow$, *then*

$$\mathrm{C}^A_{m-1}(x_1, \ldots, x_{m-1}) \in \{h_l(x_1, \ldots, x_{m-1}) : 1 \le l \le m-1\}.$$

So suppose that $\mathrm{A1}(x_1, \ldots, x_{m-1}) \downarrow$. Since step 2 terminates, we have i, j, y such that $1 \le i < j \le m$, $g_i(x_1, \ldots, x_{m-1}, y) \downarrow$, $g_j(x_1, \ldots, x_{m-1}, y) \downarrow$, and

$$g_i(x_1, \ldots, x_{m-1}, y)[1 : m-1] = g_j(x_1, \ldots, x_{m-1}, y)[1 : m-1].$$

Thus the set

$$\{g_1(x_1, \ldots, x_{m-1}, y)[1 : m-1], \ldots, g_m(x_1, \ldots, x_{m-1}, y)[1 : m-1]\}$$

has at most $m-1$ elements, and is equal to the set

$$\{h_1(x_1, \ldots, x_{m-1}), \ldots, h_{m-1}(x_1, \ldots, x_{m-1})\}.$$

Moreover, $\mathrm{C}^A_{m-1}(x_1, \ldots, x_{m-1})$ is in this set, by our assumption about *A* and our choice of $g_1, \ldots, g_m$.

We show that either algorithm A1 yields $\mathrm{C}^A_{m-1} \in \mathrm{EN}(m-1)$, or some other algorithm (A2, built out of the failure of A1 to work) yields that *A* is computable.

Case 1: $(\forall x_1, \ldots, x_{m-1})[\mathrm{A1}(x_1, \ldots, x_{m-1}) \downarrow]$. Then by the reasoning above, $\mathrm{C}^A_{m-1} \in \mathrm{EN}(m-1)$ via $h_1, \ldots, h_{m-1}$. By the induction hypothesis, *A* is computable.

Case 2: $(\exists x'_1, \ldots, x'_{m-1})[A1(x'_1, \ldots, x'_{m-1}) \uparrow]$. We use this tuple to devise a new algorithm, A2, that shows outright that *A* is computable.

Since $\mathrm{A1}(x'_1, \ldots, x'_{m-1}) \uparrow$, note that, for all $i < j \le m$ and for every y, it cannot be the case that $g_i(x'_1, \ldots, x'_{m-1}, y)$ and $g_j(x'_1, \ldots, x'_{m-1}, y)$ converge and their outputs agree on the first $m-1$ bits. Let $b'_1 \cdots b'_{m-1} = \mathrm{C}^A_{m-1}(x'_1, \ldots, x'_{m-1})$.

ALGORITHM A2

(1) Input y.

(2) Dovetail the $g_1(x'_1, \ldots, x'_{m-1}, y), \ldots, g_m(x'_1, \ldots, x'_{m-1}, y)$ computations, stopping when you find i and a bit b such that

$$g_i(x'_1, \ldots, x'_{m-1}, y) \downarrow = b'_1 \cdots b'_{m-1} b.$$

(3) (If this step is reached, then i and b were found in step 2.) Output b.

Let $y \in \mathsf{N}$. We show that $\mathrm{A2}(y) \downarrow = A(y)$. Since

$$\mathrm{C}^A_m(x'_1, \ldots, x'_{m-1}, y) \in \{g_1(x'_1, \ldots, x'_{m-1}, y), \ldots, g_m(x'_1, \ldots, x'_{m-1}, y)\},$$

we know that $(\exists i, b)[g_i(x'_1, \ldots, x'_{m-1}, y) \downarrow = b'_1 \cdots b'_{m-1} b]$. Hence $\mathrm{A2}(y) \downarrow$. If in step 2 it is discovered that $g_i(x'_1, \ldots, x'_{m-1}, y) \downarrow = b'_1 \cdots b'_{m-1} b$, then it cannot be the case that $(\exists j \neq i)(\exists b')[g_j(x'_1, \ldots, x'_{m-1}, y) \downarrow = b'_1 \cdots b'_{m-1} b']$, since this would imply that

$$g_i(x'_1, \ldots, x'_{m-1}, y)[1 : m-1] = g_j(x'_1, \ldots, x'_{m-1}, y)[1 : m-1],$$

contrary to the choice of $x'_1, \ldots, x'_{m-1}$. Hence for every $j \neq i$, either $g_j(x'_1, \ldots, x'_{m-1}, y)$ diverges, or it converges and is wrong on one of the first $m-1$ bits. It follows that $g_i(x'_1, \ldots, x'_{m-1}, y) = \mathrm{C}^A_m(x'_1, \ldots, x'_{m-1}, y)$, so $A(y) = b = \mathrm{A2}(y)$. ▮

Corollary 2.3. *Let $n \in \mathsf{N}$ and $A, X \subseteq \mathsf{N}$. If $\mathrm{C}^A_{2^n} \in \mathrm{FQ}(n, X)$, then A is computable.*

Proof: This follows from Theorems 2.2 and 1.10. ▮

This survey began by showing that you could answer 1000 queries to K with 10 *sequential* queries to K. By this next corollary we know that the sequential nature is inherent– we could not have answered 1000 queries to K with 999 *parallel* queries to K.

Corollary 2.4. *Let $n \geq 1$ and $A \subseteq \mathsf{N}$. If $\mathrm{C}^A_n \in \mathrm{FQ}_{||}(n-1, A)$, then A is computable.*

Proof: If $\mathrm{C}^A_n \in \mathrm{FQ}_{||}(n-1, A)$ then an easy induction shows that, for all $m \geq n$, $\mathrm{C}^A_m \in \mathrm{FQ}_{||}(n-1, A)$. In particular,

$$\mathrm{C}^A_{2^n-1} \in \mathrm{FQ}_{||}(n-1, A) \subseteq \mathrm{FQ}(n-1, A).$$

By Theorem 1.10, $\mathrm{FQ}(n-1, A) \subseteq \mathrm{EN}(2^{n-1})$, so we have $\mathrm{C}^A_{2^n-1} \in \mathrm{EN}(2^{n-1})$. By Theorem 2.2, A is computable. ▮

The queries in the algorithm given in the introduction (the one that showed that $C^K_{2^n-1} \in FQ(n, K)$) were made sequentially, and now we know that this is inherent. What if we use another oracle? By Theorem 1.10, there exists a Y such that $C^K_{2^n-1} \in FQ_{||}(n, Y)$. Note that the set Y is useful *not* because it has high Turing degree (in fact, $K \equiv_T Y$) but because of the way information is stored in it.

2.2 The Complexity of C^B_n for Natural Sets B

Sets that are Σ_i-complete or Π_i-complete are natural. The set K is surely natural, and we have $C^K_n \in EN(n+1)$. Are there other natural sets B for which $C^B_n \in EN(n+1)$, or at least $C^B_n \in EN(2^n-1)$. The answer is NO. We show that for every noncomputable set A, $C^{A'}_n \notin EN(2^n-1)$. This result first appeared in [BGGO93].

Definition 2.5. Let $k, n \in \mathrm{N}$ such that $1 \le k \le n$. $S(n,k) = \sum_{i=0}^{k-1} \binom{n}{i}$.

The following lemma has appeared in several places independently. It was first discovered by Vapnik and Chervonenkis [VC71], and subsequently rediscovered by Sauer [Sa72], Clarke, Owings, and Spriggs [COS75], and Beigel [Be87]. The reason why it has been discovered by so many is that it has applications in probability theory [VC71], computational learning theory [BEHW89], computational geometry [HW87], and of course bounded queries. It has also been attributed to Shelah [Sh72]; however, that paper does not contain it. I suspect that Shelah had a proof, and that people refer to that paper because its title *sounds* as if it should contain it.

Lemma 2.6. *Let $Y \subseteq \{0,1\}^n$. Let $k \le n$. Assume that for every $i_1, .., i_k$ with $1 \le i_1 < i_2 < \ldots < i_k \le n$, the projected set*

$$\{\vec{b}' \in \{0,1\}^k \mid (\exists \vec{b} \in Y)[\vec{b}' \textit{ is the projection of } \vec{b} \textit{ on coordinates } i_1, \ldots, i_k\,]\}$$

has at most $2^k - 1$ elements. Then Y has at most $S(n,k)$ elements.

The next lemma shows that if you can save just a little bit on enumerability (that is, $C^B_k \in EN(2^k-1)$ instead of the obvious $C^B_k \in EN(2^k)$), then, for large n, you can save a lot in terms of enumerability. This will enable us to show that for the jump of any noncomputable set, you cannot even save a little.

Theorem 2.7. *Let $k \ge 1$, and let B be a set. If $C^B_k \in EN(2^k-1)$, then $(\forall n \ge k)[C^B_n \in EN(S(n,k))]$. If $C^B_k \in SEN(2^k-1)$, then $(\forall n \ge k)[C^B_n \in SEN(S(n,k))]$.*

Proof: Assume that $C^B_k \in EN(2^k-1)$ via g. The proof for $SEN(2^k-1)$ is similar.

We show $C^B_n \in EN(S(n,k))$. In input $(x_1, \ldots, x_n)$ enumerate elements of $\{0,1\}^n$ as follows. Enumerate all strings $b_1 \cdots b_n$ such that, for all $i_1, \ldots, i_k$

with $1 \leq i_1 < i_2 < \ldots < i_k \leq n$, the projection $b_{i_1}b_{i_2}\cdots b_{i_k}$ shows up in $W_{g(x_{i_1},x_{i_2},\ldots,x_{i_k})}$

Let Y be the set of strings enumerated. We need to show that $C_n^B \in Y$ and that $|Y| \leq S(n,k)$.

Let $C_n^B = \vec{b}$. For every $1 \leq i_1 < i_2 < \cdots < i_k \leq n$ we clearly have the projection $b_{i_1}b_{i_2}\cdots b_{i_k}$ in $W_{g(x_{i_1},x_{i_2},\ldots,x_{i_k})}$; hence $\vec{b} \in Y$.

For every $1 \leq i_1 < i_2 < \cdots < i_k \leq n$ the number of elements in Y projected on those coordinates is at most $2^k - 1$ since $|W_{g(x_{i_1},x_{i_2},\ldots,x_{i_k})}| \leq 2^k - 1$; hence by Lemma 2.6 $|Y| \leq S(n,k)$. ∎

Theorem 2.8. *If* $(\exists k \geq 1)[C_k^{A'} \in \mathrm{EN}(2^k - 1)]$, *then A is computable.*

Proof: Assume $(\exists k \geq 1)[C_k^{A'} \in \mathrm{EN}(2^k - 1)]$. We show that $(\exists n \geq 1)[C_n^A \in \mathrm{EN}(n)]$, hence that A is computable (by Theorem 2.2).

By Theorem 2.7, $(\forall n \geq k)[C_n^{A'} \in \mathrm{EN}(S(n,k))]$. For large n, $S(n,k) = O(n^k)$, hence we write $C_n^{A'} \in \mathrm{EN}(O(n^k))$.

Since $A \leq_m A'$, we have $C_n^A \in \mathrm{EN}(O(n^k))$. By Theorem 1.10,

$$(\exists Y \equiv_{\mathrm{T}} A)[C_n^A \in \mathrm{FQ}_{||}(O(k \log n), Y) = \mathrm{FQ}_{||}(O(\log n), Y) = \mathrm{FQ}(1, C^Y_{O(\log n)})].$$

Since $Y \equiv_{\mathrm{T}} A$, we have $Y \leq_m A'$, hence

$$C_n^A \in \mathrm{FQ}(1, C^Y_{O(\log n)}) \subseteq \mathrm{FQ}(1, C^{A'}_{O(\log n)}).$$

By Theorem 2.7, $C^{A'}_{O(\log n)} \in \mathrm{EN}(S(O(\log n), k)) \subseteq \mathrm{EN}(O((\log n)^k))$. Hence $C_n^A \in \mathrm{EN}(O((\log n)^k))$. For n large, $O((\log n)^k) \leq n$, which implies that $C_n^A \in \mathrm{EN}(n)$. ∎

By a proof similar to that of Theorem 2.7, we obtain the following.

Theorem 2.9. *Let $k \geq 1$, and let A be a set. If $C_k^A \in \mathrm{SEN}(2^k - 1)$, then $(\forall n \geq k)[C_n^A \in \mathrm{SEN}(S(n,k))]$.*

3. The Complexity of $\#_n^A$

We now know that for noncomputable sets A, $C_n^A \notin \mathrm{EN}(n)$. What if we ask for less information? Realize that there are 2^n possibilities for C_n^A, which is a lot. In contrast, there are only $n+1$ possibilities for $\#_n^A$, so perhaps there are some noncomputable sets A such that $\#_n^A \in \mathrm{EN}(n)$. Alas, no such set exists!

Beigel conjectured that if $\#_n^A \in \mathrm{EN}(n)$, then A is computable. Owings [Ow89] showed that if $\#_n^A \in \mathrm{SEN}(n)$, then A is computable, and also that if $\#_2^A \in \mathrm{EN}(1)$, then A is computable. (Owings stated his theorem as $\#_n^A \in \mathrm{EN}(n) \Rightarrow A \leq_{\mathrm{T}} K$. The form we state follows from the same proof.)

Kummer [Ku92] then proved Beigel's conjecture. Kummer's proof used a Ramsey-type theorem on trees. Later, Kummer and Stephan [KS94] observed that Owings' proof could be extended to show Beigel's conjecture. We present Owings' proof followed by Kummer and Stephan's observation.

Most theorems in computability theory relativize. For the next theorem, we *need* to prove its relativized form, since we need this stronger version as a way to strengthen the induction hypothesis.

Notation 3.1. If D is a set, then $\mathcal{P}(D)$ denotes the power set of D.

If $\#_n^A \in \mathrm{SEN}^X(n)$ via f then $f \leq_{\mathrm{T}} X$ and $f(x_1, \ldots, x_n)$, outputs $\leq n$ possibilities for $\#_n^A(x_1, \ldots, x_n)$. Note that there may be other sets B such that $\#_n^B \in \mathrm{SEN}^X(n)$ via f. We will be interested in the set of all such sets. The next definition clarifies these concepts.

Definition 3.2. Let X be a set, let $n \geq 1$, and let f be a function such that

- $f\colon \mathsf{N}^n \to \mathcal{P}(\{0, \ldots, n\})$,
- $f \leq_{\mathrm{T}} X$, and
- $(\forall x_1, \ldots, x_n)[|f(x_1, \ldots, x_n)| \leq n]$.

The triple (n, f, X) is *helpful*, and

$$SE_f = \{Z \mid (\forall x_1, \ldots, x_n)[\#_n^Z(x_1, \ldots, x_n) \in f(x_1, \ldots, x_n)]\}.$$

(Note that the sets in SE_f are precisely the sets Z for which f is a strong enumerator-in-X of $\#_n^Z$.)

Theorem 3.3. *Let $n \geq 1$, and let A, X be sets such that $\#_n^A \in \mathrm{SEN}^X(n)$. Then $A \leq_{\mathrm{T}} X$.*

Proof: Since $\#_n^A \in \mathrm{SEN}^X(n)$, there exists an $f\colon \mathsf{N}^n \to \mathcal{P}(\{0, \ldots, n\})$ such that (n, f, X) is helpful and $A \in SE_f$.

We show that since (n, f, X) is helpful, we have $(\forall C \in SE_f)[C \leq_{\mathrm{T}} X]$. Our proof is by induction on n. For $n = 1$, SE_f has only one element, which is clearly computable in X.

So assume that $n \geq 2$ and, as induction hypothesis, that for every set Y and every function $g\colon \mathsf{N}^{n-1} \to \mathcal{P}(\{0, \ldots, n-1\})$,

$$(n-1, g, Y) \text{ helpful} \Rightarrow (\forall D \in SE_g)[D \leq_{\mathrm{T}} Y].$$

To prove that $A \leq_{\mathrm{T}} X$, we actually prove that the following three conditions hold.

(1) $(\forall B, C \in SE_f)[C \leq_{\mathrm{T}} B \oplus X]$.

(2) SE_f is countable.

(3) $(\forall C \in SE_f)[C \leq_{\mathrm{T}} X]$.

That $(1) \Rightarrow (2)$ is trivial. That $(2) \Rightarrow (3)$ comes from the following two facts: (a) SE_f is the set of infinite branches of a tree that is computable in f, (b) any tree with countably many infinite branches (but at least one) has some branch computable in the tree (proved by Owings, and independently by Jockusch and Soare [JS72]; see also [Od89, Prop. V.5.27, page 507]).

Hence we need prove only (1).

Let $B, C \in SE_f$. To prove that $C \leq_{\mathrm{T}} B \oplus X$, we first show that $B - C \leq_{\mathrm{T}} B \oplus X$ and $C - B \leq_{\mathrm{T}} B \oplus X$.

To show that $B - C \leq_{\mathrm{T}} B \oplus X$, we use the induction hypothesis. In particular, we show that there exists an $(n-1)$-ary function g such that $(n-1, g, B \oplus X)$ is helpful and $B - C \in SE_g$. By the induction hypothesis, this yields $B - C \leq_{\mathrm{T}} B \oplus X$.

If $B - C = \emptyset$, then clearly $B - C \leq_{\mathrm{T}} B \oplus X$ and we are done. Hence we can assume $B - C \neq \emptyset$, so choose $z_0 \in B - C$. We use z_0 in our algorithm for g.

ALGORITHM FOR g

(1) Input $(x_1, \ldots, x_{n-1})$.

(2) Make the queries "$x_1 \in B$?", ..., "$x_{n-1} \in B$?". If there is some i such that $x_i \notin B$, then $x_i \notin B - C$; hence $\#_{n-1}^{B-C}(x_1, \ldots, x_{n-1}) < n-1$, so let $g(x_1, \ldots, x_{n-1}) = \{0, \ldots, n-2\}$ and halt. Otherwise, go to the next step.

(3) (Since we have reached this step, we know that $\{x_1, \ldots, x_{n-1}\} \subseteq B$.) Note that $B - C \cap \{x_1, \ldots, x_{n-1}\} = \overline{C} \cap \{x_1, \ldots, x_{n-1}\}$. Hence $\#_{n-1}^{B-C}(x_1, \ldots, x_{n-1}) = n - 1 - \#_{n-1}^{C}(x_1, \ldots, x_{n-1})$. Therefore, we need only find $\leq n-1$ possibilities for $\#_{n-1}^{C}(x_1, \ldots, x_{n-1})$.

(4) Since $z_0 \notin C$, we have $\#_{n-1}^{C}(x_1, \ldots, x_{n-1}) = \#_{n}^{C}(x_1, \ldots, x_{n-1}, z_0)$. Hence, we need only find $\leq n-1$ possibilities for $\#_{n}^{C}(x_1, \ldots, x_{n-1}, z_0)$.

(5) Since $x_1, \ldots, x_{n-1}, z_0 \in B$ and $B \in SE_f$, $n \in f(x_1, \ldots, x_{n-1}, z_0)$. Since $z_0 \notin C$, we know that $\#_{n}^{C}(x_1, \ldots, x_{n-1}, z_0) \neq n$. Moreover, $C \in SE_f$, so $\#_{n}^{C}(x_1, \ldots, x_{n-1}, z_0) \in f(x_1, \ldots, x_{n-1}, z_0) - \{n\}$. Now note that $|f(x_1, \ldots, x_{n-1}, z_0) - \{n\}| \leq n-1$, so we have $\leq n-1$ possibilities for $\#_{n}^{C}(x_1, \ldots, x_{n-1}, z_0)$. Using this information, define $g(x_1, \ldots, x_{n-1})$ and halt.

Since (n, f, X) is helpful (hence $f \leq_T X$) and $B, C \in SE_f$, it is clear from the algorithm that $(n-1, g, B \oplus X)$ is helpful and $B - C \in SE_g$. By the induction hypothesis, $B - C \leq_T B \oplus X$.

Noting that $C - B = \overline{B} - \overline{C}$, we easily see that the proof that $C - B \leq_T B \oplus X$ is similar. Now

$$C = (C \cap B) \cup (C - B) = [\overline{(B - C)} \cap B] \cup (C - B).$$

Since we have shown that both $B - C$ and $C - B$ are computable in $B \oplus X$, we have $C \leq_T B \oplus X$. This proves that condition (1) holds. Since (1) $\Rightarrow$ (3), we also have $C \leq_T X$. ∎

By Theorem 3.3, we have that if $\#_n^A \in \mathrm{SEN}(n)$, then A is computable. We want to obtain the analogous result for $\mathrm{EN}(n)$. Extensive sets X are used, since they can turn an $\mathrm{EN}(n)$ computation into an $\mathrm{SEN}^X(n)$ computation.

Theorem 3.4. *Let $n \geq 1$, and let A be a set. If $\#_n^A \in \mathrm{EN}(n)$, then A is computable.*

Proof: Let $\#_n^A$ be n-enumerable via computable partial functions $h_1, \ldots, h_n$. Thus for all $x_1, \ldots, x_n \in \mathsf{N}$,

$$\#_n^A(x_1, \ldots, x_n) \in \{h_1(x_1, \ldots, x_n), \ldots, h_n(x_1, \ldots, x_n)\}.$$

We can assume that each h_i has finite range, namely $\{0, \ldots, n\}$.

Let X be *any* extensive set (see Definition 1.16). Since X is extensive, each h_i has a total extension that is computable in X. Using these extensions, we obtain that $\#_n^A \in \mathrm{SEN}^X(n)$. By Theorem 3.3, we have $A \leq_T X$. Since X was *any* extensive set, we have that, for *every* extensive set X, $A \leq_T X$. Since there exist minimal pairs of extensive sets (Theorem 1.18), we obtain that A is computable. ∎

4. The Complexity of ODD_n^A

We now know that if A is noncomputable, then $\#_n^A \notin \mathrm{EN}(n)$. Asking for $\#_n^A$ seems (in retrospect) like asking for a lot of information. What if we just want to know the parity of $\#_n^A$? Determining the parity of $\#_n^A$ entails finding only one bit of information. This problem is easy in terms of enumerability in a trivial way: there are only two possibilities for it.

If we return to bounded queries (rather than enumerability), interesting questions arise. For example, how hard is ODD_n^A in terms of queries to A?

The main theorem of this chapter is the following:

If A is c.e. and $(\exists n \geq 1)[\mathrm{ODD}_n^A \in \mathsf{Q}_{||}(n-1, A)]$, then A is computable.

We will give two proofs of this theorem. The first proof shows that if $\mathrm{ODD}_n^A \in \mathrm{Q}_{||}(n-1, A)$ and A is c.e. then $\mathrm{C}_n^A \in \mathrm{EN}(n)$, hence by Theorem 2.2, A is computable. This proof gives intuition for the result but involves some details that need to be done carefully. The second proof uses selective and extensive sets and is very elegant; however, it is less insightful as to why the theorem is true. I leave as an exercise the task of comparing the two proofs and determining which one is better.

The following are also known.

(1) If A and B are c.e. and $\mathrm{ODD}_n^A \in \mathrm{Q}_{||}(n-1, B)$ then A is computable.

(2) If A and B are c.e. and $\mathrm{ODD}_{2^n}^A \in \mathrm{Q}(n, B)$ then A is computable.

The first result can be proven by simple variations of the proofs given here. The second one is more complicated. The results in this section, and the two results stated below that we are not going to prove, have appeared in both [GM99] and [Be*etal* 00]. Our main result has three proofs. We present two here; one additional proof can be found in [GM99].

4.1 A Direct Proof

Theorem 4.1. *If A is computably enumerable and* $\mathrm{ODD}_n^A \in \mathrm{Q}_{||}(n-1, A)$ *then A is computable.*

Proof: In the following we assume that A has an enumeration A_s and that ODD_n^A is computed by M with $n-1$ parallel queries to A itself. We show that then C_n^A is in $\mathrm{EN}(n)$ which gives that A is computable by Theorem 2.2.
ALGORITHM FOR $\mathrm{C}_n^A \in \mathrm{EN}(n)$.

(1) Input $(x_1, \ldots, x_n)$.

(2) Run $M^{()}(x_1, \ldots, x_n)$ until the queries are made. Let them be $(y_1, \ldots, y_{n-1})$. (Do not ask them.)

(3) Enumerate a tuple $(A_s(x_1), \ldots, A_s(x_n))$ as a possibility for C_n^A iff the computation $M(x_1, \ldots, x_n)$ with query answers $(A_s(y_1), \ldots, A_s(y_{n-1}))$ terminates within s steps and its output agrees with $A_s(x_1)+\ldots+A_s(x_n)$ modulo 2.

The enumerated set contains $\mathrm{C}_n^A(x_1, \ldots, x_n)$ since for sufficiently large s the sets A_s and A coincide at all queried places and M has converged. We now show that we have enumerated at most n strings. There are two cases.
Case 1: There are two different outputs $(a_1, \ldots, a_n)$ and later $(b_1, \ldots, b_n)$ where the corresponding computation $M(x_1, \ldots, x_n)$ with query answers $(A_s(y_1), \ldots, A_s(y_n)))$ uses in both cases the same values $c_1, \ldots, c_{n-1}$ for $A_s(y_1), \ldots, A_s(y_{n-1})$. Then $a_1 + a_2 + \ldots + a_n + 2 \leq b_1 + b_2 + \ldots + b_n$

and no tuple with cardinality $a_1 + a + 2 + \ldots + a_n + 1$ is enumerated. Since for every cardinality at most one tuple is enumerated, at most n elements are enumerated.

Case 2: For every output $(a_1, \ldots, a_n)$ the corresponding computation $M(x_1, \ldots, x_n)$ with query answers $(A_s(y_1), \ldots, A_s(y_n))$ uses values $c_1, \ldots$ $\ldots, c_{n-1}$ for $A_s(y_1), \ldots, A_s(y_{n-1})$ not used for any other output. As these values $c_1, \ldots, c_{n-1}$ originate from an enumeration of A, this tuple can take at most n values and so the cardinality of the set enumerated is at most n. ∎

4.2 An Elegant Proof

We show the following.

- If A is selective and $(\exists n \geq 1)[\mathrm{ODD}_n^A \in \mathrm{Q}_{||}(n-1, A)]$, then A is computable.
- If A is c.e. and $(\exists n \geq 1)[\mathrm{ODD}_n^A \in \mathrm{Q}_{||}(n-1, A)]$, then A is computable.

In investigating the complexity of ODD_n^A, we first look at selective sets A, and we then proceed to c.e. sets A. This is because we obtain the result about c.e. sets *from* the result about selective sets.

4.2.1 ODD_n^A for Selective Sets A.

Theorem 4.2. *Let $n \geq 1$, and let A be a selective set such that* $\mathrm{ODD}_n^A \in \mathrm{Q}_{||}(n-1, A)$. *Then A is computable.*

Proof: If $n = 1$, then since $(\forall x)[\mathrm{ODD}_1^A(x) = A(x)]$, we have that A is computable. Hence we can assume that $n \geq 2$. Let A be selective via $\sqsubseteq$, and assume that $\mathrm{ODD}_n^A \in \mathrm{Q}_{||}(n-1, A)$ via M^A.

The following algorithm shows that $\mathrm{C}_{2n+1}^A \in \mathrm{FQ}_{||}(2n, A)$. By Corollary 2.4, this yields that A is computable.

(1) Input $(x_1, \ldots, x_{2n+1})$, where $x_1 \sqsubseteq \cdots \sqsubseteq x_{2n+1}$. Note that

$$\mathrm{C}_{2n+1}^A(x_1, \ldots, x_{2n+1}) \in \{1^i 0^{2n+1-i} \mid 0 \leq i \leq 2n+1\}.$$

(2) Simulate the computation of $M^A(x_2, x_4, x_6, \ldots, x_{2n})$ to obtain the queries $z_1, \ldots, z_{n-1}$ that are made in this computation. (We do not make these queries at this point. We have not yet used A in any manner.)

(3) Obtain $\mathrm{C}_{2n}^A(x_1, x_3, x_5, \ldots, x_{2n+1}, z_1, z_2, z_3, \ldots, z_{n-1})$, by making $2n$ parallel queries to A.

(4) (We know the status of both x_1 and x_{2n+1} with respect to membership in A.) If $x_1 \notin A$, then $\mathrm{C}_{2n+1}^A(x_1, \ldots, x_{2n+1}) = 0^{2n+1}$, so output 0^{2n+1}

and halt. If $x_{2n+1} \in A$, then $C^A_{2n+1}(x_1, \ldots, x_{2n+1}) = 1^{2n+1}$, so output 1^{2n+1} and halt. If $x_1 \in A$ and $x_{2n+1} \notin A$, go to the next step.

(5) There is a unique $i < n$ such that $\{x_1, x_2, x_3, \ldots, x_{2i+1}\} \subseteq A$ and $\{x_{2i+3}, x_{2i+4}, x_{2i+5}, \ldots, x_{2n+1}\} \subseteq \overline{A}$. We do not yet know whether $x_{2i+2} \in A$, but we do know that

$$C^A_{2n+1}(x_1, x_2, x_3, \ldots, x_{2n+1}) = \begin{cases} 1^{2i+1}0^{2n+1-(2i+1)}, & \text{if } x_{2i+2} \notin A; \\ 1^{2i+2}0^{2n+1-(2i+2)}, & \text{if } x_{2i+2} \in A. \end{cases}$$

Moreover, $C^A_{n+1}(x_1, x_3, x_5, \ldots, x_{2n+1}) = 1^{i+1}0^{(n+1)-(i+1)}$, so

$$C^A_n(x_2, x_4, x_6, \ldots, x_{2n}) = \begin{cases} 1^i 0^{n-i}, & \text{if } x_{2i+2} \notin A; \\ 1^{i+1}0^{n-(i+1)}, & \text{if } x_{2i+2} \in A. \end{cases}$$

Hence $x_{2i+2} \in A$ iff $\mathrm{ODD}^A_n(x_2, x_4, x_6, \ldots, x_{2n})$ and i are of opposite parity.

Using the value of $C^A_{n-1}(z_1, \ldots, z_{n-1})$ from step 3, compute

$b = M^A(x_2, x_4, x_6, \ldots, x_{2n}) = \mathrm{ODD}^A_n(x_2, x_4, x_6, \ldots, x_{2n})$.

Using i and b, compute and output $C^A_{2n+1}(x_1, \ldots, x_{2n+1})$: There are two cases.

- $b = i \bmod 2$: Then $x_{2i+2} \notin A$, so output $1^{2i+1}0^{2n+1-(2i+1)}$.
- $b \neq i \bmod 2$: Then $x_{2i+2} \in A$, so output $1^{2i+2}0^{2n+1-(2i+2)}$.

∎

4.2.2 ODD^A_n for C.E. Sets A.

Our plan is to make c.e. sets *look like* selective sets and then apply a version of Theorem 4.2.

Lemma 4.3. *Let A be a c.e. set, and let X be an extensive set. Then A is selective in X.*

Proof: Choose a computable enumeration $\{A_s\}_{s \in \mathrm{N}}$ of A. Let g be the 0,1-valued computable partial function such that $\mathrm{dom}(g) \subseteq \mathrm{N}^2$ and, for all x, y,

$$g(x,y) = \begin{cases} 1, & \text{if } (\exists s)[x \in A_s \wedge y \notin A_s]; \\ 0, & \text{if } (\exists s)[y \in A_s \wedge x \notin A_s]; \\ \uparrow, & \text{otherwise.} \end{cases}$$

Since X is extensive, there is a 0,1-valued total function $h \leq_{\mathrm{T}} X$ such that h extends g. Now define $f\colon \mathrm{N}^2 \to \mathrm{N}$ by

$$f(x,y) = \begin{cases} x, & \text{if } h(x,y) = 1; \\ y, & \text{if } h(x,y) = 0. \end{cases}$$

It is easy to show that A is selective in X via f. ∎

Theorem 4.4. *Let $n \geq 1$, and let A be c.e. If $\mathrm{ODD}_n^A \in \mathrm{Q}_{||}(n-1, A)$, then A is computable.*

Proof: Suppose that $\mathrm{ODD}_n^A \in \mathrm{Q}_{||}(n-1, A)$. Let X be *any* extensive set (see Definition 1.16). Trivially, $\mathrm{ODD}_n^A \in \mathrm{Q}_{||}^X(n-1, A)$. By Lemma 4.3, A is selective in X. By a relativized version of Theorem 4.2, we have $A \leq_{\mathrm{T}} X$. Since X was *any* extensive set, we have that, for *every* extensive set X, $A \leq_{\mathrm{T}} X$. Since there exist minimal pairs of extensive sets (Theorem 1.18), we obtain that A is computable. ∎

5. Do More Queries Help?

In prior sections, we asked 'How many queries does it take to compute BLAH?' We now ask a more abstract question: 'If I have k queries, can I compute more functions than I could if I had only $k-1$ queries?' The answer depends on what you want to compute (functions or sets) and what you are querying. The short answer is that for computing functions, it always helps to have more queries, but for deciding sets there are cases where more queries do not help.

5.1 More Queries Do Help Compute More Functions!

The results in this section are due to Beigel [Be88]. They later appeared in [GM99].

Theorem 5.1. *If $(\exists n)[\mathrm{FQ}(n, A) = \mathrm{FQ}(n+1, A)]$, then A is computable.*

Proof: By way of contradiction, suppose A is noncomputable. We exhibit a function in $\mathrm{FQ}(n+1, A) - \mathrm{FQ}(n, A)$. If $n = 0$, then $\mathrm{C}_1^A \in \mathrm{FQ}(n+1, A) - \mathrm{FQ}(n, A)$. So assume that $n \geq 1$.

By Corollary 2.3, $\mathrm{C}_{2^n}^A \notin \mathrm{FQ}(n, A)$. Since $\mathrm{C}_n^A \in \mathrm{FQ}(n, A)$, there is some $i \geq n+1$ such that $\mathrm{C}_i^A \notin \mathrm{FQ}(n, A)$ and $\mathrm{C}_{i-1}^A \in \mathrm{FQ}(n, A)$. We show that $\mathrm{C}_i^A \in \mathrm{FQ}(n+1, A)$.

Choose an oracle Turing machine $M^{()}$ so that $\mathrm{C}_{i-1}^A \in \mathrm{FQ}(n, A)$ via M^A. The following algorithm computes C_i^A with at most $n+1$ queries to A. Thus $\mathrm{C}_i^A \in \mathrm{FQ}(n+1, A) - \mathrm{FQ}(n, A)$.

(1) Input $(x_1, \ldots, x_i)$.

(2) Run $M^A(x_1, \ldots, x_{i-1})$. (By hypothesis, M^A makes at most n queries to A.)

(3) Make one additional query to A, namely, "$x_i \in A$?".

(4) Output $C_i^A(x_1, \dots, x_i)$ (by concatenating the results of steps 2 and 3).

The theorem is proved. ∎

The next theorem is an analogue of Theorem 5.1 for the bounded-query classes $\mathrm{FQC}(n, A)$, $\mathrm{FQ}_{||}(n, A)$, and $\mathrm{FQC}_{||}(n, A)$. We leave the proof to the reader.

Theorem 5.2. *Let A be a set.*

(1) If $(\exists n)[\mathrm{FQC}(n, A) = \mathrm{FQC}(n+1, A)]$, *then A is computable.*

(2) If $(\exists n)[\mathrm{FQ}_{||}(n, A) = \mathrm{FQ}_{||}(n+1, A)]$, *then A is computable.*

(3) If $(\exists n)[\mathrm{FQC}_{||}(n, A) = \mathrm{FQC}_{||}(n+1, A)]$, *then A is computable.*

5.2 More Queries Do Not Always Help Decide More Sets!

Definition 5.3. $\emptyset^{(\omega)} = \{\langle x, i\rangle \mid x \in \emptyset^{(i)}\}$.

The next theorem shows that when deciding sets by making *parallel* queries to $\emptyset^{(\omega)}$, more queries do not help. We then show that when deciding sets by making *serial* queries to $\emptyset^{(\omega)}$, allowing more queries does enable us to decide more sets. Finally, we exhibit an (unnatural) set such that allowing a greater number of serial queries (with this set as oracle) does not help.

Most of the results in this chapter are in [GM99] but were known many years earlier. They are due to Beigel and Gasarch (no reference available, but I was there). The one exception is Theorem 5.5, which Frank Stephan proved recently and appears here for the first time.

Theorem 5.4. *For all* $n \geq 1$, $\mathrm{Q}_{||}(n, \emptyset^{(\omega)}) = \mathrm{Q}(1, \emptyset^{(\omega)})$.

Proof: Let $n \geq 1$, and let $A \in \mathrm{Q}_{||}(n, \emptyset^{(\omega)})$ via $M^{()}$. Here is an algorithm for $A \in \mathrm{Q}(1, \emptyset^{(\omega)})$.

(1) Input x.

(2) Run $M^{()}(x)$ until the questions ($q_1 \in \emptyset^{(\omega)}, \dots, q_n \in \emptyset^{(\omega)}$) are asked, but do not try to answer them. Note that, for each i, there exist y_i, z_i such that the question "$q_i \in \emptyset^{(\omega)}$?" is actually the question "$y_i \in \emptyset^{(z_i)}$?" Let z be the max of the z_i.

(3) The question "Is there a set of answers for $y_1 \in \emptyset^{(z_1)}, \dots, y_n \in \emptyset^{(z_n)}$ that are true and lead to a path of the $M^{()}(x)$ computation that converges to 1?" (note that the answer to this question is yes iff $x \in A$) can be

phrased as a query to $\emptyset^{(\omega)}$ (via a query to $\emptyset^{(z+1)}$). Let "$q \in \emptyset^{(\omega)}$?" be that query, and ask it.

(4) If $q \in \emptyset^{(\omega)}$, output 1; otherwise, output 0.

∎

Theorem 5.4 uses the set $\emptyset^{(\omega)}$, which is somewhat natural. Can we use the same set as an example of an oracle for which additional serial queries do not help? As the next result shows, the answer to this question is no.

Theorem 5.5. $\mathrm{Q}(2, \emptyset^{(\omega)}) - \mathrm{Q}(1, \emptyset^{(\omega)}) \neq \emptyset$.

Proof: Let C be the set of ordered pairs (x, y) such that

- $x \in K$, and
- if s is the length of the longest string of 1's on the tape after $M_x(x)$ halts then $y \in \emptyset^{(s)}$. (We assume $s \geq 2$.)

Note that we think of x as an index of a Turing machine and y as an index of an oracle Turing machine.

Clearly, $C \in \mathrm{Q}(2, \emptyset^{(\omega)})$. We show that $C \notin \mathrm{Q}(1, \emptyset^{(\omega)})$. Assume, by way of contradiction, that $C \in \mathrm{Q}(1, \emptyset^{(\omega)})$ via $M^{()}$.

We create Turing machine M_x and oracle Turing machine $M_y^{()}$ such that $M^{\emptyset^{(\omega)}}(x, y) \neq C(x, y)$. The construction of these two machines uses the recursion theorem implicitly.

PROGRAM FOR M_x

(1) Simulate $M^{()}(x, y)$ in such a way that you never write two consecutive 1's on the tape. (E.g., use 00 for 0 and 01 for 1.) Stop the simulation when the one query is made. Let this query be "$q \in \emptyset^{(k)}$?" (Do not make the query.)

(2) Print 01^k0 and then halt. (Hence the longest string of 1's on the tape has length k.)

Note 5.6. Since $x \in K$ and prints out a sequence of k 1's we know that $(x, y) \in C$ iff $y \in \emptyset^{(k)}$.

PROGRAM FOR $M_y^{()}$

(1) Simulate $M^{\emptyset^{(\omega)}}(x, y)$. When the query, "$q \in \emptyset^{(k)}$?", is encountered, make the query. (We will be supplying y with an oracle for $\emptyset^{(k)}$ so this can be done.)

(2) If the simulation outputs 0 (so $M^{\emptyset^{(\omega)}}(x, y)$ thinks that $(x, y) \notin C$ which is equivalent to $y \notin \emptyset^{(k)}$) then halt (which causes $y \in \emptyset^{(k)}$). If the simulation outputs 1 (so $M^{\emptyset^{(\omega)}}(x, y)$ thinks that $(x, y) \in C$ which is equivalent to $y \in \emptyset^{(k)}$) then diverge (which causes $y \notin \emptyset^{(k)}$). The simulation must output something since $M^{\emptyset^{(\omega)}}(x, y)$ computes C.

The key point is that oracle program x is constructed to print out a long enough string of 1's so that oracle program y is able to simulate $M^{\emptyset^{(\omega)}}(x, y)$, make an appropriate query, and diagonalize. It is easy to see that $M^{\emptyset^{(\omega)}}(x, y) \neq C(x, y)$. ∎

The next theorem shows that when deciding sets with sequential queries, more queries do not always help. The set being queried is much less natural than the set used in the previous theorem.

Definition 5.7. $A^{\rm tt}$ is the set of (codes of) Boolean combinations of formulas of the form "v_i" that are true when you interpret v_i to be $i \in A$. For example, the (code of the) formula $(v_{12} \wedge (v_{14} \vee \neg v_9))$ is in $A^{\rm tt}$ iff $12 \in A$ and ($14 \in A$ or $9 \notin A$).

Theorem 5.8. *If B is a c.b. set (see Definition 1.19) and $n \geq 1$, then* $\mathrm{Q}(n, B^{\rm tt}) = \mathrm{Q}(1, B^{\rm tt})$. *In fact,* $\mathrm{Q}(n, B^{\rm tt}) = \mathrm{QC}(1, B^{\rm tt})$.

Proof: Let $n \geq 1$, and let $A \in \mathrm{Q}(n, B^{\rm tt})$. Clearly, $A \in \mathrm{Q}(n, B^{\rm tt}) \Rightarrow A \leq_{\rm T} B$. By Lemma 1.20, $A \leq_{\rm T} B \Rightarrow A \leq_{\rm tt} B$. By the definition of $\leq_{\rm tt}$ and $B^{\rm tt}$, we easily have that $A \leq_m B^{\rm tt}$. Clearly, $A \in \mathrm{Q}(1, B^{\rm tt})$. Note that we actually have $A \in \mathrm{QC}(1, B^{\rm tt})$. ∎

6. Does Allowing Divergence Help?

The algorithm in the introduction showed that $\mathrm{C}^K_{2^n-1} \in \mathrm{FQ}(n, K)$. For that algorithm, incorrect answers could cause divergence. Is there an algorithm where all query paths converge? More formally, can we obtain $\mathrm{C}^K_{2^n-1} \in \mathrm{FQC}(n, K)$? Also, can we obtain $\mathrm{C}^K_{2^n-1} \in \mathrm{SEN}(2^n)$? Combining the first theorem in this section (which was first proved in [BGGO93]) with Theorem 1.11, we find that the answer to both of these questions is NO—in a strong way.

However, the more general question arises as to when does allowing divergence help. We would like to know whether there are sets A such that any computation with A as an oracle can be replaced with one where all query paths converge. To accomplish this, we explore the question of whether, and to what extent, it helps to allow divergence when incorrect answers are given to one or more of the queries.

The results in this section were stated in [Be*etal* 96]. Full proofs appeared in [GM99].

6.1 A Natural Example of a Function Where Allowing Divergence Helps

We address the question that motivated the study of divergence: is there an algorithm for C_n^K which has all paths converging which asks less than n queries?

Theorem 6.1. *For all* $n \geq 1$, $C_n^K \notin \mathrm{SEN}(2^n - 1)$. *(Hence, by Theorem 1.10,* $(\forall X)[C_n^K \notin \mathrm{FQC}(n-1, X)]$.*)*

Proof: We offer two proofs. The first one uses the Recursion Theorem. The second one avoids using the Recursion Theorem. We leave as an open problem the question of which proof is better.

A Proof That Uses the Recursion Theorem

Let $n \geq 1$, and suppose, by way of contradiction, that $C_n^K \in \mathrm{SEN}(2^n - 1)$. Choose a computable function f such that $C_n^K(x_1, \ldots, x_n) \in D_{f(x_1,\ldots,x_n)}$ and $|D_{f(x_1,\ldots,x_n)}| = 2^n - 1$. By an implicit use of the recursion theorem, we construct programs $a_1, \ldots, a_n$ such that $C_n^K(a_1, \ldots, a_n) \notin D_{f(a_1,\ldots,a_n)}$.

Program a_i does the following: Compute $f(a_1, \ldots, a_n)$ and determine the set $D_{f(a_1,\ldots,a_n)}$. Find the vector $b_1 b_2 \cdots b_n \notin D_{f(a_1,\ldots,a_n)}$. Halt iff $b_i = 1$.

Programs $a_1, \ldots, a_n$ conspire to make $C_n^K(a_1, \ldots, a_n) = b_1 b_2 \cdots b_n \notin D_{f(a_1,\ldots,a_n)}$. This is the contradiction.

A Proof That Does Not Use the Recursion Theorem

It is easy to construct a c.e. set A such that, for all n, $C_n^A \notin \mathrm{SEN}(2^n - 1)$. Since K is m-complete, $A \leq_m K$. One can use this to show that if there is an n such that $C_n^K \in \mathrm{SEN}(2^n - 1)$ then, for that n, $C_n^A \in \mathrm{SEN}(2^n - 1)$. Since this is not true, we must have that, for all n, $C_n^K \notin \mathrm{SEN}(2^n - 1)$. ∎

6.2 A Natural Example of a Set Where Allowing Divergence Does Not Help

Let A, B be sets and $n \in \mathsf{N}$. If $A \in \mathrm{Q}(n, B)$, we can decide whether $x \in A$ by making n queries to B; if the wrong answers are supplied, however, the algorithm may diverge. Is there a set B such that whenever $A \in \mathrm{Q}(n, B)$ we also have $A \in \mathrm{QC}(n, B)$ (that is, even with wrong answers, the algorithm does not diverge)?

By Theorem 5.8, there exists such a set, but it is not natural. We show that K, clearly a natural set, has this property.

Theorem 6.2. *For all* $n \in \mathsf{N}$, $\mathrm{Q}(n, K) = \mathrm{QC}(n, K)$.

Proof: Let $A \in \mathrm{Q}(n, K)$ via M^K. We show that $A \in \mathrm{QC}(n, K)$.

Notation 6.3. We are using $M^{()}$ for our oracle Turing machine. We intend to run it with oracle K. To approximate this we will run it for s steps and use oracle K_s. This is denoted $M_s^{K_s}$. Do not confuse this with running Turing machine s. The subscript is the number of steps I am running the machine, not an index of a machine.

We first give an intuition behind the proof. Consider the following scenario: Given x, find $M_s^{K_s}(x)$ for $s = 1, 2, 3, \ldots$ until an s_0 is found such that $M_{s_0}^{K_{s_0}}(x) \downarrow = b \in \{0, 1\}$. (Here, the subscripts s, s_0 refer to the number of steps of the computation, not the index of the oracle machine $M^{()}$.) There is no reason to believe that $b = A(x)$; however, we can ask questions about whether at some *later* time the machine (with a better approximation to K) *changes its mind.* These are questions about *mindchanges.* If we find out that the number of mindchanges is even, then b is the answer. If the number of mindchanges is odd, then $1 - b$ is the answer. Note, however, that we never 'run a machine and see what happens' or carry out any other computation that risks diverging.

We now proceed rigorously.

Definition 6.4. Let $M^{()}$ be an oracle Turing machine. Let $x, s_0 \in \mathrm{N}$ and $b \in \{0, 1\}$. Assume that $M_{s_0}^{K_{s_0}}(x) \downarrow = b$. The phrase "there are at least m mindchanges past stage s_0" means that there exist $s_1 < s_2 < s_3 < \cdots < s_m$ such that $s_0 < s_1$, $M_{s_1}^{K_{s_1}}(x) \downarrow = 1 - b$, $M_{s_2}^{K_{s_2}}(x) \downarrow = b$, $M_{s_3}^{K_{s_3}}(x) \downarrow = 1 - b$, $M_{s_4}^{K_{s_4}}(x) \downarrow = b$, etc., and

$$M_{s_m}^{K_{s_m}}(x) \downarrow = \begin{cases} b & \text{if } m \text{ is even;} \\ 1 - b & \text{if } m \text{ is odd.} \end{cases}$$

Note that the question "Are there at least m mindchanges?" can be phrased as a query to K.

Since $M^K(x)$ makes only n queries, there can be at most $2^n - 1$ mindchanges. The following algorithm shows that $A \in \mathrm{QC}(n, K)$.

(1) Input x.

(2) Find the least s_0 such that $M_{s_0}^{K_{s_0}}(x) \downarrow$, and let $b \in \{0, 1\}$ be the output. Note that such an s_0 exists, since $M^K(x) \downarrow$.

(3) Using binary search, one can determine, in n queries to K, how many mindchanges the $M^K(x)$ computation makes past stage s_0. If this number is even, output b; otherwise, output $1 - b$.

Note that the above algorithm converges even if fed the wrong answers. The algorithm never runs any process that might not halt. ∎

6.3 An Unnatural Example of a Set Where Allowing Divergence Helps A Lot

We show that there exists a set A such that $Q(1, A) - \bigcup_{n=1}^{\infty} QC(n, A) \neq \emptyset$. In other words, there is a set B that you can decide with just one query to A, provided you allow divergence if wrong answers are given; if you insist that convergence occurs even if one or more of the queries are answered incorrectly, however, then no fixed number of queries suffices.

The results in this chapter are due to Frank Stephan. He never published them; however, they appear in [GM99].

The following easy lemma we leave to the reader.

Lemma 6.5. *For all* $A \subseteq \mathrm{N}$ *and* $n \geq 1$, $QC(n, A) \subseteq QC_{||}(2^n - 1, A)$. *Hence* $\bigcup_{n=1}^{\infty} QC_{||}(n, A) = \bigcup_{n=1}^{\infty} QC(n, A)$.

The next lemma restates the problem of getting $B \notin \bigcup_{n=1}^{\infty} QC(n, A)$ in terms of strong enumerability.

Lemma 6.6. *Let* A, B *be sets such that* $(\exists k \geq 1)[C_k^A \in SEN(2^k - 1)]$ *and* $(\forall k \geq 1)[C_k^B \notin SEN(2^k - 1)]$. *Then* $B \notin \bigcup_{n=1}^{\infty} QC(n, A)$. *(The intuition behind the statement of this lemma is that* A *is "easy" and* B *is "hard," so it is reasonable that* B *cannot be reduced to* A *in certain ways.)*

Proof: We show that $B \notin \bigcup_{n=1}^{\infty} QC_{||}(n, A)$. By Lemma 6.5, we obtain $B \notin \bigcup_{n=1}^{\infty} QC(n, A)$.

Suppose that $(\exists n_0 \geq 1)[B \in QC_{||}(n_0, A)]$. By making queries in parallel, we have that $(\forall n \geq 1)[C_n^B \in FQC(1, C_{n_0 n}^A)]$.

Since $(\exists k \geq 1)[C_k^A \in SEN(2^k - 1)]$, $C_{n_0 n}^A \in SEN(O(n^k))$ (by Theorem 2.9). Hence $C_n^B \in SEN(O(n^k))$. For large enough n, this contradicts the hypothesis on B. ∎

The next definition and lemma restate the problem of getting $B \in Q(1, A)$ and $(\forall k \geq 1)[C_k^B \notin SEN(2^k - 1)]$ in terms of fast-growing functions.

Definition 6.7. A function $f : \mathrm{N} \rightarrow \mathrm{N}$ is *computably dominated* if there is a computable g such that $(\forall x)[f(x) < g(x)]$.

Lemma 6.8. *Let* A *be a set such that there exists a function* $f \in FQ(1, A)$ *that is not computably dominated. Then there exists* $B \in Q(1, A)$ *such that* $(\forall k \geq 1)[C_k^B \notin SEN(2^k - 1)]$.

Proof: Choose $f \in FQ(1, A)$ so that f is not computably dominated. We construct a set $B \in Q(1, A)$ such that, for all $e, k \in \mathrm{N}$ with $k \geq 1$, we satisfy requirement

$$R_{\langle e,k \rangle} : \neg(C_k^B \text{ is strongly } (2^k - 1)\text{-enumerable via } \varphi_e).$$

Let $e, k \in \mathsf{N}$ with $k \geq 1$. If φ_e is not total, requirement $R_{\langle e,k\rangle}$ is automatically satisfied. If φ_e is total and we satisfy $R_{\langle e,k\rangle}$, there will be some k-tuple $(x_1, \ldots, x_k)$ of numbers such that

$$|D_{\varphi_e(x_1,\ldots,x_k)}| \geq 2^k \ \vee\ \mathrm{C}_k^B(x_1, \ldots, x_k) \notin D_{\varphi_e(x_1,\ldots,x_k)}.$$

Choose a computable partition $\{Z_{\langle e,i,k\rangle}\}_{e,i,k\in\mathsf{N}}$ of N so that, for all e, i, k, $|Z_{\langle e,i,k\rangle}| = k$. For all e, i, k with $k \geq 1$, let $\vec{z}_{\langle e,i,k\rangle}$ be the k-tuple of numbers that is formed by taking the elements of $Z_{\langle e,i,k\rangle}$ in increasing numerical order. For all e, k with $k \geq 1$, we intend to satisfy $R_{\langle e,k\rangle}$ by constructing B so that if φ_e computes a total function, then there is some i such that

$$|D_{\varphi_e(\vec{z}_{\langle e,i,k\rangle})}| \geq 2^k \ \vee\ \mathrm{C}_k^B(\vec{z}_{\langle e,i,k\rangle}) \notin D_{\varphi_e(\vec{z}_{\langle e,i,k\rangle})}.$$

We construct $B \in \mathrm{Q}(1, A)$ by giving an algorithm for it.

(1) Input x.

(2) Find e, i, k such that $x \in Z_{\langle e,i,k\rangle}$.

(3) Compute $t = f(i)$. (This requires at most one query to A.)

(4) Compute $M_{e,t}(\vec{z}_{\langle e,i,k\rangle})$.

(5) There are two cases.

   (a) $M_{e,t}(\vec{z}_{\langle e,i,k\rangle}) \uparrow$: Output 0. (We have not made progress towards satisfying requirement $R_{\langle e,k\rangle}$.)

   (b) $M_{e,t}(\vec{z}_{\langle e,i,k\rangle}) \downarrow = y$: There are two cases.

   - $|D_y| \geq 2^k$: Output 0. (Note that requirement $R_{\langle e,k\rangle}$ is automatically satisfied.)
   - $|D_y| \leq 2^k - 1$: We want to set $B(x)$, and for that matter $B(z)$ for all $z \in Z_{\langle e,i,k\rangle}$, such that $\mathrm{C}_k^B(\vec{z}_{\langle e,i,k\rangle}) \notin D_y$. For now, we can set only $B(x)$. Find σ, the lexicographically least string in $\{0,1\}^k - D_y$, and let j be such that x is the jth component of $\vec{z}_{\langle e,i,k\rangle}$. Output $\sigma(j)$. (Note that for all $z \in Z_{\langle e,i,k\rangle}$, running this algorithm on input z will get us to this same step and will yield the same σ; hence we will have $\mathrm{C}_k^B(\vec{z}_{\langle e,i,k\rangle}) = \sigma \notin D_y$, and $R_{\langle e,k\rangle}$ will be satisfied.)

Let $e, k \in \mathsf{N}$ such that $k \geq 1$. We show that $R_{\langle e,k\rangle}$ is satisfied. If φ_e is not total, then clearly $R_{\langle e,k\rangle}$ is satisfied. Assume, by way of contradiction, that φ_e is total and $R_{\langle e,k\rangle}$ is not satisfied. We use this to obtain a computable function g that dominates f, in contradiction to our assumption about f.

Since $R_{\langle e,k\rangle}$ is not satisfied, we know that, for every i, the $M_e(\vec{z}_{\langle e,i,k\rangle})$ computation does not halt within $f(i)$ steps. (Otherwise, $R_{\langle e,k\rangle}$ would have been satisfied when the elements of $Z_{\langle e,i,k\rangle}$ were input to the algorithm.) Since φ_e is total, the following computable function dominates f:

$$g(i) = \mu t[M_{e,t}(\vec{z}_{\langle e,i,k\rangle}) \downarrow].$$

Thus $R_{\langle e,k\rangle}$ *is* satisfied. ∎

To obtain our result from Lemmas 6.6 and 6.8, it suffices to have a set A such that

- $(\exists k \geq 1)[\mathrm{C}_k^A \in \mathrm{SEN}(2^k - 1)]$ and
- there exists $f \in \mathrm{Q}(1, A)$ such that f is not computably dominated.

These two properties seem hard to obtain at the same time, since the first one says that A is "easy" while the second one says that A is "hard." Even so, the following lemma allows us to obtain such sets easily.

Lemma 6.9.

(1) If A is selective, then $(\exists k \geq 1)[\mathrm{C}_k^A \in \mathrm{SEN}(2^k - 1)]$. (Actually, $(\forall k \geq 1)[\mathrm{C}_k^A \in \mathrm{SEN}(k + 1)]$.)

(2) If A is a noncomputable c.e. set, then there is a function $f \in \mathrm{FQ}(1, A)$ such that f is not computably dominated. (This is well known.)

(3) There exist noncomputable c.e. sets that are selective. (This is from [Jo68]. The proof I present uses a set defined by Dekker [De54].)

Proof:

1) This follows from Lemma 1.15.

2) Choose a computable enumeration $\{A_s\}_{s\in\mathsf{N}}$ of A, and let f be the function defined by

$$f(x) = \begin{cases} \mu s[x \in A_s], & \text{if } x \in A; \\ 0, & \text{otherwise.} \end{cases}$$

Clearly, $f \in \mathrm{FQ}(1, A)$. Suppose f is computably dominated, and choose a computable g so that $(\forall x)[f(x) < g(x)]$. Then

$$(\forall x)[x \in A \text{ iff } x \in A_{g(x)}].$$

This demonstrates that A is computable, a contradiction!

3) Let C be a noncomputable c.e. set. Choose a computable enumeration $\{C_s\}_{s\in\mathsf{N}}$ of C such that at every stage exactly one new element comes in. Let c_s be the new element that comes in at stage s, and let

$$A = \{s \mid (\exists t > s)[c_t < c_s]\}.$$

Clearly, A is c.e. Using Definition 1.13.2, it is easy to show A selective. ▮

Theorem 6.10. *There exists A such that* $\mathrm{Q}(1, A) - \bigcup_{n=1}^{\infty} \mathrm{QC}(n, A) \neq \emptyset$.

Proof: This follows from Lemmas 6.6, 6.8, and 6.9. ▮

Note 6.11. The following is known. Let A be a set.

(1) There exists $B \equiv_{\mathrm{tt}} A$ such that $(\forall n)[\mathrm{Q}(n, B) = \mathrm{QC}(n, B)]$ iff all $f \leq_{\mathrm{wtt}} A$ are computably dominated.

(2) There exists $B \equiv_{\mathrm{tt}} A$ such that $\mathrm{Q}(1, B) - \bigcup_{i=1}^{\infty} \mathrm{QC}(n, B) \neq \emptyset$ iff there exists $f \leq_{\mathrm{wtt}} A$ such that f is not computably dominated.

7. Does Order Matter?

Let $A, B \subseteq \mathsf{N}$. I am allowed to make one query to A *and* one query to B in some computation. Does the order in which I make the queries matter? The first theorem in this section is due to Beigel (unpublished) and has been generalized by McNicholl [McN00]. The second theorem is due to McNicholl [McN00]; however, the proof given here is due to Frank Stephan and has not been published previously. Questions of this type were asked in complexity theory [HHW98] before they were asked in computability theory.

Definition 7.1. Let $A, B \subseteq \mathsf{N}$. $\mathrm{QO}(A, B)$ is the set of sets that I can decide by an algorithm that makes one query to A and then one query to B. (The 'O' stands for 'order.') $\mathrm{Q}_{||}(A, B)$ is the set of sets that I can decide by an algorithm that makes one query to A and one query to B at the same time. (This differs from the use of $\mathrm{Q}_{||}(n, A)$ used earlier in this paper.)

Definition 7.2. If $\mathrm{QO}(A, B) = \mathrm{QO}(B, A)$, then A and B *commute.*

Notation 7.3. We denote an oracle Turing machine that is going to query two oracles, with one query each, by $M^{()()}$. We fill in the first () with the oracle it queries first, and the second () with the oracle it queries second. If $b_1 b_2 \in \{0, 1\}^2$ then $M^{b_1 b_2}(x)$ denotes what happens if you assume the answer to the first question is b_1 and the answer to the second question (if such a question exists) is b_2. (Note that you do not actually query either oracle.) If $B \subseteq \mathsf{N}$, then (1) $M^{(b_1)(B)}(x)$ denotes what happens if you assume the answer to the first question is b_1 but you get the true answer to the second question (if such a question exists) by querying B, and (2) $QUERY(M^{(b_1)(B)}(x))$ denotes the query to B that is encountered in the $M^{(b_1)(B)}(x)$ computation, if it exists (if it does not exist, then $QUERY(M^{(b_1)(B)}(x))$ is undefined). If we use this notation, we are implicitly assuming that the query to B exists.

We leave the proof of the following easy lemma to the reader.

Lemma 7.4. *Let $i, j, q, x \in \mathrm{N}$, $b_1, b_2, b \in \{0,1\}$, and $M^{()()}$ be an oracle Turing machine. Assume $i < j$. Then the following hold.*

(1) If $i \geq 1$, then the truth value of the statement

$$(q \in \emptyset^{(i)}) \wedge (M^{b_1 b_2}(x) \downarrow = b)$$

can be determined by a query to $\emptyset^{(i)}$.

(2) If $i \geq 2$, then the truth value of the statement

$$(q \notin \emptyset^{(i)}) \wedge (M^{b_1 b_2}(x) \downarrow = b)$$

can be determined by a query to $\emptyset^{(i)}$.

(3) The truth value of the statement

$$(q \in \emptyset^{(j)}) \wedge (M^{(b_1)(\emptyset^{(i)})}(x) \downarrow = b)$$

can be determined by a query to $\emptyset^{(j)}$.

(4) The truth value of the statement

$$(q \notin \emptyset^{(j)}) \wedge (M^{(b_1)(\emptyset^{(i)})}(x) \downarrow = b)$$

can be determined by a query to $\emptyset^{(j)}$.

(5) The truth value of the statement

$$[(q \in \emptyset^{(i)}) \wedge (QUERY(M^{(b_1)(\emptyset^{(j)})}(x)) \in \emptyset^{(j)})]$$
$$\vee[(q \notin \emptyset^{(i)}) \wedge (QUERY(M^{(b_2)(\emptyset^{(j)})}(x)) \in \emptyset^{(j)})]$$

can be determined by a query to $\emptyset^{(j)}$.

Theorem 7.5. *For all $i, j \geq 1$, $\emptyset^{(i)}$ and $\emptyset^{(j)}$ commute. In fact,*

$$\mathrm{QO}(\emptyset^{(i)}, \emptyset^{(j)}) = \mathrm{QO}(\emptyset^{(j)}, \emptyset^{(i)}) = \mathrm{Q}_{||}(\emptyset^{(j)}, \emptyset^{(i)}).$$

Proof: Assume $i < j$. Let $C \in \mathrm{QO}(\emptyset^{(i)}, \emptyset^{(j)})$ via $M^{\emptyset^{(i)},\emptyset^{(j)}}$. We show that $C \in \mathrm{Q}_{||}(\emptyset^{(j)}, \emptyset^{(i)})$. The intuition is that we can ask the question "what is the answer to the second question going to be" and the question "what is the answer to the first question" at the same time.

(1) Input x.

(2) Run the $M^{\emptyset^{(i)},\emptyset^{(j)}}(x)$ computation until the first query is encountered. Call this query q (we do not make this query).

(3) Consider the following statement:

$$\begin{array}{l}[(q \in \emptyset^{(i)}) \wedge (QUERY(M^{(1)(\emptyset^{(j)})}(x)) \in \emptyset^{(j)})] \\ \vee[(q \notin \emptyset^{(i)}) \wedge (QUERY(M^{(0)(\emptyset^{(j)})}(x)) \in \emptyset^{(j)})]\end{array}$$

By Lemma 7.4.5, this can be phrased as a query z to $\emptyset^{(j)}$. note that $z \in \emptyset^{(j)}$ iff the second query of the $M^{\emptyset^{(i)},\emptyset^{(j)}}$ computation has the answer YES.

(4) Ask "$z \in \emptyset^{(j)}$?" and "$q \in \emptyset^{(i)}$?" at the same time. This will give you all the information you need to simulate the computation of $M^{\emptyset^{(i)},\emptyset^{(j)}}(x)$.

Now let $C \in \mathrm{QO}(\emptyset^{(j)}, \emptyset^{(i)})$ via $M^{\emptyset^{(j)},\emptyset^{(i)}}$. We show that $C \in \mathrm{QO}(\emptyset^{(i)}, \emptyset^{(j)})$.

We first prove this for the case where $i \geq 2$. This is needed, since Lemma 7.4.2 does not hold when $i = 1$. We will prove the $i = 1$ case later.

The intuition is that we first find an approximation to the answer by seeing which query path converges first, and then ask about mindchanges. We first ask if there is a mindchange because of the second query, and then we ask if there is a mindchange because of the first query.

(1) Input x.

(2) Run $M^{()()}(x)$ along all query paths until one of them halts. (At least one must halt, since the correct answers yield a halting path.) Let b be the output on the halting path, let q_1 be the first query encountered (the query to $\emptyset^{(j)}$), and let q_2 be the second query encountered (a query to $\emptyset^{(i)}$). We will be asking questions about whether the computation wants to change its mind about b. There are four cases, corresponding to the four possible pairs of answers supplied to the queries.

(3) (a) Case 1: The answers 0,0 yield a halting path. By Lemma 7.4.1, we can determine the truth value of the following statement by making a query to $\emptyset^{(i)}$:

$$(q_2 \in \emptyset^{(i)}) \wedge (M^{01}(x) \downarrow = 1 - b).$$

Make the query. If the answer is YES, let $c = 1 - b$; otherwise, let $c = b$. If $q_1 \notin \emptyset^{(j)}$ (do not ask this), then the final correct answer is c. By Lemma 7.4.3, we can determine the truth value of the following statement by making a query to $\emptyset^{(j)}$:

$$(q_1 \in \emptyset^{(j)}) \wedge (M^{(1)(\emptyset^{(i)})}(x) \downarrow = 1 - c).$$

Make the query. If the answer is YES, output $1 - c$; otherwise, output c.

(b) Case 2: The answers 0,1 yield a halting path. By Lemma 7.4.2 (and $i \geq 2$), we can determine the truth value of the following statement by making a query to $\emptyset^{(i)}$:

$$(q_2 \notin \emptyset^{(i)}) \wedge (M^{00}(x) \downarrow = 1 - b).$$

Make the query. If the answer is YES, let $c = 1 - b$; otherwise, let $c = b$. If $q_1 \notin \emptyset^{(j)}$ (do not ask this), then the final correct answer is c. The rest of this case is identical to Case 1.

(c) Case 3: The answers 1,0 yield a halting path. By Lemma 7.4.1, we can determine the truth value of the following statement by making a query to $\emptyset^{(i)}$:

$$(q_2 \in \emptyset^{(i)}) \wedge (M^{11}(x) \downarrow = 1 - b).$$

Make the query. If the answer is YES, let $c = 1 - b$; otherwise, let $c = b$. If $q_1 \in \emptyset^{(j)}$ (do not ask this), then the final correct answer is c. By Lemma 7.4.4, we can determine the truth value of the following statement by making a query to $\emptyset^{(j)}$:

$$(q_1 \notin \emptyset^{(j)}) \wedge (M^{(0)(\emptyset^{(i)})}(x) \downarrow = 1 - c).$$

Make the query. If the answer is YES, output $1 - c$; otherwise, output c.

(d) Case 4: The answers 1,1 yield a halting path. By Lemma 7.4.2 (and $i \geq 2$), we can determine the truth value of the following statement by making a query to $\emptyset^{(i)}$:

$$(q_2 \notin \emptyset^{(i)}) \wedge (M^{10}(x) \downarrow = 1 - b).$$

Make the query. If the answer is YES, let $c = 1 - b$; otherwise, let $c = b$. If $q_1 \in \emptyset^{(j)}$ (do not ask this), then the final correct answer is c. The rest of the proof is identical to Case 3.

We now look at the case where $i = 1$. We need to show that if $C \in \mathrm{QO}(\emptyset^{(j)}, K)$, then $C \in \mathrm{QO}(K, \emptyset^{(j)})$. Assume $C \in \mathrm{QO}(\emptyset^{(j)}, K)$ via $M^{()()}$. We show that $C \in \mathrm{QO}(\emptyset^{(i)}, \emptyset^{(j)})$.

(1) Input x.

(2) Run all query paths of the $M^{()()}(x)$ machine. If a second query is encountered (a query to K), then before pursuing the YES path, enumerate K and wait for the element to enter. (If it never enters K, there is no point in pursuing the YES path.) Wait until one of the four query paths halts—with the caveat about the existence of a second query and pursuit of the YES path for that query. (At least one must halt, since the correct answers yield a halting path.) Let b be the answer on the halting path, let q_1 be the first query encountered (the query to $\emptyset^{(j)}$), and let q_2 be the second query encountered (a query to K). We will be asking questions about whether the computation wants to change its mind about b. Note that if there is a mindchange because of the first query, then q_2 may change (that is, the actual query made to K may change). If it does, the answer to the new q_2 may differ from the one supplied for q_2 on the original halting query path (even if that answer was verified by enumeration of K). There are four cases, corresponding to the four possible pairs of answers supplied to the queries.

(3) (a) Case 1: The answers 0,0 yield a halting path. This is identical to Case 1 in the previous algorithm.

(b) Case 2: The answers 0,1 yield a halting path. The query to K was answered correctly. Note that if $q_1 \notin \emptyset^{(j)}$ (do not ask this), then the final correct answer is b. The rest of this case is identical to Case 1.

(c) Case 3: The answers 1,0 yield a halting path. This is identical to Case 3 in the previous algorithm.

(d) Case 4: The answers 1,1 yield a halting path. The query to K was answered correctly. Note that if $q_1 \in \emptyset^{(j)}$ (do not ask this), then the final correct answer is b. The rest of the proof is identical to Case 4 in the previous algorithm.

∎

We show that, for all i, $\emptyset^{(i)}$ and $\emptyset^{(\omega)}$ do not commute. We first need a lemma of interest in its own right.

Lemma 7.6. *For all i,* $\mathrm{QO}(\emptyset^{(\omega)}, \emptyset^{(i)}) \subseteq \mathrm{Q}(1, \emptyset^{\omega})$.

Proof: Let $A \in \mathrm{QO}(\emptyset^{(\omega)}, \emptyset^{(i)})$ via $M^{\emptyset^{\omega}, \emptyset^{(i)}}$. The following algorithm shows $A \in \mathrm{Q}(1, \emptyset^{\omega})$. The intuition is that once we know the first query we can ask a complex query about answering the first one and the rest of the computation.

(1) Input x

(2) Run $M^{()()}(x)$ until a query $q \in \emptyset^{\omega}$ is encountered. Do not ask this query.

(3) Find z, k such that the query is actually of the form $z \in \emptyset^{(k)}$.

(4) Phrase the query

$$[(z \in \emptyset^{(k)}) \wedge (M^{(1)(\emptyset^{(i)})}(x)) \downarrow = 1]$$
$$\vee[(z \notin \emptyset^{(k)}) \wedge (M^{(0)(\emptyset^{(i)})}(x)) \downarrow = 1]$$

as a query y to $\emptyset^{\omega}$.

(5) Ask $y \in \emptyset^{\omega}$. If $y \in \emptyset^{\omega}$ then output YES, else output NO.

The lemma follows. ▮

Theorem 7.7. *K and $\emptyset^{(\omega)}$ do not commute.*

Proof: Let C be the set from Theorem 5.5. Clearly, $C \in \mathrm{QO}(\emptyset^{(i)}, \emptyset^{(\omega)})$. We show that $C \notin \mathrm{QO}(\emptyset^{(\omega)}, \emptyset^{(i)})$. Assume, by way of contradiction, that $C \in \mathrm{QO}(\emptyset^{(\omega)}, \emptyset^{(i)})$. By Lemma 7.6 we have $C \in \mathrm{Q}(1, \emptyset^{(\omega)})$. By Theorem 5.5 $C \notin \mathrm{Q}(1, \emptyset^{(\omega)})$. Hence we have our contradiction. ▮

8. Acknowledgments

I would like to thank Richard Beigel, Fawzi Emad, Lance Fortnow, Omer Horovitz, Georgia Martin, Jim Owings, and Frank Stephan for proofreading and commentary. I particularly want to thank Georgia Martin for her meticulous proofreading and Frank Stephan for some new proofs. In addition I would like to thank the referee, who has likely already been thanked, for helpful suggestions.

References

[Be87] Beigel, Richard (1987). ***Query-Limited Reducibilities.*** PhD thesis, Stanford University. Also available as Report No. STAN-CS-88–1221.

[Be88] Beigel, Richard (1988). When are $k+1$ queries better than k? Technical Report 88-06, The Johns Hopkins University, Dept. of Computer Science.

[BGGO93] Beigel, Richard, Gasarch, William, Gill, John, and Owings, James (1993). Terse, superterse, and verbose sets. *Information and Computation*, 103(1):68–85.

[Be*etal* 00] Beigel, Richard, Gasarch, William, Kummer, Martin, Martin, Georgia, McNicholl, Timothy, and Stephan, Frank (2000). The complexity of ODD_n^A. *Journal of Symbolic Logic*, pages 1–18.

[Be*etal* 96] Beigel, Richard, Gasarch, William, Kummer, Martin, Martin, Georgia, McNicholl, Timothy, and Stephan, Frank (1996). On the query complexity of sets. In *21st International Symposium on Mathematical Foundations of Computer Science (MFCS '96), Cracow, Poland.*

[BEHW89] Blumer, A., Ehrenfeucht, A., Haussler, D., and Warmuth, M. (1989). Learnability and the Vapnik-Chervonenkis dimension. *Journal of the ACM*, 36:929–965.

[CH89] Cai, Jin-yi and Hemachandra, Lane A. (1989). Enumerative counting is hard. *Information and Computation.* Earlier version in Structures 1988.

[COS75] Clarke, Steven, Owings, Jim, and Spriggs, James (1975). Trees with full subtrees. In *Proc. of the 6th Southeastern Conference on Combinatorics, Graph Theory, and Computing*, pages 169–172.

[De54] Dekker, J. C. E. (1954). A theorem on hypersimple sets. *Proceedings of the AMS*, 5:791–796.

[Ga85] Gasarch, William (1985). A hierarchy of functions with applications to recursive graph theory. Technical Report 1651, University of Maryland, Dept. of Computer Science.

[GM99] Gasarch, William and Martin, Georgia (1999). *Bounded Queries in Recursion Theory.* Progress in Computer Science and Applied Logic. Birkhäuser, Boston.

[HW87] Haussler, D. and Welzl, E. (1987). ϵ-nets and simplex range queries. *Discrete Computational Geometry*, 2:127–151.

[HHW98] Hemaspaandra, Hempel, and Wechsung (1998). Query order. *sicomp*, 28.

[Jo68] Jockusch, Carl (1968). Semirecursive sets and positive reducibility. *Transactions of the AMS*, 131:420–436.

[Jo89] Jockusch, Carl (1989). Degrees of functions with no fixed points. In Fenstad, J.E., Frolov, I., and Hilpinen, R., editors, *Logic, Methodology, and Philosophy of Science VIII*, pages 191–201. North Holland.

[JS72] Jockusch, Carl and Soare, Robert (1972). Π_1^0 classes and degrees of theories. *Transactions of the AMS*, 173:33–56.

[Ku92] Kummer, Martin (1992). A proof of Beigel's cardinality conjecture. *Journal of Symbolic Logic*, 57(2):677–681.

[KS94] Kummer, Martin and Stephan, Frank (1994). Effective search problems. *Mathematical Logic Quarterly*, 40:224–236.

[McN00] McNichol, Tim (2000). On the commutativity of jumps. *Journal of Symbolic Logic*, 65(4).

[MM68] Miller, Webb and Martin, Donald A. (1968). The degree of hyperimmune sets. *Zeitsch. f. math. Logik und Grundlagen d. Math.*, 14:159–166.

[Od89] Odifreddi, Piergiorgio (1989). *Classical Recursion Theory (Volume I).* North-Holland, Amsterdam.

[Ow89] Owings, Jr., James C. (1989). A cardinality version of Beigel's Nonspeedup Theorem. *Journal of Symbolic Logic*, 54(3):761–767.

[Sa72] Sauer, N. (1972). On the density of families of sets. *Journal of Combinatorial Theory (series A)*, 13:145–147.

[Sc62] Scott, Dana (1962). Algebras of sets binumerable in complete extension of arithmetic. In *Proceedings Symposium Pure and Applied Mathematics 5*, pages 117–121.

[Sh72] Shelah, S. (1972). A combinatorial problem: stability and order for models and theories of infinitary languages. *Pacific Journal of Mathematics*, 41:247–261.

[So87] Soare, Robert (1987). *Recursively Enumerable Sets and Degrees.* Perspectives in Mathematical Logic. Springer-Verlag, Berlin.

[So96] Soare, Robert (1996). Computability and recursion. *Bulletin of Symbolic Logic*, 27.

[VC71] Vapnik, V. N. and Chervonenkis, A. Y. (1971). On the uniform convergence of relative frequencies of events to their probabilities. *Theory of Probab. and its Applications*, 16(2):264–280.

FINITE END INTERVALS IN DEFINABLE QUOTIENTS OF $\mathcal{E}$

Eberhard Herrmann
Department of Mathematics
Humboldt-University at Berlin
10099 Berlin
Germany
herrmann@mathematik.hu-berlin.de

Abstract It will be shown that in a factor lattice of the lattice of c.e. sets under inclusion with respect to a special congruence relation, there exists a finite end interval which is not a Boolean algebra.

Keywords: lattice of computable enumerable sets, congruence relation, factor lattice, end intervals

Introduction

Let $\mathcal{E}$ be the lattice of computably enumerable (c.e.) sets under inclusion. For A a c.e. set, we denote by $\mathcal{L}(A)$ the family $\{B \text{ c.e.}: A \subseteq B\}$, forming trivially a sublattice of $\mathcal{E}$ under the inclusion relation. Let $\mathcal{L}^*(A)$ be $\mathcal{L}(A)$ factorized according to finite differences of sets.

One question which arose from attempts to characterize $\mathcal{E}$ was that of which finite lattices take the form $\mathcal{L}^*(A)$ for some c.e. set A.

A partial answer followed from Friedberg's proof of the existence of maximal sets. Namely, if A is chosen suitably (i.e. is equal to a finite, nonempty intersection of some number of maximal sets) then $\mathcal{L}^*(A)$ is isomorphic to a given finite Boolean algebra.

The converse was provided by Lachlan in [5]. He showed that if $\mathcal{L}^*(A)$ is finite then it is a Boolean algebra.

We will formulate the result for $\mathcal{L}^*(A)$ a little differently, but in equivalent form. Denote by $=^*$ the equality relation (between two sets) modulo finite differences, with $\mathcal{E}_{/=^*}$ the factor lattice of $\mathcal{E}$ by $=^*$ (usually denoted by $\mathcal{E}^*$) and with $[X]_{/=^*}$ the equivalence class with respect to $=^*$ including the set X (for which usually we write X^*). Let $\mathcal{E}_{/=^*}([X]_{/=^*})$ be

$$\{[B]_{/=^*} \in \mathcal{E}_{/=^*} : [A]_{/=^*} \leq [B]_{/=^*}\}.$$

Then the family

$$\{\mathcal{E}_{/=^*}([A]_{/=^*}) : A \text{ c.e. and coinfinite and } \mathcal{E}_{/=^*}([A]_{/=^*}) \text{ is finite}\} \qquad (1)$$

(which is $\{\mathcal{L}^*(A) : A - q\text{-maximal}\}$) is exactly the class of all finite Boolean algebras.

The motivation in writing the result for $\mathcal{L}^*(A)$ in this way is the following: $=^*$ belongs to a class of (binary) relations on $\mathcal{E}$ which are called congruence relations and which allow one to factorize $\mathcal{E}$ by them. The question considered in this paper is: If instead of $=^*$ we take another congruence relation in $\mathcal{E}$, say $\approx$, then which finite lattices of the form (1) occur when $=^*$ is replaced by $\approx$?

Taking the well-known congruence relations on $\mathcal{E}$, in each case we get, quite easily, either the class of all finite Boolean algebras, as before, or none. (By 'none' we mean apart from the trivial one, i.e. $\mathcal{E}_{/\approx}([\mathbb{N}]_{/\approx})$, $\mathbb{N}$ the set of all numbers, which has one element and always occurs.)
Does this hold for *every* congruence relation on $\mathcal{E}$?

The main result of this paper shows that this is not so. It will be shown that the union of two well-known congruence relations on $\mathcal{E}$ (one is defined via the q-maximal sets and the other one by the major subset relation) is a congruence relation for which (1) (with this union in place of $=^*$) properly extends the class of all finite Boolean algebras (and gives moreover the greatest possible class in which a member of a factor lattice of $\mathcal{E}$ can appear).

The rest of the paper is divided into four sections.

In the first section we give some well-known examples of congruence relations on $\mathcal{E}$ and analyse the corresponding classes (1) defined using them. At the end of this section we give an upper estimation of the class (1) for $a(n)$ an (arbitrary) congruence relation on $\mathcal{E}$ (with this in place of $=^*$), which can be done quite easily. In the second section we define and characterize the congruence relation in which we are particularly interested (i.e. the union mentioned above). The third section includes basic constructions needed later in the main theorem. It seems reasonable from the methodological point of view to give these first separately. In the last section we prove the main theorem.

The paper is a further development of the investigations in [4], in particular of Theorem 2.4, which Theorem 10 of this paper (our main theorem) essentially generalizes. But an understanding of this paper does not depend on a knowledge of [4].

Notation and symbols used throughout the paper are:

$\mathcal{E}$, c.e., $\mathcal{E}^*$, $=^*$, X^* and $\mathbb{N}$ — these having already been defined. And $\subseteq^*$ meaning $\subseteq$ (modulo $=^*$).
$2^{<\omega}$ is the set of all finite 01-sequences. $\langle\,\rangle$ is the symbol for the empty sequence, while small greek letters are used as variables over $2^{<\omega}$ (and exclusively so). For $\sigma \neq \langle\,\rangle$ we have $\sigma = (\sigma_0\sigma_1\ldots\sigma_{k-1})$ for some $k \geq 1$ (and $\sigma_i = 0$ or 1). $|\sigma|$ is the length of σ ($|\sigma| = k$ in the above example).
$<_l$ denotes the lexicographical order in $2^{<\omega}$ (i.e. $\tau <_l \sigma$ if $|\tau| < |\sigma|$ or $|\tau| = |\sigma|$ and $(\exists i)(\tau_i = 0,\ \sigma_i = 1,\ \forall j < i\,(\tau_j = \sigma_j))$). $\tau \leq_i \sigma$ means that τ is an initial part of σ.
An infinite branch is an infinite subset of $2^{<\omega}$ linearly ordered by $\leq_i$. There is an obvious $1-1$ correspondence between infinite branches and functions from $\mathbb{N}$ into $\{0,1\}$ (called *state-functions*). If f is a state-function and n a number then $f|n$ means the element $(f(0),\ldots,f(n-1))$ of $2^{<\omega}$, for $n \geq 1$ and $\langle\,\rangle$ for $n = 0$.
Given c.e. sets A and B we write $A \subset_m B$ if A is a major subset of B. If $\mathcal{A}$ and $\mathcal{B}$ are isomorphic structures we write $\mathcal{A} \cong \mathcal{B}$.
For further notation and symbols used in the paper see e.g. [So87].

1. Examples of Congruence Relations on $\mathcal{E}$, Finite End Intervals

Let $\mathcal{L}$ be a (countable, distributive) lattice. A binary relation R on $\mathcal{L}$ is called a *congruence relation on* $\mathcal{L}$ if R is an equivalence relation on $\mathcal{L}$ and

$$(\forall x_1, x_2, y_1, y_2 \in \mathcal{L})(R(x_1, y_1) \wedge R(x_2, y_2) \Rightarrow R(x_1 \wedge y_1, x_2 \wedge y_2) \wedge R(x_1 \vee y_1, x_2 \vee y_2)). \tag{2}$$

We use $\approx$ (sometimes with indices) as a variable over the congruence relations (and exclusively so).

Given $\approx$ on $\mathcal{L}$ we can form the factor lattice of $\mathcal{L}$ with respect to $\approx$, for which we write $\mathcal{L}_{/\approx}$, denoting by $[x]_{/\approx}$ the element of $\mathcal{L}_{/\approx}$ including a given $x \in \mathcal{L}$.

We consider the following examples of congruence relations on $\mathcal{E}$ (these being very far from exhaustive, but including among them the well-known ones, as well as those in which we are primarily interested).

Given c.e. sets X and Y, we have:

1.	$X = Y$	if X is equal to Y.
2.	$X =^* Y$	if $(X - Y) \cup (Y - X)$ is finite.
3.	$X \approx_{\mathrm{ms}} Y$	if $X \cap Y \subset_{\mathrm{m}} X \cup Y$ or $X \cap Y$ is cofinite in $X \cup Y$.
4.	$X \approx_{\mathcal{QM}} Y$	if $X \cap Y$ is q-maximal in $X \cup Y$ ($X \cap Y \subset_{\mathcal{QM}} X \cup Y$) or is cofinite in $X \cup Y$.
5.	$X \approx_{\mathcal{HH}} Y$	if $X \cap Y$ is hh-simple in $X \cup Y$ or is cofinite in $X \cup Y$.
6.	$X \approx_{\mathcal{S}} Y$	if $(\exists\, S$ simple$)(X \cap S = Y \cap S)$.
7.	$X \approx_i Y$	if $(X - Y) \cup (Y - X)$ is immune or finite.
8.	$X \approx_I Y$	for all sets (from $\mathcal{E}$) (i.e. $X = X \wedge Y = Y$).

We have the following inclusions (symbolized with an arrow) between the congruence relations 1. to 8.:

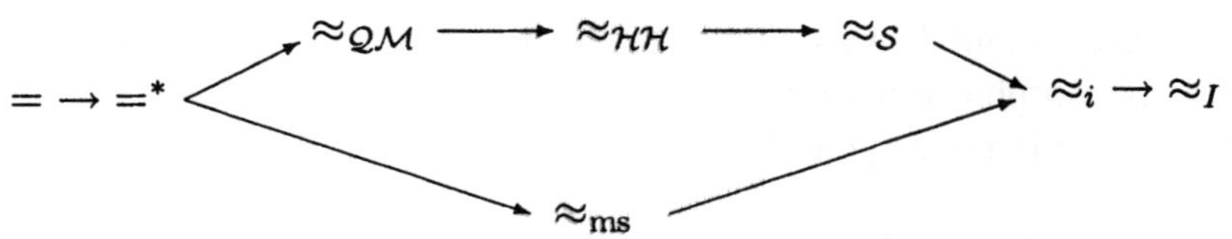

The above schema is complete, i.e. no further arrow can be drawn and no two of the above congruence relations coincide. That not $\approx_{\mathrm{ms}} \rightarrow \approx_{\mathcal{S}}$, and $\approx_{\mathcal{S}} \neq \approx_i$, follows easily from the fact that if A is a small major subset of B and B is such that

$$\neg(\exists\, C \text{ c.e.})(B \cap C = \emptyset \wedge B \cup C \text{ is simple})$$

then not $A \approx_{\mathcal{S}} B$, but $A \approx_{\mathrm{ms}} B$.

Let $\mathcal{L}$ be a lattice and x and y be in $\mathcal{L}$ with $x \leq y$. An *interval* is a sublattice of the form

$$\{z \in \mathcal{L} : z \leq x\} \qquad \text{or} \tag{3}$$

$$\{z \in \mathcal{L} : x \leq z \leq y\} \qquad \text{or}$$

$$\{z \in \mathcal{L} : x \leq z\}. \tag{4}$$

The sublattices (3) are called *initial intervals* and those in (4) *end intervals*. We write $\mathcal{L} \wedge x$, $\mathcal{L}(x, y)$ and $\mathcal{L}(x)$ respectively, for these. (It had previously become common to write $\mathcal{L}(X)$ for $\mathcal{E}(X)$, for X c.e. – see the introduction – which can lead to misunderstanding. But in this paper $\mathcal{L}(X)$, with a big X, is used only in the context of $\mathcal{E}(X)$). Thus we write

$$\mathcal{L}_{/\approx}([x]_{/\approx}) \tag{5}$$

for the end interval in $\mathcal{L}_{/\approx}$ defined by $[x]_{/\approx}$, $x \in \mathcal{L}$.

Lemma 1. *The class of all finite end intervals in the factor lattices formed by the congruence relations* 1. *to* 4. *respectively is always the class of all finite Boolean algebras, and for the relations* 5. *to* 8. *this class is empty.*

Proof. For 1. we get this at once, since all cofinite sets are c.e.
For 2. this is exactly the well-known fact given in the introduction.
For 3. we get at least all finite Boolean algebras, since $[X]_{/\approx_{\mathrm{ms}}} = X^*$ for all hh-simple sets, hence in particularly for the q-maximal sets. That there are not more follows from the fact that

$$(X \cup R)^* \in \mathcal{L}_r^*(X) \Longrightarrow [X \cup R]_{/\approx_{\mathrm{ms}}} \in \mathcal{E}_{/\approx_{\mathrm{ms}}}([X]_{/\approx_{\mathrm{ms}}})$$

($\mathcal{L}_r^*(X) =_{\mathrm{df}} \{(X \cup R)^* : R$ a computable set$\}$) is an embedding.

Hence the finiteness of $\mathcal{L}_r^*(X)$ is necessary for the finiteness of $\mathcal{E}_{/\approx_{\mathrm{ms}}}([X]_{/\approx_{\mathrm{ms}}})$. $\mathcal{L}_r^*(X)$ is finite means that X is a finite intersection of r-maximal sets (or X is cofinite). But for these X $\mathcal{E}_{/\approx_{\mathrm{ms}}}([X]_{/\approx_{\mathrm{ms}}})$ is isomorphic to $\mathcal{L}_r^*(X)$, hence is a finite Boolean algebra.
To show it for 4., i.e. for $\approx_{\mathcal{QM}}$, we first let X be a coinfinite c.e. set, which is not hh-simple. Then there is a $Y \in \mathcal{L}(X)$ with $Y - X$ splittable into $Y_1 - X, \ldots, Y_n - X$, with all $Y_i - X$ not co-c.e. Thus the sets $Y_i - X$ are not q-cohesive, which gives that $\mathcal{E}([X]_{/\approx_{\mathcal{QM}}})$ is not finite. Denote by $\oplus 2$ the Boolean algebra formed by all finite and cofinite subsets of $\mathbb{N}$. If A is a finite intersection of hh-simple sets B with $\mathcal{L}^*(B) \cong \oplus 2$ then $\mathcal{E}_{/\approx_{\mathcal{QM}}}([A]_{/\approx_{\mathcal{QM}}})$ is a finite Boolean algebra. If A is hh-simple, not of the above form and not q-maximal, then the corresponding end interval formed by A is infinite. Hence for $\approx_{\mathcal{QM}}$ we get exactly the class of all finite Boolean algebras.
For $\approx_{\mathcal{HH}}$ the same argumentation as for $\approx_{\mathcal{QM}}$ gives that the equivalence classes of the coinfinite, not hh-simple sets define infinite end intervals in $\mathcal{E}_{/\approx_{\mathcal{HH}}}$. Thus in $\mathcal{E}_{/\approx_{\mathcal{HH}}}$ there are no finite end intervals.
That $\mathcal{E}_{/\mathcal{S}}$ has no finite end intervals follows easily from the fact that if X is a coinfinite, not simple c.e. set then there is an infinite c.e. set Y with $X \cap Y = \emptyset$. Y can be decomposed into infinitely many c.e. sets Y_i, $i \geq 0$. But $X \cup Y_i$ are pairwise not $\approx_{\mathcal{S}}$-equivalent. This means that $\mathcal{E}_{/\approx_{\mathcal{S}}}([X]_{/\approx_{\mathcal{S}}})$ is infinite.
The proof that $\mathcal{E}_{/\approx_i}$ has no finite end intervals is the same as that for $\mathcal{E}_{/\approx_{\mathcal{S}}}$.
$\mathcal{E}_{/\approx_I}$ has no finite end intervals by definition.

In the following we give an upper estimation of the class of possible finite end intervals in an (arbitrary) factor lattice $a(n)$ of $\mathcal{E}$ with respect to a congruence relation.

In doing this the following two lattice-theoretic properties are important: These are the distributivity, and the so-called Reduction Principle which is defined as follows. Let $\mathcal{L}$ be a lattice with 0 (i.e. with a smallest element). We say that in $\mathcal{L}$ the *Reduction Principle* (symbolically: RP) holds if

$$(\forall x \in \mathcal{L})(\forall y \in \mathcal{L})(\exists x' \in \mathcal{L})(\exists y' \in \mathcal{L})(x' \leq x,\ y' \leq y,\qquad (6)$$
$$x' \wedge y' = \mathbf{0},\ x' \vee y' = x \vee y).$$

x' and y' are called *reduction sets* for x and y respectively.

Lemma 2. *Let $\mathcal{L}$ be a distributive lattice with 0 in which RP holds and $\approx$ a congruence relation in $\mathcal{L}$. Then the class of all finite end intervals in $\mathcal{L}_{/\approx}$ is included in the class of all finite distributive lattices in which RP holds.*

Proof. Both properties, the distributivity and RP, are positive sentences, i.e. they do not include the negation or the implication. Thus they are also satisfied in the factor lattice $\mathcal{L}_{/\approx}$, since $\mathcal{L}_{/\approx}$ is a homomorphic image of $\mathcal{L}$. The distributivity is a universal sentence, hence holds also in every sublattice. Thus the end intervals in $\mathcal{L}_{/\approx}$ are distributive.
RP is an $\forall\exists$-sentence, hence must not hold in every sublattice (of a lattice in which RP holds). But in sublattices which are intervals RP also is satisfied, as can easily be seen. Given $x, y, z \in \mathcal{L}_{/\approx}$ with $x \leq y$, $x \leq z$. If y' and z' are reduction sets for y and z resp. in $\mathcal{L}_{/\approx}$ then $x \vee y'$ and $x \vee z'$ are reduction sets for y and z resp. in $\mathcal{L}_{/\approx}(x)$.

We see that the examples 1. to 8. given before satisfy this upper estimation. The class of all finite Boolean algebras as also the one-element lattice have these properties.

Since the class of all finite distributive lattices in which RP holds is of course properly greater than the class of all finite Boolean algebras, it will be interesting to analyse whether there is a congruence relation in $\mathcal{E}$ such that the class of finite end intervals in the factor lattice defined by means of this relation is properly greater than the class of all finite Boolean algebras.[1]

Such a congruence relation exists and will be defined in the next section.

2. The Congruence Relation $\approx_{\mathrm{ms}} \vee \approx_{\mathcal{QM}}$

Let $\mathcal{L}$ be a (countable and distributive) lattice. Denote by CR($\mathcal{L}$) the family of all congruence relations in $\mathcal{L}$. In CR($\mathcal{L}$) we have the partial ordering $\leq$

[1] That every class of finite end intervals in any factor lattice of $\mathcal{E}$, if not empty, includes all finite Boolean algebras can be easily shown, at least for all congruence relations in $\mathcal{E}$ which are invariant with respect to computable permutations, and thus much more for lattice-theoretic ones. Let $\approx$ be a congruence relation in $\mathcal{E}$ with $X \approx Y \Rightarrow p(X) \approx p(Y)$, p a computable permutation. If $\mathcal{E}_{/\approx}([X]_{/\approx})$ is a finite lattice (with more than one element) then there is a $Y \in \mathcal{L}(X)$, and $[Y]_{/\approx}$ is maximal in $\mathcal{E}_{/\approx}$. (Y must be infinite, since otherwise $\emptyset \approx I\!N$, which contradicts the assumption on X).
Take a decomposition of $I\!N$ into n infinite computable sets $R_0, R_1, \ldots, R_{n-1}$ with $\bar{Y} \subseteq R_0$. Let $Y_1, \ldots, Y_{n-1}$ be images of Y under computable permutations (which map R_0 onto R_i, $i = 1, \ldots, n-1$). Then $Z = Y \cap Y_1 \cap \ldots \cap Y_{n-1}$ is a c.e. set and $\mathcal{E}_{/\approx}([Z]_{/\approx})$ is a finite Boolean algebra with n atoms.

defined by

$$\approx_1 \leq \approx_2 \text{ if } (\forall x \in \mathcal{L})(\forall y \in \mathcal{L})(x \approx_1 y \Rightarrow x \approx_2 y),$$

for $\approx_1, \approx_2 \in \mathrm{CR}(\mathcal{L})$ ($\leq$ was already used in the schema before in the form of the arrow). Further the intersection of a nonempty family of elements of $\mathrm{CR}(\mathcal{L})$ also belongs to $\mathrm{CR}(\mathcal{L})$. From this we also get that the union of two congruence relations $\approx_1$ and $\approx_2$ exists as the smallest member of $\mathrm{CR}(\mathcal{L})$ including both with respect to $\leq$. We denote it by $\approx_1 \vee \approx_2$.

Thus we have in $\mathrm{CR}(\mathcal{E})$ the element $\approx_{\mathrm{ms}} \vee \approx_{\mathcal{QM}}$. In the following we write $\approx^\circ$ instead of $\approx_{\mathrm{ms}} \vee \approx_{\mathcal{QM}}$.

At first we show that $\approx^\circ$ can be quite easily (equivalently) characterized. To do this we need the symbols: $\subset_m^0$ and $\subset_{\mathcal{QM}}^0$ with the meanings:

$$X \subset_m^0 Y \text{ if } X \subset_m Y \text{ or } X \subseteq Y \wedge X =^* Y \qquad \text{and}$$
$$X \subset_{\mathcal{QM}}^0 Y \text{ if } X \subset_{\mathcal{QM}} Y \text{ or } X \subseteq Y \wedge X =^* Y.$$

Lemma 3. *Let X and Y be c.e. sets. Then $X \approx^\circ Y$ iff there is a c.e. set Z such that*

$$(X \cap Y) \subset_m^0 Z \wedge Z \subset_{\mathcal{QM}}^0 (X \cup Y). \tag{7}$$

Proof. $\Leftarrow$ Obvious.

$\Rightarrow$ We first show that for c.e. sets Y_0, Y_1, Y_2 with $Y_0 \subseteq Y_1 \subseteq Y_2$ and

$$Y_0 \subset_{\mathcal{QM}} Y_1 \subset_m Y_2$$

there is a c.e. set Z such that

$$Y_0 \subset_m Z \subset_{\mathcal{QM}} Y_2.$$

From $X_0 \subset_{\mathcal{QM}} Y_1$ it follows that there is a computable set R such that $Y_0 \cup R = Y_1$. Now let Z be equal to

$$(\bar{R} \cap Y_2) \cup (R \cap Y_0).$$

Since R is computable, Z is c.e. We show $Y_0 \subset_m^0 Z$. Let W be c.e. with $\bar{Z} \subseteq W$ and V be equal to $W \cap \bar{R}$. Then $\bar{Y_2} \subseteq V$, since $\bar{Y_2} \subseteq \bar{R}$. Thus $\overline{Y_0 \cup R} \subseteq^* V$, since $Y_1 \subset_m Y_2$. Thus $\bar{Y_0} \subseteq^* W$. Further $Z \cup R = Y_2$ and $Y_2 - Z = R - Y_0$. Hence $Y_2 - Z$ is q-cohesive. This gives $Z \subset_{\mathcal{QM}} Y_2$.
Suppose $X \approx^\circ Y$. Then there are c.e. sets $Y_0, Y_1, \ldots, Y_n$ with $Y_0 = X \cap Y$, $Y_0 \subseteq Y_1 \subseteq \ldots \subseteq Y_n$, $Y_n = X \cup Y$ such that

$$Y_0 \subset_m^0 Y_1 \subset_{\mathcal{QM}}^0 Y_2 \subset_m^0 \ldots \subset_x^0 Y_n,$$

where $\subset_x^0$ is $\subset_m^0$ if n is odd and $\subset_{\mathcal{QM}}^0$ for n even or

$$Y_0 \subset_{\mathcal{QM}}^0 Y_1 \subset_m^0 Y_2 \subset_{\mathcal{QM}}^0 \cdots \subset_y^0 Y_n,$$

with $\subset_y^0$ conversely to $\subset_x^0$. (For this see e.g. [3, p. 183].)
Applying stepwise the previously shown "commutativity" between $\subset_{\mathcal{QM}}$ and $\subset_m$, and the transitivity of congruence relations, we get (7).

Thus the fact that d.-c.e., q-cohesive sets are co-c.e. gives an easy description of the relation $\approx^\circ$.

We see that for $\approx_{\text{ms}} \vee \approx_{\mathcal{HH}}$ a similar result holds, since hh-immune sets have the same property.

Corollary 4. *The relation $\approx^\circ$ is elementarily definable in $\mathcal{E}$.*

Proof. By using Lemma 3 we have $X \approx^\circ Y$ if

$$(\exists Z)(X \cap Y \subset_m^0 Z \wedge (\exists R \text{ computable})(Z \cup R = X \cup Y \wedge Z \cup \bar{R} \text{ is } q\text{-maximal or finite})). \tag{8}$$

Since the condition "X is q-maximal" is elementarily definable in $\mathcal{E}$, see [2], the expression (8) is elementary.

Connected with $\approx^\circ$ is the filter in $\mathcal{E}$ consisting of all c.e. sets $\approx^\circ$-equivalent to $I\!N$. It is easy to see that this is the class $\text{COF} \cup \mathcal{QM} \cup \mathcal{MS}_{\mathcal{QM}}$ (i.e. the class of all cofinite, of all q-maximal and of all major subsets of the q-maximal sets). Thus in particularly if M is a maximal set then $M \approx^\circ I\!N$.
Denote by $\mathcal{J}$ the union $\text{COF} \cup \mathcal{QM} \cup \mathcal{MS}_{\mathcal{QM}}$. $\mathcal{J}$ is a filter in $\mathcal{E}$ and thus generates a congruence relation $\approx_{\mathcal{J}}$ via

$$X \approx_{\mathcal{J}} Y \quad \text{if} \quad (\exists F_1 \in \mathcal{J})(\exists F_2 \in \mathcal{J})(X \cap F_1 = Y \cap F_2).$$

We have $\approx_{\mathcal{J}} \leq \approx^\circ$, but not equality, since $\approx_F \leq \approx_S$. Thus $\approx^\circ$ is an example of a congruence relation which is unequal to the congruence relation generated by its filter.

3. Basic Constructions

We will give here two theorems, and a few remarks which will later be slightly modified for use in Theorem 10.

The first theorem is motivated by the following observation:
Given a noncomputable c.e. set A we can find a sequence $(R_n)_{n\geq 0}$ of pairwise disjoint computable subsets of A with $\bigcup_{n\geq 0} R_n = A$, and additionally ensure that either $(R_n)_{n\geq 0}$ is a c.e. sequence or

$$(\forall R)(R - \text{ a computable subset of } A \Rightarrow (\exists n)(R \subseteq R_0 \cup \ldots \cup R_n)). \tag{9}$$

But both properties simultaneously cannot hold.

In the following theorem we construct a sequence of computable subsets of A. This sequence will be c.e. and has a property which is similar to (9), but a little weaker. As we shall see this will be sufficient for us.

To state the theorem we still need some additional notation:

- Let A be a noncomputable c.e. set. Then $\mathrm{st}_V(A)$ will denote the state-function defined by:

$$\mathrm{st}_V(A)(n) = \begin{cases} 1 & \text{if } \bar{A} \subseteq W_n, \\ 0 & \text{otherwise} \end{cases}, \quad n \geq 0.$$

- Given a computable enumeration $(a_s)_{s\geq 0}$ of A, $\mathrm{st}_V(a_s)$ will denote the state-function:

$$\mathrm{st}_V(a_s)(n) = \begin{cases} 1 & \text{if } (\forall x \leq a_s)(x \in A_s \cup W_{n,s}), \\ 0 & \text{otherwise} \end{cases}, \quad n \geq 0.$$

- Let $\sigma = (\sigma_0, \dots, \sigma_{k-1}) \in 2^{<\omega}$ and f a state-function. We write $\sigma \leq_i f$ if

$$f(i) = \sigma_i \qquad \text{for} \quad i = 0, 1, \dots, k-1^2. \tag{10}$$

(Where (10) implies $f|k = \sigma$ – see the introduction.)

Theorem 5. **[4]** *Let A be a noncomputable c.e. set and $(a_s)_{s\geq 0}$ be a computable enumeration of A. Then there exists a c.e. sequence $(R_\sigma)_{\sigma\in 2^{<\omega}}$ with the properties:*

(i) $R_\sigma \subseteq A$, $R_\sigma \cap R_\nu = \emptyset$ $(\sigma \neq \nu)$ *and* $\bigcup_{\sigma\in 2^{<\omega}} R_\sigma = A$.

(ii) $a_s \in R_\sigma \Rightarrow \sigma \leq_i \mathrm{st}_V(a_s)$ *for all* $s \geq 0$.

(iii) $\sigma \neq \langle\,\rangle \Rightarrow R_\sigma$ *is computable* [3].

(iv) $\sigma \leq_i \mathrm{st}_V(A) \Rightarrow R_\sigma$ *is infinite.*

Proof. We construct the sets R_σ, $\sigma \in 2^{<\omega}$, uniformly and stepwise. Let $R_{\sigma,0} = \emptyset$ for all $\sigma \in 2^{<\omega}$.

[2]For $\sigma = \langle\,\rangle$ and f as above we assume always $\sigma \leq_i f$.
[3]$R_{\langle\,\rangle}$ must be not computable. But this is unimportant for us.

Step $s+1$. Take the number a_s. a_s will be put into exactly one of the $R_{\sigma,s}$'s. To find this one σ consider the set

$$\{\sigma : \sigma \leq_i \mathrm{st}_V(a_s),\ \sigma \neq \langle\,\rangle,\ \max|R_{\sigma,s}| < a_s, \\ \mathrm{card}(R_{\sigma,s}) \leq \mathrm{card}(R_{P(\sigma),s}) - 2\}^{4}. \quad (11)$$

If the set (11) is empty put a_s into $R_{\langle\,\rangle,s+1}$.
Otherwise take the maximal element in (11) with respect to $\leq_i$. Let σ_0 be this and put a_s into $R_{\sigma_0,s+1}$. Let $R_\sigma = \bigcup_{s\geq 0} R_{\sigma,s}$, $\sigma \in 2^{<\omega}$.
Firstly we observe that the maximal element in (11) can be found effectively. But since for every $s \geq 0$ almost all sets $R_{\sigma,s}$ are empty, hence $\mathrm{card}(R_{\sigma,s}) = 0$, the set (11) is strongly finite in s. Thus σ_0 can be found effectively.
Every number a_s is put into exactly one of the sets R_σ (and no other numbers except elements of A come into R_σ). This gives (i).
(ii) follows from the requirement "$\sigma \leq_i \mathrm{st}_V(a_s)$" in (11).
From "$\max|R_{\sigma,s}| < a_s$" in (11) we get (iii), since the elements of R_σ are effectively enumerated in increasing order. But this means that R_σ is a computable set.
To show (iv) we prove that

$$(\forall n)(\exists^\infty a_s)(\mathrm{st}_V(A)|n = \mathrm{st}_V(a_s)|n).$$

The proof is the same as that for the existence of major subsets.
Given n, let $\gamma = \{i < n : \bar{A} \subseteq W_i\}$. Let $W_\gamma = \bigcap_{i\in\gamma} W_i$. W_γ, as an intersection of finitely many c.e. sets, is of course c.e. The set $W_\gamma \setminus A$ (the set of all numbers enumerated into W_γ earlier than they are enumerated into A) is c.e. and includes $\bar{A}$, by definition of γ. Hence $(W_\gamma \setminus A) \cap A$ must be infinite, since otherwise $\bar{A}$ would be c.e. The function F with $F(s)$ equal to

$$\max\{x : (\forall y \leq x)(x \in A_s \cup W_{\gamma,s})\}$$

is computable, (weakly) increasing and $\lim_s F(s) = \infty$, since $A \cup W_\gamma = I\!N$. The set

$$\{a_s \in W_\gamma \setminus A : F(s) < a_s\}$$

is computable by the properties of F and the condition "$F(s) < a_s$". If this were to be equal (mod $=^*$) to $W_\gamma \setminus A \cap A$ then the set $(W_\gamma \setminus A) - ((W_\gamma \setminus A) \cap A)$ would be c.e. and equal (mod $=^*$) to $\bar{A}$. But this cannot be. Thus for infinitely many elements a_s from $W_\gamma \setminus A \cap A$ for every $i \in \gamma$

$$\mathrm{st}_V(a_s)(i) = 1.$$

[4] $P(\sigma)$ for $\sigma \neq \langle\,\rangle$ is the predecessor of σ resp. to $\leq_i$.

Of these numbers a_s, with $i < n$ with $i \notin \gamma$, for only finitely many can we have $\mathrm{st}_V(a_s)(i) = 1$, by definition of $\mathrm{st}_V(a_s)$. Thus for every n we have infinitely many numbers a_s with

$$\mathrm{st}_V(a_s)|n = \mathrm{st}_V(A)|n.$$

For the sequence $(R_\sigma)_{\sigma\in 2^{<\omega}}$ we use the following properties:

1° If R is a computable set and $R \subseteq A$ then for almost all σ with $\sigma \leq_i \mathrm{st}_V(A)$ $R \cap R_\sigma = \emptyset$. Given R computable, let $W_e = \bar{R}$. Then $\bar{A} \subseteq W_e$. Hence $\mathrm{st}_V(A)(e) = 1$ and thus for all σ with $e < |\sigma|$ and $\sigma \leq_i \mathrm{st}_V(A)$ $\sigma_e = 1$. This gives $R_\sigma \subseteq W_e$, hence $R_\sigma \cap R = \emptyset$.

2° The property (ii) in Theorem 5 implies that for every σ with $\mathrm{st}_V(A)||\sigma| <_l \sigma$ the set R_σ is finite.

3° For almost all set R_σ with $\mathrm{st}_V(A)||\sigma| <_l \sigma$ $R_\sigma \cap R = \emptyset$ (R as in 1°). If $\sigma_i = 1$, but $\mathrm{st}_V(A)(i) = 0$ then only for finitely many elements $\mathrm{st}_V(a_s)(i) = 1$. Hence only for finitely many σ's with $\sigma_i = 1$ $R_\sigma \neq \emptyset$. Hence only for finitely many such σ's $R_\sigma \cap R \neq \emptyset$.

4° For us it is important that: If B and C are c.e. sets with $R_\sigma \subseteq B$ and $R_\sigma \subseteq C$ for all $\sigma <_l \mathrm{st}_V(A)||\sigma|$ and $R_\sigma \cap B =^* R_\sigma \cap C$ for all other σ's then $B \approx_{\mathrm{ms}} C$. From 1° and 3° we get for a computable subset R of A that

$$R \subseteq^* B \quad \text{iff} \quad R \subseteq^* C.$$

But this means $B \cap C \subset_m^0 B \cup C$, hence $B \approx_{\mathrm{ms}} C$.

Later we need a method for constructing finitely many c.e. sets with special properties. This in its easiest form will be given now in Theorem 6. This is a double application of Friedberg's maximal set construction. After this we generalize this via some steps till we get the needed most general form. It is given in Remark 9.

The motivation for Theorem 6 is a desire to analyse more deeply the well-known fact that if a c.e. set A is maximal in B then $B - A$ is co-c.e. (This is the exact reason why if $\mathcal{L}^*(A)$ is finite then it is a Boolean algebra.) But we can construct A such that A is maximal in B, hence $\overline{B - A} = W_e$ for some e, and every such e must be greater than a given number n.

Theorem 6. *Let $V_0, V_1, \ldots, V_{n-1}$ be c.e. sets. Then there are (effectively in $V_0, \ldots, V_{n-1}$) c.e. sets M and N such that*

(i) *M is a maximal set.*

(ii) *N is a maximal subset of M.*

(iii) *For every $i = 0, 1, \dots, n-1$*

$$\bar{M} \subseteq^* V_i \Longrightarrow M - N \subseteq^* V_i. \tag{12}$$

(Observe that in (12) we have both times $\subseteq^*$ and not $\subseteq$, this being essential for us, and making the construction a little more complicated.)

Sketch of the proof. The construction of M and N will be provided stepwise starting with $M_0 = N_0 = \emptyset$.

Denote by m_i^s, $i \geq 0$, the elements of $\bar{M}_s$, and by n_i^s, $i = 0, \dots, l_s - 1$ (with $l_0 = 0$), the elements of $M_s - N_s$. We first introduce the following state-functions:

$$\mathrm{st}_W(m_i^s)(j) = \begin{cases} 1 & \text{if } m_i^s \in W_{j,s} \\ 0 & \text{otherwise} \end{cases}, \quad i, j \geq 0,$$

with $\mathrm{st}_W(n_i^s)(j)$ for $i < l_s$ similarly defined. Further

$$\mathrm{st}_V(m_i^s)(j) = \begin{cases} 1 & \text{if } m_i^s \in V_{j,s} \\ 0 & \text{otherwise} \end{cases}, \quad i \geq 0,\ j < n$$

and

$$\mathrm{st}_V(n_i^s, m_k^s)(j) = \begin{cases} 1 & \text{if } n_i^s \in V_{j,s},\ m_k^s \in V_{j,s} \\ 0 & \text{otherwise} \end{cases}, \quad i < l_s,\ k \geq 0,\ j < n,$$

and $\mathrm{st}_V(m_i^s, m_k^s)$ similarly defined for all $i \geq 0$. We now (at stage $s+1$) look to see if there are numbers i and j with $i < j$ such that

$$\begin{array}{lll} \text{(a)} & \mathrm{st}_W(m_i^s)|i <_l \mathrm{st}_W(m_j^s)|i & \text{or} \\ \text{(b)} & i < l_s \wedge \mathrm{st}_V(n_i^s, m_i^s) <_l \mathrm{st}_V(m_j^s, m_i^s) & \text{or} \\ \text{(c)} & j < l_s \wedge \mathrm{st}_V(n_i^s, m_i^s) <_l \mathrm{st}_V(n_j^s, m_i^s) & \text{or} \\ \text{(d)} & j < l_s \wedge \mathrm{st}_V(n_i^s, m_i^s) \leq_l \mathrm{st}_V(n_j^s, m_i^s) \wedge \mathrm{st}_W(n_i^s)|i <_l \mathrm{st}_W(n_j^s)|i. & \end{array} \tag{13}$$

If "yes" then take the smallest such numbers and move m_j^s or n_j^s to m_i^s or n_i^s. (If both m_i^s and n_i^s qualify move m_j^s to position i in $\bar{M}_{s+1}$, i.e. $m_j^s = m_i^{s+1}$ and $m_i^s = m_{l_s}^{s+1}$ ($l_{s+1} = l_s + 1$).

We have the following schema for the moving of the numbers:

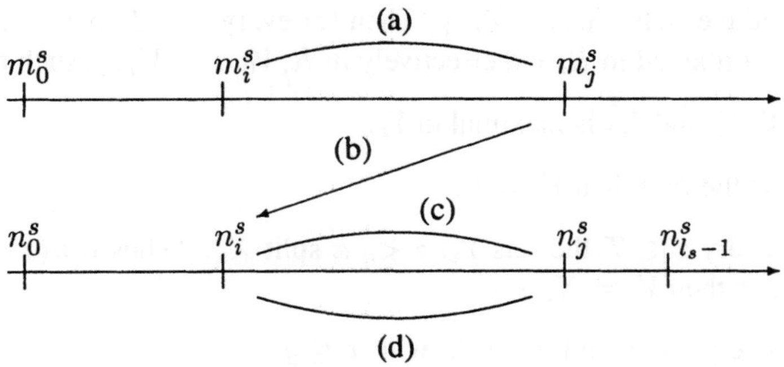

Remark 7. We see that Theorem 6 can be generalized to the construction of finitely many c.e. sets $M_0, \ldots, M_k$ such that M_j is maximal in M_{j+1}, $j = 0, \ldots, k$ ($M_{k+1} =_{\text{df}} \mathbb{N}$), and for $i = 0, \ldots, n-1$

$$M_{j+1} - M_j \subseteq^* V_i \Longrightarrow M_j - M_{j-1} \subseteq^* V_i.$$

Remark 8. Another generalization of Theorem 6 is the construction of c.e. sets M a maximal set and S and T a splitting of M and sets S_0, T_0 with S_0 maximal in S and T_0 maximal in T such that

$$\overline{M} \subseteq^* V_i \Longrightarrow (S - S_0 \subseteq^* V_i \wedge T - T_0 \subseteq^* V_i), \quad i < n. \tag{14}$$

(Here, if at some step $s+1$ m_j^s becomes m_i^{s+1}, then the m_i^s's are put alternately into $T_{s+1} - T_{0,s+1}$ and into $S_{s+1} - S_{0,s+1}$).

We see that instead of a splitting of M into two sets S and T, we can also split M into finitely many sets and ensure (14) for the corresponding differences of the sets.

*

Later in Theorem 10 we need a construction which generalizes all constructions mentioned here. In the following we describe the most general case of this:

A finite partial ordering $(\mathcal{T}, \leq)$ is called a *tree* if for all $x, y, z \in \mathcal{T}$ we have $(x \leq y,\ x \leq z) \Rightarrow (y \leq z \wedge z \leq y)$. The maximal elements in $(\mathcal{T}, \leq)$ are called *roots* (of $(\mathcal{T}, \leq)$). For $x, y \in \mathcal{T}$ we write $x <_d y$ if $x \leq y$, $x \neq y$ and there is no element of $\mathcal{T}$ between x and y with respect to $\leq$. For every $x \in \mathcal{T}$ we define the number $n(x)$ to be equal to

$$\operatorname{card}(\{z \in \mathcal{T} : z <_d x\}).$$

From the definition it follows that $n(x) = 0$ for the minimal elements x in $(\mathcal{T}, \leq)$.

Remark 9. Assume given an infinite c.e. set R, a finite tree $(\mathcal{T}, \leq)$ with only one root and c.e. sets $V_0, \ldots, V_{n-1}$. Then for every $z \in \mathcal{T}$ there are c.e. sets X_z and Y_z contained in R, and effectively in $R, V_0, \ldots, V_{n-1}$, such that

- $X_z \subseteq Y_z$ and X_z is maximal in Y_z.
- If u is the root then $Y_u = R$.
- For every $x \in \mathcal{T}$ the sets Y_z, $z <_d x$ split X_x. (Thus if $n(x) = 1$ and $z <_d x$ then $Y_z = X_x$.)
- For every $i < n$ and $x, y \in \mathcal{T}$ with $x \leq y$

$$Y_y - X_y \subseteq^* V_i \Longrightarrow Y_x - X_x \subseteq^* V_i.$$

Let X be equal to $\bigcup\{X_z : n(z) = 0\}$, X_z being the sets constructed in Remark 9 above. Then we see, since all sets $Y_z - X_z$, $z \in \mathcal{T}$ are cohesive that X is q-maximal in R. But the lattice formed by just the sets X_x, Y_x and $V_0, \ldots, V_{n-1}$ and their unions and no other c.e. sets from $\mathcal{L}^*(X, R)$ is isomorphic to the lattice corresponding to $(\mathcal{T}, \leq)$.

4. Finite End Intervals in $\mathcal{E}_{/\approx^0}$

Finite distributive lattices in which RP holds have a quite easily described structure. Every such lattice corresponds in a unique way to a finite tree. (See for this [6].)

We assume that the tree has only one root. Otherwise we can take a decomposition of $\mathbb{N}$ into the same number of infinite computable sets as there are roots. Inside each of these computable sets we provide the construction of the part of the tree with one root and then take the union of all constructed sets.

For the construction in Theorem 10 we need a set M which is Σ^0_3-maximal, i.e. is a Σ^0_3-set which, in $\mathcal{E}^3$ (the lattice of all Σ^0_3-sets under inclusion), has the same property as a maximal set in $\mathcal{E}$.

Apart from the Σ^0_3-maximality, M must have a further property given by the following: Using a computable coding of the elements of $2^{<\omega}$ by numbers, we can also talk about Σ^0_3-sets in $2^{<\omega}$ (meaning by this that the corresponding set of coding numbers belongs to Σ^0_3). If A is a noncomputable c.e. set we can consider the set

$$\{\sigma \in 2^{<\omega} : \sigma \leq_i \mathrm{st}_V(A)\}. \tag{15}$$

(where $\mathrm{st}_V(A)$ was defined before Theorem 5). We need a set M which is Σ^0_3-maximal in $2^{<\omega}$ and satisfies

$$\overline{M} \subseteq \{\sigma \in 2^{<\omega} : \sigma \leq_i \mathrm{st}_V(A)\}. \tag{16}$$

That (16) can be satisfied follows from the fact that the set (15) belongs to Δ_3. This is because "$\sigma \leq_i \mathrm{st}_V(A)$" holds if

$$\sigma = \langle\rangle \vee (\forall i < |\sigma|)((\sigma_i = 1 \Rightarrow A \cup W_i = I\!N) \wedge (\sigma_0 = 0 \Rightarrow A \cup W_i \neq I\!N)),$$

giving a set in $\Pi_2 \wedge \Sigma_2 \subseteq \Delta_3$. Just as every infinite computable set includes a complement of a maximal set, the same also holds with infinite Δ_3-sets and Σ_3^0-maximal sets in $\mathcal{E}^3$.

Since $M \in \Sigma_3^0$, there is a computable approximation function λ, such that

$$\tau \in M \iff (\exists l)(\lambda_\tau(l) = \omega).$$

(where $\lambda_\tau(l)$ means $\lim_s \lambda_\tau(l, s)$).

Theorem 10. *Every finite, distributive lattice in which RP holds is isomorphic to an end interval in the lattice $\mathcal{E}_{/\approx^\circ}$.*

Proof. Assume we are given a finite tree $(\mathcal{T}, \leq)$ with one root, a maximal set A, the c.e. sequence $(R_\sigma)_\sigma \in 2^{<\omega}$ for A (see Theorem 5), M a Σ_3^0-maximal set with the property (16) and λ a recursive approximation function for M. For every $\sigma \in 2^{<\omega}$ inside R_σ we construct for every $z \in \mathcal{T}$ c.e. subsets X_z^σ and Y_z^σ with $X_z^\sigma \subseteq Y_z^\sigma$ such that the following is satisfied:

(i) If $\sigma \in M$ and $\sigma <_l \mathrm{st}_V(A) \| \sigma|$ then $X_z^\sigma = R_\sigma$ for all $z \in \mathcal{T}$. If $\sigma \in M$ and $\mathrm{st}_V(A) \| \sigma| \leq_l \sigma$ then $X_z^\sigma =^* R_\sigma$ for all $z \in \mathcal{T}$.

(ii) If $\sigma \notin M$ then the sets X_z^σ, X_z^σ for $z \in \mathcal{T}$ satisfy Remark 9 with R_σ in place of R and the sets $W_0, \ldots, W_{|\sigma|}$ in place of the sets $V_0, \ldots, V_{n-1}$.

Construction. We start with $X_{z,0}^\sigma = Y_{z,0}^\sigma = \emptyset$ for all $z \in \mathcal{T}$.

After s steps we have for every $z \in \mathcal{T}$ the sets $X_{z,s}^\sigma$ and $Y_{z,s}^\sigma$. The elements of $Y_{z,s}^\sigma - X_{z,s}^\sigma$ are denoted by $m_z^s(i)$, $i = 0, \ldots, l_z^s - 1$.
In step $s + 1$ we do the following in the given order:

a) If $R_{\sigma,s+1} \neq R_{\sigma,s}$ put the new element into $Y_{z,s+1}^\sigma - X_{z,s+1}^\sigma$ (z-the root) at the first free place.

b) If $R_{\sigma',s+1} \neq R_{\sigma',s}$ for some σ' with $|\sigma| = |\sigma'|$ and $\sigma <_l \sigma'$ put all $m_z^s(0)$, for all $z \in \mathcal{T}$ (if defined) into $X_{z,s+1}^\sigma$. (Hence if $m_z^s(0)$ is defined and z is not minimal $m_z^s(0)$ is put into some $Y_{u,s+1}^\sigma - X_{u,s+1}^\sigma$ for $u <_d z$.)

c) If $\lambda_\sigma(l, s+1) \neq \lambda_\sigma(l, s)$ for some number l then put all numbers $m_z^s(l)$, $z \in \mathcal{T}$ (if defined) into $X_{z,s+1}^\sigma$.
(Part c) is the same as b), but with l in place of 0.)

d) Provide with the sets $X_{z,s+1}^\sigma$ and $Y_{z,s+1}^\sigma$ (which we get after actions a) to c)) the construction from Remark 9.

Let X_z^σ be $\bigcup_{s\geq 0} X_{z,s}^\sigma$, $Y_z^\sigma = \bigcup_{s\geq 0} Y_{z,s}^\sigma$ and $X_z = \bigcup_{\sigma\in 2^{<\omega}} X_z^\sigma \cup R_{\langle\rangle}$, $Y_z = \bigcup_{\sigma\in 2^{<\omega}} Y_z^\sigma \cup R_{\langle\rangle}$ for all $z \in \mathcal{T}$.
By the effectivity of the construction of X_z^σ and Y_z^σ and since $(R_\sigma)_{\sigma\in 2^{<\omega}}$ is c.e., the sets X_z and Y_z, $z \in \mathcal{T}$, are c.e. The set X (by using this we realize the end interval) is equal to

$$\bigcup\{X_z : z \in \mathcal{T},\ n(z) = 0\}.$$

Result. First we show (i). If $\sigma <_l \mathrm{st}_V(A)| |\sigma|$ then, since $R_{\sigma'}$ with $\sigma' = \mathrm{st}_V(A)| |\sigma|$ is infinite, infinitely often step b) holds for σ. But this means $R_\sigma \subseteq X$. For σ with $\mathrm{st}_V(A)| |\sigma| <_l \sigma$ we get (i), since R_σ is finite and for $\sigma = \mathrm{st}_V(A)| |\sigma|$ and $\sigma \in M$, by part c) we have $\lambda_\sigma(l) = \omega$ for some l, which gives $X_z^\sigma =^* R_\sigma$ for all $z \in \mathcal{T}$.
(ii) is satisfied, since if $\sigma \notin M$ then all $\lambda_\sigma(l)$ are finite and also all $R_{\sigma'}$ with $\sigma <_l \sigma'$ are finite. Then by induction we can show that for almost all s $m_z^s(i)$ is defined and that $\lim_s m_z^s(i)$ exists for all $z \in \mathcal{T}$ and $i \geq 0$. Thus the sets X_z^σ and Y_z^σ, $z \in \mathcal{T}$, satisfy Remark 9 (with R_σ in place of R and $W_0, \ldots, W_{|\sigma|}$ for $V_0, \ldots, V_{n-1}$).

Now we will analyse the class $\mathcal{L}(X)$ for the set X defined at the end of the construction.

- Suppose for $Y \in \mathcal{L}(X)$ $\bar{A} \subseteq^* Y$. Then there is a computable set R with $R \subseteq A$ and $\bar{R} \subseteq^* Y$. But this and $Y \in \mathcal{L}(X)$ give $Y \in \mathcal{QM}$ or Y cofinite. This follows from the facts 1° and 3° after Theorem 5 and from it that $X \cap R_\sigma$ is q-maximal or cofinite in R_σ. Thus $Y \approx^\circ I\!N$.

- Suppose $Y \in \mathcal{L}(X)$ and $X \subseteq Y \subseteq^* A$. Then for every $\sigma \in 2^{<\omega}$ and $x \in \mathcal{T}$ $Y_x^\sigma - X_x^\sigma \subseteq^* X$ or both sets are disjoint (modulo $=^*$), since all these sets if they are infinite are cohesive (by construction). Let $Y = W_e$. For all σ with $e < |\sigma|$ if $Y_x^\sigma - X_x^\sigma \subseteq^* W_e$ and $y \leq x$ then $Y_y^\sigma - X_y^\sigma \subseteq^* W_e$, since then W_e belongs to the sets $V_0, \ldots, V_{n-1}$ (for this σ). Further the set

$$\{\sigma \in 2^{<\omega} : Y_x^\sigma - X_x^\sigma \subseteq^* W_e\} \tag{17}$$

is a Σ_3^0-set. Hence, since M is Σ_3^0-maximal, (17) is equal (modulo $=^*$) to M or to $I\!N$. For $\sigma \in M$ is $R_\sigma \subseteq^* X \subseteq W_e$. Thus (17) includes M (modulo $=^*$).
Thus, by 4° after Theorem 5, there is a computable set R with $R \subseteq A$ such that $R \cup Y$ is $\approx_{\mathrm{ms}}$-equivalent to

$$\bigcup_{z\in\mathcal{T}} \bigcup_{\sigma\in 2^{<\omega}} \{Y_z^\sigma : Y_z^\sigma \subseteq^* W_e\}$$

and $Y \cup \overline{R}$ is q-maximal or cofinite in A. Both give that Y is $\approx^\circ$-equivalent to a union of sets Y_z, $z \in \mathcal{T}$ ($Y_z = \bigcup_{\sigma \in 2^{<\omega}} Y_z^\sigma$). A was chosen maximal, hence $A \approx^\circ I\!N$.

Let $x \in \mathcal{T}$ and $y \in \mathcal{T}$ with $x \neq y$. For every $\sigma \notin M$ if $x < y$ we have $Y_y^\sigma \setminus Y_x^\sigma$ is infinite (i.e. quasi-cohesive) and if $x \wedge y = 0$ then $Y_y^\sigma \cap Y_x^\sigma = \emptyset$. Thus, since $\overline{M}$ is infinite, Y_y and Y_x are not $\approx^\circ$-equivalent by 4° after Theorem 5.

These all show that $\mathcal{E}_{/\approx^\circ}([X]_{/\approx^\circ})$ is isomorphic to the lattice corresponding to $(\mathcal{T}, \leq)$.

Question. The proof of Theorem 10 would be much simpler if instead of working inside the sets R_σ, $\sigma \in 2^{<\omega}$, we could take $(R_n)_{n\geq 0}$ as uniform computable decomposition of $I\!N$ into infinite computable sets, and provide inside the sets R_n a similar construction as that in R_σ. But it is not clear then that property 3° after Theorem 5, essential in Theorem 10, also holds. Can this be ensured by a small modification of the construction in Theorem 10, with the sets R_n in place of R_σ (and a Σ_3^0-maximal set contained in $I\!N$)?

Final remarks

It is an open question whether the class

$$\{X \text{ c.e.} : \mathcal{E}_{/\approx^\circ}([X]_{/\approx^\circ}) \text{ is finite}\}$$

is elementarily definable in $\mathcal{E}$. A proof similar to that for $\mathcal{QM}$, given in [2], seems also to be possible in this case, which together with Corollary 4 would give a positive answer.

We see that the same fact regarding the finite end intervals as for $\approx^\circ$ also holds for $\approx_{\text{ms}} \vee \approx_{\mathcal{HH}}$. The sets constructed in Theorem 10 define exactly the same end intervals in $\mathcal{E}_{\approx_{\text{ms}} \vee \approx_{\mathcal{HH}}}$ as in $\mathcal{E}_{/\approx^\circ}$. Thus for all congruence relations between these two we also get the same class for their corresponding factor lattices.

Whether for others we get a class of finite end intervals properly between the class of all finite Boolean algebras and the class of all finite distributive lattices in which RP holds still remains open. A possible candidate for this is hard to see.

A generalization of the problem considered in the paper is get by taking sublattices of $\mathcal{E}$ and providing the analogous investigations inside them. Since there are known sublattices (even elementarily definable in $\mathcal{E}$) in which RP does not hold, the upper estimation given in Lemma 2 does not hold. Whether there is a sublattice of $\mathcal{E}$, and a congruence relation on it such that the factor lattice of this sublattice with respect to this congruence relation has finite end intervals in which RP does not hold, is a still open, but interesting problem.

Moreover, those questions considered above are part of a more general problem, namely that of finding which finite structures can be defined (without or with parameters) in $\mathcal{E}$. And this kind of investigation also promises a deeper knowledge of the lattice $\mathcal{E}$.

References

[1] Cenzer, D., Nies, A.: Initial segments of the lattice of Π_1^0 classes. Journ. Symb. Logic 66 (2001), 1749–1765.

[2] Harrington, L., Nies, A.: Coding in the lattice of enumerable sets. To appear.

[3] Hermes, H.: Einführung in die Verbandstheorie. Grundl. Math. Bd. 73, Springer-Verlag, Berlin, Heidelberg, New York, 1967.

[4] Herrmann, E.: On the $\forall\exists$-theory of the factor lattice by the major subset relation. To appear.

[5] Lachlan, A.H.: On the lattice of computably enumerable sets. Trans. Amer. Math. Soc. 130 (1968), 1–37.

[6] Lachlan, A.H.: The elementary theory of computably enumerable sets. Duke Math. Journal 35 (1968), 123–146.

[7] Soare, R.I.: Recursively enumerable sets and degrees. Perspectives in Math. Logic; Springer-Verlag, Berlin 1987.

A TOUR OF ROBUST LEARNING

Sanjay Jain*
School of Computing
National University of Singapore
Singapore 119260
sanjay@comp.nus.edu.sg

Frank Stephan†
Mathematisches Institut, INF 294
Universität Heidelberg
69121 Heidelberg
Germany
fstephan@math.uni-heidelberg.de

Abstract Bārzdiņš conjectured that only recursively enumerable classes of functions can be learned robustly. This conjecture, which was finally refuted by Fulk, initiated the study of notions of robust learning. The present work surveys research on robust learning and focuses on the recently introduced variants of uniformly robust and hyperrobust learning. Proofs are included for the (already known) results that uniformly robust Ex-learning is more restrictive than robust Ex-learning, that uniformly robustly Ex-learnable classes are consistently learnable, that hyperrobustly Ex-learnable classes are in Num and that some hyperrobustly BC-learnable class is not in Num.

1. Introduction

The general setting of function learning is the following: Given a set S of functions, a learner simultaneously reads data $f(0), f(1), \ldots$ on the function to be learned and outputs hypotheses $e_0, e_1, \ldots$ such that these descriptions converge in some abstract sense to an explanation of the observed function f, whenever $f \in S$. Gold [11] made this informal setting more precise by requiring syntactical convergence in the case that $f \in S$: Almost all e_k output

*Sanjay Jain is supported in part by the NUS grant number RP3992710.
†Frank Stephan is supported by the Deutsche Forschungsgemeinschaft (DFG) Heisenberg grant Ste 967/1–1; part of the work was done while he visited the University of Wisconsin – Madison.

Computability and Models: Perspectives East and West
Edited by Cooper and Goncharov, Kluwer Academic/Plenum Publishers, 2003

by the learner are the same program e for the function f.

Taking this as a starting point, learning theorists gave a large variation of learning paradigms in order to capture more precisely the intuitive notion of learning. But while the formalization of the intuitive notion of computing by recursive functions and equivalent concepts like Turing machines has turned out to be unique, the development in learning theory is more parallel to the one in complexity theory: there is a large family of important notions including Gold's notion of "explanatory learning" (Ex-learning) discussed above. One main line of research in inductive inference is to compare and evaluate these various notions of learning.

One of Bārzdiņš' main concerns with respect to the development of inductive inference was the question, to which extent the notions and also the considered learning problems are natural. Most natural learning problems S have an enumeration $f_0, f_1, \ldots$ such that $S \subseteq \{f_0, f_1, \ldots\}$ and the mapping $e, x \to f_e(x)$ is total and recursive in both parameters. Such classes can be learned by enumeration: the learner searches for the first e such that f_e is consistent with the input and outputs a canonical program for f_e.

Of course, not all learnable classes are of this form. It is almost a standard homework for computer science students to produce a program which outputs its own code. This principle is used in the definition $SD = \{f : f = \varphi_{f(0)}\}$ of the class of self-describing functions. The class SD is learnable but not in Num, that is, SD is not contained in a recursively enumerable class of total functions. As the learning part consists only mapping the input $f(0)\,f(1)\ldots f(n)$ to $f(0)$, this example looks a bit like cheating and it is obviously not very natural. One might ask, whether one can come up with a more natural example. Bārzdiņš thought that the answer is negative and that there are no naturally defined Ex-learnable classes outside Num.

Therefore, Bārzdiņš was looking for a framework of learning, which does not permit such obvious coding and enforces that learnable classes are naturally defined or at least subclasses of naturally defined learnable classes. His approach is to postulate that not only S itself but also all variants of S obtained by general recursive operators have to be Ex-learnable. The class SD does not satisfy this condition as $\Theta(SD) = REC$, where REC is the class of all total recursive functions and the operator Θ is given by the simple equation $\Theta(f)(x) = f(x+1)$. So Bārzdiņš intention is that the notion of Robust Learning captures all naturally defined learnable classes and Bārzdiņš' Conjecture states: Every robustly learnable class is already in Num. Although first incidence supported Bārzdiņš' Conjecture, Fulk [10] was able to disprove it – see Section 3 for more details.

On a philosophical side, Herman Weyl [26] started to describe the famous Erlangen program on founding geometry algebraically due to Felix Klein as follows: "If you are to find deep properties of some object, consider all the natural transformations that preserve your object (i.e. under which the object remains invariant)." Since general recursive operators (or other types of operators) can be looked upon as natural transformations, it is interesting to consider robust identification from a purely philosophical point of view too.

Case [4] was concerned whether coding tricks really occur in nature. As it turns out that they occur quite frequent not only in inductive inference but also in recursion theory and other branches of mathematics. Case thinks, that perhaps the real world may actually have some of its parts coded in other of its parts. As he essentially pointed out, such a view is consistent with both certain Eastern metaphysical principles and Leibniz' Monadology. So, one might conclude that coding is not so unnatural as it seemed to be when Bārzdiņš formulated his conjecture.

Although Fulk [10] had settled the main question, the field of robust learning has received much more attention recently. The failure of Bārzdiņš' Conjecture motivated to introduce and investigate several modified approaches.

Uniformly robust learning, as outlined in Section 4, postulates that one can compute a learner for $\Theta(S)$ from any program for Θ. Jain, Smith and Wiehagen [14] introduced this notion and showed that also for this restrictive variant, Bārzdiņš' Conjecture fails. Furthermore, the counterexample for both notions, robust and uniformly robust learning, can be taken to consist of functions f which are topologically self-describing by having that the first non-zero value at a place $x > e$ where e is the minimal program of f. This result questioned the intention that robust learning makes coding tricks impossible. On the other hand, uniformly robustly Ex-learnable classes turned out to be natural in the way that they are always consistently learnable [6]. Note that a consistent learner might be partial, but for every $f \in S$ where S is the class to be learned, and for every n, a consistent learner outputs on input $f(0)f(1)\dots f(n)$ a function which is defined and coincides with f on the arguments $0, 1, \dots, n$. Taking this result as a starting point, Section 5 gives an overview on the connections between robustness and the various notions of consistency.

Hyperrobust learning, considered in Section 6, turned out to satisfy Bāzdiņš' Conjecture for Ex-learning. Several further results support the naturalness of the definition. Hyperrobust BC-learning does not collapse to Num [20] and the corresponding class is an alternative example for Jain's result [12] that robust BC-learning is more powerful than robust EX-learning.

Section 7 compares the team hierarchies for hyperrobust and standard versions of Ex-learning. On the one hand the Non-Union Theorem (which states that there are Ex-learnable classes whose union is even not BC-learnable) cannot be generalized to hyperrobustly Ex-learnable classes. On the other hand there is a class which can be hyperrobustly Ex-learned by a team of two learners but not by a single one.

Finally Section 8 gives some sample result from applying robustness notions to the scenario where a learner receives, in addition to the data on the function to be learned, also the data on a related function.

2. Function Identification

The basic learning model of inductive inference was introduced by Gold [11]: a class S of recursive functions is learnable iff there is a partial recursive machine M which maps strings of natural numbers to numbers such that

- Given any function $f \in S$, M is defined on input $f(0)f(1)\ldots f(n)$ and outputs a number e_n;
- Almost all e_n are the same number e;
- The number e is an index for f, that is, $\varphi_e = f$.

Here the indices are chosen with respect to any fixed acceptable numbering φ. It does not matter which acceptable numbering is chosen since all acceptable numberings can be translated into another via total recursive functions. Examples of acceptable numberings are the numbering of all Turing machines and the numbering of all those programs in common programming languages with variables using natural numbers but not having a "maxint-bound" and using exactly one input and at most one output. Note that an acceptable numbering always contains also partial recursive functions; following a common convention, a "recursive function" is a function φ_e which is total and a "partial recursive function" is any function φ_e, whether total or not. So "partial recursive" indicates that the function is permitted to be undefined somewhere but it is not required to be undefined somewhere.

Gold [11] showed that the class of all total recursive functions is not learnable, that is, for every recursive learner M there is a recursive function f not learned by M. So one of the central topics of inductive inference became the search for the boundaries of learning and for further learning criteria, which were compared to the already existing ones. The main criteria are the following.

Definition 2.1 (Learning Criteria) [1, 2, 3, 8, 11, 15, 17, 19, 28, 29]. Let S be a class of total recursive functions. S is learnable with respect to one of the following criteria iff there is a partial recursive machine M with the corresponding properties.

Ex-learning: if $f \in S$ then $M(f(0)f(1)\ldots f(n))$ is defined for all n and there is an index e of f such that $M(f(0)f(1)\ldots f(n)) = e$ for almost all n.

Ex$_k$-learning: if $f \in S$ then there is an n such that $M(f(0)f(1)...f(m))$ is defined iff $m > n$ and there are at most k numbers $m' > n$ with $M(f(0)f(1)\dots f(m')) \neq M(f(0)f(1)\dots f(m'+1))$. That is, there are at most k mind changes by M during the process of learning f.

Confident learning: M is a total Ex-learner for S and M converges on every, even non-recursive, function f. That is, given any total f, there is an index e with $M(f(0)f(1)\dots f(n)) = e$ for almost all n.

Consistent learning: M Ex-learns S and, for every $f \in S$ and for every n, $M(f(0)f(1)\dots f(n))$ is defined and outputs a program e such that, for $m = 0, 1, \dots, n$, $\varphi_e(m)\downarrow = f(m)$.

Conforming learning: M Ex-learns S and, for every $f \in S$ and for every n, $M(f(0)f(1)\dots f(n))$ is defined and outputs a program e such that, for $m = 0, 1, \dots, n$, either $\varphi_e(m)\uparrow$ or $\varphi_e(m)\downarrow = f(m)$.

Reliable learning: M is a total Ex-learner for S and whenever M converges on some function f to an index e, then e is an index for f; note that reliable learners diverge on all nonrecursive functions.

BC-learning: if $f \in S$ then $M(f(0)f(1)\dots f(n))$ is defined for all n and $M(f(0)f(1)\dots f(n))$ is a program for f for almost all n; note that these programs can be different for all n.

Definition 2.2 [24]. Let I be one of the notions from Definition 2.1. Then S is *I-learnable by a team of k learners* iff there are k learners $M_1, M_2, \dots, M_k$ such that every $f \in S$ is I-learned by at least one of the learners $M_1, M_2, \dots, M_k$.

Note that one can postulate without loss of generality for the criteria of Ex-learning, confident learning and BC-learning that the learner M is even a total recursive function on the strings of natural numbers. As one can make any hypothesis of a BC-learner to be consistent (and thus also conforming) with the data seen so far without loosing any learning power, a BC-learner is without loss of generality already consistent and conforming and therefore it does not make sense to introduce these notions for the BC-case. Furthermore, there are also alternative definitions of reliable learning which are not equivalent to the above one, see for example [20] for a brief discussion of these criteria in the context of robust and hyperrobust learning.

Case and Smith [8] published the following characterizations of a restricted version of Ex-learning and of BC-learning, the first characterization is due to Leeuwen and Bārzdiņš and the second is due to Podnieks [23].

Theorem 2.3 [8, 23]. *A class S is Ex-learnable via an M outputting (on all inputs) only indices of total functions iff there is a recursive N such that, for all $f \in S$ and for almost all n, $N(f(0)f(1)\dots f(n)) = f(n+1)$.*

A class S is BC-learnable iff there is a partial recursive N such that, for all $f \in S$ and for almost all n, $N(f(0)f(1)\dots f(n))\downarrow = f(n+1)$. Note that *even for any fixed $f \in S$ the predictor N is permitted to be undefined on finitely many inputs of the form $f(0)f(1)\ldots f(n)$.*

A class S is recursively enumerable iff $S = \{f_0, f_1, \ldots\}$ for a family of total functions such that the mapping $e, x \rightarrow f_e(x)$ is recursive. A class S is defined to be in *Num* iff it is a *subclass* of a recursively enumerable class. These classes have the straight forward learning algorithm given below.

Remark 2.4 (Learning by Enumeration) [11, 23]. *Every class in Num is Ex-learnable by the following algorithm, called learning by enumeration: Given a recursive enumeration $f_0, f_1, \ldots$ of functions such that $S \subseteq \{f_0, f_1, \ldots\}$, the learner searches on any input the first f_e which is consistent with the data seen so far and translates the e to an index of the given acceptable numbering. More precisely, a recursive function h is chosen such that $\varphi_{h(e')} = f_{e'}$ for all e' and the learner outputs, for any given input $f(0)f(1)\ldots f(n)$, the value $h(e)$ where e is the first index satisfying $(\forall m \leq n)\,[f_e(m) = f(m)]$.*

Note that every class S in Num has a recursively enumerable superclass $S' = \{f_0, f_1, \ldots\}$ which is also dense. Here a class S' is called *dense* iff every data σ is extended by some $f \in S'$, denoted $\sigma \preceq f$: $(\forall x \in \text{domain}(\sigma))\,[\sigma(x) = f(x)]$. If learning by enumeration is based on this dense superclass S', then the resulting algorithm is total.

There are various learning-theoretic characterizations for S being in Num: S is Ex-learnable with a machine outputting only indices of total functions, S is learnable by enumeration, S has a total-recursive machine predicting every function in S almost always. This justifies to look upon Num as being a learning criterion and the study of robust learning was motivated by Bārzdiņš' question whether Num captures all natural learnable classes.

3. Bārzdiņš' Conjecture

Self-reference combined with coding-tricks gives an elegant way to prove many separation results in inductive inference as shown in the following example, in particular for the case of the criterion SD.

Example 3.1 (Classes SD and ASD). The following classes obtained by self-referential coding separate some of the learning criteria from Definition 2.1.

- The class $SD = \{f : \varphi_{f(0)} = f\}$ of all self-describing functions can be Ex_0-learned and can be consistently learned. Thus SD is also conformingly learnable, confidently learnable, Ex-learnable and BC-learnable. But SD cannot be learned reliably and is also not in Num.

- The class $ASD = \{f : \varphi_{f(0)} =^* f\}$ of all almost self-describing functions f (these are computed at almost all x by the index $f(0)$) is BC-learnable but not Ex-learnable.

- The class $\{f : (\exists e)\,[\varphi_e = f \wedge e = \max\{f(x) : x = 0, 1, \dots\}]\}$ of all functions f bounded by a constant and computed by the maximum of its values is Ex-learnable and therefore also BC-learnable. But it cannot be learned according to any other learning criteria from Definition 2.1.

A proof for the first statement is provided as a sample for proofs using the Recursion Theorem. An Ex_0-learner for SD just outputs on input $f(0)f(1)\dots f(n)$ the hypothesis $f(0)$. An Ex_0-learner is obviously also an Ex-learner, BC-learner and confident learner as it makes at most one hypothesis. Consistency follows from the fact that the unique guess $f(0)$ is correct whenever the function f is in SD.

To prove that SD is not in Num, assume by way of contradiction that $f_0, f_1, \dots$ is a recursive sequence of recursive functions which contains all the self-describing functions. Let $g(x) = f_x(x+1) + 1$. Now consider the function h defined by

$$\varphi_{h(e)}(x) = \begin{cases} e & \text{if } x = 0; \\ g(x-1) & \text{if } x > 0. \end{cases}$$

By Kleene's recursion theorem, this function h has a fixed point e and $\varphi_e = \varphi_{h(e)} \in SD$. Let e' be such that $f_{e'} = \varphi_e$. But then $f_{e'}(e'+1) = g(e') = f_{e'}(e'+1) + 1$, a contradiction. Thus SD is not in Num.

Based on the results by Blum and Blum [3] and Minicozzi [17], one can indirectly prove that SD is not reliably Ex-learnable. The proof looks at the union $SD \cup AEZ$ where the class AEZ from Example 4.5 below is reliably Ex-learnable. On the one hand, $SD \cup AEZ$ is not Ex-learnable [3]. On the other hand, the union of any two reliably learnable classes is reliably learnable [3, 17] and thus also Ex-learnable. So, SD cannot be reliably learnable.

Case and Smith [8] study in detail many applications of self-reference to separate several learning criteria including the use of ASD to separate BC-learning from Ex-learning.

Although these proofs are elegant, they have the property of coming up with example classes which look very artificial. The main idea is that the coding permits the less pretentious learner to learn while the more pretentious one has to transform the data obtained in a way which would enforce the learner to solve some unsolvable problem. For example a hypothetical Ex-learner for the class ASD would fail because it cannot detect the finitely many undefined places which the index $f(0)$ might have. But the BC-learner would, on input

$f(0)f(1)\dots f(n)$, output an index which looks up $f(m)$ in a table for input $m \leq n$ and computes $\varphi_{f(0)}(m)$ for $m > n$. After seeing sufficiently many examples, all places where $\varphi_{f(0)}$ is undefined or wrong are overwritten by entries in the table and from then on, each hypothesis is correct.

Besides the elegance of the proofs, a further reason to use self-describing function classes for separation results is that many candidates S for such a separation have already a self-describing subclass S' which witnesses the same separation but has the learning algorithm already "built in". This principle is just stated for the case of the second criterion being Num.

Proposition 3.2. *If S is closed under finite variants and if S is not in Num, then $S \cap SD$ is in Ex_0 but not in Num.*

Proof. Wiehagen [27] observed that for every function f there is a self-describing function g such that $f(x) = g(x)$ for all $x > 0$. Thus S satisfies

$$S = \{f : (\exists g \in SD \cap S)\,(\forall x > 0)\,[f(x) = g(x)]\}$$

and whenever $SD \cap S$ is in Num, so is S. It follows that $SD \cap S$ is not in Num. But as SD is in Ex_0, so is its subclass $SD \cap S$. ▮

Bārzdiņš' idea was to postulate that not only the class S itself but also every transformed class $\Theta(S)$ should be learnable. Taking the class of suitable permitted operators Θ, the resulting notion of robust learning was intended to meet the following requirements.

- Every class S in Num should also be robustly learnable.
- If the learnability of S depends essentially on coding, then $\Theta(S)$ should be non-learnable for a suitable Θ. That is, only natural classes should be robustly learnable.

Bārzdiņš conjectured that these two requirements would be met by permitting all general recursive operators Θ and that only classes in Num are robustly learnable.

Indeed one had also considered many other classes of operators, but none of these classes was able to satisfy both requirements. Clearly all permitted operators have to be recursive, but the question was whether to tolerate or not that some total functions are mapped to partial ones.

If one only requires that functions in S are mapped to total functions, then one could take the class S of all constant functions where the constant c is an index of a total function. If one then takes the operator defined by

$$\Theta(f)(x) = \varphi_{f(0)}(x)$$

one would have that $\Theta(S)$ is the class of all recursive functions which is not learnable. On the other hand, even this weak requirement does not destroy all coding tricks as the following example shows.

Example 3.3 (Class TSD). Jain, Smith and Wiehagen [14] introduced the class

$$TSD = \{f : (\exists e)\,[f = \varphi_e \wedge (\forall x \leq e)\,[f(x) = 0]]\,\}$$

of topologically self-describing functions. This class satisfies the following: Let Θ be a recursive operator mapping every function in TSD to a total one. Then $\Theta(TSD)$ is Ex-learnable as sketched below.

The class TSD contains 0^∞, so the learner conjectures $\Theta(0^\infty)$ until some x is found with $\Theta(f)(x) \neq \Theta(0^\infty)(x)$. Now one can compute the maximum argument y of the function 0^∞ evaluated by Θ in order to compute $\Theta(0^\infty)(x)$. It follows that $f(z) \neq 0$ for some $z \leq y$. Therefore, provided that $f \in S$, it holds that $f = \varphi_e$ for some $e \leq y$. From this information one can compute an upper bound $b_{\Theta,y}$ for the minimal index of $\Theta(f)$ and having this bound it is possible to find a program for $\Theta(f)$ in the limit.

It follows that it does not give any advantage to consider operators which fail to map some total (not necessarily recursive) functions to partial functions. So the main goals for a suitable notion of robust learning became the following one: classes in Num should be mapped to classes in Num and the concept should be as easy as possible, as the easiest concept is often the most natural one among a choice of several acceptable concepts to formalize a notion. Thus, one can define robust learning as below. Here a general recursive operator is an operator which is recursive and which maps every total function to a total one.

Definition 3.4 (Robust Learning) [31]. Fix a learning criterion I from Definition 2.1. A class S is robustly learnable with respect to the criterion I iff for every general recursive operator Θ the class $\Theta(S)$ is learnable with respect to the criterion I.

Since one can choose Θ to be the identity mapping every f to itself, every robustly learnable class is also learnable (for any criterion I of learning considered). As already mentioned in Example 3.3, the class TSD is robustly Ex-learnable but not in Num, therefore Bārzdiņš' Conjecture is refuted.

Theorem 3.5 [10, 14]. *There is a robustly Ex-learnable class which is not in Num.*

Furthermore, one can ask whether all naturally defined Ex-learnable classes are also robustly Ex-learnable. Blum and Blum [3] considered the following class which on the one hand has a natural definition but on the other hand is not robustly Ex-learnable in the case that the complexity measure used in the definition is a interpolating complexity measure. Recall that a complexity measure Φ associated to the given acceptable numbering φ is defined by the following properties.

- $\Phi_e(x)$ is defined iff $\varphi_e(x)$ is defined.
- The set $\{(e, x, t) : \Phi_e(x)\downarrow \leq t\}$ is recursive.

A complexity measure Φ is interpolating iff one can find for every i and every string σ a j such that $\Phi_j(x) = \sigma(x)$ for $x \in \text{domain}(\sigma)$ and $\Phi_j(x) = \Phi_i(x)$ for all other x.

Example 3.6 (Class BB) [25]. Let Φ be an interpolating complexity measure for the acceptable numbering φ. Using that $\Phi_e(e) < \infty$ iff e is in the diagonal halting problem K, Blum and Blum [3] defined the following approximations to K:

$$\psi_e(x) = \begin{cases} 1 & \text{if } \Phi_x(x) \leq \Phi_e(x) \text{ and } \Phi_e(x) < \infty; \\ 0 & \text{if } \Phi_x(x) > \Phi_e(x) \text{ and } \Phi_e(x) < \infty; \\ \uparrow & \text{if } \Phi_e(x) = \infty. \end{cases}$$

Note that ψ_e is total whenever Φ_e is total. Now they defined the class

$$BB = \{\psi_e : \Phi_e \text{ is total and increasing}\}$$

and showed that this class is Ex-learnable. Stephan and Zeugmann [25] reinvestigated the class and showed that BB is neither robustly Ex-learnable nor in Num.

One of the main topics of the research on robust learning was also the question whether Bārzdiņš' Conjecture holds for further criteria of learning, namely whether for a given criterion I every class learnable with respect to I is also in Num. Zeugmann [31] published Bārzdiņš' Conjecture and showed as an intermediate result that the notion of reliable learning satisfies this conjecture. The notion of reliable learning was introduced by Minicozzi [17] and intended to model that the learner is never misleading in the limit as it either converges to a correct program or diverges by outputting infinitely many different hypotheses.

Theorem 3.7 [31]. *If a class S is robustly reliably learnable then S is in Num.*

Proof. Let T be an infinite binary recursive tree without infinite recursive branches and let $\sigma_0, \sigma_1, \ldots$ be a recursive enumeration of all of its leaves. Now one applies the general recursive operator Θ to S where Θ is given by

$$\Theta(f) = \sigma_{f(0)}\sigma_{f(1)}\sigma_{f(2)}\cdots$$

which maps f to that $\{0,1\}$-valued function whose course of values is given by the infinite string $\sigma_{f(0)}\sigma_{f(1)}\sigma_{f(2)}\ldots$ and considers any reliable learner M for $\Theta(S)$. Recall that M is required to be total. Furthermore, without loss of generality, M is increasing, that is, whenever $M(\tau) \neq M(\eta)$ and $\tau \preceq \eta$, then $M(\tau) < M(\eta)$. Note that this property can be obtained by using a padding function which is a recursive function p such that $p(e,n)$ is an index of the same function as e and satisfies $p(e,n) > n$. Now consider uniformly recursive binary trees

$$U_\tau = \{\eta \in \{0,1\}^* : \tau \preceq \eta \wedge M(\tau) = M(\eta)\}$$

where τ ranges over all strings in $\{0,1\}^*$. The set of all τ where U_τ is finite is recursively enumerable since whenever U_τ is finite then there is a level $l > |\tau|$ such that $U_\tau \cap \{0,1\}^l = \emptyset$. Thus one can define the following functions g_τ by taking the value of that case which is found first to hold:

$$g_\tau(x) = \begin{cases} \varphi_{M(\tau)}(x) & \text{if } \varphi_{M(\tau)}(x)\downarrow \leq 1; \\ 0 & \text{if } U_\tau \text{ is finite.} \end{cases}$$

Every function g_τ is total. If U_τ is finite then the second condition guarantees g_τ to terminate. If U_τ is infinite then U_τ has an infinite branch g by König's Lemma. M converges on this infinite branch to $M(\tau)$ and since M is reliable, $M(\tau)$ is a program for this total function g. Furthermore, whenever M learns some function g, then there is a $\tau \preceq g$ such that $M(\tau)$ is already the final value for g and g is an infinite branch of U_τ, so $g = g_\tau$ for this τ. For every g_τ, consider the function f_τ where

$$f_\tau(x) = y_x \Leftrightarrow (\exists y_0, y_1, \ldots, y_{x-1})\,[\sigma_{y_0}\sigma_{y_1}\cdots\sigma_{y_x} \preceq g_\tau].$$

Note that the definition cannot be ambiguous since $\sigma_i \not\preceq \sigma_j$ whenever $i \neq j$. Furthermore, T has no infinite recursive branch and thus every recursive function g_τ cannot be on any shifted version σT of the tree T. Thus one can show inductively that g_τ is not on $\sigma_{y_0}\sigma_{y_1}\ldots\sigma_{y_x}T$ and therefore find an y_{x+1} with $\sigma_{y_0}\sigma_{y_1}\cdots\sigma_{y_x}\sigma_{y_{x+1}} \preceq g_\tau$. One can furthermore verify that this process can be done uniformly for all g_τ and that thus the class S' of all f_τ is a recursively enumerable family. Since for every $f \in S$, M learns $\Theta(f)$, every $f \in S$ satisfies $\Theta(f) = g_\tau$ for some $\tau \in \{0,1\}^*$. Thus $f = f_\tau$ for this τ. Thus $S \subseteq S'$ and S is in Num. ∎

4. Uniformly Robust Learning

Quite often it may be useful to consider not only whether one can robustly identify a class, but whether one can effectively get a machine to learn every "transformed" class, using the description of the transformation. For example, if one can learn a certain type of geometric figure, one might expect to be able to learn its transformations via some operations such as rotation, scaling and so on, effectively from the parameters of the transformations. This motivates the consideration of uniform robust learning.

In the context of uniform robust learning this means that the images (under general recursive operators) of S are not only learnable, but one can effectively find a machine to learn them. For this strengthened version of robust learning, one looks into two directions: (I) Are there nontrivial examples which are uniformly robustly learnable with respect to a given criterion I? (II) Is uniformly robust learning with respect to I different from classical robust learning with respect to I? Before investigating this, the formal definition is given.

Definition 4.1 [14]. Fix a learning criterion I from Definition 2.1. A class S is uniformly robustly learnable with respect to the criterion I iff there exists a recursive function g such that, for all e such that Θ_e is general recursive, $M_{g(e)}$ learns $\Theta_e(S)$ with respect to the criterion I.

Remark 4.2 [14]. *For the notion of learning with a constant bound on the number of mind changes, uniformly robust learning is very restrictive. If S is uniformly robustly Ex_k-learnable then S is finite and contains less than 2^{k+1} functions. In contrast to this, a class is robustly Ex_0-learnable iff it is finite and some infinite class is even robustly Ex_1-learnable.*

Remark 4.3 [6]. *The class of topologically self-describing functions TSD from Example 3.3 is uniformly robustly confidently learnable and thus also uniformly robustly Ex-learnable. But TSD is not in Num.*

So one knows that the uniformly robust version of confident learning is not trivial. The following result of Case, Jain, Stephan and Wiehagen [6] separates the uniform robust and robust versions of confident learning, this separation is much more involved than the previous results and the proof is therefore omitted.

Theorem 4.4 [6]. *There is a class S which is robustly confidently learnable but not uniformly robustly confidently learnable.*

Of course, the most interesting question is whether the robust and uniform robust versions of Ex-learning and BC-learning are different. Indeed, Case, Jain, Stephan and Wiehagen [6] constructed a class S which separates both notions simultaneously. Before constructing that class, the below variant of the Non-Union Theorem [2, 3] is given as it is a tool used in the proof.

Example 4.5 (Class AEZ). Let

$$AEZ = \{f : (\forall^{\infty} x)\, [f(x) = 0]\}$$

denote the class of all almost everywhere zero functions. This class is an example of a dense class in Num.

Example 4.6 (Non-Union Theorem). Let AEZ be as above and let TSD be the class of all topological self-describing functions from Example 3.1.

Both classes, AEZ and TSD, are uniformly robust Ex-learnable but their union $AEZ \cup TSD$ is not BC-learnable [2, 3, 14]. Therefore the Non-Union Theorem extends to the notions of robust and uniformly robust Ex-learning and BC-learning.

Concerning team-learning as defined in Definition 2.2, it should be noted that $AEZ \cup TSD$ is an example of a class robustly learnable by a team of two Ex-learners but not being in Ex.

Theorem 4.7 [6]. *There is a class S which is robustly Ex-learnable but not uniformly robustly BC-learnable. So the notions of robust Ex-learning and robust BC-learning are more general than their uniformly robust counterparts.*

Proof. A more detailed proof is found in the paper of Case, Jain, Stephan and Wiehagen [6].

One first constructs finite functions σ_x for all x such that one can compute the σ_x as strings in the limit from x. Then one takes a sequence $a_0, a_1, \ldots$ and chooses, for each n, finitely many recursive functions f extending σ_{a_n} and puts these functions into S. The basic idea is to take these functions in an adversary manner for uniformly robust BC-learners, while on the other hand the classes remain "almost uniformly Ex-learnable" which means that one can take an uniform Ex-learner which learns all but finitely many functions in $\Theta_e(S)$ and then makes a non-uniform upgrade in order to cover these finitely many functions as well.

The technical part in the construction of S is the inductive construction of the finite functions σ_x. The goal is that, for all $e \leq x$, Θ_e either maps all or no functions extending σ_x to $\Theta_e(0^\infty)$. The σ_x are constructed in stages s as follows.

- One uses two auxiliary partial-recursive two-place functions ψ, ϑ which are initialized as being undefined everywhere at the beginning and are used for book-keeping which Θ_e have already been addressed. Furthermore, $\sigma_{x,0} = 0^x 1$ before stage 0.

- At stage s: If there is an $e \leq x, y \leq s$ and $\tau \succeq \sigma_{x,s}$ such that $\tau \in \{0,1\}^s$, $\psi(e,x)$ is still undefined and the computations $\Theta_e(\tau)(y)$, $\Theta_e(0^\infty)(y)$ converge within s steps to different values,

- Then take least such e and let $\psi(e,x) = y$ and $\sigma_{x,s+1} = \tau$ for corresponding y and τ. Furthermore let $\vartheta(e,x)$ be the largest argument z of the function 0^∞ queried during the computation of $\Theta_e(0^\infty)(\psi(e,x))$.
- Else all values $\psi(e,x)$ and $\vartheta(e,x)$ remain unchanged and also $\sigma_{x,s+1} = \sigma_{x,s}$.

Now the sequence of the a_n is defined inductively by letting $a_0 = 0$ and letting a_{n+1} be the sum of $a_n + 1$ and all values $\vartheta(e,x)$ where $e, x \leq a_n$ and $\vartheta(e,x)$ is defined. This is a finite sum and so a_{n+1} exists.

Now let h be a function dominating every total recursive function. For every n there is an operator, whose index is called e_n, doing the following.

$$\Theta_{e_n}(f) = \begin{cases} g & \text{if } f = \sigma_{a_{2n}} \cdot g \text{ or } f = \sigma_{a_{2n+1}} \cdot g \text{ for some } g; \\ 0^\infty & \text{otherwise.} \end{cases}$$

For every $e' \leq h(e_n)$ there is by the Non-Union Theorem a function $g_{n,e'} \in AEZ \cup TSD$ which the e'-th learner $M_{e'}$ does not BC-learn. Now put for each n and each $e' \leq h(e_n)$ the following function into S:

- If $g_{n,e'} \in AEZ$ then put $\sigma_{a_{2n}} \cdot g_{n,e'}$ into S;
- If $g_{n,e'} \in TSD$ then put $\sigma_{a_{2n+1}} \cdot g_{n,e'}$ into S.

In the following it is shown that the resulting class is robustly Ex-learnable but not uniformly robustly BC-learnable.

Claim 1. S is not uniformly robust BC-learnable.

For every recursive function h' there is an n such that $h'(e_n) < h(e_n)$. It follows from the construction that $M_{h'(e_n)}$ does not BC-learn $g_{n,h'(e_n)}$ and that this function is in $\Theta_{e_n}(S)$. So there is no recursive function h' witnessing that S would be uniformly robustly BC-learnable, hence non-learnability is established.

Claim 2. The class $\Theta_e(S)$ is Ex-learnable whenever Θ_e is a general recursive operator.

Let Θ_e be a given general recursive operator and let $f \in S$ be a function such that $\Theta_e(f)$ should be learned.

Let $\{f_0, f_1, \ldots, f_k\}$ be the set of all $\Theta_e(f')$ such that either $f' = 0^\infty$ or $f' \in S \wedge f' \succeq \sigma_{a_n}$ for some $n < e$. Finding this set is the non-uniform part of the learning algorithm. The learner searches for the least l such that f_l is consistent with the data seen so far and if this finite search terminates the learner outputs a hypothesis for f_l.

Otherwise, none of the functions $\{f_0, f_1, \ldots, f_k\}$ is consistent with the input. Let n be the number with $\sigma_{a_n} \preceq f$. Note that $n \geq e$, $\psi(e, a_n)$ is defined and $\Theta_e(f)(\psi(e,a_n)) \neq \Theta_e(0^\infty)(\psi(e,a_n))$ as $\Theta_e(f) \notin \{f_0, f_1, \ldots, f_k\}$. If $e \leq m < n$ then either $\psi(e,a_m)$ is not defined or $\Theta_e(0^\infty)(\psi(e,a_m))$ is computed relative to 0^∞ by only asking queries up to $\vartheta(e,a_m)$, thus the values $\Theta_e(0^\infty)(\psi(e,a_m))$ and $\Theta_e(f)(\psi(e,a_m))$ coincide. Therefore one can find n in the limit by searching for the first number $m \geq e$ such that $\psi(e,a_m)$ is defined and witnesses that $\Theta_e(f)$ and $\Theta_e(0^\infty)$ are different.

As AEZ and TSD are uniformly robustly Ex-learnable and as one can compute an index for the general recursive operator mapping g to $\Theta_e(\tau \cdot g)$ from τ, one can, for every input $\Theta_e(f)(0)\,\Theta_e(f)(1)\ldots\Theta_e(f)(m)$, work with approximations n_m, $a_{n,m}$ and $\sigma_{a_{n,m},m}$ to n, a_n, σ_{a_n} and run as a subalgorithm the algorithm to learn $\Theta_e(\sigma_{a_{n,m},m} \cdot AEZ)$ if n_m is even and the algorithm to learn $\Theta_e(\sigma_{a_{n,m},m} \cdot TSD)$ if n_m is odd. As these approximations coincide with the correct values for almost all m, the resulting algorithm coincides with the algorithm to learn $\Theta_e(\sigma_{a_n} \cdot AEZ)$ or $\Theta_e(\sigma_{a_n} \cdot TSD)$, respectively, for almost all m. It follows that $\Theta_e(f)$ is Ex-learned. ▮

A further important question is whether the requirement of robustness is so strong that every robustly BC-learnable class is already (non-robustly) Ex-learnable. Jain [12] answered this question negatively, the class BWT from Example 6.13 is a witness for the following Theorem.

Theorem 4.8 [12]. *There is a class which is uniformly robustly BC-learnable but not Ex-learnable.*

5. Robust Learning and Consistency

This section deals with the relations between robust learning on the one hand and notions of consistency on the other hand. The main topic of this section is the surprising implication that every uniformly robustly Ex-learnable class is also consistently learnable. After that, a closer look will be given to the investigation of variants of the criterion of consistent learning.

Theorem 5.1 [6]. *If a class is uniformly robustly Ex-learnable, then it is also consistently learnable.*

Proof. Suppose that S is uniformly robustly Ex-learnable and that g is a recursive function witnessing this in the sense that whenever Θ_e is a general recursive operator then $M_{g(e)}$ is an Ex-learner for $\Theta_e(S)$. Without loss of generality one can assume that each $M_{g(e)}$ is total. Using Kleene's recursion theorem [18], there exists an operator Θ_e such that $\Theta_e(f(0)f(1)\ldots f(n))$ is defined inductively for $n = 0, 1, \ldots$ by using the preceding string $\sigma = \lambda$ for $n = 0$ and $\sigma = \Theta_e(f(0)f(1)\ldots f(n-1))$ for $n > 0$ and by taking $\Theta_e(f(0)f(1)\ldots f(n))$ according of the first of the three cases below to apply.

- $\Theta_e(f(0)f(1)\ldots f(n)) = \sigma$ if σ is already infinite.
- $\Theta_e(f(0)f(1)\ldots f(n)) = \sigma \cdot (f(n)+1) \cdot 0^s$ for the first s such that, for $e' = M_{g(e)}(\sigma \cdot (f(n)+1) \cdot 0^s)$, for all m in the domain of $\sigma \cdot (f(n)+1)$, $\varphi_{e'}(m)$ outputs $(\sigma \cdot (f(n)+1))(m)$ within s steps.
- $\Theta_e(f(0)f(1)\ldots f(n)) = \sigma \cdot (f(n)+1) \cdot 0^\infty$ if there is no such s in the previous case.

The dot is used here to denote string-concatenation in places where one might otherwise misunderstand it. Note that $\Theta_e(f(0)f(1)\ldots f(n))$ being infinite does not mean that an infinite function is computed in finite time but it means that the values of the function $\Theta_e(f)$ does not depend on the values of f beyond n. It is easy to check that Θ_e is a general recursive operator.

Claim 1. If $f \in S$ then all $\Theta_e(f(0)f(1)\ldots f(n))$ are defined according to the second case of the algorithm for Θ_e.

This is proven by induction. Assume that a number n is given such that all $\Theta_e(f(0)f(1)\ldots f(m))$ with $m < n$ are defined according to the second case. Then $\Theta_e(f(0)f(1)\ldots f(n))$ cannot be defined according to the first case since the previously defined strings are all finite.

If $\Theta_e(f(0)f(1)\ldots f(n))$ would be defined according to the third case, this would produce an infinite string. So $M_{g(e)}$ would have to converge on this string to an index e' since $M_{g(e)}$ Ex-learns this infinite string and this e' would be an index for the function given by this infinite string. So there would be an s such that $M_{g(e)}(\Theta_e(f(0)f(1)\ldots f(n-1))(f(n)+1)0^s)$ outputs e' and that the function $\varphi_{e'}$ terminates within s steps on the domain of the string $\Theta_e(f(0)f(1)\ldots f(n-1))(f(n)+1)$.

Thus the second case of the algorithm applies in contradiction to the assumption and so for $f \in S$, all strings $\Theta_e(f(0)f(1)\ldots f(n))$ are build according to the second case from the preceding strings.

Learner. Now one considers the following (consistent) learner N for S. $N(f(0)f(1)\ldots f(n))$ simulates $\Theta_e(f)$ and $M_{g(e)}$. If $\Theta_e(f(0)f(1)\ldots f(n))$ is defined by the second case in the definition of Θ_e, then N takes the index e' coming up in that second case and outputs $h(e')$ where h is defined such

$$\varphi_{h(e')}(k) = \varphi_{e'}(x_k) - 1$$

where x_k is the first x such that the set $\{y \le x : \varphi_{e'}(y)\downarrow > 0\}$ has at least $k+1$ elements.

Note that N is permitted to be undefined on input not belonging to any $f \in S$. The following Claims 2 and 3 establish that N is a consistent learner for S.

Claim 2. N is consistent, that is, if $f \in S$ then $N(f(0)f(1)\ldots f(n))$ is defined and outputs an index $h(e')$ which is defined for $m \leq n$ and outputs the value $f(m)$ there.

The index e' used in the second case of the definition of Θ_e computes a function which extends a string of the form $(f(0)+1)0^*(f(1)+1)0^*\ldots 0^*(f(n)+1)$. Thus the function computed by the index $h(e')$ satisfies $\varphi_{h(e')}(m) = f(m)$ for $m = 0, 1, \ldots, n$ and the consistency follows.

Claim 3. N Ex-learns every $f \in S$.

As $M_{g(e)}$ converges on $\Theta_e(f)$ to some index e'', the learner N almost always outputs $h(e'')$. As N is consistent for $f \in S$, it follows that $h(e'')$ extends for infinitely many n the string $f(0)f(1)\ldots f(n)$. This implies that $h(e'')$ is an index for f and N is an Ex-learner for f.

Conclusion. Claim 3 shows that N Ex-learns S and Claim 2 shows that all intermediate hypotheses for $f(0)f(1)\ldots f(n)$ with $f \in S$ exist and are consistent with the data seen so far. Thus N is a consistent Ex-learner for S. ∎

The above proof can even be made effective in the sense that one can construct for every $\Theta_d(S)$ a consistent learner $M_{g'(d)}$ and so one obtains that uniformly robustly Ex-learnable classes are also uniformly robustly consistently learnable. Furthermore, every consistent learner is conforming, so one has the following corollary.

Corollary 5.2. *If S is uniformly robustly Ex-learnable then S is conforming learnable.*

One of the main open questions in the theory of robust learning is whether the preceding result can be generalized to robust Ex-learning.

Problem 5.3. *Is every robustly Ex-learnable class also consistently learnable?*

A further interesting question is to study the relations between robust learning and the variants of consistent and conforming learning. The most restrictive variants are those of globally conforming and globally consistent learners. These learners have to be total and for every input $y_0y_1\ldots y_n$ they are defined and output an index e which satisfies for each $x \leq n$ either $\varphi_e(x)\uparrow$ or $\varphi_e(x)\downarrow = y_x$; for the criterion of global consistency the first case never and the second case always applies.

It is easy to adapt the proof of Theorem 3.7 to show that every robustly globally conforming learnable class is in Num. The key idea of this adaptation is the following. Given such a learner M for the $\Theta(S)$ defined there, every U_τ which contains for a given length more than two strings indicates that the corresponding index $M(\tau)$ does not compute a total function. Therefore one can define the g_τ by checking whether there is a unique string and outputting 0 if this fails:

$$g_\tau(x) = \begin{cases} \eta(x) & \text{if } \eta \text{ is the unique string of } U_\tau \text{ of length } |\tau| + x + 1; \\ 0 & \text{if such an } \eta \text{ does not exist or is not unique.} \end{cases}$$

The rest of the proof carries over directly. So one has the following.

Corollary 5.4 [6, 31]. *If S is robustly globally conforming learnable or robustly globally consistently learnable then S is in Num.*

Between these two extremes for consistent learning, there is the case where one adds to the standard definition the requirement that the learner has to be total although it might output inconsistent hypotheses for inputs not belonging to any function in the class S to be learned. A similar version can also be considered for conforming learning. These notions can be separated.

Theorem 5.5 [6]. *There is a class which is uniformly robustly Ex-learnable but which does not have a total conforming learner.*

Theorem 5.6 [6]. *There is a class S and a recursive function g such that $M_{g(e)}$ is a total conforming learner for every class $\Theta_e(S)$ where Θ_e is general recursive but S itself does not have a total consistent learner.*

Theorem 5.7 [6]. *There is a class S and a recursive function g such that $M_{g(e)}$ is a total consistent learner for every class $\Theta_e(S)$ where Θ_e is general recursive but S is not in Num.*

6. Hyperrobust Learning

As even uniformly robust learning does not rule out self-referential coding, Ott and Stephan [20] introduced the notion of hyperrobust learning. The key idea is that the learner has to deal not only with one image $\Theta(S)$ but with a lot of images at the same time without being able to distinguish among these. But one cannot take all general recursive operators at the same time, since every recursive function is the image of the function 0^∞ under a suitable general recursive operator. Therefore Ott and Stephan [20] defined that a class is hyperrobustly learnable iff there is *one* learner learning the image of *every* function in S under *every* primitive recursive operator.

It is shown in Theorem 6.3 below that the hyperrobustly learnable classes remain the same if one takes any larger recursively enumerable class of general recursive operators instead of the one above. Furthermore, hyperrobust learning is compatible to the standard notion of robust learning: if S is hyperrobustly learnable then S is also robustly learnable. Moreover, if S is closed under finite variants then both notions are equivalent. The set $[S]$ of all images of functions in S under primitive recursive operators is dense for every nonempty class S of functions. Thus, hyperrobust learning cannot respect any bounds on the number of mind changes. Therefore, it is not suitable to combine hyperrobust learning with the notions of Ex_k-learning and confident learning. However there are interesting results for Ex-learning, BC-learning and their team variants.

The notion of hyperrobust learning also has a more intuitive motivation: Assume that a learner M can learn all axis-parallel rectangles in the plane. Certainly, one assumes that from M one can build a learner which infers all rotated rectangles in addition. However, clearly one does not want to build, for every different rotation Θ, a learner succeeding just on rectangles mapped by the rotation Θ. Instead, one is interested in a learner which infers every image of any axis-parallel rectangle under *any* rotation Θ. The notion of hyperrobustness reflects this situation by requiring that *one* learning machine M learns every image of the functions in a class S under *all* primitive recursive operators. Now the informal definition is made more precise.

Definition 6.1. An operator Θ is called primitive recursive iff there are primitive recursive functions g_1, g_2 such that

$$(\forall \text{ total } f)\,[\Theta(f)(x) = g_1(x, f(0)f(1)\ldots f(g_2(x)))].$$

Let I be one of the criteria Ex, BC or their team versions.

A class S is *hyperrobustly learnable* with respect to the criterion I if there is one learner M which learns every function in

$$[S] = \{\Theta(f) : \Theta \text{ is primitive recursive and } f \in S\}$$

with respect to the criterion I.

As the criteria Ex and BC permit total learners, learners to witness that $[S]$ is Ex-learnable or BC-learnable will be assumed to be total in the following. Hyperrobust learning satisfies two simple observations.

Fact 6.2. *Given a non-empty class S, $[S]$ contains all primitive recursive functions since, for every primitive recursive function g, there is a primitive recursive operator Θ which maps every function f to g: $\Theta(f) = g$.*

No class S is hyperrobustly learnable with any bound on the number of mind changes, since $[S]$ is dense and dense classes cannot satisfy mind change bounds.

These two facts establish a real difference between hyperrobust learning and robust learning because there are classes of recursive functions which are robustly learnable with at most one mind change [14]. On the other hand, for hyperrobust learning, the notions Ex, BC and their team-variants are the most interesting ones.

The definition of the mapping $S \to [S]$ and thus the definition of hyperrobustness are based on the class of primitive recursive operators. The decision to choose the class of primitive recursive operators may seem to be just arbitrary and one may wonder how other choices for the class of operators affect the notion of hyperrobustness. The next result justifies the definition: If in the definition of hyperrobust learning, the class of primitive recursive operators is replaced by a larger recursively enumerable class of general recursive operators, then one still gets the same learning notion. So the definition of hyperrobust learning does not depend on the actual choice of the class of operators as long as this class is "sufficiently rich" (for example, if the class contains all primitive recursive operators, or, all polynomial time computable operators). Clearly, if the class of operators contains only the identity operator, then hyperrobust and ordinary Ex-learning coincide and so, "sufficiently rich" is a necessary and natural postulate.

Theorem 6.3 [20]. *If S is hyperrobustly Ex-learnable or hyperrobustly BC-learnable and U is any recursively enumerable class of general recursive operators, then also the class*

$$\{\Theta(f) : f \in S \wedge \Theta \in U\}$$

is Ex-learnable or BC-learnable, respectively.

Corollary 6.4. *If S is hyperrobustly learnable with respect to some criterion I, then S is also robustly learnable with respect to the same criterion I.*

Theorem 6.5 [20]. *If S is closed under finite variants then S is hyperrobustly learnable with respect to a criterion I iff S is robustly learnable with respect to the same criterion I.*

Corollary 6.4 and Theorem 6.5 show that hyperrobust learning is a natural generalization of robust learning: it is equivalent to first taking the closure under all finite variants and then applying a suitable general recursive operator Θ.

Intuitively, the notion of robustness was designed to prevent coding tricks: for example, if $f(2x)$ is a program for f for almost all x, then the general recursive operator mapping f to $f(1)f(3)f(5)\ldots$ destroys this coding trick. Such coding tricks are called *numerical* since the self-referential information is directly contained in the numerical values of the function.

Recall that Jain, Smith and Wiehagen [14] showed that the class TSD of the topologically self-describing functions is uniformly robustly Ex-learnable. So these topological coding cannot be destroyed by using a single general recursive operator. However, topological coding is destroyed by adding all finite variants of the functions in S to the class to be learned. Combining these two methods, that is, considering robust learning of classes closed under finite variants, one might hope that no coding tricks are left. Indeed, this hope is confirmed by the following characterization result which shows that the hyperrobustly Ex-learnable classes coincide with the classes in Num. So the next result gives a further learning-theoretic characterization for Num.

Theorem 6.6 [20]. *A class S is hyperrobustly Ex-learnable iff it is in Num.*

Proof. One direction is straightforward: If S is in Num, so is $[S]$. That is, if $S \subseteq \{f_0, f_1, \dots\}$ for an enumeration $f_0, f_1, \dots$ of total functions, then $[S]$ is contained in the enumeration of all $g_{e,e'} = \Psi_e(f_{e'})$, where $\Psi_0, \Psi_1, \dots$ is an enumeration of all primitive recursive operators.

For the converse direction, let M be an Ex-learner for $[S]$. Furthermore, one considers now the operator Θ_e from the proof of Theorem 5.1 where one uses M in place of the $M_{g(e)}$ there. The operator Θ_e constructed there is primitive recursive and as primitive recursive operators are closed under composition, it holds that $\Theta_e([S]) \subseteq [S]$. Thus M learns $\Theta_e([S])$ and one has that $[S]$ is consistently learnable via some N. As $[S]$ is dense, the learner N has to be total and to be consistent on every input. Therefore, N is a globally consistent learner. Furthermore, one can choose the tree T in the proof of Theorem 3.7 such that the resulting Θ is also primitive recursive. As a consequence, $\Theta([S]) \subseteq [S]$ and N is a consistent Ex-learner for $\Theta([S])$. As a globally consistent learner is also reliable, it follows from the proof of Theorem 3.7 that $\Theta([S])$ and thus also $[S]$ and S are in Num. ▌

As hyperrobust Ex-learning is equivalent to Num, it remains to investigate how hyperrobust BC-learning behaves. The next result shows that hyperrobustly BC-learnable classes share one property of classes in Num, namely that they are bounded in the following sense.

Definition 6.7. A class S is bounded if there is a total recursive function g which dominates every $f \in S$:

$$(\forall f \in S)\,(\exists x)\,(\forall y \geq x)\,[f(y) \leq g(y)].$$

Theorem 6.8 [20]. *If S is hyperrobustly BC-learnable, then S is bounded.*

Proof. It is convenient to represent the BC-learner as a machine M which predicts the function to be learned almost everywhere but which may, for each function in the class to be learned, be undefined at finitely many places, see

Theorem 2.3.

Now one defines inductively, for every f, the following function $\Theta(f) = \lim_x \sigma_x$ starting with $\sigma_0 = \lambda$ and using $M_s(\sigma_x)$ as a notation for the result of $M(\sigma_x)$ after s computational steps (which is either undefined or $M(\sigma_x)$):

$$\sigma_{x+1} = \begin{cases} \sigma_x 1 & \text{if } M_s(\sigma_x)\downarrow = 0 \text{ for some } s \leq f(0) + f(1) + \ldots + f(x); \\ \sigma_x 0 & \text{otherwise.} \end{cases}$$

Since Θ is a primitive recursive operator, M has to infer $\Theta(f)$ for every $f \in S$. But whenever $\Theta(f)(x)$ is 1, then M has made a prediction mistake and so, $\Theta(f)$ takes only finitely often a value different from 0. Since, by Fact 6.2, M has to infer every primitive recursive function, M learns, in particular, all functions of the form $\sigma 0^\infty$. Thus, the following function g is recursive:

$$g(x) = \max\{\min\{s : (\exists t < s)\,[M_s(\sigma 0^t)\downarrow = 0]\} : \sigma \in \{0,1\}^x\}.$$

Whenever $f(x) > g(x)$ then one finds within $f(x)$ steps some $t < f(x)$ such that $M(\sigma_x 0^t)\downarrow = 0$. So, if $\sigma_{x+t} = \sigma_x 0^t$, then $\sigma_{x+t+1} = \sigma_x 0^t 1$, in particular $\sigma_{x+t+1} \neq \sigma_x 0^{t+1}$. Thus there is a $y \in \{x, x+1, \ldots, x+t\}$ such that $\sigma_{y+1} = \sigma_y 1$. This happens only if $M(\sigma_y)\downarrow = 0$ and so M makes a false prediction on $\Theta(f)$ beyond x. But since M infers $\Theta(f)$, there exist at most finitely many x with $f(x) > g(x)$. Therefore, g dominates f. Since the construction of g does not depend on the actual choice of f, g dominates every function in S. ∎

Ott and Stephan [20] showed that there are classes S not in Num which are hyperrobustly BC-learnable. A major tool in this research is the use of recursively bounded recursive trees [18, vol. I, page 509]; these trees are called *bounded recursive trees* from now on. These trees are a generalization of binary recursive trees: for a bounded recursive tree T one can compute for every $\sigma \in T$ a complete list of the immediate successors in T. In general, this is impossible even if σ has only finitely many successors. But it holds when some recursive function b bounds the size of the set of successors; that is, when $a \leq b(|\sigma|)$ for all $\sigma a \in T$. So, one can define a bounded recursive tree as a recursive function c which associates with every $\sigma \in T$ a finite and explicit list of all nodes $\sigma a \in T$. If c is primitive recursive then T is called a *bounded primitive recursive tree.*

Recall that a learning machine M is said to be *reliable* if, for any input function f, M either converges to a correct program for f or outputs infinitely often a signal for divergence, which, in the case of Ex-learning, can simply be a mind change. Producing semantic mind changes alone is not sufficient to get a reliable version of BC-learning that differs from ordinary BC, as the following fact shows. This fact is based on two observations: First, behavioural correct learners can be made consistent. That is, the new consistent learner outputs for every input a hypothesis which is correct on the data seen so far [2, 9]. Second, consistent learners either converge semantically to the desired function or make infinitely many semantic mind changes.

Fact 6.9. *For every BC-learnable class S there is a BC-learner for S which, on any total function f, either converges semantically to programs for f or makes infinitely many semantic mind changes.*

Since Fact 6.9 states that BC-learners can be made semantically divergent on functions not learned, the analogue of reliable learning for BC must signal divergence more explicitly. A suitable definition is the following: The reliable BC-learner indicates divergence either by outputting a special value like "?", or, by making a definitely wrong prediction where the underlying BC-learner is given by a partial recursive predictor as in Theorem 2.3.

Definition 6.10. M is a *reliable BC-learner* if, for every total function f, either M BC-learns f by predicting almost always the correct value (that is, for almost all x, $M(f(0)f(1)\dots f(x))\downarrow = f(x+1)$) or M diverges on f by outputting infinitely often either "?" or a defined but wrong prediction.

Zeugmann [31] observed that robustly reliably Ex-learnable classes are just those in Num. Together with Theorem 6.6, one obtains the following equivalence.

Fact 6.11. *For a bounded class S the following three statements are equivalent.*

- *S is in Num;*
- *S is reliably Ex-learnable;*
- *S is hyperrobustly Ex-learnable.*

The central question of the following results in this section is: to what extent can the equivalence of the statements above be transferred to BC? The next characterization of bounded reliably BC-learnable classes is an important tool to attack this question.

Theorem 6.12 [20]. *A bounded class S is reliably BC-learnable iff there is a recursively enumerable family $T_0, T_1, \dots$ of bounded recursive trees such that every tree has only finitely many infinite branches and every $f \in S$ is an infinite branch of such a tree.*

Proof. ($\Rightarrow$): Assume that g bounds S and M is a partial recursive predictor for S which in addition signals infinitely often divergence on every function f

which M does not learn. Now let the tree T_σ contain all prefixes of σ plus all $\eta \succeq \sigma$ such that

- $\eta(x) \leq g(x)$ for all $x \in \text{domain}(\eta) - \text{domain}(\sigma)$ and
- there are no τ, a with $\sigma \preceq \tau a \prec \eta$ and $M_{|\eta|}(\tau)\downarrow \neq a$,

where, of course, the special symbol "?" is different from a.

Clearly, the trees T_σ form a recursively enumerable family of trees bounded by g. Assume now that T_σ has infinitely many infinite branches. As a consequence of König's Lemma and the fact that T_σ is finitely branching, there is an infinite branch f which is not isolated. So for every x there is an $y > x$ and a further infinite branch h of T_σ such that f and h both agree on $0, 1, \ldots, y$ but differ on $y + 1$. Then it cannot happen that $M(f(0)f(1)\ldots f(y))$ predicts $f(y+1)$ or $h(y+1)$ since otherwise either f or h would not be an infinite branch of T_σ. Furthermore, M does also not signal divergence for any input $f(0)f(1)\ldots f(x)$ with $x \geq |\sigma|$ since then f could not be an infinite branch of T_σ. This contradicts the assumption that M is a reliable BC-learner and therefore no tree T_σ has infinitely many infinite branches.

For every $f \in S$, there is a prefix $\sigma \preceq f$ such that M predicts f correctly after seeing σ and all x with $f(x) > g(x)$ are in domain(σ). Then it follows from the definition that f is an infinite branch of T_σ. Direction ($\Rightarrow$) is completed.

($\Leftarrow$): Let $T_0, T_1, \ldots$ be a recursively enumerable family of bounded recursive trees such that every tree has only finitely many infinite branches and every function in S is a branch of such a tree. Without loss of generality, the family is dense in the sense that for every σ there is a tree containing σ. This can be achieved by adding all finite trees of the form $\{\tau : \tau \preceq \sigma\}$ to the list. The new family is still recursively enumerable and the class of functions on trees in the family remains the same. Let $T[\tau]$ denote all nodes of the tree T which are comparable to τ. Now the reliable BC-learner works as follows:

- $M(\sigma)$ finds the first tree T_e with $\sigma \in T_e$.
- If there is a recent change of the tree, that is, if there is $e' < e$ with $\tau \in T_{e'}$ for all $\tau \prec \sigma$, then $M(\sigma) = ?$ in order to signal divergence.
- Otherwise $M(\sigma)$ searches for an a such that $T_e[\sigma b]$ is finite for all $b \neq a$ and $M(\sigma) = a$, if such an a is found.

The first step of the algorithm is well-defined since every σ is a node of some tree T_e.

If f is not an infinite branch of any tree T_e, then, during the inference of f, M signals infinitely often divergence, since M has to change the trees infinitely often.

If f is an infinite branch of some tree then there is a first such tree T_e in the enumeration. For sufficiently large $\sigma = f(0)f(1)\dots f(x)$, f is the only infinite branch of $T_e[\sigma]$ and $\sigma \notin T_{e'}$ for any $e' < e$. Now $f(x+1)$ is the unique value a with $T_e[\sigma a]$ being infinite. Since the trees T_e are uniformly bounded recursive trees, a suitable search algorithm finds the value $f(x+1)$. Therefore, M correctly predicts f almost everywhere; that is, $M(f(0)f(1)\dots f(x))\downarrow = f(x+1)$ for almost all x.

So, for every function f, M either BC-learns f (in the prediction model) or M signals infinitely often divergence. ∎

Example 6.13 (Class BWT). Case, Kaufmann, Kinber and Kummer [7] showed that there is a recursively enumerable family $\{T_0, T_1, \dots\}$ of binary recursive trees of width two such that the class

$$BWT = \{f : (\exists e)\,[f \text{ is infinite branch of } T_e]\,\}$$

is not Ex-learnable. This class BWT is reliably BC-learnable. So BWT witnesses that for bounded classes, the concept of reliable BC-learning is also a proper generalization of Num.

As one can show that BWT is even uniformly robustly BC-learnable, one gets an alternative proof for Jain's separation of uniformly robust BC-learning from Ex-learning in Theorem 4.8.

Together with the next result one obtains that the three notions from Fact 6.11 become all different for BC: Num is properly included in bounded reliable BC, which is properly included in hyperrobust BC.

Theorem 6.14 [20]. *If S is a bounded and reliably BC-learnable class then S is also hyperrobustly BC-learnable. But the converse implication does not hold.*

So, the main theorems of this section showed that there is a class S which is hyperrobustly BC-learnable but not hyperrobustly Ex-learnable.

Theorem 6.15 [20]. *There is a hyperrobustly BC-learnable class which is not hyperrobustly Ex-learnable.*

Zeugmann [31] showed that a class S is robustly reliably Ex-learnable iff S is in Num. One can deduce from it that S is robustly reliably Ex-learnable iff S is bounded. Zeugmann's diagonalization strategy for this result can be transferred to the case where reliably BC-learnable classes are considered. That is, an unbounded reliably BC-learnable class is not robustly reliably learnable.

Theorem 6.16 [20]. *A reliably BC-learnable class S is also robustly reliably BC-learnable iff S is bounded.*

7. Team-Learning and the Union-Theorem

Bārzdiņš [2, 3] showed that there are explanatorily learnable classes S_1, S_2 such that their union is not explanatorily learnable. This result can be easily generalized to the fact that there are unions of $n+1$ learnable classes which are not contained in the union of n learnable classes [24]. Pitt and Smith [21, 22] showed that these unions can also be characterized in terms of probabilistic learners and teams of learners. That is, the following statements are equivalent for Ex-learning as well as for BC-learning.

- S is contained in the union of n learnable classes.
- Some probabilistic machine learns S with some probability p where $p > \frac{1}{n+1}$.
- A (h, k)-team with $\frac{h}{k} > \frac{1}{n+1}$ learns S in the sense that there are k learners such that, for every $f \in S$, at least h of them learn f.

Note that the probability and the fraction $\frac{h}{k}$ of successful machines in the team have to be greater than $\frac{1}{n+1}$, since a team of $k = h \cdot (n+1)$ learners, where h learners have to succeed, can already infer the union of $n+1$ learnable classes: the first h learners follow the algorithm to learn S_1, the second h learners follow the algorithm to learn S_2, ..., the last h learners follow the algorithm to learn S_{n+1}. Recall that Example 4.5 stated that the classes AEZ and TSD are both uniformly robustly Ex-learnable while their union $AEZ \cup TSD$ is not BC-learnable. Thus the Non-Union Theorem holds for the criteria of robust and uniformly robust Ex-learning and BC-learning.

For hyperrobust Ex-learning, one can show that this connection between team-learning on the one side and unions on the other side no longer holds. The hyperrobustly Ex-learnable classes are closed under union but teams of $n+1$ hyperrobust Ex-learners are more powerful than teams of n learners. An intuitive explanation for this fact is that if $[S_1 \cup S_2]$ needs a team of two Ex-learners then so does $[S_1]$ or $[S_2]$. So, the closure operation does not permit to split a class of functions into two classes which are really easier to learn.

Fact 7.1. *If S_1 and S_2 are hyperrobustly Ex-learnable, so is $S_1 \cup S_2$.*

The result follows from the equivalence of hyperrobust Ex-learning and Num and from the fact that Num is closed under union. Note that the corresponding question for BC-learning is open.

Problem 7.2. *If S_1 and S_2 are hyperrobustly BC-learnable, is $S_1 \cup S_2$ also hyperrobustly BC-learnable?*

The next result establishes that the team hierarchies for hyperrobust Ex-learning and hyperrobust BC-learning are proper. This stands, for the Ex-case, in contrast to the collapse of the union-hierarchy.

Theorem 7.3 [20]. *The team hierarchies for hyperrobust Ex-learning and hyperrobust BC-learning are proper.*

Proof. This proof needs the notion of the rank of a tree T which measures the embeddability of complete binary trees into T as follows: The *rank of a tree* T is the maximal k for which there is a finite function g with domain $\bigcup_{k' \leq k} \{0,1\}^{k'}$ such that the range of g is a subset of T and $g(\sigma) \preceq g(\tau) \Leftrightarrow \sigma \preceq \tau$ for all strings σ, τ in the domain of g. If there is, for every k, such a function g then the rank of T is ∞.

Let S_k be the set of all functions which are infinite branches of some bounded primitive recursive tree of rank up to k; there is a uniformly recursive enumeration of these trees, in the sense that one can not only check whether $\sigma \in T_e$ but also compute an explicit list of the immediate successors of σ in T_e if $\sigma \in T_e$.

Given f, the learning algorithm first finds (in the limit) a tree T such that f is an infinite branch of T. Having found this tree T, one uses the algorithm of Case, Kaufmann, Kinber and Kummer [7]. Knowing an index of the tree and having a primitive recursive function majorizing all infinite branches, their algorithm learns the function by a team of $k+1$ Ex-learners or k BC-learners, respectively. This team-size is optimal. The class S_k is closed; that is, $[S_k] = S_k$. So, it follows that S_k is learnable by a team of hyperrobust learners of size k (BC) and $k+1$ (Ex), respectively, but not by a smaller team. ∎

Note that the above classes S_k are also uniformly robustly Ex-learnable by teams of $k+1$ Ex-learners and uniformly robustly BC-learnable by teams of k BC-learners. So these classes witness that the hierarchies are also proper for robust and uniformly robust Ex-learning and BC-learning.

Furthermore, for hyperrobust Ex-learning, one can even show that there exists a proper team hierarchy *within* the class of all hyperrobustly BC-learnable functions. The n-th level of this hierarchy is given by the class of all infinite branches of bounded primitive recursive trees of width up to n; that is, of trees which have at every depth at most n nodes.

8. Learning Aided by Context

Learning aided by context means that the learner receives not only the data of the function to be learned but also the data of some further $f \in S$ and might try to exploit this further data-source for the learning process. Case, Jain, Ott, Sharma and Stephan [5] investigated robust notions of learning aided by context. The straight forward definition would be that S is Ex-learnable aided by context iff for every $f \in S$ and every $g \in S - \{f\}$ the learner M converges on input f, g to an index for f. But it turned out that this notion coincides with standard Ex-learning. Therefore the alternative version where the context is selected is the more interesting one.

Definition 8.1. A class S is learnable aided by context iff there is a mapping assigning to every $f \in S$ a context $g_f \in S$ and S has a total Ex-learner M (taking as input, two functions) such that for all $f \in S$, M converges on input f and context g_f to an index e for f:

$$(\forall^\infty n)\, [M(f(0)f(1)\dots f(n), g_f(0)g_f(1)\dots g_f(n)) = e \wedge f = \varphi_e].$$

Furthermore, S is robustly Ex-learnable aided by context iff for every general recursive operator Θ there is an M Ex-learning every $\Theta(f)$ from input $\Theta(f)$ and context $\Theta(g_f)$.

It was shown that many but not all classes are robustly Ex-learnable aided by context.

Theorem 8.2 [5]. *If S has a dense and recursively enumerable subclass then S can be robustly Ex-learned aided by context. In particular the class of all recursive functions is robustly Ex-learnable aided by context. But on the other hand there is an infinite class which is not robustly Ex-learnable aided by context.*

The further investigations about learning aided by context dealt with the question, how learnability is affected if some requirements are imposed on the mapping $f \to g_f$. The first result shows that for many classes the context becomes useless if it is required that the context is generated by a general recursive operator.

Theorem 8.3 [5]. *If S is closed under finite variants and S is robustly Ex-learnable aided by context via a mapping $f \to g_f$ computed via a general recursive operator Θ then S is already robustly Ex-learnable.*

In the following the hyperrobust variant of the notion of Ex-learning aided by context is introduced. Note that it is important that one requires Ex-learning only if f and g_f are transformed by the same primitive recursive operator.

Definition 8.4 [20]. A class S is hyperrobustly Ex-learnable aided by context if there exist a learner M and, for every $f \in S$, a context function $g_f \in S$ such that, for every primitive recursive operator Θ, M Ex-learns the function

$\Theta(f)$ from the data-stream of the pair $(\Theta(f), \Theta(g_f))$. Similarly one defines the notion of hyperrobust BC-learning.

The next result shows that for Ex-learning this notion collapses to Num.

Theorem 8.5 [20]. *If a class S is hyperrobustly Ex-learnable from selected context then S is already hyperrobustly Ex-learnable and therefore S is in Num.*

Proof. The general idea of the proof is to split S into two classes S_1 and S_2 and to show that both $[S_1]$ and $[S_2]$ are in Num. This implies that S is in Num, too. Assume that M hyperrobustly learns S from selected context.

Let S_1 be the class of all functions $f \in S$ such that M learns f from the context f itself. Correspondingly, S_2 denotes the class of all $f \in S$ such that for all contexts $g \in S$ from which M infers f, it holds that $g \neq f$.

For any $f \in S$ and any primitive recursive operator Θ, the learner M Ex-infers $\Theta(f)$ from $\Theta(f)$ itself plus the image of the context under Θ. So, simulating M by adding the context $\Theta(f)$ to $\Theta(f)$ itself, one obtains an Ex-learner for $[S_1]$. As a consequence, $[S_1]$ and the generating class S_1 are both in Num.

For any $f \in S_2$, for the context g of f and for any primitive recursive operator Θ, there is a further primitive recursive operator Θ' such that $\Theta'(f) = \Theta(f)$ and $\Theta'(g) = 0^\infty$. This implies that M Ex-learns $\Theta(f)$ from the context 0^∞. Again one can simulate the learner by adding just the context 0^∞ to the input function. Thus, it follows that $[S_2]$ and S_2 are in Num, too.

For any two classes in Num, their union is also in Num. In particular, the union S of S_1 and S_2 is in Num, which completes the proof. ▮

For BC-learning, the corresponding question seems to be more difficult and is linked to Open Problem 7.2. If the answer to that problem is that hyperrobust BC is not closed under union, then selected context offers a real support for hyperrobust BC-learning.

Proposition 8.6 [20]. *A class is hyperrobustly BC-learnable aided by selected context iff it is contained in the union of two hyperrobustly BC-learnable classes.*

Proof. The proof of Theorem 8.5 for Ex can be transferred to BC such that $[S]$ is shown to be the union $[S_1] \cup [S_2]$ of two BC-learnable classes $[S_1]$ and $[S_2]$. Only the last argument that the union of classes in Num is again in Num cannot be transferred to the BC-case since Problem 7.2 is still open.

For the converse direction, let the hyperrobustly BC-learnable classes S_1 and S_2 be given. Now, one assigns to every $f \in S_1$ the context f itself and to every $f \in S_2 - S_1$ some fixed context $g \in S_1$. Let M_1 and M_2 be the BC-learners for the classes $[S_1]$ and $[S_2]$. Now the context-aided BC-learner M follows the output of M_1 as long as the function and the context are equal and changes to the output of M_2 when the context differs from the function itself. More precisely, for input σ, τ of the same length:

$$M(\sigma,\tau) = \begin{cases} M_1(\sigma) & \text{if } \sigma = \tau; \\ M_2(\sigma) & \text{otherwise.} \end{cases}$$

Now let $h = \Theta(f)$ be the function to be learned. If the context equals h then either $f \in S_1$ or Θ maps f and its context g to the same function. In the first case, the algorithm is correct since M_1 BC-learns $[S_1]$. In the second case, $h = \Theta(g)$ and h is again in $[S_1]$ since $g \in S_1$. Thus, M_1 is again a correct algorithm. If the context is different from h, then the original context must be different from f and $f \in S_2$. It follows that $h = \Theta(f)$ is in $[S_2]$. M outputs almost always the guesses of M_2 and thus M BC-learns h in this case, too. ∎

9. Conclusion

Although self-referential coding is an elegant proof-method, the resulting classes used as witnesses to separate some learning criteria look artificial. So Bārzdiņš asked whether there are really naturally defined classes which are learnable but not already learnable by enumeration. To formalize this, Bārzdiņš proposed the notion of robust learning and conjectured that every robustly learnable class is already in Num; as a consequence every natural learnable class would already be in Num.

Zeugmann [31] published intermediate results on the way to solve Bārzdiņš' Conjecture and showed that it holds for the notion of reliable learning. Fulk [10] refuted Bārzdiņš' Conjecture and showed that there are robustly Ex-learnable classes which are not in Num. Later Jain, Smith and Wiehagen [14] found even a class defined with a topological version of self-referential coding which is robustly Ex-learnable. Jain [12] extended this line of research by giving an example of a robustly BC-learnable class which cannot be Ex-learned. Similar robust separations for several notions of consistency followed [6].

There were several attempts to tighten the notion of robust learning such that all types of coding were ruled out. In the case of uniformly robust learning the approach failed. In the case of hyperrobust learning, the approach succeeded for Ex-learning by collapsing this notion to Num, so that Bārzdiņš' Conjecture holds for this more restrictive variant of robustness. Furthermore, hyperrobust BC-learning turns out to contain some classes outside Num without applying self-reference. It has been shown that uniformly robust learning and robust learning are different for the criteria of confident learning, Ex-learning and BC-learning.

Although much work has been done, several interesting open problems remain.

(1) Is there a more appropriate way to model learnability in a natural way? On the one hand, the notions of robust and uniform robust Ex-learning cannot prevent all coding tricks. On the other hand, the three notions of robust, uniformly robust and hyperrobust learning all fail to learn the quite natural class BB from Example 3.6 which was introduced by Blum and Blum [3, 25].

(2) Is every robustly Ex-learnable class also consistently learnable? It is a surprising result that this holds for uniformly robust Ex-learning [6] but all attempts to generalize it have failed so far.

(3) Does the Union-Theorem hold for hyperrobust BC-learning? The characterization that the hyperrobust Ex-learnable classes are exactly the classes in Num implies as a corollary the Union-Theorem. As this proof-method cannot be applied in the case of hyperrobust BC-learning, there is a certain chance that the union of some hyperrobustly BC-learnable classes is not hyperrobustly BC-learnable.

Acknowledgments. The authors would like to thank the referee for useful comments and suggestions. Further thanks go to the co-authors (John Case, Wolfgang Merkle, Matthias Ott, Arun Sharma, Carl Smith, Rolf Wiehagen and Thomas Zeugmann) of joint research reported in this survey.

References

[1] Janis Bārzdiņš. Inductive inference of automata, functions and programs. In *Int. Math. Congress, Vancouver*, pages 771–776, 1974.

[2] Janis Bārzdiņš. Two theorems on the limiting synthesis of functions. In *Theory of Algorithms and Programs*, Latvian State University, Riga, 210:82–88, 1974.

[3] Lenore Blum and Manuel Blum. Towards a mathematical theory of inductive inference. *Information and Control*, 28:125–155, 1975.

[4] John Case. Learning Machines. In Demopoulos, W. and Marras, A., editors, *Language Learning and Concept Acquisition*, pages 83–102, Ablex Publishing Company, 1986.

[5] John Case, Sanjay Jain, Matthias Ott, Arun Sharma and Frank Stephan. Robust learning aided by context. *Journal of Computer and System Sciences* (*Special Issue COLT 1998*), 60:234–257, 2000. Extended Abstract in *Proceedings of the Eleventh Annual ACM Conference on Computational Learning Theory* – COLT 1998, ACM-Press, 44–55, 1998.

[6] John Case, Sanjay Jain, Frank Stephan and Rolf Wiehagen. Robust learning – rich and poor. *Proceedings of the Fourteenth Annual Conference on Computational Learning Theory and the Fifth European Conference on Computational Learning Theory* – COLT 2001 and EuroCOLT 2001, Springer LNAI 2111:143–159, 2001. Also available as: Technical Report LSA-2000-03E, Centre for Learning Systems and Applications, Department of Computer Science, University of Kaiserslautern, 2000.

[7] John Case, Susanne Kaufmann, Efim Kinber and Martin Kummer. Learning recursive functions from approximations. *Journal of Computer and System Sciences*, 55:183–196, 1997.

[8] John Case and Carl Smith. Comparison of identification criteria for machine inductive inference. *Theoretical Computer Science*, 25:193–220, 1983.

[9] Jerome Feldmann. Some decidability results on grammatical inference and complexity. *Information and Control*, 20:244–262, 1972.

[10] Mark Fulk. Robust separations in inductive inference. In *Proceedings of the 31st Annual Symposium on Foundations of Computer Science* (FOCS), pages 405–410, St. Louis, Missouri, 1990.

[11] E. Mark Gold. Language identification in the limit. *Information and Control*, 10:447–474, 1967.

[12] Sanjay Jain. *Robust Behaviourally Correct Learning. Information and Computation*, 153:238–248, 1999.

[13] Sanjay Jain, Daniel Osherson, James Royer and Arun Sharma: *Systems that Learn, Second Edition* [of 19]. The MIT Press, Cambridge, Massachusetts, 1999.

[14] Sanjay Jain, Carl Smith and Rolf Wiehagen. Robust Learning is Rich. *Journal of Computer and System Sciences*, 62:178–212, 2001.

[15] Klaus-Peter Jantke and Hans-Rainer Beick. Combining postulates of naturalness in inductive inference. *Journal of Information Processing and Cybernetics* (EIK), 17:465–484, 1981.

[16] Wolfgang Merkle and Frank Stephan. Trees and learning. In *Proceedings of the Ninth Annual Conference on Computational Learning Theory* (COLT), pages 270–279, ACM Press, New York, 1996.

[17] Eliana Minicozzi. Some natural properties of strong-identification in inductive inference. *Theoretical Computer Science*, 2:345–360, 1976.

[18] Piergiorgio Odifreddi. *Classical Recursion Theory* I *and* II. North-Holland, Amsterdam, 1989 and 1999.

[19] Daniel Osherson, Michael Stob and Scott Weinstein. *Systems that Learn*. MIT Press, Cambridge, Massachusetts, 1986.

[20] Matthias Ott and Frank Stephan. Avoiding coding tricks by hyperrobust learning. *Proceedings of the Fourth European Conference on Computational Learning Theory* – EuroCOLT 1999. Springer LNCS 1572, 183–197, 1999.

[21] Lenny Pitt. Probabilistic inductive inference. *Journal of the Association of Computing Machinery*, 36:383–433, 1989.

[22] Lenny Pitt and Carl Smith. Probability and plurality for aggregations of learning machines. *Information and Computation*, 77:77–92, 1998.

[23] Karlis Podnieks. Comparing various concepts of function prediction, Part 1. *Theory of Algorithms and Programs, Latvian State University, Riga*, 210:68–81, 1974.

[24] Carl Smith. The power of pluralism for automatic program synthesis *Journal of the Association of Computing Machinery*, 29:1144–1165, 1982.

[25] Frank Stephan and Thomas Zeugmann. On the uniform learnability of approximations to non-recursive functions. *Proceedings of the Ninth Annual Workshop on Algorithmic Learning Theory* – ALT 1999, Springer LNCS 1720, 276–290, 1999. Also available as *Learning Classes of Approximations to Non-Recursive Functions*. Department of Informatics, Report 166, Kyushu University, Fukuoka, 1999.

[26] Hermann Weyl. *Symmetry*. Princeton University Press, 1952

[27] Rolf Wiehagen. Identification of formal languages. In *Mathematical Foundations of Computer Science, Proceedings, 6th Symposium, Tatranska Lomnica*, volume 53 of LNCS, pages 571–579. Springer, 1977.

[28] Rolf Wiehagen. *Zur Theorie der Algorithmischen Erkennung*. Dissertation B, Humboldt University of Berlin, 1978.

[29] Rolf Wiehagen and Walter Liepe. Charakteristische Eigenschaften von erkennbaren Klassen rekursiver Funktionen. *Journal of Information Processing and Cybernetics* (EIK), 12:421–438, 1976.

[30] Thomas Zeugmann. On the nonboundability of total effective operators. *Zeitschrift für mathematische Logik und Grundlagen der Mathematik* 30:169–172, 1984.

[31] Thomas Zeugmann. On Bārzdiņš' Conjecture. In K. P. Jantke, editor, *Proceedings of the International Workshop on Analogical and Inductive Inference* (AII'86), volume 265 of LNCS, pages 220–227. Springer, 1986.

ON PRIMITIVE RECURSIVE PERMUTATIONS

Iskander Kalimullin*
Department of Mathematics
Kazan State University
Kremlevskaja 18
Kazan 420008
Russia
Iskander.Kalimullin@ksu.ru

Abstract We show that each computable permutation is generated by polynomial time computable permutations via a three-placed term.

Keywords: computable functions, computable permutations, primitive recursive functions, polynomial time computable functions, computably enumerable sets

Introduction

Let $\mathcal{C}$ be the group of all computable permutations of ω. It is known (see [2]) that the the group $\mathcal{C}$ is generated by primitive recursive (indeed elementary) permutations. Moreover, in [2] Koz'minyh considered the following hierarchy of computable permutations:

$$\mathcal{P}_n^+ = \{p_1\pi_2^{-1}p_3 \dots p_n^{(-1)^{n+1}} : p_1, \dots, p_n \in \mathcal{P}\};$$
$$\mathcal{P}_n^- = \{p_1^{-1}p_2p_3^{-1} \dots p_n^{(-1)^n} : p_1, \dots, p_n \in \mathcal{P}\},$$

where $\mathcal{P}$ is the class of primitive recursive permutations. He proved that $\mathcal{P}_6^+ = \mathcal{P}_6^- = \mathcal{C}$ and $(\mathcal{P}_{n+1}^+ \cap \mathcal{P}_{n+1}^-) - (\mathcal{P}_n^+ \cap \mathcal{P}_n^-) \neq \emptyset$ for each n, $1 \leq n \leq 5$. Moreover, he found a convenient description for each class $\mathcal{P}_n^+, \mathcal{P}_n^-, 1 \leq n \leq 5$. For example, $p \in \mathcal{P}_2^+$ iff the graph of p is the range of some injective primitive recursive function; $p \in \mathcal{P}_5^+$ iff there is an infinite primitive recursive set A such that the complement of $p(A)$ is the range of some injective primitive recursive etc. In the spirit of these results we obtain descriptions of two special classes of computable permutations.

*Supported by grants RFBR 99-01-00174, RFBR-INTAS 97-01-71001 (IR-97-139).

Since $\mathcal{P}_6^+ = \mathcal{C}$, each computable permutation p is presentable in the form $p = p_1p_2^{-1}p_3p_4^{-1}p_5p_6^{-1}$, where p_i is a primitive recursive permutation for $1 \le i \le 6$. In this paper we consider the following two questions: In this presentation, can we get at least two permutations p_i and p_j, $i \neq j$, to be equal? Can we get every p_i to be a polynomial time computable permutation? We answer both questions by showing that each computable permutation p can be presented in the form $p = p_1p_2^{-1}p_3p_1^{-1}p_2p_1^{-1}$, where p_i is a primitive recursive permutation for $1 \le i \le 3$.

Our notation is standard and generally follows [3], [4] and [6]. In particular, ω is the set of all natural numbers with zero, $\langle x, y\rangle = \frac{(x+y)(x+y+1)}{2} + x$ (the standard pairing function from $\omega \times \omega$ onto ω). Let l and r be functions such that $\langle l(x), r(x)\rangle = x$ for all $x \in \omega$.

The graph of a function f is the set $graph\, f = \{\langle x, y\rangle : y = f(x)\}$. The domain and the image (range) of a (partial) function g will be abbreviated by $dom\, g$ and $rng\, g$, respectively. The notation $f(A)$ means the image of a set $A \subseteq \omega$ under a function f. If f and g are partial functions on ω then we denote by fg the composition of functions f and g, i.e., $dom\, fg = \{x \in dom\, g : g(x) \in dom\, f\}$ and $fg(x) = f(g(x))$ for all $x \in dom\, fg$. If a total function f is injective then we denote by f^{-1} its inverse, i.e. $dom\, f^{-1} = rng\, f$ and $f^{-1}(f(x)) = x$ for all $x \in \omega$.

For partial functions f and g on ω we denote by $f \otimes g$ the following function:

$$dom\, f \otimes g = dom\, f \cap dom\, g$$
$$(f \otimes g)(x) = \langle f(x), g(x)\rangle \text{ for all } x \in dom\, f \otimes g.$$

The symbol μ denotes the minimalization operator: if $R(x)$ is some predicate then $\mu x R(x)$ is equal to the least x such that $R(x)$ is true (if there is no such x then $\mu x R(x)$ is undefined).

Also we will identify the set ω of all naturals with the set of all finite strings in the alphabet $\{0, 1\}$. The length of a string x we denote by $|x|$. The notation $[c_1, \dots, c_n]$ means the string of the length n with symbols $c_1, \dots, c_n$. In particular, the notation $[0]^n$ means the string of length n which contains only symbols 0. The concatenation of strings x and y we denote by $x * y$. We write $x \preceq y$ if the string x is an initial segment of the string y. We write $x \prec y$ if $x \preceq y$ and $x \neq y$. We also assume that the usual linear ordering $<$ and the partial ordering $\prec$ are concordant, i.e., if $x \prec y$ then $x < y$. Therefore, if we write $z = (\mu x \preceq y)R(x)$ then this means that z is an initial segment of y such that $R(z)$ and the length of z is minimal.

A function f on ω is polynomial time computable if there is a Turing machine M and a polynomial function p such that for any input x written in the binary notation M gives the output $f(x)$ (in the binary notation) implementing at most $p(|x|)$ steps (see [4] for a precise definition).

1. P.t.c.e Sets and P.r.e. Sets

We begin with a useful definition.

Definition 1. 1) We say an infinite set A is *polynomial time computably enumerable (p.t.c.e.)* if $A = rng\, f$ for some injective polynomial time computable function f.
2) An infinite set A is *primitive recursively enumerable (p.r.e.)* if $A = rng\, f$ for some injective primitive recursive function f.

The following theorem gives some properties of p.r.e. sets:

Theorem 2. (Koz'minyh [2]) *1) If A is c.e. and there is a p.r.e. set $B \subseteq A$ then A is p.r.e.*
2) For every p.r.e. set A there is an increasing function f such that $rng\, f \subseteq A$.

The main our goal in this section is to obtain similar properties for the class of p.t.c.e. sets. Note that original proof of Theorem 2 is not adaptable for p.t.c.e. sets. We begin with the following theorem, which is a weak version of the statement 2) of Theorem 2 for the class of p.t.c.e. sets.

Theorem 3. *For every p.t.c.e. set A there exists a p.t.c.e. computable set $B \subseteq A$.*

Proof. Assume $A = rng\, f$ for some injective polynomial time computable function f.

Define a function h by the following recursion: $h(z) = \mu t(z \preceq t \preceq z * [0]^{2|z|}\ \&\ f(t) \geq |z|\ \&\ (\forall s \prec z)[t \neq h(s)])$. It is easy to check that h is polynomial time computable and injective. Let $g(z) = f(h(z))$ for all $z \in \omega$ and $B = rng\, g$. Then g is injective and polynomial time computable, so that B is p.t.c.e.

By the definition of h we have $g(z) \geq |z|$ for all z, giving B computable. □

The following theorem is a weak version of statement 1) of Theorem 2 for p.t.c.e. sets. However combining this theorem with Theorem 3 we are able to prove a full version of this statement for p.t.c.e. sets (see Corollary 5 below).

Theorem 4. *Let A be a p.t.c.e. set, B be a c.e set, and $A \cap B = \emptyset$. Then $A \cup B$ is p.t.c.e.*

Proof. Let $A = rng\, f$ for some injective polynomial time computable function. Since B is c.e. $B = rng\, g$ for some polynomial time computable function. Defining, if necessary, $g'(2x) = g(0)$ and $g'(2x+1) = g(x)$ we can assume that there are infinitely many x such that $g(x) = g(y)$ for some $y > x$.

Define the function h by the following recursion:

$$h(z) = \begin{cases} f(z) & \text{if the string } z \text{ contains the symbol 1,} \\ g(n) & \text{if } z = [0]^n \text{ and } (\forall t \prec z)[g(n) \neq h(t)], \\ f(\mu s \preceq z((\forall t \prec z[f(s) \neq h(t)]))) & \text{if } z = [0]^n \text{ and } (\exists t \prec z)[g(n) = h(t)]. \end{cases}$$

A routine check shows that h is polynomial time computable. It is easy to see that h is injective.

Since there exist infinitely many x such that $g(x) = g(y)$ for some $y > x$ we have $rng\, h = A \cup B$. Hence $A \cup B$ is p.t.c.e. □

Corollary 5. *If A is c.e. and there is a p.t.c.e. set $B \subseteq A$ then A is p.t.c.e.*

Proof. By Theorem 3 there exists a p.t.c.e. computable set $X \subseteq B$. Then the set $A - X$ is c.e. By Theorem 4 $A = X \cup (A - X)$ is p.t.c.e. □

2. P.t.c.e. Sets, P.r.e. Sets and Permutations

A relationship between p.r.e. sets and the hierarchy $\mathcal{P}_n^+$ and $\mathcal{P}_n^-$ was noted by Koz'minyh [2]. This relationship is obviously preserved in the case of polynomial time computable functions:

Theorem 6. *1) A partial unary function f is presentable in the form $f = ba^{-1}$ for some primitive recursive (polynomial time computable) function b and some injective primitive recursive (polynomial time computable) function a iff the graph of f is p.r.e. (p.t.c.e.).*
2) A pair of partial unary functions f and g are presentable as $f = ba^{-1}$ and $g = ca^{-1}$ for some primitive recursive (polynomial time computable) functions b, c and some injective primitive recursive (polynomial time computable) function a iff $dom\, f = dom\, g$ and the graph of the function $f \otimes g$ is p.r.e. (p.t.c.e.).
3) A total unary function f is presentable in the form $f = ba^{-1}$ for some primitive recursive (polynomial time computable) function b and some primitive recursive (polynomial time computable) permutation a iff the graph of f is p.r.e. (p.t.c.e.).
4) A pair of total unary functions of f and g are presentable as $f = ba^{-1}$ and $g = ca^{-1}$ for some primitive recursive (polynomial time computable) functions b, c and some primitive recursive (polynomial time computable) permutation a iff the graph of the function $f \otimes g$ is p.r.e. (p.t.c.e.).

Proof. It is easy to see that 3) follows from 1) and that 4) follows from 2). Note also that 1) follows from 2) since *graph* $f \otimes f$ is p.r.e. (p.t.c.e.) iff *graph* f is p.r.e (p.t.c.e.).

Proof of 2). Let $f = ba^{-1}$ and $g = ca^{-1}$, for some injective primitive recursive (polynomial time computable) function a and some primitive recursive (polynomial time computable) functions b, c. It follows that $dom\, f = rng\, a = dom\, g$.

Define $u(x) = \langle a(x), \langle b(x), c(x)\rangle\rangle$, all $x \in \omega$. Then $rng\, u = graph\, f \otimes g$ and u is an injective primitive recursive (polynomial time computable) function.

To prove 2) in the reverse direction suppose that $graph\, f \otimes g = rng\, u$ for some injective primitive recursive (polynomial time computable) function u and $dom\, f = dom\, g$. Define $a(x) = l(u(x))$, $b(x) = l(r(u(x)))$ and $c(x) = r(r(u(x)))$ for all $x \in \omega$. Then $dom\, f = dom\, g = rng\, a = dom\, a^{-1}$, $f = ba^{-1}$, $g = ca^{-1}$. It is easy to see that functions a, b and c are primitive recursive (polynomial time computable) and a is injective. □

Now we can easily extend a result of Robinson[5]. Note also that Cannonito and Finkelstein [1] proved the following theorem for the class of elementary functions.

Corollary 7. *1) For each partial computable unary function f there are polynomial time computable functions a and b such that a is injective and $f = ba^{-1}t$, where t(x)=2x for all $x \in \omega$.*

2) For each (total) computable unary function f there are polynomial time computable functions a and b such that a is a permutation and $f = ba^{-1}t$, where t(x)=2x for all $x \in \omega$.

Proof. First note that 2) follows from 1). We prove 1). Let f be a partial computable unary function. Define a partial computable function h such that $dom\, h = \{2x : x \in dom\, f\} \cup \{2x+1 : x \in \omega\}$, and $h(2x+1) = 0$ for all $x \in \omega$, and $h(2x) = f(x)$ for all $x \in dom\, f$. By the definition the graph of f it contains the p.t.c.e. subset $\{\langle 2x+1, 0\rangle : x \in \omega\}$. Hence, by Corollary 5 the graph of f is p.t.c.e. By Theorem 6 $h = ba^{-1}$ for some polynomial time computable functions a and b such that a is injective. It remains to recall that $f = ht$, where $t(x) = 2x$ for all $x \in \omega$. □

3. A Presentation of Computable Permutations via Three Polynomial Time Computable Permutations

Theorem 8. *For each partial computable unary function f with an infinite domain there exist a computable permutation u and a partial computable function v such that $f = u^{-1}vu$ and the graph of the function $u \otimes v$ is p.t.c.e.*

By Theorems 6 and 8 we have

Corollary 9. *1) For each partial computable unary function f with an infinite domain there exist polynomial time computable functions a, b, d such that a and b are permutations, d is injective and $f = ab^{-1}cd^{-1}ba^{-1}$.*

2) For each computable unary function f there exist polynomial time computable functions a, b and c such that a and b are permutations and $f = ab^{-1}ca^{-1}ba^{-1}$.
3) For each computable permutation f there exist polynomial time computable permutations a, b and c such that $f = ab^{-1}ca^{-1}ba^{-1}$.

Proof of Theorem 8. Let $A_k = \{4x + k : x \in \omega\}$, $D_k = \{y \in dom\, f : y \notin A_k \ \&\ f(y) \notin A_k\}$ and $C_k = \{y \in D_k : f(y) \neq y\}$ for each integer k, $0 \leq k < 4$. Choose an integer k_0 such that $A_{k_0} \cap dom\, f$ is infinite. Choose an integer $k_1 \neq k_0$ such that D_{k_1} is infinite. Let k_2 and k_3 be such integers that $0 \leq k_2 < k_3 < 4$ and $\{k_2, k_3\} \cap \{k_0, k_1\} = \emptyset$.

Consider three possible cases:

Case 1. The set $f(C_{k_1})$ is infinite. Then there is an increasing function s such that the function fs is increasing, $B_1 \subseteq C_{k_1}$, and $B_1 \cap B_2 = \emptyset$, where $B_1 = rng\, s$ and $B_2 = rng\, fs$. Moreover, we can choose the function s such that the computable set $B_3 = \overline{A_{k_1} \cup B_1 \cup B_2}$ is infinite. Let $B_3 = rng\, t$ for some injective computable function t.

Define a computable permutation u as a function for which $u(4x + k_1) = 4x + k_0$, $u(s(x)) = 4x + k_1$, $u(f(s(x))) = 4x + k_2$, and $u(t(x)) = 4x + k_3$ for all $x \in \omega$. Define a partial computable function $v = ufu^{-1}$. Then $f = u^{-1}vu$. It is easy to see that $v(4x + k_1) = 4x + k_2$. Hence $graph\, u \otimes v$ contains the p.t.c.e. subset $\{\langle 4x + k_1, \langle 4x + k_0, 4x + k_2\rangle\rangle : x \in \omega\}$. Thus, by Theorem 5 the set $graph\, u \otimes v$ is p.t.c.e.

Case 2. The set $f(C_{k_1})$ is finite but the set C_{k_1} is infinite. Therefore, there is an element y_0 such that $f(y) = y_0$ for infinitely many $y \in C_{k_1}$. Then there exists an increasing computable function s such that $B_1 = rng\, s \subseteq C_{k_1}$ and $f(s(x)) = y_0$ for all $x \in \omega$. Again we can choose the function s such that the computable set $B_2 = \overline{A_{k_1} \cup B_1}$ is infinite. Let $rng\, t = B_2$ and $rng\, w = A_{k_2} \cup A_{k_3}$ for some injective functions t and w.

Let $u(4x + k_1) = 4x + k_0$, $u(s(x)) = 4x + k_1$ and $u(t(x)) = w(x)$ for all $x \in \omega$. Then u is a computable permutation. Let $v = u^{-1}fu$. Note that $v(4x + k_1) = v(y_0)$ for all $x \in \omega$. Thus, $f = uvu^{-1}$ and $graph\, u \otimes v$ is a p.t.c.e. set, since $graph\, u \otimes v$ contains the p.t.c.e. subset $\{\langle 4x + k_1, \langle 4x + k_0, v(y_0)\rangle\rangle : x \in \omega\}$.

Case 3. The set C_{k_1} is finite. This means that $f(y) = y$ for almost all $y \in D_{k_1}$. By the same arguments as in previous cases we can choose injective computable functions s and t such that $rng\, s \subseteq D_{k_1}$, $f(s(x)) = s(x)$ for all $x \in \omega$, and $rng\, t = \overline{A_{k_1} \cup rng\, s}$.

Define u as in Case 2: $u(4x + k_1) = 4x + k_0$, $u(s(x)) = 4x + k_1$ and $u(t(x)) = w(x)$ for all $x \in \omega$, where w is an injective computable function such that $rng\, w = A_{k_2} \cup A_{k_3}$. The function u is a computable permutation so that we can define $v = ufu^{-1}$. Then $f = u^{-1}fu$ and $v(4x + k_1) = 4x + k_1$ for all $x \in \omega$. Hence, $graph\, u \otimes v$ contains the p.t.c.e.

subset $\{\langle 4x + k_1, \langle 4x + k_0, 4x + k_1 \rangle\rangle : x \in \omega\}$. By Theorem 5 $graph\, u \otimes v$ is p.t.c.e. □

Using Theorem 5 it is easy to prove that each infinite c.e. set is computably isomorphic to a p.t.c.e. set. The next corollary shows that this isomorphism can be chosen in the form ab^{-1}, where a and b are some polynomial time computable permutations.

Corollary 10. *For each infinite computably enumerable set A there are polynomial time computable permutations a and b and a p.t.c.e. B such that $A = ab^{-1}(B)$.*

Proof. Since A is infinite and c.e. then there exists an injective computable function f such that $A = rng\, f$. By Corollary 9 there exist polynomial time computable permutations a, b and an injective polynomial time computable function c such that $f = ab^{-1}ca^{-1}ba^{-1}$. Let $B = rng\, c$. Then the set B is p.t.c.e. and $A = ab^{-1}(B)$. □

4. Generating Group Terms

Definition 11. Let $\mathbf{T}(a_1, \ldots, a_n)$ be a group-theoretic term.
1) Let $\mathcal{P}_{\mathbf{T}}$ be the class of permutations $p \in \mathcal{C}$ for which there are primitive recursive permutations $a_1, \ldots, a_n$ such that $p = \mathbf{T}(a_1, \ldots, a_n)$
2) We say that the term $\mathbf{T}$ is *generating* if $\mathcal{P}_{\mathbf{T}} = \mathcal{C}$.

By Theorem 9 the term $\mathbf{T}(a, b, c) = ab^{-1}ca^{-1}ba^{-1}$ is generating. By results of Koz'minyh [2] there are no generating terms with less than six compositions.

It would be interesting to know how many polynomial time computable (or at least primitive recursive) permutations are sufficient to generate any computable permutations. By Theorem 9 three permutations are sufficient. Is it possible to decrease the number of permutations to two? In particular, is there some two-placed generating term $\mathbf{T}(a, b)$? Of course the number of compositions in such term must be greater than six, otherwise $\mathbf{T}(a, b) = ab^{-1}ab^{-1}ab^{-1}$ but not every computable permutation is a cube. An indirect argument for the existence of such a term is that there are two primitive recursive (even polynomial time computable) permutations which form together an automorphism base for the group $\mathcal{C}$.

Definition 12. *Let*

$$\mathbf{T}_1(a, b, c, d) = a^{-1}bc^{-1}db^{-1}a$$

and

$$\mathbf{T}_2(b, c, d) = bc^{-1}db^{-1}.$$

Note that by Theorem 6 a permutation p belongs to $\mathcal{P}_{\mathbf{T}_2}$ iff $p = u^{-1}v$ for some permutations u and v such that $graph\, u \otimes v$ is p.r.e.

Koz'minyh [2] proved that a computable permutation p belongs $\mathcal{P}_4^+$ iff there are p.r.e. sets A and B such that $p(A) \cap B = \emptyset$. By the following theorem $p \in \mathcal{P}_{\mathbf{T}_2}$ if (but not only if) $p(B) \cap B = \emptyset$ for some p.r.e. set B.

Theorem 13. *A computable permutation p belongs to $\mathcal{P}_{\mathbf{T}_2}$ iff there exist an injective primitive recursive function α and a primitive recursive set A such that for all $n \in \omega$*

$$n \in A \implies p(\alpha(n)) = \alpha(n)$$
$$n \notin A \implies (\forall m \in \omega) p(\alpha(n)) \neq \alpha(m).$$

Proof. Let $p = u^{-1}v$ for some permutations u and v such that $graph\, u \otimes v = rng\, w$, where w is an injective primitive recursive function. Let s be the injective primitive recursive function such that $s(n)$ is equal to the least number t such that

$$\{l(r(t)), r(r(t))\} \cap (\{l(r(m)) : m < n\} \cup \{r(r(m)) : m < n\}) = \emptyset$$

for every $n \in \omega$. Then the second condition of the theorem is satisfied if we set $\alpha = \lambda n[l(r(s(n)))]$ and $A = \{n \in \omega : l(r(s(n))) = r(r(s(n)))\}$.

Conversely, suppose that there exist an injective primitive recursive function α and a primitive recursive set A such that $n \in A \implies p(\alpha(n)) = \alpha(n)$ and $n \notin A \implies (\forall m \in \omega) p(\alpha(n)) \neq \alpha(m)$ for all n.

Note that without loss of generality we can assume that α is increasing and that the set $X = p^{-1}(\overline{B}) \cap \overline{B}$ is infinite, where $B = rng\, \alpha$ (otherwise we can define $\alpha'(n) = \lambda n[\alpha(2q(n))]$ and $A' = \{n : 2q(n) \in A\}$, where $q(n) = (\mu t)[(\forall m < n)[\alpha(t) > \alpha(q(m))]]$). Therefore we can assume that the set X is computable so that $X = rng\, \kappa$ for some injective computable function κ.

Let

$$\delta(n) = \begin{cases} 3n & \text{if } n \in A \\ 3n+1 & \text{if } n \notin A \end{cases}$$

and let ν be an injective computable function such that $rng\, \nu = \{3n+2 : n \in \omega\} \cup \{3n+1 : n \in A\}$. Then $p = u^{-1}v$, where

$$v(\alpha(n)) = u(p(\alpha(n))) = 3n$$
$$v(p^{-1}(\alpha(n))) = u(\alpha(n)) = \delta(n)$$
$$v(\kappa(n))) = u(p(\kappa(n)) = \nu(n).$$

It is easy to see that u and v are permutations and $graph\, u \otimes v$ is p.r.e. since $\{\langle \alpha(n), \langle \delta(n), 3n \rangle \rangle : n \in \omega\} \subseteq graph\, u \otimes v$. □

Theorem 14. *A computable permutation p belongs to* $\mathcal{P}_{\mathbf{T}_1}$ *iff the set of fixed points* $\{x \in \omega : p(x) = x\}$ *or the set* $\{x \in \omega : p(x) \neq x\}$ *contains an infinite primitive recursive subset.*

Proof. Let $p \in \mathcal{P}_{\mathbf{T}_1}$. Then $p = a^{-1}qa$ for some primitive recursive permutation a and some $q \in \mathcal{P}_{\mathbf{T}_2}$. Then by Theorem 13 there exist an injective primitive recursive function α and a primitive recursive set A such that

$$n \in A \implies p(\alpha(n)) = \alpha(n),$$
$$n \notin A \implies p(\alpha(n)) \neq \alpha(n)$$

for all n. By the proof of Theorem 13 we can assume that α is an increasing function. Consider primitive recursive sets $C = \{x : (\exists n \in A)[a(x) = \alpha(n)]\}$ and $D = \{x : (\exists n \notin A)[a(x) = \alpha(n)]\}$. Clearly, one of the sets C and D is infinite since a is a permutation. To prove the left-to-right direction of the theorem it remains to note that $p(x) = x$ for all $x \in C$ and $p(x) \neq x$ for all $x \in D$.

Let us prove the theorem in the reverse direction. Suppose that there exists an infinite primitive recursive set C such that $p(x) = x$ for all $x \in C$. Then there exists a permutation a such that for all x

$$a(x) \text{ is an even number } \implies x \in C.$$

Therefore, there exists a computable permutation q such that $p = a^{-1}qa$ and $q(2n) = 2n$ for all n. By Theorem 6 $q = bc^{-1}$ for some primitive recursive permutations b, c. Then $p = a^{-1}bc^{-1}a = \mathbf{T}_1(a, b, c, b) \in \mathcal{P}_{\mathbf{T}_1}$.

Suppose now that there exists an infinite primitive recursive set D such that $p(x) \neq x$ for all $x \in D$. In this case there exists a permutation a such that for all x and y

$$a(x) \text{ and } a(y) \text{ are both even } \implies x \in D \text{ and } y \in D \text{ and } p(x) \neq y.$$

Indeed, let $Ev = \{2n : n \in \omega\}$ and $Od = \{2n+1 : n \in \omega\}$. By Kleene's Normal Form Theorem there is a primitive recursive function U and primitive recursive predicates T_1 and T_2 such that

$$p(x) = U((\mu y)T_1(x, y)), p^{-1}(x) = U((\mu y)T_2(x, y))$$

for any $x \in \omega$. Define $a(0) = 0$. Suppose that $x > 0$ and the values $a(z)$ for $z < x$ are already defined. Then define $a(x) = (\mu x)[x \in Ev - \{a(z) : z < x\}]$ if

1) $x \in D$, and

2) for all $z < x$ if $a(z)$ is even then there are integers $y_1 < x$ and $y_2 < x$ such that

$$T_1(z, y_1) \,\&\, T_2(z, y_2) \,\&(\forall t < y_1)[\neg T_1(z,t)] \,\&(\forall t < y_2)[\neg T_2(z,t)]$$
$$\&\, U(y_1) \neq x \,\&\, U(y_2) \neq x,$$

and otherwise define $a(x) = (\mu x)[x \in Od - \{a(y) : y < x\}]$.

Consider a computable permutation q such that $p = a^{-1}qa$. Then $q(2n) \neq 2m$ for all n and m. By Theorem 13 (with $\alpha(n) = 2n$ and $A = \emptyset$) we have $q \in \mathcal{P}_{\mathbf{T}_2}$. Hence $p \in \mathcal{P}_{\mathbf{T}_1}$. □

In contrast with Theorem 9 the term $\mathbf{T}_1$ is not generating. Indeed, let A be any computable set such that both A and $\overline{A}$ do not contain infinite primitive recursive sets. Consider now any computable permutation such that for any $x \in \omega$

$$x \in A \iff p(x) = x.$$

Then by Theorem 14 $p \notin \mathcal{P}_{\mathbf{T}_1}$.

References

[1] F. B. Cannonito, M. Finkelstein, *On primitive recursive permutations and their inverses*, J. Symbolic Log., 34, N4, 1969, 634-638.

[2] V. V. Koz'minyh, *On a presentation of partial recursive functions by compositions*, Algebra i Logika, 11, N3, 1972, 270–294 (In Russian).

[3] P. G. Odifreddi, *Classical Recursion Theory*, vol. I, North-Holland, Amsterdam, New York, Oxford, Tokyo, 1989.

[4] P. G. Odifreddi, *Classical Recursion Theory*, vol. II, North-Holland, North-Holland, Amsterdam, New York, Oxford, 1999.

[5] J. Robinson, *General recursive functions*, Proc. Amer. Math. Soc., 1, 1950, 703–718.

[6] H. Rogers, Jr., *Theory of Recursive Functions and Effective Computability*, McGraw-Hill, New York, 1967.

ON SELF-EMBEDDINGS OF COMPUTABLE LINEAR ORDERS

Steffen Lempp*
Department of Mathematics, University of Wisconsin
Madison, WI 53706-1388, USA
lempp@math.wisc.edu

Andrei S. Morozov†
Sobolev Institute of Mathematics
Russian Academy of Sciences, Siberian Branch
630090 Novosibirsk
Russia
morozov@math.nsc.ru

Charles F. D. McCoy‡
Department of Mathematics, University of Wisconsin
Madison, WI 53706-1388, USA
mccoy@math.wisc.edu

D. Reed Solomon§
Department of Mathematics, University of Wisconsin
Madison, WI 53706-1388, USA
rsolomonmath.wisc.edu

Abstract We show the existence of a computable linear order without nontrivial $\mathbf{0}'$-computable self-embedding.

Keywords: computable linear order, self-embedding

*The first author's research was partially supported by NSF grant DMS-9732526.
†The second author's research was partially supported by binational NSF grant DMS-0075899 and by RFBR (Russian Fund for Basic Research) grant No. 99-01-00485.
‡The third author's research was partially supported by an NSF VIGRE fellowship.
§The fourth author's research was partially supported by NSF Fellowship DMS-0071586.

1. Background and Theorems

A *self-embedding* of a linear ordering $\mathcal{L}$ is an order-preserving injection from $\mathcal{L}$ into itself. We call a self-embedding *nontrivial* if it is not the identity function.

One of the fundamental classical theorems on countable linear orderings is the

Dushnik-Miller Theorem on Countable Linear Orderings [DM40]. *Let $\mathcal{L}$ be a countably infinite linear ordering. Then there is a nontrivial self-embedding of $\mathcal{L}$.*

One of the long-standing open questions on computable linear orders is the characterization of those computable linear orders which admit a nontrivial computable self-embedding in the following sense.

Definition Let $\mathcal{L} = \langle L, < \rangle$ be a linear ordering.

(i) An *interval* (or *convex subset*) of $\mathcal{L}$ is any subset $I \subseteq L$ such that $a, b \in I$ and $a < c < b$ implies $c \in I$. (So, in particular, an interval in our sense need *not* have endpoints.)

(ii) An interval I is *discrete* if every $a \in I$ (except a possible left endpoint of I) has an immediate predecessor in I, and every $a \in I$ (except a possible right endpoint of I) has an immediate successor in I.

(iii) $\mathcal{L}$ is *η-like* if L is infinite and no infinite interval of $\mathcal{L}$ is discrete, i.e., if L is infinite but has no interval of order-type ω or ω^*. (Here ω^* is ω under the reverse ordering, and η is the order-type of the rational numbers.)

(iv) $\mathcal{L}$ is *strongly η-like* if L is infinite and there is a fixed finite bound on the size of all discrete intervals of $\mathcal{L}$.

The open question is now contained in the following

Conjecture (folklore, see Downey [Do98], Conjecture 5.2). *Let $\mathcal{L}$ be an infinite computable linear ordering. Then $\mathcal{L}$ contains a strongly η-like interval iff for any computable copy $\mathcal{M} \cong \mathcal{L}$, there is a nontrivial computable self-embedding of $\mathcal{M}$.*

We note right away that the left-to-right direction is trivial: Given any computable copy $\mathcal{M} = \langle M, < \rangle$ of a linear ordering with strongly η-like interval I, we can find a computable subset $J \subset I$ of order-type η; we can now effectively embed $\mathcal{M}$ into the subset J.

The other direction, however, has defied researchers for decades, and only partial results have been obtained thus far. These partial results seem to indicate that the solution to this question is very hard:

Theorem 1.

(i) (Downey [Do98], remark after Conjecture 5.2) *The above conjecture holds for computable η-like linear orderings (i.e., the equivalence holds when restricted to η-like linear orderings).*

(ii) (Downey [Do98], Theorem 5.23) *The above conjecture does not hold uniformly: There is no effective procedure which, given (an index for) a computable linear ordering $\mathcal{L}$, produces (an index for) a computable copy $\mathcal{M} \cong \mathcal{L}$ such that if $\mathcal{L}$ contains no strongly η-like interval then there is no nontrivial computable self-embedding of $\mathcal{M}$.*

(iii) (Downey [Do98], Theorem 5.24) *The above conjecture does not hold if we replace "isomorphic" by "Δ_2^0-isomorphic"; in fact, there is an infinite computable linear ordering $\mathcal{L}$ without η-like interval such that for any computable copy $\mathcal{M} \cong_{\Delta_2^0} \mathcal{L}$, there is a nontrivial computable self-embedding of $\mathcal{M}$.*

The following older result was the first theorem to show that the Dushnik-Miller Theorem on Countable Linear Orderings does indeed not hold effectively.

Theorem 2 (Hay/Rosenstein, see Rosenstein [Ro82])**.** *There is a computable copy of ω without nontrivial computable self-embedding.*

This result was extended as follows, which also allows the proof-theoretic characterization of the Dushnik-Miller Theorem on Countable Linear Orderings in the sense of reverse mathematics (see Simpson [Si99] for background on reverse mathematics).

Theorem 3 (Downey/Lempp [DL99])**.** *There is a computable copy $\mathcal{L}$ of ω such that any nontrivial self-embedding of $\mathcal{L}$ computes the halting problem $\emptyset'$. Thus, over the base theory RCA_0, the Dushnik-Miller Theorem on Countable Linear Orderings is equivalent to the theory ACA_0.*

Since it is easy to see that any computable linear order has a nontrivial self-embedding computable in $\emptyset''$, the question naturally arises as to the exact upper bound on the possible complexity of a nontrivial self-embedding. The main result of the present paper partially answers this question as follows:

Theorem 4. *There is a computable linear ordering $\mathcal{L}$ such that no nontrivial self-embedding of $\mathcal{L}$ is computable from the halting problem $\emptyset'$.*

It remains open whether Theorem 3 can be improved to coding $\emptyset''$ (in place of $\emptyset'$) into any nontrivial self-embedding of a computable linear order. At the same time, Theorem 4 is further evidence that the solution to the above conjecture is very difficult.

2. The Proof of Theorem 4

We build a computable linear order $\mathcal{L} = \langle \omega, <_{\mathcal{L}} \rangle$ and ensure that no function computable in $\emptyset'$ is a nontrivial self-embedding of $\mathcal{L}$. (Our notation generally follows Soare [So87].)

For this, we first fix an effective listing $\{\Psi_e^{\emptyset'}\}_{e\in\omega}$ of all functions partial computable in $\emptyset'$ and an effective listing of partial computable functions $\{\psi_e\}_{e\in\omega}$ such that for all $e \in \omega$, $\Psi_e^{\emptyset'} = \lim_u \psi_{e,u}$. Without loss of generality, we may assume that each $\psi_{e,u}$ is a total function (although the limit in u may diverge for some arguments).

We need to meet the following

Requirement:

$$\mathcal{R}_e : \Psi_e^{\emptyset'} \text{ is not a nontrivial self-embedding of } \mathcal{L}.$$

Rather than meeting this requirement all at once, we will try to meet, for all $e, x \in \omega$, the following subrequirement

$$\mathcal{R}_{e,x} : \Psi_e^{\emptyset'} \text{ is not a self-embedding of } \mathcal{L} \text{ moving } x.$$

Remark. Note that, for any $e \in \omega$, we actually need to meet the subrequirement $\mathcal{R}_{e,x}$ only for at least one x in each maximal discrete interval of $\mathcal{L}$. (This will be crucial to our construction later.)

Strategy for $\mathcal{R}_{e,x}$. For simplicity, abbreviate $\psi_{e,u}(\psi_{e,u}(x))$ as $\psi^2_{e,u}(x)$. As the construction proceeds, we consider $\psi_{e,u}(x)$ and $\psi^2_{e,u}(x)$ for larger and larger u and proceed as follows:

1. Whenever $\psi_{e,u}(x) \neq \psi_{e,u-1}(x)$ or $\psi^2_{e,u}(x) \neq \psi^2_{e,u-1}(x)$, we restart, canceling any restraint (since $\Psi_e^{\emptyset'}(x)$ or $\Psi_e^{\emptyset'}(\Psi_e^{\emptyset'}(x))$ appear undefined).

2. Unless $x <_{\mathcal{L}} \psi_{e,u}(x) <_{\mathcal{L}} \psi^2_{e,u}(x)$ or $x >_{\mathcal{L}} \psi_{e,u}(x) >_{\mathcal{L}} \psi^2_{e,u}(x)$, we do nothing (since $\Psi_e^{\emptyset'}$ does not appear to move x, or does not appear to preserve $<_{\mathcal{L}}$).

3. By symmetry, we now assume that $x <_{\mathcal{L}} \psi_{e,u}(x) <_{\mathcal{L}} \psi^2_{e,u}(x)$. We then ensure that $|[x, \psi_{e,u}(x)]| > |[\psi_{e,u}(x), \psi^2_{e,u}(x)]|$ by *restraining the interval* $[\psi_{e,u}(x), \psi^2_{e,u}(x)]$ (which is finite at this stage, of course) i.e., not allowing any new points into this interval; and by adding enough points into the interval $[x, \psi_{e,u}(x)]$.

Outcomes of the $\mathcal{R}_{e,x}$-strategy. An individual $\mathcal{R}_{e,x}$-strategy may have the following outcomes (keeping in mind that we assume that each $\psi_{e,u}$ is total):

∞ The $\mathcal{R}_{e,x}$-strategy restarts infinitely often in Step 1: Then at least one of $\Psi_e^{\emptyset'}(x)$ and $\Psi_e^{\emptyset'}(\Psi_e^{\emptyset'}(x))$ is not defined, the requirement $\mathcal{R}_e$ is met, and the $\mathcal{R}_{e,x}$-strategy has no permanent effect on our construction.

id The $\mathcal{R}_{e,x}$-strategy is eventually stuck waiting in Step 2 forever: Then $\Psi_e^{\emptyset'}$ does not move x or does not preserve $<_{\mathcal{L}}$, the subrequirement $\mathcal{R}_{e,x}$ is met, and the $\mathcal{R}_{e,x}$-strategy has no permanent effect on our construction.

fin The $\mathcal{R}_{e,x}$-strategy is eventually stuck waiting in Step 3 forever: Then $\Psi_e^{\emptyset'}$ cannot preserve $<_{\mathcal{L}}$ since $||[x, \Psi_e^{\emptyset'}(x)]|| > ||[\Psi_e^{\emptyset'}(x), \Psi_e^{\emptyset'}(\Psi_e^{\emptyset'}(x))]||$, the requirement $\mathcal{R}_e$ is met, and the $\mathcal{R}_{e,x}$-strategy has only finite effect on our construction.

Thus under outcomes ∞ and *fin*, the overall requirement $\mathcal{R}_e$ is met, while under outcome *id*, only the subrequirement $\mathcal{R}_{e,x}$ is met. We agree that the apparent outcome of the $\mathcal{R}_{e,x}$-strategy at a given stage is ∞ if the strategy restarts at that stage under Step 1; and *id* or *fin* if the strategy is waiting at Step 2 or 3, respectively, at that stage.

Outcomes of the $\mathcal{R}_e$-requirement. For each $e \in \omega$ and along any path p of the tree of strategies T (to be defined precisely later), either (i) we have an $\mathcal{R}_{e,x}$-strategy α_x for at least one x in each maximal discrete interval of $\mathcal{L}$ such that each α_x has outcome *id* along the path p; or (ii) there is a longest $\mathcal{R}_{e,x}$-strategy α for some x such that α has outcome ∞ or *fin* along the path p. In the former case, if $\Psi_e^{\emptyset'}$ is indeed a self-embedding of $\mathcal{L}$, then it is the identity; in the latter case, $\Psi_e^{\emptyset'}$ is partial or does not preserve $<_{\mathcal{L}}$.

Conflicts between strategies. The only possible conflict is between an $\mathcal{R}_{e',x'}$-strategy $\alpha' \in T$, wanting to restrain an interval, and an $\mathcal{R}_{e,x}$-strategy $\alpha > \alpha'$, wanting to insert numbers into that interval. Note that for a fixed α along the true path, there can be at most finitely many such α' wanting to restrain an interval $[a_{\alpha'}, b_{\alpha'}]$, say. This means that x is in one (or several) of these intervals $[a_{\alpha'}, b_{\alpha'}]$, and that as long as α is not initialized, these intervals do not increase in size. Let $[a, b]$ be the smallest interval containing x and either containing, or disjoint from, all intervals $[a_{\alpha'}, b_{\alpha'}]$. (This interval is the restraint on points near x that α has to deal with.)

We now recall our remark at the beginning of this section: We only need to meet subrequirement $\mathcal{R}_{e,x}$ for at least one x in each maximal discrete interval of $\mathcal{L}$. So, to meet subrequirement $\mathcal{R}_{e,x}$, we essentially "replace" x by a and b and proceed as follows:

1. Whenever $\psi_{e,u}(a) \neq \psi_{e,u-1}(a)$, $\psi^2_{e,u}(a) \neq \psi^2_{e,u-1}(a)$, $\psi_{e,u}(b) \neq \psi_{e,u-1}(b)$, or $\psi^2_{e,u}(b) \neq \psi^2_{e,u-1}(b)$, we restart, canceling any restraint (since $\Psi_e^{\emptyset'}(a)$, $\Psi_e^{\emptyset'}(\Psi_e^{\emptyset'}(a))$, $\Psi_e^{\emptyset'}(b)$ or $\Psi_e^{\emptyset'}(\Psi_e^{\emptyset'}(b))$ appear to be undefined).

2. Unless $b <_{\mathcal{L}} \psi_{e,u}(b) <_{\mathcal{L}} \psi^2_{e,u}(b)$ or $a >_{\mathcal{L}} \psi_{e,u}(a) >_{\mathcal{L}} \psi^2_{e,u}(a)$, we do nothing (since $\Psi_e^{\emptyset'}$ does not appear to move b to the right, or a to the left, or does not appear to preserve $<_{\mathcal{L}}$). (Note here that if the restraint on $[a, b]$ is permanent then if $\Psi_e^{\emptyset'}$ moves x to the right, it must also move b to the right, and if it moves x to the left, it must also move a to the left.)

3. By symmetry, we now assume that $b <_{\mathcal{L}} \psi_{e,u}(b) <_{\mathcal{L}} \psi^2_{e,u}(b)$. We then ensure that $|[b, \psi_{e,u}(b)]| > |[\psi_{e,u}(b), \psi^2_{e,u}(b)]|$ by restraining the interval $[\psi_{e,u}(b), \psi^2_{e,u}(b)]$ (which is finite at this stage, of course), and adding enough points into the interval $[b, \psi_{e,u}(b)]$. (Note here that by our assumption on b, there is no restraint on points immediately to the right of b, so there is no conflict with higher-priority strategies.)

Construction. It is now easy to describe the full construction. We first define the *set of outcomes* $\Lambda = \{\infty <_\Lambda \mathit{id} <_\Lambda \mathit{fin}\}$ and the *tree of strategies* $T = \Lambda^{<\omega}$ (with the lexicographical order $<$ induced by the order $<_\Lambda$). We then assign subrequirements to nodes of T by recursion as follows: Fix any effective ordering of all subrequirements $\mathcal{R}_{e,x}$ of order-type ω and assign the highest-priority subrequirement to the root of T. Call a subrequirement $\mathcal{R}_{e,x}$ *satisfied along* $\alpha \in T$ if there is some $\mathcal{R}_{e,x}$-strategy $\alpha' \subset \alpha$, or if, for some $x' \neq x$, there is an $\mathcal{R}_{e,x'}$-strategy α' with $\alpha' \hat{\ } \langle \infty \rangle \subseteq \alpha$ or $\alpha' \hat{\ } \langle \mathit{fin} \rangle \subseteq \alpha$. (The intuition here is that then either $\mathcal{R}_{e,x}$ or $\mathcal{R}_e$ is satisfied along α by this α'.) We then assign to α the highest-priority subrequirement not satisfied along α.

A strategy $\alpha \in T$ is *initialized* by setting its parameter u to 0 and canceling its restraint. At stage 0, all strategies are initialized.

Each *stage* $s > 0$ consists of *substages* $t \leq s$, at which a strategy $\alpha \in T$ of length t is *eligible to act*. At substage t, the $\mathcal{R}_{e,x}$-strategy α eligible to act then proceeds as described above (incrementing its parameter u by 1 at each stage at which α is eligible to act), taking into account the restraint of all the finitely many strategies $\alpha' < \alpha$ currently restraining intervals. (If x has not yet appeared in $\mathcal{L}$ by this stage, we simply end the stage.) At the end of stage s, we add an additional point to the right of all current points in $\mathcal{L}$ (to ensure that $\mathcal{L}$ is infinite) and initialize all strategies $>$ the last strategy eligible to act at stage s.

Verification. By the action at the end of each stage, the linear ordering $\mathcal{L}$ is infinite (and we may assume that its domain is all of ω). Thus, for each t, stage s is eventually not ended before substage t due to lack of points. So we can define the *true path of the construction* $f \in [T]$ as the leftmost infinite path through T along which all strategies are eligible to act infinitely often.

For each $e \in \omega$, there are now three possibilities:

(1) There is a longest $\mathcal{R}_{e,x}$-strategy α (for some $x \in \omega$) and $\alpha\hat{}\langle\infty\rangle \subset f$: Then, by definition of initialization and of the true path, α is not initialized after some stage s_α, say. So, after stage s_α, α meets the requirement $\mathcal{R}_e$ as in the basic module above by showing that $\Psi_e^{\emptyset'}$ is partial.

(2) There is a longest $\mathcal{R}_{e,x}$-strategy α (for some $x \in \omega$) and $\alpha\hat{}\langle fin\rangle \subset f$: Then, by definition of initialization and of the true path, α is not initialized after some stage s_α, say. So, after stage s_α, α meets the requirement $\mathcal{R}_e$ as in the basic module above by showing that $\Psi_e^{\emptyset'}$ does not preserve $<_\mathcal{L}$, handling the conflict with the finitely many higher-priority strategies $\alpha' < \alpha$ as explained above.

(3) Otherwise: Then, by definition of the tree of strategies, for each x, there is an $\mathcal{R}_{e,x}$-strategy $\alpha_x \subset f$, and for each x, $\alpha_x\hat{}\langle id\rangle \subset f$. By definition of initialization and of the true path, α_x is not initialized after some stage s_x, say. Thus, for each x, after stage s_x, α_x shows that $\Psi_e^{\emptyset'}$ does not preserve $<_\mathcal{L}$ or that $x = \Psi_e^{\emptyset'}(x)$, which implies that $\Psi_e^{\emptyset'}$ is not a nontrivial computable self-embedding of $\mathcal{L}$.

This concludes the proof of Theorem 4.

References

[Do98] Downey, R. G., *Computability theory and linear orderings*, Handbook of recursive mathematics, Vol. 2, Stud. Logic Found. Math., 139, North-Holland, Amsterdam, 1998, pp. 823–976.

[DL99] Downey, Rodney G. and Lempp, Steffen, *The proof-theoretic strength of the Dushnik-Miller theorem for countable linear orders*, Recursion theory and complexity (Kazan, 1997), de Gruyter Series Logic Appl. No. 2, de Gruyter, Berlin, 1999, pp. 55–57.

[DM40] Dushnik, B. and Miller, E. W., *Concerning similarity transformations of linearly ordered sets*, Bull. Amer. Math. Soc. **46** (1940), 322–326.

[Ro82] Rosenstein, J. G., *Linear Orderings*, Academic Press, New York, 1982.

[Si99] Stephen G. Simpson, *Subsystems of Second Order Arithmetic*, Springer-Verlag, Berlin, 1999.

[So87] Soare, R. I., *Recursively enumerable sets and degrees*, Perspectives in Mathematical Logic, Springer-Verlag, Berlin, New York, 1987.

DEFINABLE RELATIONS ON THE COMPUTABLY ENUMERABLE DEGREES

Angsheng Li*
Institute of Software, Chinese Academy of Sciences
P.O. Box, 8718, Beijing
P. R. of CHINA
liang@ox.ios.ac.cn

Abstract We review recent developments in the study of definable relations on the computably enumerable (c.e.) degrees, including the following aspects:

(1) Structure and hierarchies
(2) Definable relations
(3) Continuity in the c.e. degrees
(4) Automorphisms
(5) Defining $\mathcal{E}$ in $\mathcal{E}_n$

Keywords: Computably enumerable degrees, definability, continuity, automorphism

The *computably enumerable* (c.e.) sets naturally occur in various branches of mathematics and computer science, and the existence of noncomputable c.e. sets led to early, fundamental, undecidability results such as the undecidability of the solution of certain classes of Diophantine equations, the unsolvability of the word problem for finitely presented groups, and of course the unsolvability of the *halting problem*.

Let K be the set of all natural numbers x such that program P_x with input x halts. Then K is a noncomputable c.e. set, produced by *diagonalisation*.

Notions of *reducibility* provide one of the main approaches to analysing the noncomputable universe. Various reducibilities have yielded rich theories for

*The author was partially supported by an EPSRC Research Grant, "Turing Definability", No. GR/M 91419 (UK) and by NSF Grant No. 69973048 and NSF Major Grant No. 19931020 (P. R. CHINA). I am grateful to Barry Cooper for help, encouragement and suggestions during the preparation of this paper. Thanks to Yue Yang for helpful comments on the preliminary draft of the paper.

the noncomputable universe. Among which *Turing reducibility* has proven to be the most natural and successful one, and has become one of the main tools in computability theory.

Given sets $A, B \subseteq \omega$, we say that A is *computable in* (or *Turing reducible to*) B, if there is an algorithm to decide for each $x \in \omega$, whether or not $x \in A$, when given answers to all questions of the form "Is $y \in B$?". We use $A \leq_T B$ to denote A Turing reducible to B, and write $A \equiv_T B$ if $A \leq_T B$ and $B \leq_T A$. A (Turing) *degree* is an equivalence class of A under $\equiv_T$ for some set $A \subseteq \omega$. A degree is called *computably enumerable* (c.e.) if it contains a c.e. set. This can be intuitively understood via *oracle Turing machines*, i.e., Turing machines with an (additional) oracle tape: say that A is Turing reducible to B, if there is an oracle Turing machine Φ say, which computes A when oracle B is given, denoted by $A = \Phi(B)$. Using Turing reducibility, we can define a diagonalisation set K^A to be the set of all x such that $\Phi_x(A; x)$ *converges*, where Φ_x is the x-th oracle Turing machine. We say that K^A is the *Turing jump* of A. We also use A' to denote the Turing jump of A.

For $A, B \subseteq \omega$, we define the *join* of A and B by $A \oplus B = \{2x \mid x \in A\} \cup \{2x + 1 \mid x \in B\}$.

Post [1944] observed that there is a greatest c.e. degree $\mathbf{0}'$, i.e., the degree of K, and asked whether or not there is a c.e. degree other than $\mathbf{0}$ (the least degree), and $\mathbf{0}'$. Friedberg [1957] and Muchnik [1956] independently solved this problem by constructing two incomparable c.e. degrees. In their proof, the *priority method* was first introduced. Over the years, this method has been expanded and developed, while the study of the c.e. degrees has prospered, leaving a small number of important open problems which seems out of reach of the existing technical resources. Shore [1999] gave a comprehensive survey of important results in the history of the study of the c.e. degrees. In this article, the emphasis is on new developments and related aspects of the study of the c.e. degrees. We try to avoid reiteration of the content of Shore [1999], but we may have to return to some of the more fundamental achievements.

1. Structure and Hierarchies

We use lower-case boldface letters $\mathbf{a}, \mathbf{b}, \mathbf{c}, \cdots$ to denote degrees, and $\mathcal{E}$ to denote the set of all c.e. degrees. For a degree $\mathbf{a}$, the *Turing jump* of $\mathbf{a}$, denoted by $\mathbf{a}'$, is the Turing degree of A' for some $A \in \mathbf{a}$. For degrees $\mathbf{a}, \mathbf{b}$, the *join* of $\mathbf{a}, \mathbf{b}$, denoted by $\mathbf{a} \vee \mathbf{b}$, is the degree of $A \oplus B$ for some sets $A, B \in \mathbf{a}, \mathbf{b}$ respectively. Then $\mathcal{E}$ is an upper semi-lattice with least element $\mathbf{0}$ and greatest element $\mathbf{0}'$. Extending Friedberg [1957] and Muchnik [1956], Sacks [1963] proved the following:

THEOREM 1.1. (Sacks Splitting Theorem) For any nonzero c.e. degree $\mathbf{a}$, there are c.e. degrees $\mathbf{b}, \mathbf{c} < \mathbf{a}$ such that $\mathbf{b} \vee \mathbf{c} = \mathbf{a}$.

The main new feature of the proof of this theorem is referred to as *Sacks preservation.* Using this method, Sacks [1964] proved the density theorem by introducing a new strategy, which is known as the *Sacks coding strategy.*

THEOREM 1.2. (Sacks Density Theorem) Given any c.e. degrees $\mathbf{a} < \mathbf{b}$, there is a c.e. degree $\mathbf{c}$ such that $\mathbf{a} < \mathbf{c} < \mathbf{b}$.

The proof of this theorem is in fact an *infinite injury argument,* in which a strategy to satisfy a requirement may be injured infinitely many times. However for a given requirement $\mathcal{R}$ say, there is a strategy α such that the *injury set* of α is a computable set which is recognised by α so that α is still successful in satisfying $\mathcal{R}$. Consequently, every requirement is eventually satisfied.

Based on this progress, Shoenfield [1965] suggested an algebraical approach to understanding the structure $\mathcal{E}$.

CONJECTURE 1.3. (Shoenfield [1965]) Given finite partial orderings $P \subseteq Q$, then any embedding of P into $\mathcal{E}$ (the set of all c.e. degrees), in which degree theoretic joins and $\mathbf{0}, \mathbf{0}'$ are consistent with Q, can be extended to an embedding of Q into $\mathcal{E}$.

In particular, Shoenfield listed two consequences of the conjecture:

C1. For any c.e. degrees $\mathbf{a}, \mathbf{b}$, if $\mathbf{a}, \mathbf{b}$ are incomparable, then the greatest lower bound of $\mathbf{a}, \mathbf{b}$ does not exist.

C2. For any c.e. degrees $\mathbf{0} < \mathbf{b} < \mathbf{c}$, there is a c.e. degree $\mathbf{a} < \mathbf{c}$ such that the least upper bound of $\mathbf{a}, \mathbf{b}$, denoted by $\mathbf{a} \vee \mathbf{b}$, equals to $\mathbf{c}$.

Conjecture 1.3 is not true, because consequence C1 of the conjecture was refuted by the minimal pair theorem of Lachlan [1966] and Yates [1966].

THEOREM 1.4. (Minimal Pair Theorem) There exist nonzero c.e. degrees $\mathbf{a}, \mathbf{b}$ such that $\mathbf{a} \wedge \mathbf{b} = \mathbf{0}$, where $\mathbf{a} \wedge \mathbf{b}$ is the greatest lower bound of $\mathbf{a}$ and $\mathbf{b}$.

The proof is another kind of infinite injury argument. It played an important role in the refinement of the *priority tree method* which has become the main framework for theorem proving in the theory of the c.e. degrees. Also this theorem introduced the *meet operator* $\wedge$ and initiated the investigation of lattice embeddings. However, the meet operator $\wedge$ is not always defined.

THEOREM 1.5. (Lachlan [1966], Yates [1966]). $\mathcal{E}$ is not a lattice.

Lachlan [1966] also discovered some connections between the Turing jump and the operators $\vee$ and $\wedge$ in $\mathcal{E}$.

THEOREM 1.6. (Lachlan Nondiamond Theorem) There are no c.e. degrees $\mathbf{a}, \mathbf{b}$ such that $\mathbf{0} < \mathbf{a} < \mathbf{0}'$, $\mathbf{a} \vee \mathbf{b} = \mathbf{0}'$ and $\mathbf{a} \wedge \mathbf{b} = \mathbf{0}$.

Using the Turing jump, Cooper [1972] and Soare [1974] defined a useful hierarchy of the (c.e.) degrees, namely, the *high/low hierarchy.*

DEFINITION 1.7. (Cooper [1972], Soare [1974]) The *high/low hierarchy* is defined by

$$\mathbf{H}_n = \{\mathbf{a} \in \mathcal{E} \mid \mathbf{a}^{(n)} = \mathbf{0}^{(n+1)}\}$$

$$\mathbf{L}_n = \{\mathbf{a} \in \mathcal{E} \mid \mathbf{a}^{(n)} = \mathbf{0}^{(n)}\}$$

where $n \geq 0$, $\mathbf{x}^{(n+1)}$ is the Turing jump of $\mathbf{x}^{(n)}$, and $\mathbf{x}^{(0)} = \mathbf{x}$ for each $\mathbf{x}$.

An element of $\mathbf{H}_n(\mathbf{L}_n)$ is called *high*$_n$ (*low*$_n$), and for $n = 1$, we also call an element of $\mathbf{H}_1$ $(\mathbf{L}_1)$ *high* (*low*).

By the minimal pair theorem, a natural problem is that of characterising the finite lattices which can be embedded into $\mathcal{E}$. This question relates to the decidability problem for fragments of the elementary theory of $\mathcal{E}$. Progress on this topic is referred to in Shore [1999].

In contrast to the minimal pair theorem, Yates [1966] showed:

THEOREM 1.8. (Yates [1966]) There exists a c.e. degree $\mathbf{a} \neq \mathbf{0}$ such that for any c.e. degree $\mathbf{x}$, $\mathbf{a} \wedge \mathbf{x} = \mathbf{0}$ if and only if $\mathbf{x} = \mathbf{0}$.

Theorem 1.8 was extended later by Ambos-Spies [1984a]: There exists a nonzero (low) c.e. degree $\mathbf{a}$ such that for any c.e. degree $\mathbf{x}$, $\mathbf{a} \wedge \mathbf{x}$ exists if and only if $\mathbf{a}$ and $\mathbf{x}$ are comparable. The degree $\mathbf{a}$ is called a *strongly noncappable* (s.n.c.) degree. Lempp [1988] constructed a high strongly noncappable degree. This leaves an interesting open question: (Lempp [1988]) whether or not for each degree $\mathbf{c}$, c.e.a. in $\mathbf{0}'$, there is a strongly noncappable degree $\mathbf{a}$ such that $\mathbf{a}' = \mathbf{c}$?

More importantly, by theorems 1.4 and 1.8, we can define the notion of *cappable* and *noncappable* c.e. degrees. Say a c.e. degree $\mathbf{a}$ is *cappable*, if there is a c.e. degree $\mathbf{x} \neq \mathbf{0}$ such that $\mathbf{a} \wedge \mathbf{x} = \mathbf{0}$, and *noncappable*, otherwise. And dually, we say that a c.e. degree $\mathbf{a}$ is *cuppable*, if there is an incomplete c.e. degree $\mathbf{x}$ such that $\mathbf{a} \vee \mathbf{x} = \mathbf{0}'$, and *noncuppable* (a nontrivial notion as we will see), otherwise. The study of cappable and noncappable degrees leads to some elegant characterisations and descriptions of subclasses of the c.e. degrees, while the study of cupping and noncupping of c.e. degrees provides significant ingredients for undecidability and definability results for the c.e. degrees.

In addition, the minimal pair theorem provides an instance of close relationship between structural properties and the high/low hierarchy for $\mathcal{E}$.

THEOREM 1.9. (Cooper [1974a] Minimal Pair Theorem) For any high c.e. degree $\mathbf{h}$, there are c.e. degrees $\mathbf{a}, \mathbf{b}$ such that $\mathbf{0} < \mathbf{a}, \mathbf{b} \leq \mathbf{h}$ and $\mathbf{a} \wedge \mathbf{b} = \mathbf{0}$.

In fact, **a**, **b** here can be chosen to be high. Also Robinson [1971a] had found a similar kind of property related to lowness:

THEOREM 1.10. (Robinson Low Splitting Theorem) For any c.e. degrees $\mathbf{l} < \mathbf{a}$, if $\mathbf{l}$ is low, then there are c.e. degrees $\mathbf{a}_0$, $\mathbf{a}_1$ such that $\mathbf{l} \leq \mathbf{a}_0, \mathbf{a}_1 < \mathbf{a}$ and $\mathbf{a}_0 \vee \mathbf{a}_1 = \mathbf{a}$.

Theorems 1.9 and 1.10 suggested a natural question: that of characterising the high/low hierarchy via structural properties of $\mathcal{E}$.

In contrast to the results on the minimal pairs and noncappable degrees, Cooper [1974b] and Yates proved a remarkable theorem:

THEOREM 1.11. (Cooper-Yates Noncupping Theorem) There exists a c.e. degree $\mathbf{a} \neq \mathbf{0}$ such that for any c.e. degree $\mathbf{x}$, $\mathbf{a} \vee \mathbf{x} = \mathbf{0}'$ if and only if $\mathbf{x} = \mathbf{0}'$.

Clearly the set of all noncuppable degrees is an ideal of $\mathcal{E}$. This is the first example of a nontrivial definable ideal of $\mathcal{E}$. So this theorem suggests the question of characterising (at least, finding more) definable ideals of $\mathcal{E}$.

Harrington [1976] noted that the noncuppable degree in theorem 1.11 can be chosen to be high. And using the *Cooper's high permitting method* (in the proof of theorem 1.9), Harrington [1976] extended the Cooper-Yates noncupping theorem: For every high c.e. degree $\mathbf{h}$, there is a high c.e. degree $\mathbf{a} < \mathbf{h}$ such that for any c.e. degree $\mathbf{x}$, $\mathbf{a} \vee \mathbf{x} = \mathbf{h}$ if and only if $\mathbf{x} = \mathbf{h}$ (see Miller [1981]). A consequence of this theorem is:

COROLLARY 1.12. (Harrington [1976]) Every high c.e. degree bounds a high noncuppable degree.

And more extensions of the Harrington's noncupping result have been provided by Ding and Yu [ta]. On the other hand, Harrington [1978] also proved a *plus cupping theorem* that there is a nonzero c.e. degree $\mathbf{a}$ such that for any c.e. degrees $\mathbf{x}, \mathbf{y}$, if $\mathbf{0} < \mathbf{x} \leq \mathbf{a} \leq \mathbf{y}$, then there is a c.e. degree $\mathbf{z} < \mathbf{y}$ such that $\mathbf{x} \vee \mathbf{z} = \mathbf{y}$. We say that the $\mathbf{a}$ is a *plus cupping degree*. And one can prove the following:

THEOREM 1.13. (Li [ta1]) There is a high$_2$ plus cupping degree.

By the Robinson [1971b] interpolation theorem, for any incomplete c.e. degree $\mathbf{a}$, there is an incomplete high c.e. degree $\mathbf{h}$ such that $\mathbf{a} \leq \mathbf{h}$. We say that $\mathbf{a} \in \mathbf{H}_n$ is *cuppable in* $\mathbf{H}_n$, if there is an incomplete $\mathbf{b} \in \mathbf{H}_n$ such that $\mathbf{a} \vee \mathbf{b} = \mathbf{0}'$, and *noncuppable in* $\mathbf{H}_n$, otherwise. By theorem 1.13, for each $n > 1$, there is a c.e. degree $\mathbf{a} \in \mathbf{H}_n$ which bounds no degrees noncuppable in $\mathbf{H}_n$. By corollary 1.12, every high degree bounds a degree which is noncuppable in $\mathbf{H}_1$. This gives the following:

THEOREM 1.14. (Li [ta1]) For each natural number $n > 1$, $\mathrm{TH}(\mathbf{H}_1) \neq \mathrm{TH}(\mathbf{H}_n)$ holds, where $\mathrm{TH}(\mathbf{H}_n)$ is the elementary theory of upper semi-lattice $\mathbf{H}_n$.

However we do not know whether or not there are $m \neq n$ such that $\mathbf{H}_m$ and $\mathbf{H}_n$ are elementarily equivalent.

An important technical breakthrough is due to Lachlan [1975]:

THEOREM 1.15. (Lachlan Nonsplitting Theorem) There exist c.e. degrees $\mathbf{b} < \mathbf{a}$ such that for any c.e. degrees $\mathbf{x}, \mathbf{y}$, if $\mathbf{b} \leq \mathbf{x}, \mathbf{y} < \mathbf{a}$, then $\mathbf{x} \vee \mathbf{y} \neq \mathbf{a}$.

In the proof of this theorem, Lachlan introduced a new method, in which a strategy to satisfy a requirement may fail to satisfy the requirement due to infinitary injury, so a *backup strategy for the strategy* will be introduced. In so doing, for a fixed requirement $\mathcal{R}$ say, there are finitely many $\mathcal{R}$-strategies on the *true path*, among which one of them works in a *well defined environment* provided by higher priority strategies, so that the strategy will be successful in satisfying its requirement eventually. Consequently, every requirement will be satisfied. This method was categorised as $\mathbf{0}'''$*-priority tree method* by Harrington.

Harrington [1980] extended theorem 1.15: there exists an incomplete c.e. degree $\mathbf{a}$ such that $\mathbf{0}'$ is not splittable over $\mathbf{a}$. The degree $\mathbf{a}$ is called a *Harrington nonsplitting base*. In this proof, Harrington abstracted some basic principles from the Lachlan nonsplitting proof which have become "golden rules" of various priority tree arguments. Also Harrington introduced some new features such as *Harrington capricious destruction*, and *Harrington honestification* which have proven to be useful in the proof of many theorems since then.

Lachlan [1979] also proved that the highness condition in the Cooper minimal pair theorem is necessary.

THEOREM 1.16. (Lachlan Nonbounding Theorem) There exists a c.e. degree $\mathbf{a} \neq \mathbf{0}$ which bounds no minimal pairs.

Degree $\mathbf{a}$ in the theorem is called a *(Lachlan) nonbounding degree*. Downey, Lempp and Shore [1993] showed that there is a high$_2$ nonbounding degree. However this does not give any elementary difference among the theories of $\mathbf{H}_n$ for $n > 0$.

The cappable and noncappable degrees were very well understood through a dynamic notion of *the promptly simple sets* which was introduced by Maass [1982].

DEFINITION 1.17. (i) A coinfinite c.e. set A is *promptly simple* if there is a computable function p and a computable enumeration $\{A_s\}_{s\in\omega}$ of A such that for every e, if W_e is infinite, then there are s and x such that $x \in W_{e,\text{at } s} \cap A_{p(s)}$.

(ii) A c.e. degree **a** is called *promptly simple*, if it contains a promptly simple set.

Then:

THEOREM 1.18. (Ambos-Spies, Jockusch, Shore and Soare [1984]) Let **M** be the set of all cappable degrees; let **NC** be the set of all noncappable degrees; let **PS** be the set of all promptly simple degrees, and let **LC** be the set of all low-cuppable degrees, the degrees which join to $\mathbf{0}'$ with some low c.e. degrees. Then:

(i) **NC** = **PS** = **LC**.

(ii) **M** is a definable ideal of $\mathcal{E}$.

(iii) **NC** is a strong filter in $\mathcal{E}$, i.e., it is closed upward in $\mathcal{E}$, and if $\mathbf{a}, \mathbf{b} \in \mathbf{NC}$, then there is a $\mathbf{c} \in \mathbf{NC}$ with $\mathbf{c} \leq \mathbf{a}, \mathbf{b}$.

This provides not only another example of definable ideals of $\mathcal{E}$, but an elegant algebraic decomposition of $\mathcal{E}$.

Against a background of seemingly inexorable technical progress, Slaman and Soare [1995, 2001] were able to solve the extensions of embeddings problem.

THEOREM 1.19. (Slaman-Soare Extensions of Embeddings Theorem) There is an effective characterisation of the pairs $P \hookrightarrow Q$ of partial orderings with 0, 1 such that for every embedding $f : P \to \mathcal{E}$, there is an extension g of f to an embedding of Q into $\mathcal{E}$.

A number of new and useful features such as *soft link, hard link* are introduced in the proof of this theorem. Also interesting is that a simplified presentation of parts of the proof of the theorem has been given by Wang [2000]. The theorem indicates that the approach suggested by Shoenfield's conjecture has been exhausted. New research depends on new ideas.

In addition, it is easy to see that every noncappable degree bounds a minimal pair. So every nonbounding degree is cappable, and every nonzero c.e. degree bounds a nonzero cappable degree. However there is a relative nonbounding theorem for cappable degrees.

THEOREM 1.20. (Relative Nonbounding Theorem, Li [2000a], [2000b]) There exist c.e. degrees $\mathbf{b} < \mathbf{a}$ such that **a** and **b** bound the same class of cappable degrees.

By theorem 1.18, the set of noncappable degrees equals to the set of all low-cuppable degrees. Extending low-cuppability, we define:

DEFINITION 1.21. (Li, Wu and Zhang [2000]) Say a c.e. degree **a** is *low$_n$-cuppable*, if there is a low$_n$ c.e. degree **l** such that $\mathbf{a} \vee \mathbf{l} = \mathbf{0}'$.

Let $\mathbf{LC}_n$ be the set of all low$_n$-cuppable c.e. degrees, and $\mathbf{NLC}_n = \mathcal{E} - \mathbf{LC}_n$, for each $n > 0$, and let $\mathbf{CUP}$ be the set of all cuppable degrees. Then:

$$\mathbf{LC}_1 \subseteq \mathbf{LC}_2 \subseteq \cdots \subseteq \mathbf{CUP}.$$

Then we have a basic result:

THEOREM 1.22. (Li, Wu and Zhang [2000]) $\mathbf{LC}_1 \subset \mathbf{LC}_2$.

This yields *the Harrington cap and cup theorem*: there is a c.e. degree which are both cappable and cuppable. It is interesting that the theorem indicates that there is a relation $F \subset \mathcal{E}$ which is naturally definable in $\mathcal{E}$ such that $F \cap \mathbf{L}_1 = \emptyset$, and $F \cap \mathbf{L}_2 \neq \emptyset$.

More importantly, the hierarchy of the cuppable degrees collapses at an interesting level:

THEOREM 1.23. (Hierarchy Theorem of Cups, Li [ta2]) $\mathbf{LC}_3 = \mathbf{CUP}$.

However we do not know whether or not $\mathbf{LC}_2 \neq \mathbf{LC}_3$ holds. Given that both $\mathbf{NLC}_1$ (the cappable degrees by theorem 1.18), and $\mathbf{NLC}_3$ (by theorem 1.23) are ideals of $\mathcal{E}$, in any case, we conjecture that $\mathbf{NLC}_2$ is an ideal of $\mathcal{E}$. Finally we end this section with some further open questions.

OPEN QUESTION 1.24. (Soare, 1987) (i) Characterise the finite lattices which can be embedded into $\mathcal{E}$.

(ii) (Soare, 1987) Is the Σ_2-theory of $\mathcal{E}$ decidable?

(iii) Characterise the decidable fragments of the elementary theories of $\mathbf{L}_n$, $\mathbf{H}_n$ for $n > 0$.

(iv) Can the high/low hierarchy be characterised by algebraic properties of $\mathcal{E}$?

[Some conjectures related to this can be found in Shore [1999].]

(v) Characterise the definable ideals and/or filters of $\mathcal{E}$. In particular, are there infinitely many definable ideals of $\mathcal{E}$?

(vi) Are there any $m \neq n$ such that $\mathrm{TH}(\mathbf{H}_m) = \mathrm{TH}(\mathbf{H}_n)$? Are there any $m \neq n$ such that $\mathrm{TH}(\mathbf{L}_m) = \mathrm{TH}(\mathbf{L}_n)$? In particular, are the $\mathbf{L}_1$ and $\mathbf{L}_2$ elementarily equivalent?[1]

2. Definable Relations

One of the central problems about $\mathcal{E}$ is to characterise the definable relations of $\mathcal{E}$. The solution of this problem may rule out certain automorphisms of $\mathcal{E}$. The problem remains open, but some partial results are known.

[1] Jockusch, Li and Yang [ta] have shown that $\mathrm{TH}(\mathbf{L}_1) \neq \mathrm{TH}(\mathbf{L}_2)$.

Using first order representation of standard models of arithmetic within $\mathcal{E}$, Nies, Shore and Slaman [1998] have shown:

THEOREM 2.1. For every $n > 0$, $\mathbf{H}_n$ and $\mathbf{L}_{n+1}$ are definable in $\mathcal{E}$.

Furthermore, Cholak, Groszek and Slaman [2001] showed:

THEOREM 2.2. There exists a nonzero c.e. degree $\mathbf{a}$ such that for any c.e. degree $\mathbf{x}$, $\mathbf{x}$ is low if and only if $\mathbf{a} \vee \mathbf{x}$ is low.

Consequently, $\mathbf{L}_1$ is definable in $\mathcal{E}$ with finitely many parameters.

While Nies [ta] discovered a definable ideal of $\mathcal{E}$ other than the ideal **M** of cappable degrees, and the ideal **NCUP** of noncuppable degrees:

THEOREM 2.3. (Nies) (i) Any definable subset of $\mathcal{E}$ generates a definable ideal of $\mathcal{E}$.

(ii) The nonbounding degrees generate a definable ideal of $\mathcal{E}$ other than **M** and **NCUP**.

However a sharper question is:

OPEN QUESTION 2.4. (Slaman, 1985) Is there any nontrival principal ideal of $\mathcal{E}$ which is definable in $\mathcal{E}$?

The recent work of Cholak, Downey and Walk [ta] has provided a very interesting partial solution, based on the definability of the relation "$\mathbf{x}$ is contiguous".

THEOREM 2.5. (i) (Downey and Lempp) A c.e. degree $\mathbf{a}$ is contiguous iff it is strongly contiguous iff it is locally distributive, i.e., if $\mathbf{b} \vee \mathbf{c} = \mathbf{a} > \mathbf{d}$, then there are $\mathbf{b}_0, \mathbf{c}_0$ such that $\mathbf{b}_0 < \mathbf{b}, \mathbf{c}_0 < \mathbf{c}$ and $\mathbf{b}_0 \vee \mathbf{c}_0 = \mathbf{d}$.

(ii) (Ambos-Spies and Fejer) A c.e. degree $\mathbf{a}$ is contiguous iff it is not the top of an embedding of the N_5 into $\mathcal{E}$.

This leads to the following definability theorem:

THEOREM 2.6. (Cholak, Downey and Walk [ta]) (i) There is a maximal contiguous degree, and furthermore, there is a c.e. degree $\mathbf{a} < \mathbf{0}'$ such that there is a unique maximal contiguous degree $> \mathbf{a}$.

(ii) There is a definable infinite anti-chain of $\mathcal{E}$.

Although Nies, Shore and Slaman [1998] (theorem 2.1) proved that $\mathbf{H}_n$ and $\mathbf{L}_{n+1}$ are definable in $\mathcal{E}$ for each $n > 0$, their proof uses coding of models of arithmetic in $\mathcal{E}$, so the role of the jump in $\mathcal{E}$ is still not clear. On the other hand, the existing priority methods developed in the study of the c.e. degrees give us very little about higher levels of the high/low hierarchy. Therefore there is a special interest in obtaining more degree theoretic definitions of the high/low hierarchy, i.e. the so called *natural definitions*, as stated in Shore [2000]:

OPEN QUESTION 2.7. Are any of the jump classes $\mathbf{H}_n$ and $\mathbf{L}_{n+1}$ naturally definable in $\mathcal{E}$?

Clearly finding a natural definition of any of the jump classes would illuminate the role of the Turing jump in $\mathcal{E}$, and a positive solution of this problem would challenge the existing priority methods. It could be even more interesting to show that there is no logically simple formula which defines any of the jump classes in $\mathcal{E}$, which involves a precise definition for the notion "natural". Therefore in any case, the significance of a solution for the problem is much more than another definition of the jump hierarchy. Very significant attempts have already been made by many people.

DEFINITION 2.8. We say that a c.e. degree $\mathbf{a}$ is a *diamond base*, if $\mathbf{a}$ is the bottom of a diamond of c.e. degrees with top $\mathbf{0}'$.

Let **DB** be the set of all diamond bases.

Lachlan [1966] (theorem 1.6) showed that $\mathbf{0} \notin \mathbf{DB}$. More importantly to this discussion, Lachlan also observed that $\mathbf{DB} \cap \mathbf{L}_1 = \emptyset$ (see Ambos-Spies [1984b]).

THEOREM 2.9. (Ambos-Spies [1984b]) (i) $\mathbf{DB} \cap \mathbf{L}_1 = \emptyset$, that is, every low$_1$ c.e. degree bounds no diamond base.

(ii) (Li and Yang [2000]) $\mathbf{DB} \cap \mathbf{H}_1 \neq \emptyset$.

And Li, Slaman and Yang [ta] proved the following:

THEOREM 2.10. (Li, Slaman and Yang [ta]) There exists a nonlow$_2$ c.e. degree which bounds no diamond base.

This progress leads us to ask the following:

OPEN QUESTION 2.11. (i) Is there any naturally definable relation F of $\mathcal{E}$ such that $F \cap \mathbf{L}_n = \emptyset$ and $F \cap \mathbf{L}_{n+1} \neq \emptyset$ for any $n > 2$?

(ii) Does **DB** have minimal elements?

(iii) Characterise the c.e. degrees which bound elements of **DB**. Characterise the atomic jump classes of the diamond bases.

A solution for 2.11 (i) could be a challenge of the methodology of the constructions in the study of the c.e. degrees. 2.11 (ii) is interesting in the sense of either continuity of the c.e. degrees (see the next section) or more definable antichains of $\mathcal{E}$. 2.11 (iii) is a question about the connection of algebraic properties of $\mathcal{E}$ and the Turing jump.

As we have seen in section 1, the phenomena of splitting and nonsplitting are basic to the study of the c.e. degrees. Recent progress on this topic has greatly increased our understanding on these phenomena and their relations with some fundamental areas of the study of the c.e. degrees. We first introduce the following:

DEFINITION 2.12. (Cooper and Li [ta1]) (i) We say that a c.e. degree $\mathbf{a}$ is a *Lachlan nonsplitting base*, if there is a c.e. degree $\mathbf{b} > \mathbf{a}$ such that $\mathbf{b}$ is not splittable above $\mathbf{a}$.

If $\mathbf{a}$ is incomplete and $\mathbf{0}'$ is not splittable over $\mathbf{a}$, then $\mathbf{a}$ is called a *Harrington nonsplitting base*.

(ii) We say that a pair $(\mathbf{b}, \mathbf{a})$ of c.e. degrees $\mathbf{a}, \mathbf{b}$ is a *Lachlan nonsplitting pair*, if $\mathbf{b} < \mathbf{a}$ and $\mathbf{a}$ is not splittable above $\mathbf{b}$.

An interesting approach to giving a natural definition of low$_2$ c.e. degrees in the c.e. degrees $\mathcal{E}$ had been provided by Harrington (unpublished) and independently by Bickford and Mills (unpublished) (see Shore and Slaman [1990]):

THEOREM 2.13 (Harrington-Bickford-Mills) Every low$_2$ c.e. degree bounds no Lachlan nonsplitting pair.

Could this theorem be half of a proof that low$_2$ is definable in $\mathcal{E}$, by, such as, $\mathbf{x}$ is low$_2$ if and only if $\mathbf{x}$ bounds no Lachlan nonsplitting pair? Cooper, Li and Yi [ta] have shown that it is not the case.

THEOREM 2.14. There exists a nonlow$_2$ c.e. degree which does not bound any Lachlan nonsplitting base.

Also interesting is that Shore and Slaman (see Shore and Slaman [1993]) have observed that the existing technical resources could provide the following:

THEOREM 2.15. There exists a Lachlan nonsplitting pair $\mathbf{a} < \mathbf{l}$ such that $\mathbf{l}$ is low$_3$.

This theorem together with theorem 2.13 shows that $\mathrm{Th}(\mathbf{L}_2) \neq \mathrm{Th}(\mathbf{L}_3)$, i.e. low$_2$ and low$_3$ are not elementarily equivalent.

Let **LNSB** be the set of all Lachlan nonsplitting bases, and let **HNSB** be the set of all Harrington nonsplitting bases. By the Robinson low splitting theorem (theorem 1.10), $\mathbf{LNSB} \cap \mathbf{L}_1 = \emptyset$. Naturally we ask:

COOPER'S QUESTION. (Cooper, 1973) Does $\mathbf{LNSB} \cap \mathbf{L}_2 = \emptyset$ hold?

On one side, Arslanov, Cooper and Li [2000] have improved the Robinson low splitting theorem:

THEOREM 2.16. (Generalised Low Splitting Theorem) For any c.e. set A, any Δ^0_2 set L, if $L <_{\mathrm{T}} A$, and L has low degree, then there are c.e. sets X_0, X_1 with the following properties,

(1) $X_0 \cup X_1 = A$, $X_0 \cap X_1 = \emptyset$,

(2) for each $i \in \{0, 1\}$, $A \not\leq_{\mathrm{T}} X_i \oplus L$.

On the other hand, Cooper and Li [ta1] showed that the usual nonsplitting construction can be adapted to provide a low$_3$ Harrington nonsplitting base. This has led to the answer of the above question of Cooper:

THEOREM 2.17. (Cooper and Li [ta2] Nonsplitting Theorem) There exists a low$_2$ Harrington nonsplitting base.

The theorem pushes the nonsplitting to the limit. It indicates that low$_2$ is sufficiently complicated, and that low$_2$ is quite different from low$_1$ (for instance, let L_1^+ be the union of low and $\{\mathbf{0}'\}$, and let L_2^+ be the union of low$_2$ and $\{\mathbf{0}'\}$, then the theory of the partial ordering L_1^+ is different from the theory of the partial ordering L_2^+). And the new techniques such as *multi-threshold method*, *wake up method*, and *slow down method* introduced in the proof of the theorem are quite general and potentially useful.

Another interesting topic is the study of the definable ideals of $\mathcal{E}$. Bickford and Mills introduced the notion of *deep degree*, the c.e. degree $\mathbf{a}$ such that for any c.e. degree $\mathbf{x}$, $\mathbf{x}' = (\mathbf{a} \vee \mathbf{x})'$, $\mathbf{y}'$ is the Turing jump of $\mathbf{y}$. Lempp and Slaman [1989] showed that the only deep degree is $\mathbf{0}$. Entending the notion of the deep degree, a c.e. degree $\mathbf{a}$ is called *deep$_n$*, if for any c.e. degree $\mathbf{x}$, $\mathbf{x}^{(n)} = (\mathbf{a} \vee \mathbf{x})^{(n)}$. Let $\mathbf{DP}_n$ be the set of all deep$_n$ c.e. degrees. Then: $\{\mathbf{0}\} = \mathbf{DP}_1 \subseteq \mathbf{DP}_2 \subseteq \cdots$. Naturally we ask: Does this hierarchy collapse?

In addition, we define $\mathbf{PL}_n$ and $\mathbf{PH}_n$ by

$$\mathbf{PL}_n = \{\mathbf{a} \mid (\forall \mathbf{x})[\mathbf{x}^{(n)} = \mathbf{0}^{(n)} \leftrightarrow (\mathbf{a} \vee \mathbf{x})^{(n)} = \mathbf{0}^{(n)}]\}$$

$$\mathbf{PH}_n = \{\mathbf{a} \mid (\forall \mathbf{x})[\mathbf{x}^{(n)} = \mathbf{0}^{(n+1)} \leftrightarrow (\mathbf{a} \vee \mathbf{x})^{(n+1)}]\}.$$

By theorem 2.1, for each $n \geq 1$, both $\mathbf{PL}_{n+1}$ and $\mathbf{PH}_n$ are definable ideals of $\mathcal{E}$. While by theorem 2.2, $\mathbf{PL}_1 \neq \{\mathbf{0}\}$. However, Jockusch, Li and Yang [ta] have shown that:

THEOREM 2.18. (Jockusch-Li-Yang Join Theorem) For any c.e. degree $\mathbf{x}$, if $\mathbf{x} \neq \mathbf{0}$, then there is a c.e. degree $\mathbf{a}$ such that $(\mathbf{a} \vee \mathbf{x})' = \mathbf{a}'' = \mathbf{0}''$.

By the existence of nonzero noncuppable degrees of Cooper [1974] and Yates (theorem 1.11), the theorem is the best join theorem for the c.e. degrees. Consequently, we have the following:

COROLLARY 2.19. (Jockusch, Li and Yang) (i) For each $n \geq 1$, $\mathbf{DP}_n = \mathbf{PL}_{n+1} = \mathbf{PH}_n = \{\mathbf{0}\}$.

(ii) $\mathrm{TH}(\mathbf{L}_1) \neq \mathrm{TH}(\mathbf{L}_2)$.

(i) answers the question about the deep$_n$ degrees, and indicates that every definable ideal $\mathbf{PL}_{n+1}$ or $\mathbf{PH}_n$ of $\mathcal{E}$ is trivial, (ii) gives the first elementary difference between the partial orderings of low$_1$ and low$_2$ c.e. degrees.

This progress leads us to ask the following:

OPEN QUESTION 2.20. (i) Is there any naturally definable relation F of $\mathcal{E}$ such that F contains no high$_n$ c.e. degrees, but does contain high$_{n+1}$ c.e. degrees for any $n > 1$. (For $n = 1$, the answer is yes by existing results.)

(ii) Is there a naturally definable relation $F \subset \mathcal{E}$ of $\mathcal{E}$ which contains no low_{n+1} or high_n c.e. degrees for each $n > 0$?

(iii) Characterise the finite relations of $\mathcal{E}$ which are definable in $\mathcal{E}$.

(iv) Characterise all definable relations of $\mathcal{E}$.

3. Continuity in the C.E. Degrees

By analogy with the notion of *major subset* in the context of the lattice of c.e. sets, Lachlan suggested a corresponding notion for the c.e. degrees:

DEFINITION 3.1. (Lachlan, 1967) Given c.e. degrees $\mathbf{a} < \mathbf{b}$, we say that $\mathbf{a}$ is a *major subdegree of* $\mathbf{b}$, if for every c.e. degree $\mathbf{x}$, we have that $\mathbf{b} \vee \mathbf{x} = \mathbf{0}'$ holds if and only if $\mathbf{a} \vee \mathbf{x} = \mathbf{0}'$ holds.

Given that every noncomputable c.e. set B has a major subset (Lachlan [1968]), it is natural to ask:

MAJOR SUBDEGREE PROBLEM. (Lachlan, 1967) Does every c.e. degree $\mathbf{b} \neq \mathbf{0}$ or $\mathbf{0}'$ have a major subdegree?

Attempts to solve the problem unexpectedly led to an appreciation of the importance of the notion of *continuity* for the c.e. degrees, and a number of results have been proved. The first progress with the major subdegree problem is related to consequence C2 of Shoenfield's conjecture (Conjecture 1.3). Cooper [1974b] and Yates (theorem 1.11) showed that there exists a noncuppable c.e. degree $\mathbf{a}$, i.e. $\mathbf{a} \neq \mathbf{0}$ and for any c.e. degree $\mathbf{x}$, $\mathbf{a} \vee \mathbf{x} = \mathbf{0}'$ if and only if $\mathbf{x} = \mathbf{0}'$. By definition, if $\mathbf{a}$ is noncuppable, then any nonzero c.e. degree $\mathbf{b} < \mathbf{a}$ is a major subdegree of $\mathbf{a}$. So major subdegrees exist, since every noncuppable c.e. degree has one by the downward density of the c.e. degrees. And Jockusch and Shore [1983] gave the first instance of a cuppable c.e. degree which has a major subdegree: there exist c.e. degrees $\mathbf{l} < \mathbf{h} < \mathbf{0}'$ such that $\mathbf{l}$ is low, $\mathbf{h}$ is high and such that $\mathbf{h}$ is cuppable and $\mathbf{l}$ is a major subdegree of $\mathbf{h}$.

Furthermore, Cooper and Slaman [1988] showed:

THEOREM 3.2. (Cooper and Slaman) (i) Every high c.e. degree $\mathbf{h} \neq \mathbf{0}'$ has a major subdegree.

(ii) Every low c.e. degree $\mathbf{l} \neq \mathbf{0}$ has a major subdegree.

However, general continuity results were first achieved in the dual of the major subdegree problem.

THEOREM 3.3. (Continuity of Capping Theorem, Harrington and Soare [1989]) For any c.e. degrees $\mathbf{a}, \mathbf{b} \neq \mathbf{0}$, if $\mathbf{a} \wedge \mathbf{b} = \mathbf{0}$, then there exists a c.e. degree $\mathbf{c} > \mathbf{a}$ such that $\mathbf{c} \wedge \mathbf{b} = \mathbf{0}$.

Theorem 3.3 was then extended by Seetapun [1991], giving a full solution to the dual of the major subdegree problem (see Giorgi [2001]).

THEOREM 3.4. (Seetapun Continuity Theorem) For any c.e. degree $\mathbf{b} \neq \mathbf{0}, \mathbf{0}'$, there exists a c.e. degree $\mathbf{a} > \mathbf{b}$ such that for any c.e. degree $\mathbf{x}$, $\mathbf{b} \wedge \mathbf{x} = \mathbf{0}$ if and only if $\mathbf{a} \wedge \mathbf{x} = \mathbf{0}$.

A consequence of theorem 3.4 is the continuity of Lachlan nonbounding degrees: there is no maximal Lachlan nonbounding degree.

Also a partial result relating to the major subdegree problem was provided by Ambos-Spies, Lachlan and Soare [1993]:

THEOREM 3.5. (Continuity of Cupping Theorem) For any c.e. degrees $\mathbf{a}, \mathbf{b}$, if $\mathbf{0} < \mathbf{a} < \mathbf{0}'$ and $\mathbf{a} \vee \mathbf{b} = \mathbf{0}'$, then there is a c.e. degree $\mathbf{c} < \mathbf{a}$ such that $\mathbf{c} \vee \mathbf{b} = \mathbf{0}'$.

This theorem together with the continuity of capping theorem (theorem 3.3) gives a general continuity result.

COROLLARY 3.6. (Harrington and Soare [1989]) Let $\mathbf{F}(\mathbf{x}, \mathbf{y})$ be an open formula in the language $\{<, \vee, \wedge, \mathbf{0}, \mathbf{0}'\}$ and $\mathbf{a}, \mathbf{b}$ be distinct c.e. degrees such that $\mathbf{F}(\mathbf{a}, \mathbf{b})$ holds in the upper semilattice $\mathcal{E}$ of c.e. degrees. Then there exist c.e. degrees $\mathbf{a}_0, \mathbf{a}_1, \mathbf{b}_0, \mathbf{b}_1$ such that

(i) $\mathbf{a}_0 < \mathbf{a} < \mathbf{a}_1$, $\mathbf{b}_0 < \mathbf{b} < \mathbf{b}_1$, and

(ii) for any c.e. degrees $\mathbf{x}, \mathbf{y}$, if $\mathbf{a}_0 < \mathbf{x} < \mathbf{a}_1$ and $\mathbf{b}_0 < \mathbf{y} < \mathbf{b}_1$, then $\mathbf{F}(\mathbf{x}, \mathbf{y})$ holds.

By modifying the proof of theorem 3.5, Seetapun [1991] also improved Cooper-Slaman's partial result of theorem 3.2 (ii): every nonzero low_2 c.e. degree has a major subdegree. And Seetapun (1992, unpublished) has shown, based on joint work with Lachlan and Soare, how the proof of theorem 3.5 can be modified to prove the following:

THEOREM 3.7. (Lachlan, Seetapun and Soare) For any c.e. degree $\mathbf{b} \neq \mathbf{0}, \mathbf{0}'$, there exists a c.e. degree $\mathbf{a} \not\geq \mathbf{b}$ such that for any c.e. degree $\mathbf{x}$, if $\mathbf{0}' = \mathbf{b} \vee \mathbf{x}$, then $\mathbf{0}' = \mathbf{a} \vee \mathbf{x}$.

This progress led to Cooper and Li's full solution of the major subdegree problem. Cooper and Li [ta3] have shown:

THEOREM 3.8. (Cooper-Li Continuity Theorem) Every c.e. degree $\mathbf{b} \neq \mathbf{0}, \mathbf{0}'$ has a major subdegree, i.e., for any c.e. degree $\mathbf{b} \neq \mathbf{0}, \mathbf{0}'$, there is a c.e. degree $\mathbf{a} < \mathbf{b}$ such that for any c.e. degree $\mathbf{x}$, $\mathbf{b} \vee \mathbf{x} = \mathbf{0}'$ if and only if $\mathbf{a} \vee \mathbf{x} = \mathbf{0}'$.

In fact, the construction given to prove this theorem is a uniform one. Therefore we have:

THEOREM 3.9. (Cooper and Li [ta3]) There is a computable function f such that for every e, if $W_e \not\leq_T \emptyset$ and $K \not\leq_T W_e$, then $W_{f(e)} <_T W_e$ and $\deg_T(W_{f(e)})$ is a major subdegree of $\deg_T(W_e)$.

The proof of this theorem incorporated a number of new and potentially useful features such as *multi-permitting, dynamic priority* and *expanding outcomes*.

An immediate consequence of the Cooper-Li continuity theorem (theorem 3.9) is a continuity result for Harrington nonsplitting bases.

COROLLARY 3.10. (Cooper and Li [ta3]) There is no minimal Harrington nonsplitting base.

Proof. Suppose that $\mathbf{b}$ is a Harrington nonsplitting base. By the Harrington nonsplitting theorem and the Sacks splitting theorem, $\mathbf{b} \neq \mathbf{0}, \mathbf{0}'$. By theorem 3.9, let $\mathbf{a}$ be a major subdegree of $\mathbf{b}$. Then, for any c.e. degrees $\mathbf{x_0}, \mathbf{x_1}$, if $\mathbf{x_0} \vee \mathbf{x_1} = \mathbf{0}'$ but $\mathbf{x_i} \vee \mathbf{a} \neq \mathbf{0}'$ for both $i = 0$ and 1, then $\mathbf{b} \vee \mathbf{x_i} \neq \mathbf{0}'$ holds for both $i = 0$ and 1. Let $\mathbf{y_i}$ be a c.e. degree such that $\mathbf{b} \vee \mathbf{x_i} \leq \mathbf{y_i} < \mathbf{0}'$ for $i = 0, 1$. Since $\mathbf{x_0} \vee \mathbf{x_1} = \mathbf{0}'$, we have that $\mathbf{y_0} \vee \mathbf{y_1} = \mathbf{0}'$, but $\mathbf{b} \leq \mathbf{y_0}, \mathbf{y_1} < \mathbf{0}'$, $\mathbf{b}$ is not a Harrington nonsplitting base, contradicting the choice of $\mathbf{b}$. Therefore for any c.e. degrees $\mathbf{x_0}, \mathbf{x_1}$, if $\mathbf{x_0} \vee \mathbf{x_1} = \mathbf{0}'$, then $\mathbf{x_i} \vee \mathbf{a} = \mathbf{0}'$ for some $i \leq 1$, $\mathbf{a}$ is a Harrington nonsplitting base. So $\mathbf{b}$ is not a minimal Harrington nonsplitting base, and the corollary follows. □

Intuitively speaking, continuity results limit the potential for definability results on $\mathcal{E}$, in the sense that continuity results may rule out certain relations on $\mathcal{E}$ being definable in $\mathcal{E}$. Therefore it is important to analyse the continuity of the basic phenomena of the c.e. degrees. A question is:

OPEN QUESTION 3.11. Characterise the relations on $\mathcal{E}$ which are definable in $\mathcal{E}$ and which have maximal and/or minimal elements. In particular, is there a relation on $\mathcal{E}$ which is definable in $\mathcal{E}$ and which has a nontrivial greatest or least element? (If yes, Slaman's definability problem (open question 2.4) has positive solution.)

The only result about this question is the result of Cholak, Downey and Walk [ta]: the relation "$\mathbf{x}$ is contiguous" is definable in $\mathcal{E}$ and has maximal elements (theorem 2.6).

4. Automorphisms

One of the central topics on the structure $\mathcal{E}$ is the characterisation of the automorphisms of $\mathcal{E}$, and the study of the automorphism group of $\mathcal{E}$. The basic approaches are on the one hand to analyse the automorphism bases and on the other to analyse the limitations on definable relations on $\mathcal{E}$. By the Sacks splitting theorem (theorem 1.1), any dense set of degrees is an automorphism base for $\mathcal{E}$. And every jump class $\mathbf{H}_n$ and $\mathbf{L}_n$ is an automorphism base for $\mathcal{E}$. And furthermore, Ambos-Spies [1993] showed that every downward cone in $\mathcal{E}$ is an automorphism base for $\mathcal{E}$.

If $\mathbf{F} \subset \mathcal{E}$ is definable in $\mathcal{E}$, then for any automorphism $\psi \in \mathrm{Aut}(\mathcal{E})$, we have that $\psi(\mathbf{F}) \subseteq \mathbf{F}$. Therefore it is specially interesting to analyse the relations on $\mathcal{E}$ which are both definable in $\mathcal{E}$ and automorphism bases for $\mathcal{E}$.

And more importantly to this discussion, Cooper [1997] has described a construction to prove the following:

THEOREM 4.1. (Cooper Automorphism Theorem) $\mathrm{Aut}(\mathcal{E}) \neq \{id\}$. And furthermore, there is an automorphism ψ of $\mathcal{E}$ such that $\psi(\mathbf{L}_1) \nsubseteq \mathbf{L}_1$.

An immediate consequence of this theorem is that low_1 c.e. degrees are not definable in $\mathcal{E}$.

It is possible that more general nondefinability results within $\mathcal{E}$ may be approachable via analysis of the Cooper automorphisms. However we do not even know:

OPEN QUESTION 4.2. (Cooper, 1997) Is $\mathrm{Aut}(\mathcal{E})$ countable?

More conjectures about this question can be found in Shore [1999].

5. Defining $\mathcal{E}$ in $\mathcal{E}_n$

The n-c.e. Turing degrees provide a natural generalisation of the class of the computably enumerable degrees. Recent years have seen much research centred on the comparisons between the structures of the c.e. degrees and the n-c.e. degrees for $n > 1$, and various similarities and differences have been found. A basic problem, closely related to other longstanding questions such as Downey's conjecture is:

OPEN QUESTION 5.1. (Cooper, 1999) At what levels of the Ershov hierarchy does Turing definability of given lower levels of the hierarchy occur? In particular, are the computably enumerable (c.e.) degrees $\mathcal{E}$ definable within the class of d.c.e. (= 2-c.e.) degrees $\mathcal{E}_2$?

One of the first significant results concerning the structure of the n-c.e. degrees for $n > 1$ was provided by Arslanov [1985].

THEOREM 5.2. (Arslanov Cupping Theorem) Every nonzero c.e. degree can be joined to $\mathbf{0}'$ by an incomplete 2-c.e. degree.

By the noncupping theorem of Cooper [1974] and Yates (theorem 1.11), this gives a difference between the elementary theories of the c.e. and the n-c.e. degrees for every $n > 1$.

Following this, Downey [1989] gave another difference:

THEOREM 5.3. (Downey Diamond Theorem) The diamond lattice can be embedded into the d.c.e. degrees preserving both $\mathbf{0}$ and $\mathbf{0}'$.

Jiang [1993] showed that $\mathbf{0}'$ in the theorem can be replaced by any high c.e. degree.

Extending both Arslanov's cupping theorem and Downey's diamond theorem, Li and Yi [1999] proved the following:

THEOREM 5.4. (Li-Yi Cupping Theorem) There exist incomplete d.c.e. degrees $\mathbf{a}_0, \mathbf{a}_1$ such that for any c.e. degree $\mathbf{x}$, if $\mathbf{x} \neq \mathbf{0}$, then $\mathbf{0}' = \mathbf{a}_i \vee \mathbf{x}$ holds for some $i \leq 1$.

And the Sacks splitting theorem (theorem 1.1) can be generalised to the d.c.e. degrees:

THEOREM 5.5. (Cooper [1992] Splitting Theorem) For any d.c.e. degree $\mathbf{a}$, if $\mathbf{a} \neq \mathbf{0}$, then there are d.c.e. degrees $\mathbf{b}, \mathbf{c}$ such that $\mathbf{b} < \mathbf{a}$, $\mathbf{c} < \mathbf{a}$ and $\mathbf{b} \vee \mathbf{c} = \mathbf{a}$.

However this splitting theorem cannot be combined with cone avoidance, unlike the Sacks splitting theorem for the c.e. degrees. Cooper and Li [ta4], [ta6] have shown:

THEOREM 5.6. (Cooper and Li [ta6]) Given $n > 1$, there exists an n-c.e. degree $\mathbf{a}$ and a c.e. degree $\mathbf{b}$ such that $\mathbf{0} < \mathbf{b} < \mathbf{a}$ and such that for any n-c.e. degrees $\mathbf{x}, \mathbf{y}$, if $\mathbf{x} \vee \mathbf{y} = \mathbf{a}$, then either $\mathbf{b} \leq \mathbf{x}$ or $\mathbf{b} \leq \mathbf{y}$.

A basic analysis of the proof of theorem 5.6 establishes the following:

THEOREM 5.7. (Cooper and Li [ta6]) For any $n > 1$, the elementary theory of the low$_3$ c.e. degrees is different from that of the low$_3$ n-c.e. degrees.

Another basic and related topic is the density problem. For the c.e. degrees, Sacks [1964] proved the density theorem (theorem 1.2). But the density problem for the n-c.e. degrees has led to a more interesting situation. Cooper, Lempp and Watson [1989] proved a *weak density theorem* that for any $n > 0$, and any c.e. degrees $\mathbf{c} < \mathbf{a}$, there exists a properly n-c.e. degree $\mathbf{b}$ such that $\mathbf{c} < \mathbf{b} < \mathbf{a}$. Cooper and Yi (unpublished, for $n = 2$) and Arslanov, LaForte and Slaman [1998] (for $n > 2$) proved another *weak density theorem* which says that for $n > 1$, n-c.e. degree $\mathbf{b}$, and c.e. degree $\mathbf{a}$, if $\mathbf{a} < \mathbf{b}$, then there exists a d.c.e. degree $\mathbf{c}$ with $\mathbf{a} < \mathbf{c} < \mathbf{b}$.

While Cooper, Harrington, Lachlan, Lempp and Soare [1991] proved the nondensity of n-c.e. degrees for each $n > 1$ that:

THEOREM 5.8. (Nondensity Theorem of n-c.e. Degrees) For every $n > 1$, there is a maximal n-c.e. degree in the n-c.e. degrees.

A limitation on the scope of this result has been provided by Arslanov, Cooper and Li [2000] (theorem 2.16).

Compared with Harrington-Bickford-Mills low$_2$ splitting theorem (theorem 2.13), Cooper [1991] showed that for each $n > 0$, each n-c.e. degrees $\mathbf{a} < \mathbf{l}$, if $\mathbf{l}$ is low$_2$, then there are n-c.e. degrees $\mathbf{x}, \mathbf{y}$ such that $\mathbf{a} \leq \mathbf{x}, \mathbf{y} < \mathbf{l}$ and $\mathbf{l} = \mathbf{x} \vee \mathbf{y}$.

Considering splitting in the d.c.e. degrees, Ding and Qian (unpublished, see Ding, Lu and Qian [2000]) and independently Ishmukhametov [1996] showed:

THEOREM 5.9. For any c.e. sets $B <_T A$, there exist d.c.e. sets X_0, X_1 with the following properties:
(i) $X_0 \oplus X_1 \equiv_T A$, and
(ii) $A \not\leq_T X_i \oplus B$ holds for each $i \leq 1$.

Furthermore, Cooper and Li [ta5] showed:

THEOREM 5.10. (Cooper-Li Splitting Theorem) For any d.c.e. set A, c.e. set B, if $B <_T A$, then there are d.c.e. sets X_0, X_1 with the following properties,
(1) $X_0 \oplus X_1 \equiv_T A$, and
(2) for any $i \leq 1$, if $A \leq_T X_i \oplus B$, then there is a c.e. set I such that both $I \leq_T A$ and $A \leq_T I \oplus B$ hold.

In fact, Cooper and Li's construction in the proof of this theorem works when A is given to be c.e., in which case, the construction gives a proof for theorem 5.9.

Theorems 5.9 and 5.10 together give the following:

THEOREM 5.11. (Cooper and Li [ta5]) For any c.e. set B, d.c.e. set A, if $B <_T A$, then there are d.c.e. sets X_0, X_1 with the following properties,
(1) $X_0 \oplus X_1 \equiv_T A$, and
(2) for each $i \in \{0, 1\}$, $A \not\leq_T X_i \oplus B$ holds.

Proof. By theorem 5.10, there exist d.c.e. sets X_0, X_1 such that either X_0, X_1 satisfy the conditions of the theorem or there is a c.e. set I such that $A \equiv_T I \oplus B$. In the latter case, by theorem 5.9, there are d.c.e. sets Y_0, Y_1 satisfying the conditions of theorem 5.11. Theorem 5.11 follows. □

The degree theoretic statement of theorem 5.11 is then:

THEOREM 5.12. (Cooper and Li [ta5]) For any d.c.e. degree $\mathbf{a}$, c.e. degree $\mathbf{b}$, if $\mathbf{b} < \mathbf{a}$, then there are d.c.e. degrees $\mathbf{x_0}, \mathbf{x_1}$ such that $\mathbf{b} < \mathbf{x_0}, \mathbf{x_1} < \mathbf{a}$ and $\mathbf{x_0} \vee \mathbf{x_1} = \mathbf{a}$.

As well as extending Cooper's splitting theorem (theorem 5.5), we have an extension of the weak density result of Cooper and Yi (unpublished) as mentioned above. The theorem gives a nontrival d.c.e. Turing approximation to the class of the c.e. degrees, which is then an important first step to define $\mathcal{E}$ in $\mathcal{E}_2$. However the following questions are still open.

OPEN QUESTION 5.13. (i) Are the elementary theories of low$_m$ c.e. degrees and of low$_m$ n-c.e. degrees different for each $n > 1$, and for $m = 1$ or 2?
(ii) Are there any jump classes high$_m$, low$_m$ n-c.e. degrees definable in $\mathcal{E}_n$ for $n > 1$?

References

[1] K. Ambos-Spies [1984a], On pairs of recursively enumerable degrees, Trans. Amer. Math. Soc. 283 (1984), 507-537.

[2] K. Ambos-Spies [1984b], An extension of the non-diamond theorem in classical and α-recursion theory, J. Symbolic Logic 49 , 586-607.

[3] K. Ambos-Spies [1993], Automorphism bases for the recursively enumerable degrees, preprint.

[4] K. Ambos-Spies, C. G. Jockusch, Jr., R. A. Shore and R. I. Soare [1984], An algebraic decomposition of the recursively enumerable degrees and the coincidence of several degree classes with the promptly simple degrees, Trans. Amer. Math. Soc. 281 (1984), 109–128.

[5] K. Ambos-Spies, A. H. Lachlan and R. I. Soare [1993], The continuity of cupping to $\mathbf{0}'$, Annals of Pure and Applied Logic 64 (1993) 195–209.

[6] M. M. Arslanov [1985], Structural properties of the degrees below $\mathbf{0}'$, Dokl. Akad. Nauk. SSSR **283**, 270–273.

[7] M. M. Arslanov, S. B. Cooper and A. Li [2000], There is no low maximal d.c.e. degree, Math. Log. Quart. 46 (2000) 3, 409–416.

[8] M. M. Arslanov, G. L. LaForte and T. A. Slaman [1998], Relative enumerability in the difference hierarchy, J. Symb. Logic 63, 411–420.

[9] M. Bickford and C. F. Mills, Lowness properties, to appear.

[10] P. Cholak, R. Downey and S. Walk [ta], Maximal contiguous degrees, to appear in J. Symbolic Logic.

[11] P. Cholak, M. Groszek and T. A. Slaman [2001], An almost deep degree, J. Symbolic Logic, 66 (2001), No. 2, 881–901.

[12] S. B. Cooper [1972], Jump equivalence of the Δ^0_2 hyperhyperimmune sets, J. Symbolic Logic 37 (1972) 598–600.

[13] S. B. Cooper [1974a], Minimal pairs and high recursively enumerable degrees, J. Symbolic Logic 39 (1974), 655-660.

[14] S. B. Cooper [1974b], On a theorem of C.E.M. Yates, handwritten notes, 1974.

[15] S. B. Cooper [1991], The density of the Low_2 n-r.e. degrees, Archive for Math. Logic **31**, 19–24.

[16] S. B. Cooper [1992], A splitting theorem for the n-r.e. degrees, Proc. Amer. Math. Soc. **115** (2), 461–471.

[17] S. B. Cooper [1997], Beyond Gödel's theorem- the failure to capture information content, in Complexity, Logic and Recursion Theory (ed. A. Sorbi), Lecture Notes in Pure and Applied Mathematics, Vol. 187, Marcel Dekker, New York, 1997, pp. 93-122.

[18] S. B. Cooper [1999], Local degree theory, in Handbook of Computability Theory (ed. Edward R. Griffor), Studies in Logic, Vol. 140, Amsterdam, Laussanne, New York, Oxford, Shannon, Singapore, Tokyo, 1999.

[19] S. B. Cooper, L. Harrington, A. H. Lachlan, S. Lempp and R. I. Soare [1991], The d-r.e. degrees are not dense, Ann. Pure and Appl. Log. **55** (2), 125–151.

[20] S. B. Cooper, S. Lempp and P. Watson [1989], Week density and cupping in the d-r.e. degrees, Israel J. Math. **67**, (1989), 137–152.

[21] S. B. Cooper and A. Li [ta1], Splitting and Nonsplitting, I: A low_3 Harrington Nonsplitting Base, to appear.

[22] S. B. Cooper and A. Li [ta2], Splitting and Nonsplitting, II: A Low_2 c.e. degree above which $\mathbf{0}'$ is not splittable, to appear.

[23] S. B. Cooper and A. Li [ta3], On Lachlan's Major Subdegree Problem, to appear.

[24] S. B. Cooper and A. Li [ta4], Non-uniformity and generalised Sacks splitting, to appear in Acta Mathematica Sinica.

[25] S. B. Cooper and A. Li [ta5], Turing definability in the Ershov hierarchy, Journal of London Mathematical Society, to appear.

[26] S. B. Cooper and A. Li [ta6], Splitting and cone avoidance in the d.c.e. degrees, to appear.

[27] S. B. Cooper, A. Li and X. Yi [ta], On the distribution of Lachlan nonsplitting bases, Archive for Mathematical Logic, To appear.

[28] S. B. Cooper and T. A. Slaman [1988], Major subdegrees and cup classes, Recursive Function Theory Newsletter, Item 371, Issue No. 37, February 1988.

[29] S. B. Cooper and X. Yi, Isolated d.r.e. degrees, unpublished.

[30] D. Ding, H. Lu and L. Qian [2000], A splitting with infimum in the d.c.e. degrees, Math. Log. Quart. 46 (2000) 1, 53–76.

[31] D. Ding and L. Yu [ta], To appear.

[32] R. Downey [1989], D.r.e. degrees and the nondiamond theorem, Bull. London Math. Soc. **21**, 43–50.

[33] R. G. Downey, S. Lempp and R. A. Shore [1993], Highness and bounding minimal pairs,Math. Logic Quarterly, Vol. 39, pp 475-491.

[34] R. M. Friedberg, Two recursively enumerable sets of incomparable degrees of unsolvability, Proc. Natl. Acad. Sci. SUA 43 (1957), 236-238.

[35] M. B. Giorgi [2001], Continuity properties of degree structures, PhD dissertation, University of Leeds, 2001.

[36] L. Harrington [1976], On Cooper's proof of a theorem of Yates, Part I, (handwritten notes), 1976.

[37] L. Harrington [1978], Plus cupping in the recursively enumerable degrees (handwritten notes).

[38] L. Harrington [1980], Understanding Lachlan's monster paper (handwritten notes), 1980.

[39] L. Harrington and R. I. Soare [1989], Games in recursion theory and continuity properties of capping degrees, in Proceedings of the workshop on set theory and the continuum, Oct. 16–20, 1989, Mathematical Sciences Research Institute, Berkeley.

[40] S. Ishmukhametov [1996], Splitting above r.e. degrees, Fundamental Mathematics and Mechanics, 1996, Ulynovsk University Press, Vol. 1, 139–143.

[41] Z. Jiang [1993], Diamond lattice embedded into d.r.e. degrees, Science in China (Series A) 36, 803–811.

[42] C. G. Jockusch, Jr., A. Li and Y. Yang [ta], A join theorem for the computably enumerable degrees, to appear.

[43] C. G. Jockusch and R. A. Shore [1983], Pseudo jump operators I: The R.E. case, Trans. Amer. Math. Soc. 275 (1983), 599-609.

[44] A. H. Lachlan [1966], Lower bounds for pairs of recursively enumerable degrees, Proc. London Math. Soc. 16 (1966), 537–569.

[45] A. H. Lachlan [1968], On the lattice of recursively enumerable sets, Trans. Amer. Math. Soc. 130 (1968), 1–37.

[46] A. H. Lachlan [1975], A recursively enumerable degree which will not split over all lesser ones, Ann. Math. Logic 9 (1975), 307–365.

[47] A. H. Lachlan [1979], Bounding minimal pairs, J. Symbolic Logic 44 (1979), 626–642.

[48] S. Lempp [1988], A high strongly noncappable degree, J. Symbolic Logic.

[49] S. Lempp, A. Nies and T. A. Slaman [1998], The Π_3-theory of the enumerable Turing degrees is undecidable, Trans. Amer. Math. Soc. 350, 2719-2736.

[50] S. Lempp and T. A. Slaman [1989], A limit on relative genericity in the recursively enumerable sets, J. Symbolic Logic, 54 (1989), No. 2 376–395.

[51] A. Li [2000a], On a Conjecture of Lempp, Archive for Mathematical Logic (2000) 39: 281–309.

[52] A. Li [2000b], Bounding cappable degrees, Archive of Mathematical Logic (2000) 39: 311–352.

[53] A. Li [ta1], Elementary differences among jump hierarchies, to appear.

[54] A. Li [ta2], A hierarchy characterisation of cuppable degrees, to appear.

[55] A. Li, T. A. Slaman and Y. Yang [ta], A nonlow$_2$ c.e. degree which bounds no diamond bases, to appear.

[56] A. Li, G. Wu and Z. Zhang [2000], A hierarchy for the cuppable degrees, Illinois Journal of Mathematics, Vol. 44, No. 3, Winter 2000, 619–632.

[57] A. Li and D. Yang [2000], A high diamond theorem, ISSN 1000-9825 Journal of Software, 2000, 11(1): 23–39.

[58] A. Li and X. Yi [1999], Cupping the recursively enumerable degrees by d.r.e. degrees, Proc. London Math. Soc. **78** (3) (1999), 1–21.

[59] D. Miller [1981], High recursively enumerable degrees and the anti-cupping property, in M. Lerman, J. H. Schmerl and R. I. Soare (editors), Logic Year 1979–80: University of Connecticut, Lecture Notes in Mathematics No. 859, Springer-Verlag, Berlin, Heidelberg, Tokyo, New York, 1981, 230–245.

[60] A. A. Muchnik [1956], On the unsolvability of the problem of reducibility in the theory of algorithms, Dokl. Akad. Nauk SSSR, N.S. 108 (1956), 194–197 (Russian).

[61] A. Nies [ta], Model theoretic properties of structures from computability theory, to appear.

[62] A. Nies, R. A. Shore and T. A. Slaman [1998], Interpretability and definability in the recursively enumerable degrees, Proc. London Math. Soc. (3), 77, 241–249.

[63] E. L. Post [1944], Recursively enumerable sets of positive integers and their decision problems, Bull. Amer. Math. Soc. 50 284-316.

[64] P. Odifreddi [1989], *Classical Recursion Theory*, North-Holland, Amsterdam, New York, Oxford.

[65] R. W. Robinson [1971a], Interpolation and embedding in the recursively enumerable degrees, Ann. of Math. (2) 93 (1971), 285–314.

[66] R. W. Robinson [1971b], Jump restricted interpolation in the recursively enumerable degrees, Ann. of Math. (2) 93 (1971), 586–596.

[67] G. E. Sacks [1963], On the degrees less than $\mathbf{0}'$, Ann. of Math. (2) 77 (1963), 211–231.

[68] G. E. Sacks [1964], The recursively enumerable degrees are dense, Ann. of Math. (2) 80 (1964), 300–312.

[69] D. Seetapun [1991], Contributions to Recursion Theory, Ph.D. thesis, Trinity College, Cambridge, 1991.

[70] D. Seetapun [1992], Defeating Red, (handwritten notes), 1992.

[71] J. R. Shoenfield [1965], Application of model theory to degrees of unsolvability, in J. W. Addison, L. Henkin and A. Tarski, Symposium on the theory of models (editors), North-Holland, Amsterdam, 1965.

[72] R. A. Shore [1999], The recursively enumerable degrees, in Handbook of Computability Theory (ed. E. R. Griffor), Amsterdam, Lausanne, New York, Oxford, Shannon, Singapore, Tokyo, 1999, 169–197.

[73] R. A. Shore [2000], Natural definability in degree structures, in M. Lerman, P. Cholak, S. Lempp and R. A. Shore, eds., Computability Theory and Its Applications, Contemporary Mathematics Vol. 257, Providence RI, 2000, AMS, 255–271.

[74] R. A. Shore and T. A. Slaman [1990], Working below low_2 recursively enumerable degrees, Archive for Math. Logic, 29 (1990), 201–211.

[75] R. A. Shore and T. A. Slaman [1993], Working below high recursively enumerable degrees, 58, J.Symbolic Logic , 824-859.

[76] T. A. Slaman and R. I. Soare [1995], Algebraic aspects of the recursively enumerable degrees, Proc. Nat. Acad. Sci. USA, 92, (1995), 617–621.

[77] T. A. Slaman and R. I. Soare [2001], Extension of embeddings in the recursively enumerable degrees, Annals of Math., 153: 1–43, 2001.

[78] R. I. Soare [1974], Automorphisms of the lattice of recursively enumerable sets, Bull. Amer. Math. Soc.80 (1974), 53–58.

[79] R. I. Soare [1987], Recursively Enumerable Sets and Degrees, A study of computable function and computably generated sets, Springer-Verlag Berlin Heidelberg New York London Pairs Tokyo, 1987.

[80] D. Wang [2000], Saturation properties in the computably enumerable degrees, Ph. D. Dissertation, University of Wisconsin, 2000.

[81] C. E. M. Yates [1966], A minimal pair of recursively enumerable degrees, J. Symbolic Logic 31 (1966), 159–168.

QUASI-DEGREES OF RECURSIVELY ENUMERABLE SETS

Roland Sh. Omanadze
I. Vekua Institute of Applied Mathematics
Tbilisi State University
University str. 2.
Tbilisi, 380043
Georgia
omanr@viam.hepi.edu.ge

Contents

Introduction

1. Some classes of recursively enumerable sets and quasi-reducibility
2. Complexity properties of recursively enumerable sets and quasi-degrees
3. On the upper semilattices of recursively enumerable quasi-degrees
4. Relations between degrees
5. Complete degrees

Introduction

The basis of the modern theory of degrees of unsolvability is established in the article by Post [46] in which the notions of many-one (m-) reducibility, truth-table (tt-), bounded truth-table (btt-) and Turing (T-) reducibility are introduced.

Besides these reducibilities, other authors have defined other notions of reducibility. In particular, Tennenbaum (see [47, p. 159]) defined the notion of *quasi-reducibility* or *Q-reducibility* as follows:

A set A is *Q-reducible to* a set B ($A \leq_Q B$) if there is a recursive function f such that for every $x \in \omega = \{0, 1, 2, \dots\}$,

$$x \in A \Leftrightarrow W_{f(x)} \subseteq B.$$

Computability and Models: Perspectives East and West
Edited by Cooper and Goncharov, Kluwer Academic/Plenum Publishers, 2003

In this case we say that $A \leq_Q B$ via f. It is not hard to see that this relation is reflexive and transitive, so that it gives rise to a degree structure on 2^ω. Q-reducibility is a natural weakening of m-reducibility, where singletons are replaced by recursively enumerable (r.e.) sets. It is not hard to show that on the recursively enumerable sets Q-reducibility is a strictly stronger reducibility than T-reducibility.

The notion of Q-reducibility is very natural and important for the Theory of Algorithms. With the help of this notion a series of interesting results were obtained. This notion plays a key role in the Marchenkov solution (see [28]) of the well-known Post problem using Post's methods. Q-degrees have applications in several fields of the Theory of Algorithms, for instance in the study of word problems and in computational complexity.

The relationship of Q-reducibility to word problems is most easily perceived by considering the theorem of Dobritsa [11] stating that for every set of natural numbers A there is a word problem with the same Turing degree as that of A. The proof of this fact depends, for the harder direction, on producing a group G in which every element has a normal form with an associated finite list of numbers, which are a subset of A if and only if the element is the identity. Thus, the proof actually shows that A has the same Q-degree as the word problem of G.

A further relationship between Q-degrees and groups is exhibited in the relationship of a recursively presented group to its algebraically closed extensions. Macintyre in [26] showed that whenever G and H are recursively presented groups with word problems of Turing degrees $\mathbf{g}$ and $\mathbf{h}$, respectively, then if $\mathbf{g} < \mathbf{h}$, then G must be a subgroup of every algebraically closed group of which H is a subgroup. Belegradek [2] showed that the converse fails, but becomes true if Turing reducibility is replaced by Q-reducibility. That is, for any recursively presented groups G and H, if G is a subgroup of every algebraically closed group of which H is a subgroup, then G's word problem must be Q-reducible to that of H.

Blum and Marques in [5] introduced the notions of subcreative and effectively speedable sets and they proved that a recursively enumerable set is subcreative if and only if it is effectively speedable. Gill and Morris in [16] gave a simple and interesting characterization of effectively speedable sets in terms of Q-complete sets. They proved that an r.e. set is effectively speedable if and only if it is Q-complete.

In this article we give a survey of results on the quasi-degrees of r.e. sets. All the notions and notations used without definition can be found in [30, 47, 50].

1. Some Classes of Recursively Enumerable Sets and Quasi-reducibility

If a set A is Q-reducible to a set B via f and there is a recursive function g such that

$$(\forall x)(\forall y)(y \in W_{f(x)} \Rightarrow y < g(x)),$$

then we say that the set A is *strongly Q-reducible* (or *sQ-reducible*) to the set B ($A \leq_{sQ} B$), via f and g.

If a set A is sQ-reducible to a set B via f, g and there is some $n \in \omega$ such that

$$(\forall x)(|W_{f(x)}| < n),$$

then we say that the set A is *bounded sQ-reducible* to the set B ($A \leq_{bsQ} B$).

Finally, we say that a set A is *bounded Q-reducible* to a set B ($A \leq_{bQ} B$) if $A \leq_Q B$ via some f and there is some $n \in \omega$ such that

$$(\forall x)(|W_{f(x)}| < n).$$

Let $r \in \{bQ, bsQ, sQ, Q\}$. A set A is $r-complete$ if A is r.e. and $W \leq_r A$ for every r.e. set W.

It is not hard to see that the relations $\leq_{sQ}$, $\leq_{bsQ}$, $\leq_{bQ}$ are reflexive and transitive, so that they give rise to corresponding degree structures on 2^ω.

We note here that there is a least bsQ-degree, containing exactly the Π^0_1 sets. In fact, let $A \in \Pi^0_1$, let $B \neq \omega$ be any set, let $b \notin B$ and let f be a recursive function such that

$$W_{f(x)} = \begin{cases} \{b\}, & \text{if } x \in \overline{A} \\ \emptyset, & \text{otherwise.} \end{cases}$$

Then $A \leq_{bsQ} B$ via f, g and n, where g is constant function $g(x) = b$ and $n = 1$.

An r.e. set A is *hyperhypersimple* if it is a coinfinite set and there is no recursive function f such that

1. $(\forall x)(W_{f(x)} \cap \overline{A} \neq \emptyset)$,
2. $(\forall x)(\forall y)(x \neq y \Rightarrow W_{f(x)} \cap W_{f(y)} = \emptyset)$,
3. $(\forall x)(|W_{f(x)}| < \infty)$.

Solov'ev [51] has obtained an interesting characterization of the hyperhypersimple sets in terms of Q- complete r.e. sets. The same result was obtained by Gill and Morris [16] independently.

Theorem 1.1 (Solov'ev [51]) *A coinfinite r.e. set is hyperhypersimple if and only if it is not contained in any Q-complete r.e. set.*

A coinfinite r.e. set A is *finitely strongly hypersimple* (fsHS) if there is no recursive function f such that

1. $(\forall x)(W_{f(x)} \cap \overline{A} \neq \oslash)$,
2. $(\forall x)(\forall y)(x \neq y \Rightarrow W_{f(x)} \cap W_{f(y)} = \oslash)$,
3. $(\forall x)(|W_{f(x)}| < \infty)$,
4. $\bigcup_{x \in \omega} W_{f(x)} = \omega$.

The author [38] has obtained an unexpected connection between the class of Q-complete r.e. sets and the sets not belonging to the class fsHS.

Theorem 1.2 (Omanadze [38]) *Let A be a coinfinite r.e. set such that A does not belong to the class* fsHS. *Then there is an r.e. set C, $A \subseteq C$, such that every coinfinite superset of the set C is Q-complete.*

Before giving the proof, we introduce some notions and a result which are used in the proof.

A set $A \subseteq \omega$ is *retraceable* if there is a partial recursive function g such that $A \subseteq dom(g), g(a_0) = a_0$ and

$$(\forall n)(g(a_{n+1}) = a_n),$$

where $A = \{a_0 < a_1 < \cdots < a_n < \cdots\}$.

Lemma 1.3 (Arslanov [1]) *If A is an infinite retraceable set and B is an infinite subset of the set A, then $\overline{A} \leq_Q \overline{B}$.*

Proof of the Theorem 1.2. Let A be a coinfinite r.e. set, $A \notin$fsHS, and let f be a recursive function satisfying 1.-4. of the definition of finitely strongly hypersimple sets. Let B be a Q-complete coretraceable r.e. set.

There is such a set: For instance, [47, Theorem 9.XVI] (see also [47, Exercise 9.44(vi)]), every r.e. nonrecursive Turing degree contains a coretraceable hypersimple set.

Define the set

$$C = A \cup \left(\bigcup_{x \in B} W_{f(x)}\right).$$

Then $A \subseteq C$, $|\overline{C}| = \infty$ and

$$(\forall x)(x \in B \Leftrightarrow W_{f(x)} \subseteq C).$$

Let us show that C is the desired set. In fact, let C_1 be any coinfinite superset of the set C. Define the set

$$B_1 = \{x : W_{f(x)} \subseteq C_1\}.$$

It is obvious that $B \subseteq B_1$ and from 1.-4. we have $|\overline{B_1}| = \infty$. Then by Lemma 1.3 $B \leq_Q B_1$.

By definition of the set B_1 we have $B_1 \leq_Q C_1$. Hence $B \leq_Q C_1$, i.e. C_1 is a Q-complete set. ⊠

Corollary 1.4 *Every nonrecursive r.e. T-degree contains an r.e. set A having a coinfinite superset B such that every coinfinite superset of the set B is Q-complete.*

Given r.e. sets $A \subseteq B$, A is a *major subset* of B ($A \subset_m B$) if $B \setminus A$ is infinite and for every r.e. set W

$$\overline{B} \subseteq^* W \Rightarrow \overline{A} \subseteq^* W.$$

(For sets X, Y we use $X \subseteq^* Y$ to denote that $X \subseteq Y$ except for finitely many elements (i.e. $X \setminus Y$ is finite).

Theorem 1.5 (Lachlan [22]) *For every nonrecursive r.e. set B there exists an r.e. set A such that $A \subset_m B$.*

By using Theorem 1.5 and Theorem 1.2 the following statement is proved.

Theorem 1.6 (Omanadze [45]) *Every nonrecursive r.e. set A has a major subset B such that each superset C of the set B, satisfying the condition*

$$C \subset_m A,$$

is Q-complete.

Proof. Let A be a nonrecursive r.e. set, let $E \subset_m A$, and let f be a 1:1 recursive function such that range$(f) = A$ and $E_1 = f^{-1}(E)$. It is not hard to see that E_1 is not a finitely strongly hypersimple set. Then by Theorem 1.2 there is an r.e. set F, $E_1 \subseteq F$, such that every r.e. coinfinite superset of the set F is Q-complete. Now in order to complete the proof of Theorem 1.6 it is sufficient to consider the set $B = f(F)$. ⊠

An r.e. set A is *nowhere simple* if for every r.e. B with $B \setminus A$ infinite there is an infinite r.e. set $W \subseteq B \setminus A$. Shore [48] has proved that if **a** is an r.e. T-degree then there is a nowhere simple set $A \in$ **a**.

If a set A is Q-reducible to a set B via f and

$$(\forall x)(|W_{f(x)}| < \infty),$$

then we say that the set A is Q_k ÷ *reducible* to the set B ($A \leq_{Q_k} B$).

Solov'ev [51] has proved that if A and B are r.e. sets then

$$A \leq_Q B \Rightarrow A \leq_{Q_k} B.$$

By using this result, we can prove the following statement.

Proposition 1.7 *Let A be an r.e. set, B a nowhere simple set and $A \leq_Q B$. Then A is a nowhere simple set.*

Corollary 1.8 *Every r.e. set contained in the Q-degree of a nowhere simple set is nowhere simple.*

Proposition 1.9 *Let A, B be nowhere simple sets. Then $A \oplus B = \{2x : x \in A\} \cup \{2x + 1 : x \in B\}$ is a nowhere simple set.*

Corollary 1.10 *The Q-degrees of nowhere simple sets form an ideal in the upper semilattice of the r.e. Q-degrees.*

We will need the following statement.

Lemma 1.11 (Omanadze [33]) *Let A, C be r.e. sets and let B be a set such that*

$$A \leq_Q B \leq_Q C.$$

Then $A \leq_{Q_k} B$.

The following question naturally arises: Is a simple or a nowhere simple set contained in every r.e. Q-degrees? A negative answer to this question is given by the following theorem which is a consequence of Lemma 1.11.

Theorem 1.12 (Omanadze [34]) *Every nonrecursive r.e. T-degree contains an r.e. set A such that the Q-degree of A contains neither simple nor nowhere simple sets.*

Sketch of the proof. The proof of this theorem is based on the following result, which is of independent interest.

Theorem 1.13 (Omanadze [34]) *Let A be a simple set, B an arbitrary set, C a nowhere simple set, K a Q-complete r.e. set and suppose that*

$$A \leq_Q B \oplus C \leq_Q K.$$

Then $A \leq_Q B$.

By using Proposition 1.7 and Theorem 1.13 one can prove the following statement.

Proposition 1.14 *Suppose that A is not a nowhere simple set, and B is nowhere simple. Further let A and B have incomparable Q-degrees. Then the Q-degree of $A \oplus B$ contains neither simple nor nowhere simple sets.*

Now in order to complete the proof of the Theorem 1.12, let **a** be an arbitrary nonrecursive r.e. T-degree, let $A \in \mathbf{a}$, and let B and C are r.e. sets such that

$$A = B \cup C \,\&\, B \cap C = \oslash$$

and B, C are T-incomparable. Let B_1 be an r.e. set such that B_1 is not nowhere simple and $B_1 \equiv_T B$, and let C_1 be a nowhere simple set such that $C_1 \equiv_T C$. Then the Q-degree of the set $B_1 \oplus C_1$ contains neither simple nor nowhere simple sets. It is clear that $A \equiv_T B_1 \oplus C_1$. ⊠

Denote by L_Q the upper semilattice of all r.e. Q-degrees, by J_Q the ideal of the Q-degrees of all nowhere simple sets. For every r.e. Q-degree $\mathbf{a}$ let

$$[\mathbf{a}] = \{\mathbf{b} : \mathbf{b} \in L_Q \& (\exists \mathbf{m}, \mathbf{n} \in J_Q)(\mathbf{a} \cup \mathbf{m} = \mathbf{b} \cup \mathbf{n})\},$$
$$L_Q/J_Q = \{[\mathbf{a}] : \mathbf{a} \in L_Q\},$$
$$[\mathbf{a}] \leq [\mathbf{b}] \Leftrightarrow (\exists \mathbf{m} \in J_Q)(\mathbf{a} \leq \mathbf{b} \cup \mathbf{m}).$$

In the quotient semilattice L_Q/J_Q it is possible to define the operation $\cup$ such that the following condition is satisfied

$$[\mathbf{a}] \cup [\mathbf{b}] = [\mathbf{a} \cup \mathbf{b}].$$

Proposition 1.15 (Omanadze [34]) *There is an incomparable pair in the quotient semilattice L_Q/J_Q of the r.e. Q-degrees modulo the nowhere simple Q-degrees.*

Proposition 1.16 (Omanadze [34]) *In the quotient semilattice L_Q/J_Q of the r.e. Q-degrees modulo the nowhere simple Q-degrees, for every element $[\mathbf{a}] \neq \oslash$ there is a Q-degree $\mathbf{b}$ of a simple set such that $[\mathbf{b}] \leq [\mathbf{a}]$.*

2. Complexity Properties of Recursively Enumerable Sets and Quasi-degrees

One of the most interesting and important aspects of the theory of computational complexity of recursive functions is the speed-up phenomenon, such as in the theorem of Blum [3] which asserts the existence of a 0, 1-valued recursive function with arbitrary large speed-ups.

Blum and Marques [5] extended speed-up definitions from recursive functions to partial recursive functions, or equivalently to r.e. sets and introduced the notions of speedable, nonspeedable, effectively speedable and subcreative sets. In the same work they proved that an r.e. set is subcreative if and only if it is effectively speedable. Blum and Marques [5] remarked that: "An important goal of complexity theory, as we see it is to characterize those partial recursive functions and recursively enumerable sets having some given complexity properties, and to do so in terms which do not involve the notion of complexity." Such a characterization of the effectively speedable sets was given Gill and Morris [16]. Namely they proved that an r.e. set is effectively speedable if and only if it is Q-complete. Soare [49] has found an interesting characterization of nonspeedable sets in terms of index sets.

Since Blum's papers [3,4] first appeared in 1967, methods and results from "pure" Recursive Function Theory (especially the Recursion Theorem and the priority method) have played an increasingly important role in the investigation of the computational complexity of recursive functions.

Suppose that $\{\varphi_i\}_{i\in\omega}$ is an effective numbering of all partial recursive functions and let $\{\Phi_i\}_{i\in\omega}$ be a sequence of partial recursive functions satisfying Blum's axioms, [3]:

(1) the function $\varphi_i(x)$ is defined if and only if the function $\Phi_i(x)$ is defined;

(2) the function

$$M(i,n,m) = \begin{cases} 1, & \text{if } \Phi_i(n) = m, \\ 0, & \text{otherwise} \end{cases}$$

is recursive.

Definition 2.1 (Blum and Marques [5]) An r.e. set A is called *subcreative*, if there exists a recursive function f such that for each i one can find a number a_i such that

$$W_{f(i)} = A \cup \{a_i\}$$

and

$$a_i \in (W_i \cap A) \cup (\overline{W_i \cup A}).$$

Definition 2.2 (Blum and Marques [5]) An r.e. set A is called *effectively speedable* if there exists a recursive function τ such that for all natural numbers i and l, if $W_i = A$ and φ_l is a recursive function, then

1. $W_{\tau(i,l)} = A$ (i.e. $\varphi_{\tau(i,l)} = \chi_A$),
2. $(\exists^\infty x)(x \in A \Rightarrow \Phi_i(x) > \varphi_l(x, \Phi_{\tau(i,l)}(x)))$.

Theorem 2.1 (Blum and Marques [5]) *Let A be an r.e. set. Then A is effectively speedable if and only if A is subcreative.*

Definition 2.3 (Blum and Marques [5]) An r.e. A set is called *speedable* if for all integer i such that $\varphi_i = \chi_A$ and for all recursive functions g there is an integer j such that

1. $\varphi_j = \chi_A$ (i.e. $W_j = A$),
2. $(\exists^\infty x)(x \in A \Rightarrow \Phi_i(x) > g(x, \Phi_j(x)))$.

Definition 2.4 (Blum and Marques [5]) An r.e. set A is called *nonspeedable* if there is a number i and a recursive function h such that $W_i = A$ and

$$(\forall j)[W_j = A \Rightarrow (\forall^\infty x)[x \in A \Rightarrow \Phi_i(x) \leq h(x, \Phi_j(x))]].$$

Let A be an r.e. set and χ_A its semicharacteristic function. For any $i \in \omega$, we say that $\varphi_i = \chi_A$ is *minimal* if there is a recursive function h such that for all $j \in \omega$

$$\varphi_j = \chi_A \Rightarrow (\forall^\infty x)[x \in A \Rightarrow \Phi_i(x) \leq h(x, \Phi_j(x))].$$

We say then that φ_i is *minimal modulo h*.

Now it is possible to define the notion of a nonspeedable (or speedable) set in terms of minimal programs. An r.e. set A is nonspeedable if and only if its semicharacteristic function χ_A has a minimal program modulo some recursive function h.

In other words, a nonspeedable set has a program, which is the best with respect to some recursive function, but a speedable set has no such program.

Marques [29] has proved that nonspeedable sets are contained in every r.e. T-degrees. Soare [48] has shown that an r.e. set A is nonspeedable if and only if

$$\{e : W_e \cap \overline{A} \neq \emptyset\} \leq_T \emptyset',$$

(i.e., A is semilow) and an r.e. T-degree $\mathbf{a}$ contains a speedable set A if and only if $\mathbf{a}' > \mathbf{0}'$, and if $\mathbf{a}' > \mathbf{0}'$ then A can be chosen to be simple.

Gill and Morris have given a useful characterization of the subcreative sets.

Theorem 2.2 (Gill and Morris [16]) *An r.e. set A is subcreative if and only if there is a recursive function f such that if* $W_i \cap A = \emptyset$, *then*

$$A \subset W_{f(i)} \subseteq \overline{W}_i.$$

(We denote proper inclusion by $\subset$).

Corollary 2.3 *Post's simple set is subcreative.*

Using Theorem 2.2 Gill and Morris gave a simple and interesting characterization of the subcreative sets and, hence, by Theorem 2.1, of the effectively speedable sets in terms of Q-complete sets.

Theorem 2.4 (Gill and Morris [16]) *An r.e. set A is subcreative if and only if it is Q-complete.*

Definition 2.5 (Gill and Morris [16]) A set A of natural numbers is *subproductive* if there is a recursive function f such that, for every i,

$$W_{f(i)} \cap W_i \neq \emptyset \Leftrightarrow W_{f(i)} \cap A = \emptyset;$$

f is called a *subproductive function* for A.

Theorem 2.5 (Gill and Morris [16]) *An r.e. set A is subcreative if and only if* $\overline{A}$ *is subproductive. Moreover, the subproductive function f may be chosen so that* $\{W_{f(i)}\}_{i \in \omega}$ *is a disjoint sequence of (nonempty) finite sets.*

Theorem 2.6 (Gill and Morris [16]) *An r.e. set A is subcreative if and only if there exists an r.e. subset B of A and a weak array* $\{W_{f(i)}\}_{i \in \omega}$ *of disjoint, finite, nonempty sets such that*

(i) $W_{f(i)} \setminus B \neq \emptyset$;

(ii) $A = B \cup \bigcup_{i \in I} W_{f(i)}$ *where* $I = \{i : W_{f(i)} \bigcap W_i \neq \emptyset\}$.

Corollary 2.7 *There exist hypersimple (dense simple, strongly hypersimple) subcreative sets.*

It is natural to introduce the following notions.

Definition 2.6 (Omanadze [43]) An r.e. set A is called *effectively subcreative* if there exist recursive functions f and g such that for each i one can find a number a_i such that

1. $W_{f(i)} = A \cup \{a_i\}$,
2. $a_i \in (W_i \cap A) \cup \overline{(W_i \cup A)}$,
3. $a_i \in \overline{W_i \cup A} \Rightarrow a_i < g(i)$.

Definition 2.7 (Omanadze [43]) An r.e. set A is called *strongly effectively speedable* if there exist recursive functions h and g such that for all natural numbers i and l, if $W_i = A$ and φ_l is a recursive function, then

1. $W_{h(i,l)} = A$,
2. $(\exists^\infty x)(x \in A \Rightarrow \Phi_i(x) > \varphi_l(x, \Phi_{h(i,l)}(x)))$,
3. $(\forall i)(\forall l)(\forall y)(y \in W_{h(i,l)} \setminus A \Rightarrow y < g(\langle i, l \rangle))$.

Theorem 2.8 (Omanadze [43]) *Let A be an r.e. set. Then the following statements are equivalent:*

I. *A is an effectively subcreative set;*

II. *A is an sQ-complete set;*

III. $(\exists f)(\exists g)(\forall x)(\forall y)((W_x \cap A = \oslash \Rightarrow A \subset W_{f(x)} \subseteq \overline{W}_x)\ \&$
$(y \in W_{f(x)} \setminus A \Rightarrow y < g(x))$;

IV. *A is a strongly effectively speedable set.*

Corollary 2.9 *There exists an effectively speedable (subcreative) set which is not strongly effectively speedable (effectively subcreative).*

Proposition 2.10 (Omanadze [33]) *Let A be an r.e. set, B be a nonspeedable set and $A \leq_Q B$. Then A is nonspeedable.*

From this proposition it follows that each r.e. set contained in a Q-degree of a nonspeedable set is nonspeedable. Therefore we shall name a Q-degree of nonspeedable set a *nonspeedable* Q-degree.

Proposition 2.11 (Omanadze [33]) *Let A, B be nonspeedable sets. Then $A \oplus B$ is a nonspeedable set.*

Corollary 2.12 *The Q-degrees of nonspeedable sets induce an ideal in the upper semilattice of the r.e. Q-degrees.*

We now point out an important property of bsQ-reducibility.

Proposition 2.13 (Omanadze [39]) *Let A, B, C be r.e. sets, $B \cap C = \oslash$ and $A = B \cup C$. Then $B \leq_{bsQ} A$ and $C \leq_{bsQ} A$.*

Corollary 2.14 *The upper semilattice of r.e. bsQ- (bQ-, sQ-, Q-) degrees contains no minimal elements.*

Soare (see [49, Corollary 2.3] or [50, p. 72]) has shown that an r.e. set A is nonspeedable if and only if there exists a recursive enumeration $\{A_s\}_{s\in\omega}$ of the set A such that for each e

$$(\exists^\infty s)(W_{e,s} \setminus A_s \neq \oslash) \Rightarrow W_e \setminus A \neq \oslash.$$

Using this very useful characterization of the nonspeedable sets and the previous proposition the following statement is proved.

Theorem 2.15 (Omanadze [37]) *Let* $\mathbf{a}$, $\mathbf{b}$ *be nonspeedable Q-degree such that* $\mathbf{a} <_Q \mathbf{b}$. *Then there exist infinitely many pairwise Q-incomparable nonspeedable Q-degrees* $\{\mathbf{c_i}\}_{i\in\omega}$ *such that*

$$(\forall i)(\mathbf{a} <_Q \mathbf{c_i} <_Q \mathbf{b}).$$

Sketch of the proof. Let $\mathbf{a}$, $\mathbf{b}$ be nonspeedable Q-degrees, $\mathbf{a} <_Q \mathbf{b}$, and let A, B be r.e. nonspeedable sets such that $A \in \mathbf{a}$ and $B \in \mathbf{b}$. We shall split effectively the set B into an infinite number of pairwise disjoint r.e. sets $\{B_i\}_{i\in\omega}$ which satisfy the condition

$$(\forall i)(\forall j)(i \neq j \Rightarrow B_i \nleq_Q A \oplus B_j)$$

and such that each $A \oplus B_i$ is nonspeedable. Then, by using Proposition 2.3, it is not hard to check that as suitable Q-degrees $\mathbf{c}_i$, $i = 0, 1, 2, \ldots$, we may take the Q-degrees of the sets $A \oplus B_i$, $i = 0, 1, 2, \ldots$.

For each $i \in \omega$ consider a B_i-list which consists of all natural numbers ordered according to the inverse of the natural ordering. At any stage of the computation we will use only a finite part of each list. Step by step we build r.e. sets B_i, $i = 0, 1, 2, \ldots$. By $B_{i,s}$ we will denote the part of the set B_i built at the end of step s.

In the process of the computation we use an infinite set of movable markers Γ^i_{jk}, $i \neq j$, $i, j, k \in \omega$. The markers Γ^i_{jk} are used in the B_i-list for ensuring that the set B_i is not Q-reducible to the set $A \oplus B_j$, $i \neq j$, via φ_k. In the construction it turns out that at each step s, if $\langle i, j, k\rangle < \langle i', j', k'\rangle$, then the number to which the marker Γ^i_{jk} is assigned is smaller than the number to which the marker $\Gamma^{i'}_{j'k'}$ is assigned. Denote by z^s_{ijk} the number to which the marker Γ^i_{jk} is assigned at step s.

We shall say that a natural number is *free* in a given list at a given time if neither this number nor any other number situated below it in any list have markers associated with them.

Let $\{A_s\}_{s\in\omega}$ and $\{B_s\}_{s\in\omega}$ be recursive enumerations of the sets A and B respectively such that for each e

$$(\exists^\infty s)[W_{e,s} \setminus A_s \oplus B_s \neq \oslash] \Rightarrow W_e \setminus A \oplus B \neq \oslash.$$

There are such enumerations by the fact that the set $A \oplus B$ is nonspeedable and by Soare's previously mentioned result.

Step 0. For every i, let $B_{i,0} = \oslash$.

Step s, $s > 0$. Let $x \in B_s \setminus B_{s-1}$. (If $B_s \setminus B_{s-1} = \oslash$, then immediately move to next step.) We make s steps in the computation of all one-place partial recursive functions $\{\varphi_k\}_{k\in\omega}$, and in the enumeration of all r.e. sets $\{W_k\}_{k\in\omega}$. We seek a least number $\langle i,j,k\rangle$, $i \neq j$, $\langle i,j,k\rangle \leq s$, such that either $\langle i,j,k\rangle$ has not been considered at any step t, $t < s$, or for each $u \in [0, z^{s-1}_{ijk}]$ $\varphi_k(u)$ is defined and

$$u \in B_{i,s-1} \Leftrightarrow W_{\varphi_k(u),s} \subseteq A_s \oplus B_s.$$

Moreover at step s no $\Gamma^{i'}_{j'k'}$, $\langle i,j,k\rangle > \langle i',j',k'\rangle$, is prohibiting that we enumerate into B_i the number $x \in B_s \setminus B_{s-1}$.

If there exists such a number $\langle i,j,k\rangle$, then we take off the prohibitions imposed by $\Gamma^{i'}_{j'k'}$, $\langle i',j',k'\rangle > \langle i,j,k\rangle$, and we act in the following way:

a) We enumerate $x \in B_s \setminus B_{s-1}$ into B_i.

b) For all $\langle i',j',k'\rangle \geq \langle i,j,k\rangle$ we associate the marker $\Gamma^{i'}_{j'k'}$ with the number $z^{s-1}_{i'j'k'} + 1$. If the number $\langle i,j,k\rangle$ has not been considered at any step t, $t < s$, then we associate the marker Γ^{i}_{jk} with the first free number in the B_i-list, and in addition, for all $\langle i',j',k'\rangle > \langle i,j,k\rangle$, if the marker $\Gamma^{i'}_{j'k'}$ at step $s-1$ has state $z^{s-1}_{i'j'k'}$, then we associate it with the number $z^{s-1}_{i'j'k'} + s$.

c) We prohibit the enumeration of the numbers z, $z < z^{s}_{ijk} + 1$, into B_y for all y, $y \neq i$.

If there exists no such $\langle i,j,k\rangle$ then we shall find the least number $\langle i_0,j_0,k_0\rangle$ such that the marker $\Gamma^{i_0}_{j_0k_0}$ allows us to use the number $x \in B_s \setminus B_{s-1}$, we take off the prohibitions imposed by $\Gamma^{i'}_{j'k'}$, $\langle i',j',k'\rangle > \langle i_0,j_0,k_0\rangle$, we enumerate the number x_0 into B_{i_0} and for all $\langle i,j,k,\rangle \geq \langle i_0,j_0,k_0\rangle$ we associate the marker Γ^{i}_{jk} with the number $z^{s-1}_{ij} + 1$.

Lemma 2.16 *For each* e

$$(\forall i)(\exists^\infty s)((W_{e,s} \setminus A_s \oplus B_{i,s} \neq \oslash) \Rightarrow W_e \setminus A \oplus B_i \neq \oslash).$$

Lemma 2.17 *All the assigned markers move finitely often.*

Lemma 2.18 $(\forall i)(\forall j)(i \neq j \Rightarrow B_i \not\leq_Q A \oplus B_j)$. ⊠

A set A is *semirecursive* [19] if there exists a recursive function f such that

1. $f(x, y) = x \vee f(x, y) = y$,
2. $x \in A \vee y \in A \Rightarrow f(x, y) \in A$.

Corollary 2.19 *1) Let A, B be nonspeedable semirecursive sets such that $A <_Q B$. Then there is a nonspeedable semirecursive set C such that $A <_Q C <_Q B$.*

2) Let A, B be nonspeedable nonsemirecursive sets such that $A <_Q B$. Then there are infinitely many pairwise Q-incomparable nonspeedable nonsemirecursive sets $\{C_i\}_{i \in \omega}$ such that

$$(\forall i)(A <_Q C_i <_Q B).$$

This corollary is proved by using the following statement.

Lemma 2.20 (Marchenkov [23]) *Let B be a semirecursive r.e. set, $B \neq \omega$, $B \neq \oslash$, and let A be an r.e. set with $A \leq_T B$. Then A is Q-reducible to B.*

Recently Jahn has proved the following interesting statement.

Theorem 2.21 (Jahn [18]) *Given any speedable r.e. set A, there are disjoint r.e. sets A_0, A_1, such that $A_0 \cup A_1 = A$ and both A_0, A_1 are speedable.*

From this theorem and from Propositions 2.10 and 2.13 we have the following result.

Theorem 2.22 *An r.e. set A is nonspeedable if and only if for all splitting $A = A_0 \cup A_1$, $A_0 \cap A_1 = \oslash$, with A_0, A_1 r.e. sets, both A_0 and A_1 are nonspeedable sets.*

Corollary 2.23 *An r.e.set A is speedable if and only if there is a splitting $A = A_0 \cup A_1$, $A_0 \cap A_1 = \oslash$ such that both A_0 and A_1 are speedable sets.*

Definition 2.8 (Omanadze [44]) An r.e. set is called *bounded strongly effectively speedable* if there exist recursive functions h, g and a natural number $m \in \omega$ such that for all natural numbers i and l, if $W_i = A$ and φ_l is recursive, then

1. $W_{h(i,l)} = A$,
2. $(\exists^\infty x)(x \in A \Rightarrow \Phi_i(x) > \varphi_l(x, \Phi_{h(i,l)}(x)))$,
3. $(\forall i)(\forall l)(\forall y)(y \in W_{h(i,l)} \setminus A \Rightarrow y < g(\langle i, l \rangle))$,
4. $(\forall i)(\forall l)(|W_{h(i,l)} \setminus A| < m)$.

Theorem 2.24 (Omanadze [44]) *Let A be an r.e. set. Then the following statements are equivalent:*

I. A is a creative set;

II. A is a bsQ-complete set;

III. $(\exists f)(\exists g)(\exists m)(\forall x)(\forall y)((W_x \cap A = \oslash \Rightarrow W_{f(x)} \cap \overline{A} \neq \oslash \,\&$
$\& \, W_{f(x)} \cap W_x = \oslash) \,\&\, (y \in W_{f(x)} \Rightarrow y < g(x)) \,\&\, (|W_{f(x)}| < m));$

IV. $(\exists \widetilde{f})(\exists \widetilde{g})(\exists m_1)(\forall x)(\forall y)((W_x \cap A = \oslash \Rightarrow A \subset W_{\widetilde{f}(x)} \subseteq \overline{W}_x) \,\&$
$\& \, (y \in W_{\widetilde{f}(x)} \setminus A \Rightarrow y < \widetilde{g}(x)) \,\&\, (|W_{\widetilde{f}(x)} \setminus A| < m_1));$

V. A is a bounded strongly effectively speedable set.

Corollary 2.25 *There exists a strongly effectively speedable set which is not bounded strongly effectively speedable.*

Definition 2.9 (Omanadze [44]) An r.e. set A is called *bounded effectively speedable* if there exist a recursive function h and a natural number $m \in \omega$ such that for all natural numbers i and l, if $W_i = A$ and φ_l is recursive, then

1. $W_{h(i,l)} = A$,
2. $(\exists^\infty x)(x \in A \Rightarrow \Phi_i(x) > \varphi_l(x, \Phi_{h(i,l)}(x)))$,
3. $(\forall i)(\forall l)(|W_{h(i,l)} \setminus A| < m)$.

Theorem 2.26 (Omanadze [44]) *Let A be an r.e. set. Then the following statements are equivalent:*

I. A is a creative set;

II. A is a bQ-complete set;

III. $(\exists f)(\exists m)(\forall x)(\forall y)((W_x \cap A = \oslash \Rightarrow W_{f(x)} \cap \overline{A} \neq \oslash \,\&$
$W_{f(x)} \cap W_x = \oslash) \,\&\, (|W_{f(x)}| < m));$

IV. $(\exists \widetilde{f}_1(\exists m_1)(\forall x)((W_x \cap A = \oslash \Rightarrow A \subset W_{\widetilde{f}_1(x)} \subseteq \overline{W}_x) \,\&$
$(|W_{\widetilde{f}_1(x)} \setminus A| < m));$

V. A is a bounded strongly effectively speedable set.

Corollary 2.27 *An r.e. set A is bounded effectively speedable if and only if A is bounded strongly effectively speedable.*

From Theorem 2.26 we have

Theorem 2.28 *An r.e. set A is bounded effectively speedable if and only if there are disjoint r.e. sets A_0, A_1 such that $A = A_0 \cup A_1$ and both A_0, A_1 are bounded effectively speedable.*

Problem 1. Is there an analog of Theorem 2.28 for effectively speedable and strongly effectively speedable sets?

3. On the Upper Semilattices of Recursively Enumerable Quasi-degrees

Fisher and Ambos-Spies [15] have shown that the upper semilattice of r.e. Q-degrees is not distributive and is not a lattice. Interesting results concerning the upper semilattice of the r.e. Q-degrees have been obtained by Downey, LaForte and Nies [13]. In particular, they have proved that the upper semilattice of r.e. Q-degrees is densely ordered. The fact that minimal pairs of r.e. T-degrees exist implies that there are minimal pairs of r.e. Q-degrees.

Theorem 3.1 (Downey, LaForte, Nies [13]) *There exist a nonrecursive r.e. set A and an r.e. B with $A \equiv_T B$ such that A and B form a minimal pair in the Q-degrees.*

Theorem 3.2 (Downey, LaForte, Nies [13]) *For every nonrecursive r.e. set C, there exists an r.e. set A, $C \not\leq_Q A$ such that A is non-branching in the r.e. Q-degrees.*

Theorem 3.3 (Downey, LaForte, Nies[13]) *For every pair of r.e. sets $B <_Q A$, there exists an r.e. set C with $B <_Q B \oplus C <_Q A$.*

The proofs of these last two theorems, particularly that of the density theorem, are significantly more difficult than in the case of the Turing degrees. This shows up particularly in the more subtle methods that have to replace standard permitting techniques for ensuring that a constructed set will be recursive in some set given in advance. The added complexity serves to illustrate the ideas needed to modify standard finite injury and infinite injury constructions in order to deal with r.e. Q-degrees.

An r.e. set M is *maximal* if $\overline{M}$ is infinite and there is no r.e. set W such that $|W \cap \overline{M}| = |\overline{W} \cap \overline{M}| = \infty$.

An r.e. set M is *r-maximal* if $\overline{M}$ is infinite and there is no recursive set R such that $|R \cap \overline{M}| = |\overline{R} \cap \overline{M}| = \infty$.

The well-known Sacks Splitting Theorem does not hold for the r.e. Q-degrees.

Theorem 3.4 (Omanadze [33])*The Q-degree of a maximal set is not the least upper bound of any pair of incomparable Q-degrees.*

The proof of this theorem is based on the following statement.

Theorem 3.5 (Omanadze [33]) *Among all m-degrees contained in the Q-degree of a maximal set there exists a least m-degree.*

Sketch of the proof. Let M be a maximal set, A an arbitrary set and $M \equiv_Q A$. Then by Lemma 1.11 there is a recursive function f such that

$$(\forall x)(x \in M \Leftrightarrow W_{f(x)} \subseteq A)$$

and

$$(\forall x)(|W_{f(x)}| < \infty).$$

By using recursive enumerability of the set M it is possible to assume that there is a recursive function g such that

$$(\forall x)(x \in A \Leftrightarrow W_{g(x)} \subseteq M),$$

$$(\forall x)(x \in \overline{A} \Rightarrow |W_{g(x)} \cap \overline{M}| < \infty),$$

$$(\forall x)(M \subseteq W_{g(x)}).$$

From this last relation, in particular, it follows that

$$\bigcup_{x \in \omega} W_{g(x)} =^* \omega,$$

where $X =^* Y$ stands for

$$|(X \setminus Y) \cup (Y \setminus X)| < \infty.$$

Lemma 3.6 *For every partial recursive function φ we have*

$$(|\varphi(\overline{M})| = \infty \;\&\; \varphi(\overline{M}) \subseteq \overline{M}) \Rightarrow | \{x : x \in \overline{M} \;\&\; \varphi(x) \text{ is defined } \& \; \varphi(x) \neq x\} | < \infty.$$

Now let us define a partial recursive function $\tilde{\varphi}$ as follows:
We compute simultaneously

$$\{W_{g(i)}\}_{i \in \omega} \quad \text{and} \quad \{W_{f(j)}\}_{j \in \omega}$$

and, for given number z, we seek the first integers x and y (if they exist) such that

$$z \in W_{g(y)} \;\;\&\;\; y \in W_{f(x)}.$$

If we can find such x and y, then we let

$$\tilde{\varphi}(z) = x.$$

It is clear that if $z \in \overline{M}$, and $\tilde{\varphi}(z)$ is defined, then $\tilde{\varphi}(z) \in \overline{M}$.

Lemma 3.7 *The function $\tilde{\varphi}$ is defined almost everywhere in $\overline{M}$.*

By Lemma 3.6 we have

$$|\{x : x \in \overline{M} \;\;\&\;\; \tilde{\varphi}(x) \quad \text{is defined} \quad \& \;\; \tilde{\varphi}(x) \neq x\}| < \infty.$$

Therefore for almost all x we get

$$x \in \overline{M} \Rightarrow | \{y : y \in W_{f(x)} \;\;\&\;\; x \in W_{g(y)}\} | \geq 1.$$

Then, for almost all x, we have

$$x \in \overline{M} \Rightarrow x \in W_{g(y)} \;\;\&\;\; y \in W_{f(x)} \Rightarrow y \in \overline{A},$$

$$x \in M \Rightarrow x \in W_{g(y)} \;\;\&\;\; y \in W_{f(x)} \Rightarrow y \in A.$$

For all x let $\tilde{f}(x)$ be the first element which appears in the computation of the set

$$\{y : y \in W_{f(x)} \;\;\&\;\; x \in W_{g(y)}\}.$$

With the help of $\tilde{f}$ it is possible to construct a recursive function which m-reduces the set M to the set A.⊠

Corollary 3.8 *The Q-degree of a maximal set contains no r.e. semirecursive set.*

Proof of Theorem 3.4. It is not hard to show that if M is a maximal set, and A and B are sets such that $M \leq_m A \oplus B$ then $M \leq_m A$ or $M \leq_m B$. Now assume that A and B are Q-incomparable sets and

$$M \equiv_Q A \oplus B,$$

where M is a maximal set. Then by Theorem 3.5

$$M \leq_m A \oplus B.$$

Hence we have

$$M \leq_m A \text{ or } M \leq_m B.$$

Then from relations

$$M \equiv_Q A \oplus B \quad \text{and} \quad M \leq_m A \text{ or } M \leq_m B$$

it follows that A and B are Q-comparable, which is a contradiction. ⊠

It is well-known that if A, B, C are r.e. sets and

$$A = B \cup C \;\&\; B \cap C = \oslash,$$

then $A \equiv_{bW} B \oplus C$. Therefore from Theorem 3.4 it follows:

Corollary 3.9. *Let be* $r \in \{Q, sQ, bsQ, bQ\}$, $r_1 \in \{T, W, bW\}$. *Then*

$$Th(L_r) \neq Th(L_{r_1}).$$

Lachlan's Nonsplitting Theorem [24] is well-known. This says that there are r.e. T-degrees $\mathbf{a}$ and $\mathbf{b}$ such that $\mathbf{b} < \mathbf{a}$ and for all r.e. T-degrees $\mathbf{a}_0$ and $\mathbf{a}_1$ we have

$$[\mathbf{b} \leq \mathbf{a}_0, \mathbf{a}_1 \leq \mathbf{a} \;\&\; \mathbf{a} \leq \mathbf{a}_0 \cup \mathbf{a}_1] \Rightarrow [\mathbf{a} \leq \mathbf{a}_0 \text{ or } \mathbf{a} \leq \mathbf{a}_1].$$

From Theorem 3.4 we get that the analog to Lachlan's Nonsplitting Theorem, mentioned above, holds. Namely:

Corollary 3.10 *Let* $\mathbf{a}$ *be the Q-degree of a maximal set, and let* $\mathbf{c}$ *be an r.e. Q-degree, such that* $\mathbf{c} <_Q \mathbf{a}$. *Then there are no r.e. Q-incomparable degrees* $\mathbf{b}_0$ *and* $\mathbf{b}_1$ *such that*

$$\mathbf{c} <_Q \mathbf{b}_0 \ \& \ \mathbf{c} <_Q \mathbf{b}_1 \ \& \ \mathbf{a} = \mathbf{b}_0 \cup \mathbf{b}_1.$$

The following statement shows that the maximal set in Theorem 3.5 cannot be replaced by an r-maximal one.

Proposition 3.11 *There exists* r*-maximal* Q*-complete sets* M_1 *and* M_2 *such that* $M_1 \mid_m M_2$.

Proof. By Theorem 1.6 there exits an r-maximal Q-complete set M_1 and the set M_1 has a Q-complete major subset M_2. It is known [50] that a major subset of a r-maximal set is r-maximal. Hence the set M_2 is r-maximal. It is known [8] that if M_1, M_2 are r-maximal sets $M_2 \subset M_1$ and $|M_1 \setminus M_2| = \infty$ then $M_1 \mid_m M_2$. ⊠

Problem 2. *Is there a greatest r.e. m-degree contained in the Q-degree of a maximal set?*

In [13] the decidability of the elementary theory $Th(L_Q)$ is investigated, and the following is proved.

Theorem 3.12 (Downey, LaForte, Nies [13]) $Th(L_Q)$ *has an undecidable first order theory.*

It is not hard to show that a set A is sQ-reducible to a set B if and only if there is a recursive function f such that

$$(\forall x)(x \in A \Leftrightarrow W_{\pi_1 f(x)} \cap D_{\pi_2 f(x)} \subseteq B.)$$

Let A and B be r.e. sets which are built or enumerated step by step s, with $s = 1, 2, \ldots$ and let

$$f_{A,B}(n,s) = \max\{y \leq s : (\forall x \leq y)(\varphi_n(x) \text{ is defined } \& \\ (x \in A_s \Leftrightarrow W_{\pi_1\varphi_n(x),s} \cap D_{\pi_2\varphi_n(x)} \subseteq B_s))\}.$$

Let also

$$F_{A,B}(n,0) = 0$$
$$F_{A,B}(n,s+1) = \max\{F_{A,B}(n,s),\ f_{A,B}(n,s+1)\}.$$

From this definition it follows that $F_{A,B}$ is a recursive function which is non-decreasing with respect to the second argument. The function $F_{A,B}$ is called *counter*. The fundamental property of the counter is formulated in the next lemma (see [8]).

Lemma on the counter 3.13 *For any r.e. sets* A, B *and for each* $n \in \omega$ *we have*

$$\lim_{s\to\infty} F_{A,B}(n,s) = \infty \Leftrightarrow \varphi_n \text{ is recursive } \& \ A \leq_{sQ} B \ \text{ via } \varphi_n.$$

For $r \in \{sQ, bsQ, bQ\}$ the following theorem holds and is proved with the help of Lemma 3.13.

Theorem 3.14 (Omanadze [39]) *Let* $\mathbf{a}$ *and* $\mathbf{b}$ *be r.e. r-degrees with* $\mathbf{a} <_r \mathbf{b}$. *Then there exist infinitely many pairwise r-incomparable r.e. r-degrees* $\{\mathbf{c}_i\}_{i\in\omega}$ *such that*

$$(\forall i)(\mathbf{a} <_r \mathbf{c}_i <_r \mathbf{b}).$$

Sketch of the proof. We will prove this theorem for sQ-reducibility. The proofs for bsQ- and bQ-reducibility are similar.

Let $\mathbf{a}$ and $\mathbf{b}$ be r.e. sQ-degrees, $\mathbf{a} <_{sQ} \mathbf{b}$, and let A and B be r.e. sets, $A \in \mathbf{a}, B \in \mathbf{b}$. We shall split effectively the set B into an infinite number of pairwise disjoint r.e. sets $\{B_i\}_{i\in\omega}$ which satisfies the condition

$$(\forall i)(\forall j)(i \neq j \Rightarrow B_i \not\leq_{sQ} A \oplus B_j).$$

Then it is not hard to check that as suitable sQ-degrees $\mathbf{c}_i$, $i = 0, 1, 2, \ldots$, we can take respectively the sQ-degrees of the sets $A \oplus B_i$, $i = 0, 1, 2, \ldots$.

Step by step we shall build r.e. sets B_i, $i = 0, 1, 2, \ldots$. Let f be a 1:1 recursive function which enumerates B and let

$$\Gamma(i, j, n, s) = F_{B_i, A\oplus B_j}(n, s) \quad (i \neq j).$$

At step s we act as follows.

Step s**.** We compute $f(s)$ and if

$$f(s) > \max\{\Gamma(i, j, n, s) : \langle i, j, n\rangle \leq s\},$$

then put $f(s)$ into B_s and move to next step. Otherwise let $\langle i_0, j_o, n_0\rangle \leq s$ be such that

$$\langle i_0, j_0, n_0\rangle = \min\{\langle i, j, n\rangle : f(s) \leq \Gamma(i, j, n, s)\}$$

and put $f(s)$ into B_{i_0}. Then move to next step. This completes the construction.

Lemma 3.15 $(\forall i)(\forall_j)(\forall n)(\lim_{s\to\infty} \Gamma(i, j, n, s) < \infty)$. ⊠

We shall name the r-degree of a nonspeedable (nowhere simple) set a *non-speedable (nowhere simple)* r-degree.

Corollary 3.16. 1) Let $\mathbf{a}$ and $\mathbf{b}$ be nonspeedable r-degrees. Then there exist infinitely many pairwise r-incomparable nonspeedable r-degrees $\{\mathbf{c}_i\}_{i\in\omega}$ such that

$$(\forall i)(\mathbf{a} <_r \mathbf{c}_i <_r \mathbf{b}).$$

2) Let $\mathbf{a}$ and $\mathbf{b}$ be nowhere simple r-degrees. Then there exist infinitely many pairwise r-incomparable nowhere simple r-degrees $\{\mathbf{c}_i\}_{i\in\omega}$ such that

$$(\forall i)(\mathbf{a} <_r \mathbf{c}_i < \mathbf{b}).$$

Denisov [9] has proved that if A is a nonrecursive r.e. non m-complete set then it is possible to construct an r.e. set B which is m-incomparable with the set A. Marchenkov [27] has shown that for truth-table type reducibilities the following result is valid: For every nonrecursive r.e. noncomplete degree **a** there is a degree which is incomparable with **a** and which contains a maximal set.

It is known (see [5, Theorem 4.1]), that any maximal set is speedable. Therefore by analyzing the proof of the previously mentioned theorem by Marchenkov we obtain that for every nonrecursive r.e. noncomplete sQ- (bsQ-, bQ-) degrees there is a speedable sQ-(bsQ-, bQ-) degree which is incomparable with it.

We recall Soare's result that an r.e. set A is nonspeedable if and only if there exists a recursive enumeration $\{A_s\}_{s\in\omega}$ of the set A such that

$$(\exists^\infty s)[W_{e,s} \setminus A_s \neq \oslash] \Rightarrow W_e \setminus A \neq \oslash.$$

With the help of this characterization of the nonspeedable sets and Denisov's construction [9] the following statement is proved.

Theorem 3.17 (Omanadze [39]) *For every nonrecursive r.e. noncomplete r-degree* **a** *there exists a nonspeedable r-degree which is incomparable with it.*

Sketch of the proof. We will prove this theorem for sQ-reducibility. The proofs for bsQ-reducibility and bQ-reducibility are similar. Let **a** be a nonrecursive and noncomplete r.e. sQ-degree, K an sQ-complete r.e. set, f a $1:1$ recursive function such that range$(f) = K$, and let A be r.e. and $A \in$ **a**. In the construction we use markers Γ_k^1, Γ_2^k with the priority ordering $\Gamma_0^1, \Gamma_0^2, \Gamma_1^1, \Gamma_1^2, \ldots$, and markers Γ_k^{10}, $k \in \omega$. The markers Γ_k^1 and Γ_k^{10} will move in the A- list, and the markers Γ_k^2 will move in the B- list. At every step the sign $+$ or the sign $-$ is put near some numbers, and B_s denotes the set of numbers in the B-list which are placed in B by the end of step s.

We say that the set A is *sQ-reducible to the set B via f on the set E* if for each $x \in E$ we have

$$x \in A \Leftrightarrow W_{\pi_1 f(x)} \cap D_{\pi_2 f(x)} \subseteq B.$$

Let

$$d_e^s = \begin{cases} \mu x(x \in W_{e,s} \setminus B_s), & \text{if } W_{e,s} \setminus B_s \neq \oslash, \\ \oslash, & \text{if } W_{e,s} \setminus B_s = \oslash. \end{cases}$$

Step $2s$. Let $k = \pi_1(s)$. If the marker Γ_k^2 stands at a point n, $\varphi_{k,2s}$ is defined on the set $\{0, 1, \ldots, n\}$ and for each x, $0 \leq x \leq n$,

$$x \in B_{2s-1} \Leftrightarrow W_{\pi_1 \varphi_k(x), 2s} \cap D_{\pi_2 \varphi_k(x)} \subseteq A_{2s},$$

or the marker Γ_k^2 has not yet been assigned, then we compute successively $f(0), f(1), \ldots$ until we find s_0 such that near $f(s_0)$ in the B-list stands neither the symbol $-$, posed by a marker with priority greater than the priority of Γ_k^2, nor $+$, and

$$(\forall e)(e \leq k \Rightarrow f(s_0) \neq d_e^{2s}).$$

In the B-list we place $+$ near $f(s_0)$, move the marker Γ_k^2 to $f(s_0)$ (or place it there if not yet assigned). We put the number $f(s_0)$ into B_{2s} and move to step $2s+1$.

Step $2s + 1$. Let $k = \pi_1(s)$.

I. 1) If the marker Γ_k^{10} stands at a point n and we have the relation

$$n \in A_{2s+1} \Leftrightarrow W_{\pi_1\varphi_k(n),2s+1} \cap D_{\pi_2\varphi_k(n)} \not\subseteq B_{2s},$$

then we pass to step $2s+2$.

I. 2) If the marker Γ_k^{10} either does not stand, or stands at a point n and we have the relation

$$n \in A_{2s+1} \Leftrightarrow W_{\pi_1\varphi_k(n),2s+1} \cap D_{\pi_2\varphi_k(n)} \subseteq B_{2s},$$

then we check if the set A_{2s+1} is sQ-reducible to the set B_{2s} via $\varphi_{k,2s+1}$ on the set $\mathrm{dom}(\varphi_{k,2s+1})$, or not. If yes, then we pass to point II. If not, then we seek the first number m such that $\varphi_{k,2s+1}(m)$ is defined and

$$m \in A_{2s+1} \Leftrightarrow W_{\pi_1\varphi_k(m),2s+1} \cap D_{\pi_2\varphi_k(m)} \not\subseteq B_{2s},$$

we place the marker Γ_k^{10} on (or move it to) this number and we pass to point II.

II. 1) If the marker Γ_k^1 already stands at a point n and $\varphi_{k,2s+1}$ is not everywhere defined on the set $\{0, 1, \ldots, n\}$ or the following condition holds:

$$(\forall b \in \mathrm{dom}(\varphi_{k,2s+1}))(b \notin A_{2s+1} \text{ \& the marker } \Gamma_k^1 \text{ has not yet been assigned to the point } b \Rightarrow W_{\pi_1\varphi_k(b),2s+1} \cap D_{\pi_2\varphi_k(b)} \subseteq B_{2s}),$$

then we pass to the step $2s+2$.

II. 2) If the marker Γ_k^1 either has not been assigned or stands near a number n and $\varphi_{k,2s+1}$ is everywhere defined on the set $\{0, 1, \ldots, n\}$ then we seek the first point which satisfies the condition:

$$b \in \mathrm{dom}(\varphi_{k,2s+1}) \text{ \& } b \text{ is not marked by the marker } \Gamma_k^1 \text{ \& } W_{\pi_1\varphi_k(b),2s+1} \cap D_{\pi_2\varphi_k(b)} \not\subseteq B_{2s}.$$

We denote this point by b_0. We move (or assign) the marker Γ_k^1 to this number, and near the least number u such that

$$u \in W_{\pi_1\varphi_k(b_0),2s+1} \cap D_{\pi_2\varphi_k(b_0)} \cap \overline{B}_{2s},$$

we place the sign $-$ and we move to step $2s+2$.

This completes the construction.

Lemma 3.18 *All the assigned markers of the form Γ^1_k, Γ^2_k move finitely often.*

Lemma 3.19 $A \not\leq_{sQ} B \ \& \ B \not\leq_{sQ} A$.

Lemma 3.20 *The set B is a nonspeedable set.* ⊠

Problem 3. *Let $r_1, r_2 \in \{Q, sQ, bsQ, bQ\}$. Is $Th(L_{r_1}) \neq Th(L_{r_2})$ for $r_1 \neq r_2$?*

4. Relations Between Degrees

Let $\leq_R$ and $\leq_r$ be reducibilities. The reducibility $\leq_R$ is said to be *weaker than* the reducibility $\leq_r$ (and $\leq_r$ is said to be *stronger than* $\leq_R$) if we have that $A \leq_r B \Rightarrow A \leq_R B$ for all $A, B \subseteq \omega$. It is clear that if $\leq_r$ is stronger than $\leq_R$ then each R-degree consists of a single or several (possibly infinitely many) r-degrees. One of the main approaches to studying various reducibilities concerns the question of how many r-degrees exist, and how they are arranged, in the degrees relative to the weaker reducibility. Degtev [6] has showed that each r.e. nonrecursive T-degree contains a countable number of pairwise incomparable r.e. tt-degrees. Ladner and Sasso [25] have proved that there exists a nonrecursive r.e. T-degree which contains a single r.e. W-degree, and Kobzev [20] has showed that every nonrecursive r.e. W-degrees consists of an infinite number of pairwise incomparable r.e. tt- and bW-degrees. Degtev [73] has proved that there exist a nonrecursive r.e. tt-degree in which all r.e. sets are m-equivalent.

From the above mentioned results we may have the impression that if $\leq_r$ is stronger than $\leq_R$ then each nonrecursive r.e. R-degree contains either a single r.e. r-degree or infinitely many pairwise incomparable r.e. r-degrees. However this is not the case. Downey [12] has obtained an unexpected result concerning the r.e. tt-degrees. Namely he has proved that there exists a nonrecursive r.e. tt-degree which contains exactly three r.e. m-degrees.

In this section we investigate the relations between r.e. T- and Q-degrees, r.e. W- and sQ-degrees, r.e. bW- and bsQ-degrees, as well as other questions.

Let

$$A^Q = \{x : W_x \subseteq A\}.$$

It is obvious that $A \equiv_Q A^Q$ and it is not hard to show that

$$A \leq_Q B \Leftrightarrow A^Q \leq_1 B^Q.$$

Theorem 4.1 (Hay [17]) *Let* $\mathbf{b}$ *be a nonrecursive r.e. T-degree and let us consider the following classes of T-degrees:*

$$B_1 = \{\mathbf{c} : (\exists A)(A \in \mathbf{b} \ \&\ A^Q \in \mathbf{c})\},$$
$$B_2 = \{\mathbf{c} : (\exists A)(A \in \mathbf{b} \ \&\ A \text{ is r.e.} \ \&\ A^Q \in \mathbf{c})\},$$
$$B_3 = \{\mathbf{c} : \mathbf{c} \geq \mathbf{0}' \ \&\ \mathbf{c} \text{ is r.e. in } \mathbf{b}\}.$$

Then $B_1 = B_2 = B_3$.

From this theorem, in particular, it follows that no nonrecursive r.e. non-complete T-degree can entirely contain a Q-degree. Therefore the following question naturally arises. Does $\mathbf{0}'$ contain a Q-degree? We will show that $\mathbf{0}'$ contains an infinite number of pairwise Q-incomparable nonspeedable Q-degrees.

Theorem 4.2. (Omanadze [33]*Let A be an r.e. set. Then*

$$A \in \mathbf{0}' \ \&\ A^Q \in \mathbf{0}' \Leftrightarrow (\forall B)(B \equiv_Q A \Rightarrow B \equiv_T A).$$

Corollary 4.3 *Let A be an r.e. set such that the Q-degree of A is entirely contained in* $\mathbf{0}'$. *If B is an r.e. set,* $B \in \mathbf{0}'$ *and* $B <_Q A$ *then the Q-degree of the set B is entirely contained in* $\mathbf{0}'$.

Denote by $L_Q(\mathbf{0}')$ the upper semilattice of the Q-degrees of those r.e. sets which are contained in $\mathbf{0}'$.

Proposition 4.4 (Omanadze [33] 1) *Let A be an r.e. set and* $A \in \mathbf{0}'$. *Then the Q-degree of the set A is contained in* $\mathbf{0}'$ *only if A is nonspeedable.*

2) *The Q-degrees which are contained in* $\mathbf{0}'$ *induce an ideal in the semilattice* $L_Q(\mathbf{0}')$.

From this proposition and from the fact that in every r.e. T-degree is contained a nonspeedable set it follows:

Corollary 4.5 *The degree* $\mathbf{0}'$ *entirely contains the Q-degree of a nonspeedable r.e. set.*

After having shown that $\mathbf{0}'$ entirely contains an r.e. Q-degree a question naturally arises. Are there infinitely many pairwise Q-incomparable r.e. Q-degrees which are entirely contained in $\mathbf{0}'$? An affirmative answer to this question is given in the following theorem which is proved by using the following statement which is of independent interest.

Proposition 4.6 (Omanadze [35]) *For every r.e. nonspeedable Q-degree* $\mathbf{a}$, $\mathbf{a} > 0$, *there exists a nonspeedable Q-degree which is incomparable with* $\mathbf{a}$.

The proof of this proposition is similar to the proof of Theorem 3.17.

Theorem 4.7 (Omanadze [36]) 1) *Every nonrecursive r.e. T-degree contains infinitely many pairwise Q-incomparable nonspeedable sets.*

2) *The degree* $\mathbf{0}'$ *contains infinitely many pairwise* Q*-incomparable r.e.* Q*-degrees.*

3) *Among all r.e.* Q*-degrees entirely contained in* $\mathbf{0}'$ *there is no maximal r.e.* Q*-degree.*

Proof. 1) By Theorem 3.8 [49] for every simple set B and for all nonrecursive r.e. T-degree $\mathbf{a}$ there exists a simple subset $A \subseteq B$ such that $A \in \mathbf{a}$. Therefore let $\mathbf{a}$ be a nonrecursive r.e. T-degree, and let A be a nonspeedable simple set such that $A \in \mathbf{a}$.

Let $A = A_0 \cup A_1$, $A_0 \cap A_1 = \varnothing$ and both A_0 and A_1 are nowhere simple sets. There is such a representation by [48, Theorem 2]. We know that in this case

$$A_0 \leq_Q A \;\&\; A_1 \leq_Q A$$

and if B, C are r.e. sets, C is a nonspeedable set and $B \leq_Q C$ then B is a nonspeedable set. Therefore both A_0 and A_1 are nonspeedable sets.

By Proposition 1.9 the set $A_0 \oplus A_1$ is a nowhere simple set and by Proposition 1.7 the property of being nowhere simple is preserved downwards. On the other hand a simple set is not nowhere simple and if A_0, A_1 are nonspeedable sets then $A_0 \oplus A_1$ is a nonspeedable set. Therefore $A_0 \oplus A_1$ is a nonspeedable nowhere simple set and is satisfied the following relation

$$A \equiv_T A_0 \oplus A_1 \;\&\; A_0 \oplus A_1 <_Q A.$$

Now to complete the proof of point 1) it is sufficient to use Theorem 2.15.

2) The proof of this point immediately follows from point 1) and from Proposition 4.4.

3) Assume that among all nonspeedable Q-degrees contained in $\mathbf{0}'$ there is a maximal Q-degree and this is a nonspeedable (by Proposition 4.4) Q-degree $\mathbf{a}$. Then by Proposition 4.6 there exists a nonspeedable Q-degree $\mathbf{b}$ such that $\mathbf{a} \mid_Q \mathbf{b}$. Consider the Q-degree $\mathbf{a} \cup \mathbf{b}$. This is a nonspeedable Q-degree, contained in $\mathbf{0}'$ and $\mathbf{a} < \mathbf{a} \cup \mathbf{b}$, which is a contradiction. ⊠

Theorem 4.8 (Omanadze [39]) *Every nonrecursive r.e.* W*-degree contains infinitely many pairwise* sQ*-incomparable r.e. sets.*

Proof. Let $\mathbf{a}$ be a nonrecursive r.e. W-degree. It is known that every nonrecursive r.e. W-degree contains a simple, not hypersimple set, and a nowhere simple set. Therefore let S be a simple set, A be a nowhere simple set and $S, A \in \mathbf{a}$. Then

$$A \leq_{sQ} S \oplus A \;\&\; A \equiv_W S \equiv_W A \oplus S.$$

By Theorem 3.7 there exist infinitely many pairwise sQ-incomparable r.e. sets $\{C_i\}_{i \in \omega}$ such that

$$(\forall i)(A <_{sQ} C_i <_{sQ} S \oplus A).\ ⊠$$

We call an r.e. T-degree *contiguous* if it contains a single r.e. W-degree.

Proposition 4.9 (Omanadze [41]) *An r.e. T-degree* $\mathbf{a}$ *is contiguous if and only if all r.e. semirecursive sets contained in* $\mathbf{a}$ *are sQ-equivalent.*

Theorem 4.10 *Every nonrecursive r.e. bW-degree contains infinitely many pairwise bsQ-incomparable r.e. sets.*

Sketch of the proof. For the proof it is sufficient to show that if A is an r.e. nonrecursive set then there exists an r.e. nonrecursive set B such that

$$A \equiv_{bW} B \; \& \; A \not\leq_{sQ} B.$$

This last relation is not difficult to obtain by a finite priority argument.

Now to complete the proof let us consider the sets A and $A \oplus B$. Obviously

$$A \equiv_{bW} A \oplus B \; \& \; B <_{sQ} A \oplus B.$$

Then using Theorem 3.14 we obtain the required statement. ⊠

Theorem 4.11 (Omanadze [32]) *Relatively to the class of r.e. sets, the complete m-degrees and the complete bQ-degrees coincide.*

Corollary 4.12 *R.e. bsQ-complete and r.e. m-complete sets coincide.*

Let $r \in \{sQ, bsQ, bQ\}$.

Theorem 4.13 (Omanadze [42]) *Let A, B be r.e. sets, $A \equiv_r B$ and $A <_m B$. Then the r-degree of the set A contains infinitely many pairwise m-incomparable r.e. sets $\{C_i\}_{i\in\omega}$.*

These theorems immediately yield:

Theorem 4.14 (Omanadze [42]) *Let $r \in \{bsQ, bQ\}$. Every r.e. r-degree contains either a single r.e. m-degree or infinitely many pairwise m-incomparable r.e. m-degrees.*

Theorem 4.15 (Omanadze [44]) *A bsQ-degree of a simple set contains infinitely many pairwise m-incomparable r.e. m-degrees.*

Proof. Let S be a simple, but not hypersimple set. A set A is *hereditary* if and only if there is a recursive function g such that $D_x \subseteq A \iff g(x) \in A$. It is known (see, for instance, [8]) that a simple not hypersimple set cannot be hereditary. It is not hard to show that the bsQ-degree of a nonhereditary r.e. set contains infinitely many pairwise m-incomparable r.e. m-degrees.

Now suppose that S is a hypersimple set, A, B are nonrecursive r.e. sets, $S = A \cup B$, $A \cap B = \oslash$, $A \mid_T B$ and A, B are recursively inseparable sets. Then $A \leq_{bsQ} S$. Let us show that $A \not\leq_m S$. Suppose that $A \leq_m S$. Then $A^\omega \leq_c S$, where $A^\omega = \{x : D_x \subseteq A\}$. A set A is *perfect* if it is η-closed, for some nontrivial r.e. equivalence relation η whose only recursive η-closed sets are $\oslash$ and ω. It is known [14] that A^ω is a perfect set, but no perfect set is tt-reducible to a hypersimple set [10]. Therefore, $A^\omega \not\leq_c S$. A contradiction.

Consequently, $A \not\leq_m S$. Consider the set $S \oplus A$. Then $S \equiv_{bsQ} S \oplus A$ and $S <_m S \oplus A$. By Theorem 4.13 the bsQ-degree of the set S contains infinitely many pairwise m-incomparable r.e. m-degrees. ⊠

Problem 4. *Is there a nonrecursive noncomplete r.e. bsQ-degree which contains a single r.e. m-degree?*

5. Complete Degrees

Theorem 5.1 (Omanadze [41] 1) *There exists an r.e. set which is simultaneously Q- and W- complete but not sQ-complete.*

2) *There exists an r.e. set which is simultaneously sQ- and bW- complete but not bsQ-complete.*

3) *There exists an r.e. set which is simultaneously Q- and W- complete but not bW-complete.*

Proof. 1) Similarly to Theorem 1.13 it is proved that if A is a simple set, B is an arbitrary set, C is a nowhere simple set, then

$$A \leq_{sQ} B \oplus C \Rightarrow A \leq_{sQ} B.$$

Let A be a Q-complete coretraceable hypersimple set, and let B be a W-complete nowhere simple set. Consider the set $A \oplus B$. This is simultaneously Q- and W-complete. Assume that the set $A \oplus B$ is sQ-complete and S is an sQ-complete simple set. Then $S \leq_{sQ} A \oplus B$. From here we have $S \leq_{sQ} A$, i.e. the set A is sQ-complete. But it is known that no hypersimple set can be W-complete.

2) Let A be an sQ-complete but not bsQ-complete r.e. set; let B be a bW-complete but not sQ-complete r.e. set. Then $A \oplus B$ is simultaneously sQ-complete and bW-complete. The set $A \oplus B$ is not bsQ-complete. In fact, if $A \oplus B$ is a bsQ-complete set, then $A \oplus B$ is m-complete. Then by a theorem of Lachlan, [21], either A is m-complete or B is m-complete. A contradiction.

3) Lachlan, [23], proved that r.e. bW-complete and bd-complete sets coincide and it is known that there is no d-complete simple set. From these results it follows that there is an sQ-complete but not bW-complete r.e. set. ⊠

If in the definition of Q- (sQ-, bsQ-, bQ-) reducibility we require that the recursive function f satisfy the additional condition

$$(\forall x)(\forall y)(x \neq y \Rightarrow W_{f(x)} \cap W_{f(y)} = \oslash),$$

then we obtain a reducibility which we denote by $Q_1-(sQ_1-, bsQ_1-, bQ_1-)$.

The well-known classical theorem of Myhill says that the m-complete and the 1-complete r.e. degrees coincide. In [31] it is proved that in general the analog of Myhill's theorem does not hold.

Theorem 5.2 (Gill and Morris [16]) *An r.e. set A is Q-complete if and only if A is Q_1-complete.*

Let $r \in \{sQ, bsQ, bQ\}$, $r_1 \in \{sQ_1, bsQ_1, bQ_1\}$.

Theorem 5.3 (Omanadze [40]) *An r.e. set A is r-complete if and only if A is r_1-complete.*

Sketch of the proof. From the result that r.e. m-complete and r.e. bQ-complete sets coincide it follows that for $r \in \{bsQ, bQ\}$ the theorem holds. Therefore let us prove the claim for sQ-reducibility.

It is not hard to show that an r.e. set A is sQ-complete if and only if there are recursive functions f and g such that

$$(\forall x)(W_{f(x)} \cap W_x \neq \oslash \Leftrightarrow W_{f(x)} \subseteq A), \tag{1}$$

$$(\forall x)(\forall y)(y \in W_{f(x)} \Rightarrow y < g(x)). \tag{2}$$

Analogously, an r.e. set A is sQ_1-complete if and only if (1), (2) are hold and

$$(\forall x)(\forall y)(x \neq y \Rightarrow W_{f(x)} \cap W_{f(y)} = \oslash). \tag{3}$$

Let τ be a recursive function such that

$$W_{\tau(i,n)} = \begin{cases} W_i \setminus \{a_1, \dots, a_n\}, & \text{if } n \text{ distinct elements} \\ & a_1, \dots, a_n \text{ can be computed in } W_i \cap A, \\ \oslash & \text{otherwise} \end{cases}$$

Step by step we construct f_1 and g_1.

Step n. Let λ, γ, δ be recursive functions such that

$$W_{\lambda(n)} = \bigcup_{i<n} W_{f_1(i)},$$
$$W_{\gamma(n)} = W_{\lambda(n)} \cap W_n,$$
$$\delta(n) = \sum_{i<n} g_1(i) + g\gamma(n) + g\tau(\gamma(0), 0).$$

Suppose

$$W_{f_1(n)} = \begin{cases} W_{f\gamma(n)} \cup (\bigcup\{W_{f\tau(\gamma(n),j)} \ : \ 0 \leq j \leq |W_{\lambda(n)} \cap A|\}), & \text{if } W_n \neq \oslash, \\ \bigcup\{W_{f\tau(\gamma(n),j)} \ : \ 0 \leq j \leq |W_{\lambda(n)} \cap A|\}, & \text{if } W_n = \oslash, \end{cases}$$

$$g_1(n) = g\gamma(n) + \sum_{j \leq \delta(n)} g\tau(\gamma(n), j),$$

where f and g are recursive functions which satisfy the conditions (1) and (2).

Lemma 5.4 *If $k < n$, then*

$$W_{f_1(k)} \cap [(\bigcup\{W_{f\tau(\gamma(n),j)} : 0 \leq j \leq |W_{\lambda(n)} \cap A|\}) \cap \overline{A}] = \oslash.$$

Lemma 5.5 *For each $k < n$*

$$W_{f_1(n)} \cap \overline{A} \cap W_{f_1(k)} = \oslash.$$

Lemma 5.6 *If $W_n = \omega$, then*

$$(\bigcup\{W_{f_T(\gamma(n),j)} : 0 \leq j \leq |W_{\lambda(n)} \cap A|\} \subseteq A.$$

Let α be a recursive function such that

1. $(\forall i)(\forall j)(i \neq j \Rightarrow W_{\alpha(i)} \cap W_{\alpha(j)} = \oslash)$,
2. $(\forall i)(W_{\alpha(i)} \subseteq W_{f_1(i)})$,
3. $\bigcup_{i \in \omega} W_{\alpha(i)} = \bigcup_{i \in \omega} W_{f_1(i)}$,

and let β is 1:1 be a recursive function such that

$$W_{\beta(i)} = \begin{cases} \omega, & \text{if } W_{\alpha\beta(i)} \cap W_i \neq \oslash, \\ \oslash, & \text{if } W_{\alpha\beta(i)} \cap W_i = \oslash. \end{cases}$$

Then the recursive functions μ, ν, where

$$\mu(i) = \alpha\beta(i),$$

$$\nu(i) = g_1\beta(i),$$

satisfy the conditions (1), (2) and (3). ⊠

The author is grateful to Prof. B. Cooper and the anonymous referee for their remarks and valuable suggestions.

References

[1] M.M. Arslanov, *On a class of hypersimple incomplete sets.* Mat. Zametki **38**(1985), 872-874 (Russian); Math. Notes **38**(1985), 984-985 (English translation).

[2] O.V. Belegradek. *On Algebraically closed groups*, Algebra i Logika, **13**, No. 3 (1974), 813-816 (Russian).

[3] M. Blum, *A machine-independent theory of the complexity of recursive functions*, J. Assoc. Comp. Mach. **14,** No. 2 (1967), 322-336.

[4] M. Blum, *On the size of machines,* Inform. and Control, **11,** No. 3 (1967), 257-265.

[5] M. Blum, I. Marques,*On complexity properties of recursively enumerable sets,* J. Symb. Logic, **38,** No. 4 (1973), 579-593.

[6] A. N. Degtev, *Hereditary sets and truth table reducibility,* Algebra i Logika, **11,** No. 3 (1972), 257-279 (Russian); Algebra and Logic **11** (1972), 145-152 (English translation).

[7] A. N. Degtev, *tt-and m-degrees,* Algebra i Logika **12,** No. 2 (1973), 143-161 (Russian); Algebra and Logic **12** (1973), 78-89 (English translation).

[8] A. N. Degtev, *Recursively Enumerable Sets and Truth Table Type Reducibilities.* Nauka, Moscow, 1998 (Russian).

[9] S. D. Denisov, *On m-degrees of recursively enumerable sets,* Algebra i Logika, **9,** No. 4 (1970), 422-427 (Russian); Algebra and Logic, **9** (1970), 254-256 (English Translation).

[10] S. D. Denisov, *Three theorems on elementary theories and tt-reducibility,* Algebra i Logika, **13,** No. 4 (1974), 5-8 (Russian); Algebra and Logic, **13**(1974), 1-2 (English translation).

[11] V. P. Dobritsa, *On word problem of recursively definable groups,* in: Proc. 3th. All-Union Conf. Math. Logic, Novosibirsk, (1974), 63-65 (Russian).

[12] R. G. Downey, *Recursively enumerable m- and tt-degrees I. The quantity of m-degrees,* J. Symb. Logic, **54,** No. 2 (1989.)

[13] R. G. Downey, G. Laforte, A. Nies, *Computably enumerable sets and quasi-reducibility,* Annals of Pure and Appl. Logic, **95** (1998), 1-35.

[14] Y. L. Ershov, *Positive equivalences,* Algebra i Logika, **10** (1971), 620-650 (Russian); Algebra and Logic, **10** (1971), 378-394 (English translation).

[15] P. Fisher, K. Ambos-Spies, *Q-degrees of r.e. sets,* J. Symb. Logic, **52,** No. 1 (1987), 317.

[16] J. T. Gill, P. H. Morris, *On subcreative sets and S-reducibility,* J. Symb. Logic,**39,** No. 4 (1974), 669-677.

[17] L. Hay, *The class of recursively enumerable subsets of a recursively enumerable set,* Pacif. J. Math. **46,** No. 1 (1973), 167-183.

[18] M. A. Jahn, *Implicit measurements of dynamic complexity properties and splittings of speedable sets,* J. Symb. Logic, **64,** No. 3 (1999), 1037-1064.

[19] C. G. Jockusch, Jr., *Semirecursive sets and positive reducibility,* Trans. Amer. Math. Soc. **131** (1968), 420-436.

[20] G. N. Kobzev, *Relationships between recursively enumerable tt-degrees and W-degrees.* Soobshch. Akad. Nauk. Gruz. SSR (Bull. Acad. Sci. Georgian SSR) **84** (1976), 585-587 (Russian).

[21] A. H. Lachlan, *A note on universal sets,* J. Symb. Logic, **31,** No. 4 (1966), 573-574.

[22] A. H. Lachlan, *On the lattice of recursively enumerable sets,* Trans. Amer. Math. Soc., **130** (1968), 1-37.

[23] A. H. Lachlan, *Wtt-complete sets are not necessarily tt-complete,* Proc. Amer. Math. Soc., **48,** No. 2 (1975), 429-434.

[24] A. H. Lachlan, *A recursively enumerable degree which will not split over all lesser ones,* Ann. Math. Logic, **9** (1975), 307-365.

[25] R. E. Ladner, L. P. Sasso, *The weak truth table degrees of recursively enumerable sets,* Ann. Math. Logic, **8** (1975), 429-448.

[26] A. Macintyre, *Omitting quantifier free types in generic structures,* J. Symb. Logic, **37** (1972), 512-520.

[27] S. S. Marchenkov, *Truth-table degrees of maximal sets,* Mat. Zametki, **20** (1976), 373-381 (Russian); Math. Notes **20** (1976), 766-770 (English translation).

[28] S. S. Marchenkov, *A class of incomplete sets,* Mat. Zametki **20** (1976), 473-478 (Russian); Math. Notes **20** (1976), 823-825 (English translation).

[29] I. Marques, *On degrees of unsolvability and complexity properties,* J. Symb. Logic, **40,** No. 3 (1975), 529-540.

[30] P. Odifreddi, *Classical Recursion Theory,* Amsterdam; North-Holland, 1989.

[31] R. Sh. Omanadze, *On reducibilities for the class of recursively enumerable sets,* Soobshch. Akad. Nauk Gruz. SSR (Bull. Acad. Sci. Georgian SSR), **91,** No. 3 (1976), 549-552 (Russian).

[32] R. Sh. Omanadze, *On bounded Q- reducibility,* Soobshch. Akad. Nauk Gruz. SSR (Bull. Acad. Sci. Georgian SSR), **100,** No. 1 (1980), 57-60 (Russian).

[33] R.Sh. Omanadze, *On the upper semilattice of recursively enumerable Q-degrees,* Algebra i Logika, **23,** No. 2 (1984), 175-184 (Russian); Algebra and Logic **23** (1985), 124-129 (English translation).

[34] R. Sh. Omanadze, *Q-reducibility and nowhere simple sets,* Soobshch. Akad. Nauk Gruz. SSR (Bull. Acad. Sci. Georgian SSR), **127,** No. 1 (1987), 29-32 (Russian).

[35] R. Sh. Omanadze, *Q-degrees of nonspeedable sets,* Proc. of I. Vekua Inst. of Appl. Math. **20** (1987), 21-32 (Russian).

[36] R. Sh. Omanadze, *Relation between recursively enumerable Q- and T-degrees,* VIII Intern. Congress of Logic, Meth. and Phil. of Science. Moscow, Abstracts, **1** (1987), 157-158.

[37] R. Sh. Omanadze, *On the density of nonspeedable Q-degrees,* Soobshch. Akad. Nauk Gruz. SSR (Bull. Acad. Sci. Georgian SSR), **131,** No. 3 (1988), 484-485 (Russian).

[38] R. Sh. Omanadze, *Classes of recursively enumerable sets and Q-reducibility,* Mat. Zametki, **42**(1989), 79-82 (Russian); Math. Notes, **45** (1989), 141-143 (English translation).

[39] R.Sh. Omanadze, *On the upper semilattice of recursively enumerable sQ-degrees,* Algebra i Logika, 30, No. 4(1991), 405-413 Russian; Algebra and Logic, **30**(1992), 265-271 (English translation).

[40] R.Sh. Omanadze, *On sQ-completeness of recursively enumerable sets,* Mat. Zametki, **52,** No. 3 (1992), 102-107 (Russian); Math. Notes, **52** (1993), 948-952 (English translation).

[41] R. Sh. Omanadze, *Relation between some reducibilities,* Algebra i Logika, **33,** No. 6 (1994), 79-87 (Russian); Algebra and Logic, **33,** No. 6 (1994), 381-385 (English translation).

[42] R.Sh. Omanadze, *On one strengthening of Q-reducibility,* Algebra i Logika, **34,** No. 1 (1995), 79-87 (Russian); Algebra and Logic, **34,** No. 6 (1995), 45-49 (English translation).

[43] R.Sh. Omanadze, *Complexity properties of recursively enumerable sets and sQ-completeness,* Mat. Zametki, **62,** No. 3 (1997), 425-429 (Russian); Math. Notes **62**(1997), 356-359 (English translation).

[44] R. Sh. Omanadze, *Complexity properties of recursively enumerable sets and bsQ-completeness,* Mat. Zametki, **68,** No. 4 (2000), 554-559 (Russian); Math. Notes, **68,** No. 4 (2000), 476-480 (English translation).

[45] R.Sh. Omanadze, *Major sets, classes of simple sets and Q- complete sets,* Mat. Zametki, 71, No 1 (2002) (Russian); Math. Notes, 71, No 1 (2002), 90-97 (English translation).

[46] E.L. Post, *Recursively enumerable sets of positive integers and their decision problems,* Bull. Amer. Math. Soc., **50** (1944), 284-316.

[47] H. Rogers Jr., *Theory of Recursive Functions and Effective Computability,* New York, McGraw-Hill, 1967.

[48] R. A. Shore, *Nowhere simple sets and the lattice of recursively enumerable sets*, J. Symb. Logic, **43**, No. 2 (1978), 322-330.

[49] R.I. Soare, *Computational complexity, speedable and levelable sets,* J. Symb. Logic, **42,** No. 4 (1977), 545-563.

[50] R. I. Soare, *Recursively Enumerable Sets and Degrees,* Berlin, Springer-Verlag, 1987.

[51] V. D. Solov'ev, *Q-reducibility and hyperhypersimple sets,* Veroyatn. Metod. i Kibern. (Probabilistic Math. and Cybern.) **10-11,** (1974), 121-288 (Russian).

POSITIVE STRUCTURES

Victor Selivanov*
Novosibirsk Pedagogical University
A. P. Ershov Institute of Informatics Systems,
Siberian Division of Russian Academy of Sciences
vseliv@nspu.ru

Abstract The paper reviews the current state of the theory of positive (or computably enumerable) structures. This theory is relevant to several branches of algebra, logic and computer science, e.g. to algorithmic problems in algebra, theory of quasivarieties, logic programming, algebraic specifications of data types. Along with the general theory of positive structures we consider also positive structures of special kind (the most interesting results are obtained for positive boolean algebras) and some appliations of positive structures to logic and computability theory.

Keywords: Positive, structure, quasivariety, hierarchy, boolean algebra, ideal, numbering, index set.

1. Introduction

Effective algebra was initiated in the fiftees by A. Mostowski [Mo52, Mo53], A. V. Kuznetsov [Ku56, Ku58], and A. Fröhlich and J. C. Shepherdson [FS56]. The subject was strongly influenced by the work of M. O. Rabin [Ra60] and A. I. Mal'cev [Ma61]. The seminal paper of A. I. Mal'cev was fundamental for the extensive subsequent work in effective algebra by Siberian school of algebra and logic led by Yu. L. Ershov. In parallel, active research in this area was conducted in the West, most systematically by the school of A. Nerode.

The resulting rich theory was summarized in monographs [Er80, Go96, EG99] and in handbook [EGNR98]. The main attention in this work was paid to constructive structures (known also as computable or recursive structures). Along with the constructive structures, A. I. Mal'cev defined and studied also two more general classes of effective structures: positive (or computably enumerable) and negative (co-computably enumerable) structures. To the best

*Partially supported by RFBR-INTAS Grant IR-97-139. Thanks to Barry Cooper and Sergei Goncharov for inviting me to write this survey.

of my knowledge, results on positive and negative structures were till now not presented in a systematic way.

In this paper we try to review the current stage of the theory of positive stuctures. Only from time to time we mention also some relevant facts on constructive and on negative structures. More information on negative structures and some their generalizations may be found in [Kas96].

Constructive structures seem to be insufficient for a complete treatment of effectiveness in algebra for the following reasons: many interesting finitely presented algebras (algebras are structures without predicate symbols in the signature) are positive but not constructivizable (this was shown in classical works of A. A. Markov for semigroups and of P. S. Novikov and W. W. Boone for groups); the least model of a logic program (see [Ll84]) is positive but not always constructivizable; natural classes of constructive structures may have no "acceptable" numberings.

We will discuss some quite general facts about positive structures. E.g., the class of positive structures in a given recursively axiomatizable quasivariety always has an "acceptable" numbering, and this numbering often has a nice characterization in terms of complete numberings. Among positive models of any recursively axiomatizable quasivariety there always exists the most complicated one (called universal positive structure), and it is unique in some natural sense. These general facts provide some specific flavour to the theory of positive structures and enable to consider natural decision problems on positive structures in a way similar to the study of index sets in computability theory.

Along with the above-mentioned general theory of positive structures we will also consider some special classes of positive structures, e.g. positive partial orderings and semilattices. Special attention is paid to positive boolean algebras (BA's) because in this case one can obtain rather deep facts having interesting applications. Applications of the theory of positive structures are also discussed below.

Because of the space bound, it was impossible to include proofs of theorems. Almost all proofs are published, hence it was sufficient to give references to the source papers. This is exactly what we tried to do.

The rest of the paper is organized as follows. In Section 2 we present some notation and terminology used throughout the paper. In Section 3 we present several general facts on positive structures relating them to some classical algebraic notions, to the theory of abstract data types and to the theory of quasivarieties. In Section 4 we discuss some particular kinds of positive structures, e.g. BA's, linear orderings, abelian groups and positive precomplete partial orderings. In Section 5 we consider some applications of the material of preceding sections to logic, to degree structures for strong reducibilities, and to classification of index sets.

2. Preliminaries

Fix a signature $L = \{f_i, P_j | i < i_0, j < j_0\}$ with $i_0, j_0 \leq \omega$, functional symbols f_i and predicate symbols P_j, the arity of any symbol being effectively computable from its index. The case of arity 0 is possible. For technical reasons we assume that $j_0 > 0$ and P_0 is always a binary symbol. This symbol is usually represented as the equality relation (often understood intensionally).

By *an L-rule* we mean a formula of the form $\theta_1 \wedge \cdots \wedge \theta_k \rightarrow \theta$, where all θ_l, θ are atomic L-formulas. By an *L-identity* (*L-quasiidentity*) we mean the universal closure of an atomic L-formula (resp. of an L-rule). Natural examples of quasiidentities are the so called axioms of equality, i.e. sentences stating that $P_0(x, y)$ is an equivalence relation respected by all other functions and relations. By *a quasivariety* we mean a class of structures axiomatizable by quasiidentities. We will use some well–known facts on quasivarieties which may be found in any standard text on universal algebra, e.g. [Ma73, CK73, Co81, Gor99].

By an *extensional* (resp. *intensional*) L-structure we mean an L-structure interpreting P_0 as the equality relation (resp. as a relation satisfying the axioms of equality). It is well–known that the natural factorization of an intensional L-structure is an extensional L-structure.

Definition 2.0.1. *By a* positive L-structure *we mean an extensional L-structure* $\mathbf{A} = (A; f_i^A, P_j^A)$ *having a numbering* α *such that any function* f_i^A *is represented in* α *by a recursive function uniformly in* i*, and any predicate* P_j^A *is computably enumerable (c.e.) in the numbering* α *uniformly in* j*. The pair* $(\mathbf{A}; \alpha)$ *is called* a numbered positive structure.

Numbered *negative* structures are defined similarly, only the predicates are required to be co-c.e.

Our main objects are the numbered positive structures. For technical reasons it is sometimes more convenient to consider them in the form of intensional positive L-structures $(\omega; f_i, P_j)$ with f_i uniformly recursive and P_j uniformly c.e. Note that any numbered positive structure $(\mathbf{A}; \alpha)$ induces such an intensional structure $(\omega; f_i, P_j)$ by taking f_i to be the recursive function representing f_i^A in α, and P_j to be the index set of P_j^A in the numbering α. Conversely, any intensional positive structure $(\omega; f_i, P_j)$ induces the unique numbered positive structure, namely the quotient–structure by the congruence relation P_0, with the numbering α being the natural epimorphism on the quotient–structure. Simplifying notation, we usually identify these two presentations of numbered positive structures and denote the numbering of a numbered positive structure by a "similar" letter (α for $\mathbf{A}$, β for $\mathbf{B}$, and so on).

We will sometimes use elementary notions from category theory, all of them are broadly known and may be found in any text on category theory (see e.g. [BD70]). Let $\mathcal{R}$ be the category formed by the numbered positive L-structures as objects, and by homomorphisms representable by recursive functions in the respective numberings, as morphisms. By $\simeq_r$ we denote the $\mathcal{R}$-isomorphism, and by $\simeq$ — the abstract isomorphism of the corresponding structures. Note that any numbered positive structure is $\mathcal{R}$-isomorphic to the corresponding intensional positive structure.

For any c.e. set Q of L-quasiidentities, let $\mathcal{R}_Q$ be the full subcategory of $\mathcal{R}$ formed by the models of Q. Due to our use of intensional structures, it is convenient to think that Q always contains the equality axioms (otherwise, just add them to Q). The category $\mathcal{R}_Q$ has the initial object which is known in algebra as the free Q-model with countably many generators and in the data type theory as the initial Q-model. The category $\mathcal{R}_Q$ has also the terminal object $\mathbf{E}$ which is the singleton L-structure all of whose predicates are true.

We use some well known notions and results of computability theory which may be found in [Ma70, Ro67, So87]. The standard numbering of c.e. subsets of ω is denoted by $\{W_n\}$. Levels of the arithmetical hierarchy are denoted by $\Sigma^0_n, \Pi^0_n, \Delta^0_n$, levels of the analytical hierarchy by $\Sigma^1_n, \Pi^1_n, \Delta^1_n$, and levels of the Ershov difference hierarchy [E77] — by $\Sigma^{-1}_n, \Pi^{-1}_n, \Delta^{-1}_n$.

Positive (negative, constructive) structures are sometimes also called Σ^0_1- (respectively Π^0_1-, Δ^0_1-) structures. For any oracle f, $f : \omega \to \omega$, one can define f-positive, f-negative and f-constructive sructures by relativizing the introduced notions in a strightforward way. For $f = \emptyset^n$, were $\emptyset^n$ is the n-th iterate of the Turing jump, the corresponding relativized notions are often called Σ^0_{n+1}- (Π^0_{n+1}-, Δ^0_{n+1}-) structures. These relativized versions are important for some applications.

We will also use a few notions and facts from the theory of numberings [Er77]. Recall some of them. By $dom(f)$ and $rng(f)$ we denote respectively the domain and range of a function f. By *a numbering* we mean any function ν with $dom(\nu) = \omega$. So numberings may be written also as sequences $\{\nu_n\}_{n<\omega}$. Let numberings μ, ν be given. Then μ is *reducible* to ν (in symbols $\mu \leq \nu$), if $\mu = \nu \circ g$ for a recursive function g; μ is *equivalent* to ν, if $\mu \leq \nu$ and $\nu \leq \mu$; μ is *recursively isomorphic* to ν, if $\mu = \nu \circ g$ for some recursive permutation g; ν is *a factorization (or quotient)* of μ, if ν is equivalent to $f \circ \mu$ for some $f : rng(\mu) \to rng(\nu)$. A numbering ν is *precomplete*, if for any partial recursive function ψ there is a recursive function g such that $\nu\psi(x) = \nu g(x)$ for $x \in dom(\psi)$. It is well–known that any factorization of a precomplete numbering is precomplete, and if a numbering is equivalent to a precomlete numbering, then these numberings are recursively isomorphic. A numbering ν is *complete* (with respect to an element a), if for any partial recursive function ψ there is a recursive function g such that $\nu\psi(x) = \nu g(x)$ for $x \in dom(\psi)$ and $\nu\psi(x) = a$ for $x \notin dom(\psi)$. The theory of precomplete numberings is rather well developed and useful. More information about this interesting theory may be found in [Er77, Se94].

3. General Facts

In this section we briefly recall a few general facts about positive structures relating them to some algebraic notions and to the theory of quasivatieties.

3.1 Constructivizability

Here we cite some known facts about the following natural question: under what conditions a positive structure is constructive (or, more generally, constructivizable). An (abstract) structure is *constructivizable* if it is constructive for a suitable numbering of its universe.

The next result is taken from [Ma61].

Theorem 3.1.1. *The standard numbering of any finitely generated positive algebra, all of whose nonzero congruences are of finite index, is constructive.*

In contrast to this theorem, in [Kas89] a positive nonconstructivizable algebra was found all of whose nonzero congruences are of finite index.

Recall that a lattice is called *Artin* (*Nöther*), if any ascending (respectively, descending) sequence of its elements is finite. In [Ba74] it is shown that any positive algebra with countable Artin conguence lattice is constructive. In contrast to this, in [Kas92] a positive nonconstructivizable algebra with Nöther congruence lattice was consructed.

In [Ma61], with the use of [McK45], the following relationship of constructivity to the algebraic notion of finite approximability was established:

Theorem 3.1.2. *If the initial algebra of a finitely axiomatizable quasivariety K is approximated by finite K-algebras then it is constructive.*

In [Er80] the following criterion of consructiveness of a positive algebra was obtained. In the formulation, $\langle\,\rangle$ denote the usual computable pairing function on the natural numbers.

Theorem 3.1.3. *A positive algebra $(\mathbf{A};\alpha)$ is constructive if and only if there exists a c.e. set S so that $\alpha x \neq \alpha y$ for all $\langle x,y\rangle \in S$, and for any nonzero congruence η of $\mathbf{A}$ there is $\langle x,y\rangle \in S$ with $(\alpha x,\alpha y) \in \eta$.*

In [Er80] one can find some consequences of this criterion including the following generalization of an earlier result of A. V. Kuznetsov.

Corollary 3.1.4. *If algebra $\mathbf{A}$ has only finitely many congruences then any positive numbering of $\mathbf{A}$ is constructive.*

3.2 Algebraic Specifications

Positive algebras are closely related to a branch of theoretical computer science devoted to the theory of data types (see e.g. [EM85, EM90, Go85]).

Attempts to formulate a rigorous and general mathematical notion of a data type (which is of course a central concept in computer science) resulted in several notions turned out to be relevant to initial structures in varieties and quasivarieties. We fix for a short discussion here one of the existing definitions.

Definition 3.2.1. *(i)* Abstract data type *is a positive many-sorted algebra of a finite signature generated by the values of constant symbols.*

(ii) Abstract data type $\mathbf{A}$ *is* specifiable by identities (quasiidentities, positive sentences,...) *if there is an enrichment* $\mathbf{A}^*$ *of* $\mathbf{A}$ *and a finite set* Q *of identities (quasiidentities, positive sentences,...) such that* $\mathbf{A}^*$ *is the initial* Q*-algebra.*

A natural question is to describe the algebraically specifiable abstract data types. The next general fact is proved in [BT80].

Theorem 3.2.2. *Any consructive abstract data type is specifiable by identities.*

The following result from [Kas96] shows that the last theorem cannot be generalized to positive structures.

Theorem 3.2.3. *There is an abstract data type not specifiable by positive sentences.*

B. Khoussainov recently (in 2001 at a workshop in Novosibirsk) announced the existence of a positive data type not specifiable by quasiidentities.

The theory of data types is of course not reducible to abstract results similar to those mentioned above (e.g. practically realizible data types seem to have more constructive nature) but such results may show an important barrier for realizability, just in the same way as results on undecidability are interesting for people developing practically important algorithms.

3.3 Acceptable Numberings

Here we state the existence of some natural numberings of positive structures from $\mathcal{R}_Q$, for a c.e. set Q of quasiidentities (see Section 2).

We call a numbering $\{(\omega; f_i^n, P_j^n)\}_n$ of positive L-structures *c.e.*, if the sequence $\{f_i^n\}$ is recursive and the sequence $\{P_j^n\}$ is c.e. uniformly in i, j, n. We will consider such numberings of structures always modulo $\mathcal{R}$-isomorphism, not mentioning this explicitly. We will apply to the numberings of structures the notions from the numbering theory cited in Section 2 always under this agreement. E.g., a c.e. sequence of structures $\{\mathbf{A}_n\}$ is *reducible* to a c.e. sequence of structures $\{\mathbf{B}_n\}$, if there is a recursive function g such that $\mathbf{A}_n \simeq_r \mathbf{B}_{g(n)}$ for all $n < \omega$. By *an acceptable numbering of* $\mathcal{R}_Q$ we mean a c.e. numbering $\{\mathbf{A}_n\}$ of Q-models such that any c.e. numbering of Q-models is reducible to $\{\mathbf{A}_n\}$.

We are ready to state the main result of this section obtained in [Se96].

Theorem 3.3.1. *For any c.e. set Q of L-quasiidentities there is an acceptable numbering of $\mathcal{R}_Q$, and this numbering is unique up to recursive isomorphism.*

Examples. 1. The class of positive effectively dense BA's in the signature $\cup, \cap, ^-, 0, 1, f$, where f is the function witnessing the density (for details see Section 4.3 below) has an acceptable numbering, because it is axiomatizable by quasiidentities.

2. The class of positive dense (i.e., atomless) BA's has no acceptable numbering. This follows from a result in [Go83, Go96] formulated in the next section.

3. The class of positive boolean pairs (i.e. BA's with a distinguished subalgebra) has an acceptable numbering.

Note that the problem of existence of acceptable numberings is a popular topic in constructive model theory (the definition of an acceptable numbering of a class of constuctive structures looks quite similar to the definition above). For constructive structures the situation with the existence of acceptable numberings is more intricate. Acceptable numberings are known to exist for several natural classes of constructive structures, e.g. for the class of constructive fields lying in between two given constructive fields [Er68], for BA's (Goncharov and Peretyat'kin, unpublished), for the class of all structures of a signature without function symbols, for distributive lattices and for torsion-free abelian groups of a finite rank [N74]. But they do not exist for some other natural classes of structures, say for lattices and for torsion-free abelian groups [N74]. For a survey of computable numberings of constuctive structures see [Dob98].

3.4 Universal Positive Structures

In the last section we discussed numberings of structures which are in some sense similar to the standard numbering of c.e. sets. Here we try to find an analog of the "universal", i.e. creative set. Let again Q be a c.e. set of quasiidentities. We call a positive Q-model $\mathbf{B}$ *a component* of a positive Q-model $\mathbf{A}$, if $\mathbf{A} \simeq_r \mathbf{B} \times \mathbf{C}$ for some positive Q-model $\mathbf{C}$.

Remark. For the case of BA's, $\mathbf{B}$ is a component of $\mathbf{A}$ if and only if $\mathbf{B}$ is isomorphic to an ideal of $\mathbf{A}$.

We call a positive Q-model $\mathbf{U}$ *universal*, if any positive Q-model is a component of $\mathbf{U}$. Somewhat unexpectedly, it turns out that such structures exist in many natural cases, as established in [Se96].

Theorem 3.4.1. *For any c.e. set Q of quasiidentities there exists a universal positive Q-model, and it is unique up to $\mathcal{R}$-isomorphism.*

The construction of the universal structure in [Se96] seems to have no analogs in algebra. Let us show that for some particular natural classes of structures the universal structure can be obtained by a standard algebraic construction. This will also be useful for some further considerations. Assume that the signature L contains a unique constant symbol denoted by 0, and the quasiidentities from Q imply that the terminal structure $\mathbf{E}$ specified in Section 2 is a substructure of any Q-model. We call such a theory Q *special*. Examples of such theories are: semigroups with 0, lattices with 0, rings, as well as any extension of these theories by quasiidentities.

It is well known that any sequence $\{\mathbf{B}_n\}$ of Q-models for any special set Q has *a direct sum* $\coprod_n \mathbf{B}_n$, which is also a Q-model. This is the substructure of $\prod_n \mathbf{B}_n$ formed by the sequences almost all of whose elements are 0. If the numbering $\mathbf{B}_n = (\omega; f_i^n, P_j^n)$ is c.e., then the structure $\coprod_n \mathbf{B}_n$ is clearly positive. It is easy to show that if $\{\mathbf{A}_n\}$ is the acceptable numbering of positive Q-models then the structure $\coprod_n \mathbf{A}_n$ is universal.

Remark. The construction of the universal structure by means of the direct sum operation is of course easy and natural, but it is applicable only to special sets Q. The construction of $\mathbf{U}$ from $\{\mathbf{A}_n\}$ in the proof of Theorem 3.4.1 in [Se96] is applicable only to the acceptable numbering $\{\mathbf{A}_n\}$ (if the numbering $\{\mathbf{A}_n\}$ is c.e. but not acceptable, then one cannot guarantee that the resulting structure $\mathbf{U}$ is positive).

For the case of BA's, the universal positive structure may also be obtained using a natural algebraic construction, as follows. For the acceptable numbering $\{\mathbf{B}_n\}$ of positive BA's, let $U_0(U_1)$ be the set of all sequences $\{b_n\}$ such that $b_n \in \mathbf{B}_n$ for all n and $b_n = 0_{B_n}$ (resp. $b_n = 1_{B_n}$) for almost all n. Define the boolean operations on $U = U_0 \cup U_1$ componentwise and let ν be the natural numbering for which $\mathbf{U}$ is a positive BA and both sets $\nu^{-1}(U_i)$ are recursive. It is easy to see that the structure $\mathbf{U} = \coprod_n \mathbf{B}_n$ is a universal positive BA.

Its uniqueness is checked as follows. We may assume that $\mathbf{B}_n$ is the quotient of the free countable BA $(A; \alpha)$ by the smallest ideal containing the set $\{\alpha x | x \in W_n\}$. Suppose that $\mathbf{U}'$ is another universal positive BA, then $\mathbf{U}$ is r-isomorphic to a principal ideal $\hat{c}$ of $\mathbf{U}'$. Then

$$\mathbf{U}' \simeq_r \hat{\bar{c}} \sqcup \hat{c} \simeq_r \hat{\bar{c}} \sqcup \mathbf{B}_0 \sqcup \mathbf{B}_1 \sqcup \cdots .$$

For some $a < \omega$ we have $\hat{\bar{c}} \simeq_r \mathbf{B}_a$, hence $\mathbf{U}' \simeq_r \mathbf{B}_a \sqcup \mathbf{B}_0 \sqcup \mathbf{B}_1 \sqcup \cdots$. Define the acceptable numbering $\{V_n\}$ of c.e. sets by $V_0 = W_a$ and $V_{n+1} = W_n$. By uniqueness of the acceptable numbering, $V_n = W_{p(n)}$ for some recursive permutation p. This permutation induces the evident r-isomorphism of $\mathbf{B}_a \sqcup \mathbf{B}_0 \sqcup \mathbf{B}_1 \sqcup \cdots$ and $\mathbf{B}_0 \sqcup \mathbf{B}_1 \sqcup \cdots$. Hence, $\mathbf{U}' \simeq_r \mathbf{U}$.

Remarks. 1. The existence of the universal positive BA was established in [H75]. To the best of my knowledge, the uniqueness of this algebra was not noticed till 1990 when I observed that it easily follows from the representation $\mathbf{U} = \coprod_n \mathbf{B}_n$, as explained above. This fact was reported at several conferences and seminars in Russia and abroad and was included in [Se92a]. Soon I have noticed that the universal structures exist for the special sets Q. A subsequent analysis led to the general result 3.4.1.

2. The operation of "direct sum" of BA's maybe in the evident way generalized to any family $\{B_i\}_{i \in I}$ of BA's. As noticed by Yu. L. Ershov, the BA $\coprod_i B_i$ is isomorphic to the BA generated in a natural way from the usual direct sum of B_i's considered as relatively complemented distributive lattices with 0 (see [Er80]) which are known also as Ershov lattices.

3. For some natural classes of constructive structures, the universal constructive structures also exist. E.g., this is the case for constructive BA's (to see this, just take the "direct sum" of the acceptable numbering of constructive BA's mentioned in the last section). In contrast to this, in [Go83] it was shown that there is no universal structure for the class of constructive atomic BA's.

4. Theorem 3.3.1 is not generalizable to some natural classes of positive structures. E.g., in [Go83] it was proved that there is no universal positive atomic BA, as well as there is no universal positive atomless BA.

3.5 Properties of Acceptable Numberings

Here we consider the acceptable numberings of structures from the point of view of the numbering theory. In [Se87, Se88] we have shown that some natural numberings in computability theory are universal h-2-complete, for a suitable oracle $h : \omega \rightarrow \omega$. This notion is relevant also to the numberings of structures, so we start with recalling some definitions from [Se87,Se88]. A numbering ν is *h-2-complete* (with respect to given elements $a, b \in rng(\nu)$), if it has two following properties:

for any function ψ partial recursive in h there is a recursive function g such that $\nu g(x) = \nu\psi(x)$ for $x \in dom(\psi)$ and $\nu g(x) = a$ otherwise (this is a strong relativization of the notion of a complete numbering);

for any set S c.e. in h there is a recursive function f such that $\nu f\langle x, y\rangle = \nu x$ for $y \notin S$ and $\nu f\langle x, y\rangle = b$ for $y \in S$.

Examples. The numbering $\{W_n^h\}$ is universal h-2-complete with respect to $\emptyset, \omega$. The quotient numbering $\{W_n^*\}$ of the numbering $\{W_n\}$ modulo finite sets is universal $\emptyset'$-2-complete with respect to $\omega^*, \emptyset^*$.

Let again $\mathbf{E}$ be the trivial singleton structure and $\mathbf{U}$ be the universal structure from Section 3.3. The next result from [Se96] shows a special role of these structures. The notion of a special set of quasiidentities was defined in the previous section.

Theorem 3.5.1. *For any special set Q of quasiidentities, the acceptable numbering $\{\mathbf{A}_n\}$ of positive Q-models is $\emptyset'$-2-complete with respect to $\mathbf{E}$ and $\mathbf{U}$.*

This theorem does not generalize to arbitrary c.e. set Q of quasiidentities. E.g., let Q be the set of axioms of BA's in the signature $\{\cup, \cap, ^-, 0, 1\}$. Then $\mathbf{A}_n \simeq_r \mathbf{E}$ iff $P_0^n(0, 1)$ (without loss of generality we assume that 0,1 are interpreted in any BA A as $0_A, 1_A$). So $\{n|\mathbf{A}_n \simeq_r \mathbf{E}\}$ is a nontrivial c.e. index set. This (by a relativized version of the Rice theorem) implies that the numbering $\{\mathbf{A}_n\}$ is not $\emptyset'$-2-complete. Note that the same argument applies also to some other classes of structures: lattices with 0 and 1, rings with 1, and so on.

Next we will show that in some natural cases the numbering $\{\mathbf{A}_n\}$ have a universality property among the $\emptyset'$-2-complete numberings. Call a set Q of quasiidentities in a finite signature L *good* if it is special and there exists a sequence $\{\mathbf{B}_n\}$ of finite Q-models such that $\mathbf{B}_n$ is not embeddable in $\mathbf{B}_m$ for $m \neq n$ and there is an algorithm which computes from n the diagram D_n of the structure $\mathbf{B}_n$ (note that D_n is in this case a finite set of formulas). The good sets of axioms are not rare in algebra. For instance, all examples of special sets mentioned in the last section are good. From [Se88a] it follows that the set of axioms of distributive lattices with 0 (and even of distributive lattices of width 2) is also good.

We say that numberings μ and ν are *q-equivalent* (q reminds on quotients), if any of them is a quotient of the other (note that this is an equivalence relation). A numbering μ is called *universal h-2-complete*, if it is h-2-complete, and any h-2-complete numbering is a quotient of μ. It is clear that any two universal h-2-complete numberings are q-equivalent.

The next result is also from [Se96].

Theorem 3.5.2. *For any good set Q of quasiidentities, the acceptable numbering $\{\mathbf{A}_n\}$ of positive Q-models is universal $\emptyset'$-2-complete.*

From [Se88] and from remarks before the formulation of the last theorem it follows that any numbering from 3.5.2 is q-equivalent to the quotient numbering $\{W_n^*\}$, and also to the standard numbering $\{W_n^{\emptyset'}\}$ of the class Σ_2^0.

4. Some Paricular Topics

In this section we give several examples of rather deep results about some particular classes of positive structures. Special attention is paid to positive BA's because results in this field have several interesting applications.

4.1 Arithmetical Hierarchy of Structures

Here we consider the natural question on inclusions between classes of BA's connected with arithmetical hierarchy. We say that an abstract BA B is Σ_n^0 ($B \in \Sigma_n^0$) if there is a numbering β of B such that the numbered BA $(B;\beta)$ is Σ_n^0, and analogously for other classes Π_n^0, Δ_n^0 of BA's. Of course, a similar definition applies to other classes of structures.

The next result completely describes the relationships between the introduced classes of BA's. The assertion (i) below is due to Feiner [F70], the assertion

(ii) — to S. P. Odintsov [OS90], and the assertion (iii) — to V. L. Selivanov [OS90] (according to the author of [Ha93], the assertion (iii) was independently obtained by L. Harrington). For proofs see also [Go96, Se92a].

Theorem 4.1.1. *(i) There is a Σ^0_{n+1}-BA which is not a Δ^0_{n+1}-BA.*
(ii) Every Π^0_1-BA is a Δ^0_1-BA.
(iii) Every Π^0_{n+2}-BA is a Σ^0_{n+1}-BA.

The Feiner's theorem (i) was contrasted in [T94] by showing that any Low_2-BA is constructivizible.

A relativization of the construction of universal BA's in Section 3.4 shows that the universal BA's exist in all classes $\Delta^0_1, \Sigma^0_1, \Sigma^0_2, \ldots$. From the elementary classification of BA's follows that all these universal BA's are elementarily equivalent to one another (they all have the infinite elementary charachteristic [Er80]). A relativization of Feiner's theorem shows all these universal BA's are pairwise not isomorphic.

In [F70] and [Ha93] it is shown that for positive linear orderings the complete analog of Theorem 4.1.1 holds true. The survey [Dow98] contains much additional interesting information on constructive and positive linear orderings.

In [Kh88, Kh98] N. G. Khisamiev proves among other facts the following result on the arithmetical hierarchy of abelian groups.

Theorem 4.1.2. *(i) For abelian groups, only trivial inclusions between the classes of arithmetical hierarchy hold true. In other words, for abelian groups we have $\Delta^0_{n+1} \not\subseteq \Sigma^0_n \cup \Pi^0_n$, $\Sigma^0_n \not\subseteq \Pi^0_n$ and $\Pi^0_n \not\subseteq \Sigma^0_n$.*
(ii) For abelian torsion-free groups, $\Delta^0_1 \subset \Pi^0_1 \subset \Pi^0_2 \subset \cdots$, $\Sigma^0_1 = \Delta^0_1$ and $\Sigma^0_{n+2} = \Pi^0_{n+1}$.

Hence, for the torsion-free abelian groups the situation is in a sense dual to the case of BA's. To the best of my knowledge, for other natural classes of structures the problem discussed in this section was not considered in the literature.

4.2 Complexity of Ideals of Boolean Algebras

Let us recall some well known constructions on BA's. Element a of a BA B is *an atom* if $a \neq 0$ and $x \leq a$ impllies $x \in \{0, a\}$. Element a is called *atomless* if there is no atom $x \leq a$. Element a is called *atomic* if $0 \neq x \leq a$ implies that x is not atomless. Element a is called *decomposable* if it is representable in the form $a = b \cup c$ where b is atomless and c is atomic. Algebra B is called atomless (atomic, decomposable) if the element 1_B is atomless (atomic, decomposable).

The ideal of a BA B generated by atoms is called *Frechet ideal* of B. It consists of all finite unions of atoms and is denoted by $F(B)$. The ideal of B consisting of all decomposable elements is called *Tarski-Ershov* ideal of B

and is denoted by $T(B)$. The quotient-structure $B/F(B)$ ($B/T(B)$) is called Frechet (respectively, Tarski-Ershov) derivative of B.

One could define some other similar constructions, say the quotient-structure $B/I(B)$ of B by the ideal $I(B)$ generated by atoms and atomless elements. For the sake of brevity, we sometimes call the quotient-stιucture $B/I(B)$ the I-derivative of a BA B.

The next result gives in a sense a complete characterisation of algorithmic complexity of the introduced constructions. In this result the assertions (i) and (iii) are due to V. L. Selivanov [Se88, OS89], while the assertion (ii) — to S. P. Odintsov [OS89]. See also [O98, Se92a].

Theorem 4.2.1. *(i) The Frechet derivative of any Δ_1^0-BA is a Σ_2^0-BA, and every Σ_2^0-BA is isomorphic to the Frechet derivative of a suitable Δ_1^0-BA.*

(ii) The I-derivative of any Δ_1^0-BA is a Σ_3^0-BA, and every Σ_3^0-BA is isomorphic to the I-derivative of a suitable Δ_1^0-BA.

(iii) The Tarski-Ershov derivative of any Δ_1^0-BA is a Σ_4^0-BA, and every Σ_4^0-BA is isomorphic to the Tarski-Ershov derivative of a suitable Δ_1^0-BA.

Let B^n be the n-th Tarski-Ershov derivative of a BA B, i.e. $B^0 = B$ and $B^{n+1} = (B^n)'$, where C' denote the Tarski-Ershov derivative of a BA C. The next result is an effective version of (iii) which follows immediately from the proofs in [Se88, OS89]. Analogs of this result hold true also for the other two kinds of derivatives.

Corollary 4.2.2. *There is an algorithm which computes from a Σ_{4n+1}^0-index of a numbered BA C a Σ_1^0-index of a numbered BA B such that $B^n \simeq C$.*

We conclude this section with an interesting property of universal BA's [Se91a, Se92a] which follows easily from the last theorem und the representation of the universal BA by means of the "direct sum" operation in Section 3.4.

Theorem 4.2.3. *If U is a universal positive BA then $U/F(U)$ is a universal Σ_3^0-BA, $U/I(U)$ is a universal Σ_4^0-BA, and $U/T(U)$ is a universal Σ_5^0-BA.*

The last theorem immediately implies the following assertion (to see this, one has only to note that, in the notation of the assertion, $B/F(B) \simeq U/I(U)$).

Corollary 4.2.4. *Let U be a universal positive BA and $J(B)$ be its ideal consisting of all atomless elements. Then $B = U/J(U)$ is an atomic BA whose Frechet derivative is a universal Σ_4^0-BA.*

It would be rather interesting to prove analogs of results of this section for other well-known mathematical constructions (say, to estimate the complexity of the torsion subgroup of an abelian group). It seems plausible that the results of this section may be generalized to the Ershov lattices [Er80].

4.3 Effective Ideals of Positive Boolean Algebras

In the literature (see e.g. [MN77, Re78, Re89, Od86, CR98]) one could find many interesting facts on the lattice of c.e. substructures or the lattice of c.e. ideals of a positive structure. Such questions were investigated for vector spaces, commutative rings and BA's. This investigation was often conducted in parallel with the study of the lattice of c.e. sets in computability theory. In this section we cite only few results of this kind for BA's which are related to some other topics dicussed in the paper.

Let us start with a couple of results from [CR98] were the authors relate the theory of positive BA's to the theory of Π^0_1-classes.

Theorem 4.3.1. *If a positive BA B has only finitely many prime ideals then every prime ideal of B is recursive.*

Theorem 4.3.2. *(i) There is a positive BA which has no recursive prime ideal.*

(ii) There is a positive BA B such that any two distinct prime ideals of B are Turing incomparable, were distinct means having infinite symmetric difference.

Next we discuss the lattice of c.e. ideals of a positive BA or, more generally, the lattice $\mathcal{I}(B)$ of Σ^0_n-ideals of a Σ^0_n-BA B. It is easy to see [Ni00] that the lattice $\mathcal{I}(B)$ is distributive, has the reduction property [So87] and contains all principal ideals.

It can occur that $\mathcal{I}(B) \simeq B$, even for the atomless positive BA B: one can construct an atomless BA B such that every c.e. ideal is principal [MP70]. In such cases the elementary theory of $\mathcal{I}(B)$ is of course decidable.

As shown in [Ni98, Ni00], the situation is quite different for effectively dense Σ^0_n-BA's. A nontrivial positive BA $(\mathbf{B};\beta)$ is effectively dense if for a suitable recursive function f the following conditions hold true: $\beta f(x) \leq \beta x$ and $0 < \beta x \to 0 < \beta f(x) < \beta x$. A Σ^0_n-*effectively dense* BA is a numbered Σ^0_n-BA $(\mathbf{B};\beta)$ such that for a suitable $\emptyset^{n-1}$-computable function f the conditions above hold true.

As usual, $Th(\mathbf{A})$ denote the elementary theory of a structure $\mathbf{A}$. The following result from [Ni98, Ni00] characterizes the algorithmic complexity of the elementary theory of the ideal lattice defined above.

Theorem 4.3.3. *If $(\mathbf{B};\beta)$ is a Σ^0_n-effectively dense BA then the set $Th(\mathcal{I}(\mathbf{B}))$ is recursively isomorphic to the true arithmetic $Th(N;+,\cdot)$, i.e. to $\emptyset^\omega$.*

4.4 Precomplete Positive Structures

Here we discuss positive structures whose numberings are precomplete. Interest to such structures is explained by a result from [La87] that the nontrivial precomplete positive numbering is unique up to recursive isomorphism.

The next definition is taken from [MS85, Se89].

Definition 4.4.1. *Let f be any oracle. By a* special f-partial ordering *we mean an f-positive partial ordering $\mathbf{A} = (A; \alpha, \leq, 0, 1)$, where α is f-precomplete, 0 is the least element, 1 is the greatest element and $0 \neq 1$. The notions of a* special f-upper semilattice *and of a* special f-lattice *look similarly, only the corresponding operations $\cup$ (resp. $\cup, \cap$) are required as usual to be representable by computable functions in the numbering α.*

We call a structure $\mathbf{A}$ from the definition *effectively broad,* if there is a computable function g such that $0 < \alpha g(x) < 1$ for all x, and if $0 < \alpha y < 1$ for all $y \in W_x$ then $\alpha g(x)$ is $\leq$-incomparable with αy for all $y \in W_x$. Note that for the effectively broad ordering there always exists an element $a \in A$ with $0 < a < 1$ (consider the empty W_x), and for every element $0 < \alpha b < 1$ there is an element incomparable with αb (consider $W_x = \{b\}$).

The next assertion from [Se89] states some results about the α-index sets and the algebraic properties of such orderings.

Proposition 4.4.2. *For any special f-partial ordering $\mathbf{A}$ we have:*

(i) $\mathbf{A}$ is effectively broad.

(ii) If $0 < a$ then $\{x|a \leq \alpha x\}$ is $\Sigma_1^{0,f}$-complete.

(iii) If $a < 1$ then $\{x|\alpha x \leq a\}$ is $\Sigma_1^{0,f}$-complete.

(iv) If $0 < a < 1$ then $\{x|a \not\leq \alpha x \wedge \alpha x \not\leq a\}$ is $\Pi_1^{0,f}$-complete.

(v) For all $a_0, \ldots, a_n \in A$, the set $\{x|\alpha x = a_0 \vee \cdots \vee \alpha x = a_n\}$ is $\Sigma_1^{0,f}$-complete.

We call $\mathbf{A}$ *effectively upward dense*, if there is a computable function g such that $\alpha x < 1$ implies $\alpha x < \alpha g(x) < 1$. Recall that $\Sigma_k^{-1,f}$ denote the k-th level of Ershov hierarchy relativized to f. The next result is also from [Se89].

Theorem 4.4.3. *For any special f-upper semilattice $\mathbf{A}$ we have:*

(i) $\mathbf{A}$ is effectively upward dense.

(ii) For any $n < \omega$ there exist $a_0, \ldots, a_n \in A$ such that $a_k \not\leq \cup_{l \neq k} a_l$ for all $k \leq n$.

(iii) If $0 < a < 1$ then $\{x|a < \alpha x\}$ is $\Sigma_2^{-1,f}$-complete.

(iv) If $0 < a_0 < a_1 < a_2 < \cdots$ then the sets $\{x|a_0 \leq \alpha x \wedge a_1 \not\leq \alpha x\}$, $\{x|(a_0 \leq \alpha x \wedge a_1 \not\leq \alpha x) \vee a_2 \leq \alpha x\}, \ldots$ are m-complete respectively in $\Sigma_2^{-1,f}, \Sigma_3^{-1,f}, \ldots$.

Note that the last theorem does not generalize to the special f- orderings.

We say that $\mathbf{A}$ is *effectively dense*, if there is a computable function g such that $\alpha x < \alpha y$ implies $\alpha x < \alpha g\langle x, y\rangle < \alpha y$. The next result does not generalize to the special f-upper semilattices. In this result the assertions (i) and (ii) follow from results of [MS85] while the assertions (iii)—(v) are taken from the paper [Se89].

Theorem 4.4.4. *For any special f-lattice* $\mathbf{A}$ *we have:*

(i) Any f*-constructive partial ordering is effectively embeddable in* $\mathbf{A}$.

(ii) Any f*-positive linear ordering is effectively embeddable in* $\mathbf{A}$.

(iii) $\mathbf{A}$ *is effectively dense.*

(iv) If $0 < a < 1$ *then* $\{x|\alpha x < a\}$ *is* $\Sigma_2^{-1,f}$*-complete.*

(v) If $a, b \in \mathbf{A}$ *and* $0 < a$ *or* $b < 1$ *then* $\{x|a < \alpha x < b\}$ *is* $\Sigma_2^{-1,f}$*-complete.*

Actually, the assertions (i) and (ii) were proven in [MS98] for a more general notion of effectively inseparable (e.i.) positive bounded lattice which means the effective inseparability of the index sets $\alpha^{-1}(0)$ and $\alpha^{-1}(1)$. Any precomplete positive lattice is of course e.i., but the converse inclusion does not hold.

It is easy to see that there is no nontrivial precomplete positive BA (as noticed by Yu. L. Ershov, the same is true for groups), but there are natural and important examples of e.i. positive BA's. Let us cite some interesting results on such BA's from [MS85] (the proof uses an earlier paper [KP67])

Theorem 4.4.5. *(i) Any positive partial ordering is effectively embeddable in any e.i. positive BA.*

(ii) Any two e.i. positive BA's are recursively isomorphic.

The next result from [MS85, Se89] shows how to construct new special f-orderings from given ones extending thus the applicability of the above assertions.

Proposition 4.4.6. *(i) The classes of special f-orderings, f-upper semilattices and f-lattices are closed under finite direct products.*

(ii) If $\mathbf{A}$ *is a special f-upper semilattice and* $a < 1$ *then the interval structure* $([a, 1]; \alpha^*, \cup, a, 1)$ *with the induced numbering* α^* *is a special f-upper semilattice.*

(iii) If $\mathbf{A}$ *is a special f-lattice and elments* $a, b \in A$ *satisfy* $a < b$ *then the interval structure* $([a, b]; \alpha^*, \cup, \cap, a, b)$ *with the induced numbering* α^* *is a special* f*-lattice.*

There are several natural open questions conserning the objects discussed above. E.g., we do not know whether there is a special f-upper semilattice with the decidable elementary theory. Another interesting open question is: does there exist a nontrivial positive precomplete finitely presented algebra?

4.5 Index Sets of Positive Structures

The existence of the acceptable numberings of structures enables to consider many natural algorithmic problems on such structures of the following type: given a "natural" property of structures, what is the degree of unsolvability of the (index set of the) class of structures having this property? Note that m-degrees seem to be most suitable for this problem. From Theorem 3.3.1 it follows that the degree of a property does not depend on the acceptable numbering: index sets of any property in any two given acceptable numberings are recursively isomorphic.

Questions of this type are similar to the popular topic of the classification of index sets in computability theory. The aim of this section is to discuss some particular cases of the raised problem trying to show the richness of the subject and its relatedness to some more traditional decision problems (as wil be shown in Section 5.5 below).

We start with a Rice theorem for the numberings of positive structures. It follows immediately from Theorem 3.5.1 and the properties of 2-complete numberings established in [Se82, Se83].

Corollary 4.5.1. *Let $\{\mathbf{A}_n\}$ be the acceptable numbering of positive Q-models of a special set Q of quasiidentities. Then any nontrivial index set of the numbering $\{\mathbf{A}_n\}$ is either Σ^0_2-hard or Π^0_2-hard.*

Now consider the class of all index sets in the numbering of positive structures. The next result from [Se96] follows immediately from Theorem 3.5.2 and the properties of 2-complete numberings established in [Se82, Se83].

Corollary 4.5.2. *Any index set of the numbering from Theorem 3.5.2 is recursively isomorphic to an index set of the numbering $\{W^*_n\}$, and vice versa.*

Let $(I_\nu; \leq_m)$ be the structure of m-degrees containing index sets of a given numbering ν. This structure for different numberings ν was considered by several authors. By the last corollary, this structure is the same for any numbering from 4.5.1 as well as for the numbering $\{W^*_n\}$. Results in [Se82, Se83] imply the following intriguing property of this structure.

Corollary 4.5.3. *Let $\nu = \{\mathcal{A}_n\}$ be any numbering from 4.5.1. For all $n < \omega$ and $a_0, \dots, a_n \in I_\nu$ there are $b_0, b_1, b_2, b_3 \in I_\nu$ such that: $a_i \leq_m b_j$ for all $i \leq n, j \leq 3$; if $x \in I_\nu$ and $a_i \leq_m x$ for all $i \leq n$, then $b_j \leq_m x$ for some $j \leq 3$; if $x \in I_\nu$ and $x \leq_m b_j$ for all $j \leq 3$, then $x \leq_m a_i$ for some $i \leq n$.*

We proceed with the classification of some concrete index sets in the acceptable numbering $\{\mathbf{A}_n\}$ of positive distributive lattices with 0 in the signature $\{\cup, \cap, 0\}$. By ω we denote the lattice $(\omega; max, min, 0)$, and by $k+1$ for $k < \omega$ — its sublattice with the universe $\{0, \dots, k\}$. By LO (WO) we denote the class of lattices which are linearly ordered (respectively well ordered) by the induced partial ordering.

The next result is proved in [Se96].

Theorem 4.5.4. *(i) The sets $M_k = \{n | \exists l \leq k(\mathbf{A}_n \simeq l+1)\}$ for $k < \omega$ and $\{n | \mathbf{A}_n \in LO\}$ are Π^0_2-complete.*

(ii) The set $M = \{n | \mathbf{A}_n \, is \, finite\}$ is Σ^0_3-complete.

(iii) The set $C = \{n|\mathbf{A}_n \simeq \omega\}$ *is* Π^0_4*-complete.*
(iv) The set $\{n|\mathbf{A}_n \in WO\}$ *is* Π^1_1*-complete.*

Now let us discuss results on index sets of positive boolean algebras. Many results of this kind were obtained in [Se90,Se91,Se92]. Recall definitions of some objects relevant to the elementary classification of BA's due to A. Tarski and Yu. L. Ershov, see [CK73] or [Er80]. Let $\mathcal{A}^n_k(\mathcal{B}^n_k, \mathcal{A}^n)$ be the class of BA's B such that B^n has $\leq 2^k$ elements (resp. has $\leq k$ atoms, is an atomic BA). Here B^n is again the n-th Tarski-Ershov derivative of a BA B (see Section 4.2).

The next result from [Se90, Se92a] describes the complexity of all introduced classes of BA's in the acceptable numbering $\{B_n\}$ of positive BA's. The proof makes heavy use of results from Section 4.2 and of the tree representation of positive BA's. For a class $\mathcal{X}$ of BA's, let $\theta(\mathcal{X}) = \{n|B_n \in \mathcal{X}\}$.

Theorem 4.5.5. *For all* $n, k < \omega$ *the sets* $\theta(\mathcal{A}^n_0)$, $\theta(\mathcal{A}^n_{k+1})$, $\theta(\mathcal{B}^n_k)$ *and* $\theta(\mathcal{A}^n)$ *are m-complete respectively in* Σ^0_{4n+1}, Π^0_{4n+2}, Π^0_{4n+3} *and* Π^0_{4n+4}.

Remarks. 1. In [Se90, Se91] some further universality conditions for the index sets from the last theorem were established.

2. For the ideals $F(B)$ and $I(B)$ discussed in Section 4.2 one could define in a natural way analogs of some classes from the last theorem. In [Le87, Se90, Se91, Se92, Se92a] the index sets of all such classes were estimated.

One could try to classify more index sets of classes of BA's, e.g. index sets of the form $\{n|\mathbf{A}_n \models \varphi\}$, for any given sentence φ of signature $\{\subseteq\}$. Any such index set is arithmetical, so one could try to use the arithmetical hierarchy as a scale for this classification. It turns out that it is indeed insufficient: one should find a suitable refinement of the arithmetical hierarchy. In some cases, the sequence $\{\Sigma_n^{-1,\emptyset^k}\}_{n<\omega}$, formed by the levels of the Ershov difference hiearchies relativized to the jumps $\emptyset^k (k < \omega)$, is sufficient; by the *long difference hierarchy* we mean the scale formed by these levels and by their duals.

The next result from [Se91, Se92a] is an example of a positive decision of the raised problem. The proof uses the observation in [Se90] that finitely axiomatizable classes $\mathcal{M}_\varphi = \{B \in \mathcal{U}|B \models \varphi\}$ of BA's (where $\mathcal{U}$ is the class of all BA's), are exactly the finite boolean combinations of the classes $\mathcal{A}^n_k, \mathcal{B}^n_k, \mathcal{A}^n (n, k < \omega)$ inside $\mathcal{U}$, and there is an algorithm which computes from a given sentence φ a boolean combination coinciding with $\mathcal{M}_\varphi$.

Theorem 4.5.6. *For any sentence* φ *of signature* $\{\subseteq\}$*, the index set* $\{n|\mathbf{B}_n \models \varphi\}$ *is m-complete in one of levels of the long difference hierarchy, and the level is effectively computable from the sentence* φ.

Analogs of the last theorem hold true also for the finite boolean combinations of classes of BA's mentioned in the Remark 2 above.

In [Se91] the following generalization of the last theorem was proved. It requires a further refinement of the long difference hierarchy, namely the fine hierarchy introduced in [Se83] and systematized in [Se95]. It looks like $\{\Sigma_\alpha\}_{\alpha<\varepsilon_0}$, the levels being well-ordered under inclusion with the well known order type ε_0. Relate to any sentence φ of signature $\{\subseteq\}$ the unary predicate $A \models \varphi$ on the class $\mathcal{U}$ of all BA's.

Theorem 4.5.7. *For any boolean combination $Q(A_1, \dots, A_n)$ of the predicates $A_i \models \varphi$, the index set $\theta(Q) = \{\langle x_1 \dots, x_n\rangle | Q(B_{x_1}, \dots, B_{x_n})\}$ is m-complete in one of levels $\Sigma_\alpha, \Pi_\alpha$ ($\alpha < \varepsilon_0$) of the fine hierarchy, and all the possibilities are realized.*

Remarks. 1. Analogs of the above-mentioned results on positive BA's may be proved also for the acceptable numbering of constructive BA's. The estimations in this case are "one step lower" but the proofs from [L87, Se90, Se91, Se92] generalize to this case in a straightforward way.

2. One cannot expect to prove analogs of the last two theorems for broad enough classes of structures. E.g. the analog of Theorem 4.5.6 for the numbering $\{\mathbf{A}_n\}$ of positive distributive lattices is false (otherwise one could decide the set of sentences φ for which $\{n|\mathbf{A}_n \models \varphi\}$ is m-complete in $\Pi_0^{-1} = \{\omega\}$, i.e. decide the elementary theory of positive distributive lattices, which is impossible). We see that the discussed problem is closely related to the decidability of elementary theories. The possibility of positive results for BA's is explained by the simplicity and constructive nature of the Tarski-Ershov elementary classification of BA's.

It seems interesting to try to get analogs of results of this section for the classes of structures with the effective elementary classification, e.g. for the Ershov lattices or for abelian groups.

Another natural question of this type is to get analogs of the described results for "complex" classes of structures (e.g. distributive lattices) but for "simple enough" (say, boolean combinations of Σ_2^0) sentences. This is closely related to decidability of restricted theories of such structures.

5. Applications

In this section we consider applications of several above-mentioned results to logic and computability theory.

5.1 Universal Boolean Algebras in Logic

As we know from Section 3.4, for many natural classes of structures the universal positive structures exist. Here we discuss the question on whether such structures really occur in mathematical practice. It turns out that this is the case, at least for the positive universal BA's.

If an abstract BA B is Σ_n^0-universal (i.e. it has a numbering β such that $(B;\beta)$ is Σ_n^0-universal), let us denote this by $B \sim \Sigma_n^0$. If a BA B is atomic and its Frechet derivative is a Σ_n^0-universal BA, let us denote this by $B \sim (1, \Sigma_n^0)$ (here 1 points on the first derivative, one could also consider any $k \in \omega$ in place of 1). As is well known (see e.g. [Er80]), the condition $B \sim (1, \Sigma_n^0)$ determines the BA B uniquely up to isomorphism.

For a distributive lattice L, let (L) denote the BA generated by L. As is well-known [Gr71], the algebra (L) is determined uniquely, up to isomorphism; the same is true for distributive upper semilattices L.

Let $\mathcal{E}$ be the lattice of c.e. sets and, more generally, let Σ_n^0 be the lattice of Σ_n^0-sets, $\Sigma_1^0 = \mathcal{E}$. Let $\mathcal{R}_m$ be the distributive upper semilattice of c.e. m-degrees (for simplicity we assume that all computable sets form a single m-degree). Let $(\mathcal{E})$, (Σ_n^0) and $(\mathcal{R}_m)$ be the BA's defined as in the previous paragraph. Finally, let $(\Sigma_n^0)^*$ be the Frechet derivative of (Σ_n^0).

According to a well-known result of A. Lachlan (see e.g. [So87]), any Σ_3^0-BA is isomorphic to the lattice $L^*(A)$ of c.e. supersets (modulo finite sets) of a suitable c.e. set A (which is of course hyperhypersimple). From another well-known result of A. Lachlan [La72] easily follows (see e.g. [Se88, Se92a]) that any Σ_3^0-distributive lattice with 0 and 1 is isomorphic to an initial segment of $\mathcal{R}_m$.

From these results, from Theorem 3.4.1 and from some easy estimations by the Tarski-Kuratowski algorithm (details may be found in [Se92a]) one can easily deduce the following

Corollary 5.1.1. $(\mathcal{E})^* \sim \Sigma_3^0$, $(\Sigma_n^0)^* \sim \Sigma_{n+2}^0$, $(\Sigma_n^0) \sim (1, \Sigma_{n+2}^0)$ *and* $(\mathcal{R}_m) \sim \Sigma_3^0$.

This provides some natural examples of universal BA's (and some BA's closely related to them) in computability theory. We will give also a couple of examples from logic.

Call a finite signature Ω *rich* if it contains a symbol of arity > 1 or at least two unary functional symbols. Let $L(T)$ be the Lindenbaum algebra of sentences of Ω modulo equivalence in the theory T (for an exact definition of this object see e.g. [CK73]). Note that the Gödel numbering of sentences induces a numbering of the BA $L(T)$, and that $L(T)$ is positive provided that T is recursively axiomatizable. For the empty theory T the BA $L(T)$ is denoted simply by L and is called *the Lindenbaum algebra* of Ω.

The next theorem is due to W. Hanf [H75]. Actually, this paper by W. Hanf (and some other his relevant papers) was the origin of the concept of a universal positive BA.

Theorem 5.1.2. *The Lindenbaum algebra of any finite rich signature is a universal positive BA.*

Recently, S. Lempp, M. G. Peretyat'kin and R. Solomon characterized several other natural Lindenbaum algebras. Let us cite a couple of typical examples from the manuscript [Pe00]. The assertion (i) below is proved by a very complicated construction while the assertions (ii) and (iii) follow easily from Corollary 4.2.4 and from the case of Frechet ideal in Theorem 4.2.3, respectively.

Theorem 5.1.3. *(i) Let F be the theory of all finite structures of a finite rich signature. Then $L(F) \sim (1, \Sigma_2^0)$.*

(ii) Let K be the theory of all structures of a finite rich language, whose elementary theories are finitely axiomatizable. Then $L(K) \sim (1, \Sigma_4^0)$.

(iii) Let N be the theory of all structures of a finite rich language, whose elementary theories are not finitely axiomatizable. Then $L(N) \sim \Sigma_3^0$.

From results above and results in Sections 3.4 and 4.2 one can deduce several unexpected curious relationships between some natural BA's. Let us formulate a couple of such examples. Recall that by $\simeq$ we denote the (abstract) isomorphism of structures.

Corollary 5.1.4. *(i) $L/F(L) \simeq (\mathcal{E})^* \simeq (\mathcal{R}_m) \simeq L(N)$, where L is the Lindenbaum algebra of a finite rich signature.*

(ii) $B/F(B) \simeq (\Sigma_2^0)^* \simeq L/I(L)$, where $B = L(K)$ and I is the ideal generated by atoms and by atomless elements.

There are many natural questions on characterization of Lindenbaum algebras, e.g.: is the Lindenbaum algebra $L(G)$ of the elementary group theory a universal positive BA? This question, as well as many other interesting questions on Lindenbaum algebras, was raised by M. G. Peretyat'kin, see e.g. [Pe00].

It seems also interesting to seek for examples of other types of universal positive structures (say, abelian groups) in mathematics.

5.2 Effectively Dense Lattices in Logic

Here we consider some positive structures naturally arising in logic and closely related to precomplete positive structures discussed in Section 4.4.

For any $n, k < \omega$, let $\alpha_{n,k}$ be the Gödel numbering of Σ_{n+1}^0-formulas of language $\{+, \cdot, 0, 1\}$ with free variables among $x_0, \ldots, x_{k-1}$, modulo equivalence in Peano arithmetic (which is assumed to be consistent). Let $\mathbf{A}_{n,k}$ denote the numbered structure with the universe $A_{n,k} = rng(\alpha_{n,k})$, the numbering $\alpha_{n,k}$, and the operations induced by conjunction and disjunction. For any k, let $\mathbf{L}_k$ denote the Lindenbaum algebra of formulas with free variables among $x_0, \ldots, x_{k-1}$, modulo Peano arithmetic (hence, $L_k = \cup_n A_{n,k}$).

The next result, in formulation of which some notions from Section 4.4 are used, follows from results in [Vi80].

Theorem 5.2.1. *(i) The structures $\mathbf{A}_{n,k}$ are special lattices.*

(ii) The positive BA's $\mathbf{L}_k$ are effectively inseparable.

Applying to $\mathbf{A}_{n,k}$ and $\mathbf{L}_k$ results from Section 4.4, we get several nontrivial results about these structures. In particular, the BA's $\mathbf{L}_k$ embed all positive partial orderings. In [MS85] the following strengthening of this property was obtained. In this assertion $\bar{i}$ is a numeral denoting the number i.

Theorem 5.2.2. *For every positive preorder* $P = (\omega; \leq_P)$ *there is a* Σ^0_2-*formula* $F(v)$ *such that, for all* i *and* j, *we have* $i \leq_P j$ *iff* $F(\bar{i}) \rightarrow F(\bar{j})$ *is provable in Peano arithmetic. If* P *is a distributive lattice with operations* $\cup, \cap$ *then formula* F *may be choosen to satisfy additionally the property that* $F(\overline{i \cup j}) \leftrightarrow F(\bar{i}) \vee F(\bar{j})$ *is provable in Peano arithmetic, and similarly for* $\cap$ *and* $\wedge$.

The BA $\mathbf{L}_k$ is effectively dense, so one can apply Theorem 4.3.3. In this way, the following result was obtained in [Ni00].

Theorem 5.2.3. *Let* T *be a recursively axiomatizable, consistent theory containing Peano arithmetic, and let* $\mathcal{L}_T$ *denote the lattice of all c.e. extensions of* T *under inclusion. Then the elementary theory* $Th(\mathcal{L}_T)$ *is recusively isomorphic to* $\emptyset^\omega$.

Remark. Results of this section hold true for a strong enough finitely axiomatizable theory in place of Peano arithmetic.

5.3 Computably Enumerable Sets and Π^0_1-Classes

Here we consider applications of positive structures to such classical objects of computability theory as the lattice $\mathcal{E}$ of c.e. sets and the lattice $\mathcal{L}$ of Π^0_1-classes of sets.

For $A, B \in \mathcal{E}$, $A \subseteq B$, let $[A, B]$ denote the interval sublattice of $\mathcal{E}$ defined by A, B. If this interval is a BA then, by the well known result of A. Tarski, the elementary theory $Th([A, B])$ is decidable. Using Theorem 4.3.3, A. Nies [Ni97,Ni00] was able to prove the following

Theorem 5.3.1. *Suppose* $A, B \in \mathcal{E}$ *and* $[A, B]$ *is not a BA, then* $Th([A, B])$ *is recursively isomorphic to* $\emptyset^\omega$. *A similar statement holds for* $\mathcal{E}^*$.

This implies that the elementary theory of the lattice $\mathcal{E}$ is also recursively isomorphic to $\emptyset^\omega$ (hence, is certainely undecidable), which was earlier proved in [He84,HN98] with the use of quite different methods.

Feiner [Fe70] deduced from his theorem cited in Section 4.1 the following result on relativized versions of $\mathcal{E}$.

Theorem 5.3.2. *For any oracle* f, *the lattices* $\mathcal{E}^f$ *and* $\mathcal{E}^{f'}$ *are not isomorpic. In particular,* $(\Sigma^0_m; \subseteq) \not\simeq (\Sigma^0_n; \subseteq)$ *for* $m \neq n$.

In [Ha93] some strengthenings of this last theorem are proved, also using similar methods. Using different methods, L. Harrington and A. Nies proved in [HN98] that if $m \neq n$ then the lattices $(\Sigma^0_m; \subseteq)$ and $(\Sigma^0_n; \subseteq)$ are not elementarily equivalent.

In [CN99] some interesting results on initial segments $[\emptyset, P] = \mathcal{L}(P)$ of the lattice of Π^0_1-classes were obtained. We cite a result proved again with the use of Theorem 4.3.3. A Π^0_1-class P is said to be *decidable* if it is the set of paths through a computable tree T with no dead ends.

Theorem 5.3.3. *If P is a decidable Π^0_1-class and $\mathcal{L}(P)$ is not a BA, then the elementary theory of $\mathcal{L}(P)$ is recursively isomorphic to $\emptyset^\omega$.*

5.4 Degree Structures

Here we apply results on positive structures to some degree structres in computability theory and complexity theory.

For c.e. sets A and B, let $A \leq^\omega_T B$ denote that $A^n \leq_T B^n$ for some n, where A^n is the n-th iterate of the Turing jump. By a *strong reducibility* we mean any of the usual computability-theoretic reducibilities [Ro67,So87] lying in between (and possibly equal to) m and wtt.

For any $n < \omega$, let K_n be the class of sequences $(A_0, \ldots, A_{n+1})$ of nonempty pairwise disjoint c.e. sets. Any K_n has a natural numbering which is the greatest one among the uniform c.e. (or computable) numberings, see [Er77]. This numbering induces the numberings of $K_n/\equiv_m$ and $K_n/\equiv_{sm}$, where $\equiv_m$ is the equivalence under the uniform m-reducibility of sequences, and $\equiv_{sm}$ is the equivalence corresponding to the following reducibility: a sequence $(A_0, \ldots, A_{n+1})$ is *sm-reducible* to a sequence $(B_0, \ldots, B_{n+1})$, if there is a computable function g such that $A_i \subseteq g^{-1}(B_i)$ for all $i \leq n+1$. The sm-reducibility is related to the notion of separability and was used in [Er77] for a classification of tuples of index sets.

Let $\oplus$ be the usual join operation and let $A \otimes B = \{\langle a, b\rangle | a \in A, b \in B\}$. The operations $\oplus$ and $\otimes$ act also on tuples of sets (componentwise).

In the next theorem we assemble together some results from [Dv84, JLSS89, A89, Se89]. We identify preorderings with the corresponding partial orderings and denote by $\leq_r$ any strong reducibility.

Theorem 5.4.1. *(i) The structure $(\Sigma^0_1; \leq^\omega_T)$ is a special $\emptyset^\omega$-partial ordering.*

(ii) For all n and k, $k > 0$, structures $(\Sigma^{-1}_k; \leq_r, \oplus)$ and $(K_n; \leq_m, \oplus)$ are special $\emptyset^2$-upper semilattices.

(iii) For any n, the structure $(K_n; \leq_{sm}, \oplus, \otimes)$ is a special $\emptyset^2$-lattice.

Applying results of Section 4.4 to structures from the last theorem, we can deduce several facts on degree structures, as described in [Se89]. Below are a few examples.

Corollary 5.4.2. *The structure $(\Sigma^0_1; \leq^\omega_T)$ is effectively broad, in particular it satisfies the sentence $\forall a(0 < a < 1 \rightarrow \exists b(a \not\leq b \wedge b \not\leq a))$.*

In [JLSS89] it was shown that the structure $(\Sigma^0_1; \leq^\omega_T)$ is dense.

Corollary 5.4.3. *For any $k > 0$ and any strong reducibility $\leq_r$ we have:*
(i) The structure $(\Sigma_k^{-1}; \leq_r)$ is effectively upward dense.
(ii) For any $n < \omega$ there exist $A_0, \ldots, A_n \in \Sigma_k^{-1}$ such that $A_i \not\leq_r \oplus_{l \neq i} A_l$ for all $i \leq n$.

For the case of c.e. m- and tt-degrees the assertions (i) and (ii) were proved in [De70] and [Kal73]. The last corollary generalizes these results to all levels of the difference hierarchy and to all strong reducibilities.

Corollary 5.4.4. *For any $n < \omega$, the structure $(K_n; \leq_{sm})$ is effectively dense and embeds any Δ_2^0-partial ordering.*

It seems that only few isolated facts are known about the following question: are the elementary theories of all structures discussed above undecidable?

Remark. All results from the corollaries are proved quite elementarily (without the priority argument), using the recursion theorem for precomplete numberings. It seems to be of some methodological interest because the direct proofs of some results should probably use a priority argument.

We conclude this section with an application of Theorem 4.3.3 to a fairly general result from [DN98] on complexity-theoretic reducibilities. Let $\leq_r^p$ denote a polynomial time reducibility in between (and possibly equal to) $\leq_m^p$ and $\leq_T^p$ (for undefined notions from complexity theory see any textbook on the subject). For any $h : \omega \to \omega$, let $DTIME(h)$ denote the set of languages computable in deterministic time $O(h)$.

Theorem 5.4.5. *Elementary theory of any structure $(DTIME(2^n); \leq_r^p)$ is undecidable.*

The paper [Ni00] contains a further strengthening of the last theorem.

5.5 Index Sets

Probably, the most popular topic in the investigaton of index sets is the estimation (of m-degrees) of index sets of concrete interesting classes and predicates. Here we discuss applications of results described above to this area.

First we consider some corollaries of results from Section 4.4. Please recall the examples of special structures from Theorem 5.4.1.

Let ν be an acceptable numbering of a level Σ_k^{-1} of the difference hierarchy, C be a Σ_k^{-1}-complete set, and $\leq_r$ be a strong reducibility. Interpreting Proposition 4.4.2 in this situation we get the following result from [Se87,Se88,Se89].

Corollary 5.5.1. *For all $k > 0$ and $A, A_0, \ldots, A_n \in \Sigma_k^{-1}$, we have:*
(i) if $\emptyset <_r A$ then $\{x | A \leq_r \nu x\}$ is Σ_3^0-complete;
(ii) if $A <_r C$ then $\{x | \nu x \leq_r A\}$ is Σ_3^0-complete;
(iii) if $\emptyset <_r A <_r C$ then $\{x | A \not\leq_r \nu x \wedge \nu x \not\leq_r A\}$ is Π_3^0-complete;

(iv) the set $\{x|\nu x = A_0 \vee \cdots \vee \nu x = A_n\}$ *is* Σ_3^0*-complete.*

Note that $\Sigma_1^{-1} = \Sigma_1^0$, so for $k = 1$ and $r = m$ we get the well known estimations for the index sets of c.e. m-degrees due to C. Yates and S. Kallibekov, see [Ka71]. Corollary 5.5.1 generalizes these results to all levels of the difference hierarchy and to all strong reducibilities. Of some methodological interest is the fact that the proof of these generalizations use only the recursion theorem (in the proof of Proposition 4.4.2) and some elementary observations (in the proof of Theorem 5.4.1). It does not use the priority method that seems inevitable in a direct proof of such results. Similar remarks are true for other results in this section below.

Applying Proposition 4.4.2 to the structure $(\Sigma_1^0; \leq_T^\omega)$ from the last section, we get the analog of Corollary 5.5.1 with W in place of ν and Σ_ω^0 (the class of sets c.e. in $\emptyset^\omega$) in place of Σ_3^0. A particular case of this is the following estimation from [So87] and [JLSS89] in which L (H) is the class of all c.e. sets A such that $A^n \equiv_T \emptyset^n$ (resp. $A^n \equiv_T \emptyset^{n+1}$) for some $n < \omega$.

Corollary 5.5.2. *The* W*-index sets of* L*,* H *and* $L \cup H$ *are* Σ_ω^0*-complete.*

To the structure $(\Sigma_k^{-1}; \leq_r)$ we may apply also Theorem 4.4.3. Then we get the following estimations which were probably not known before [Se89].

Corollary 5.5.3. *(i) If* $\emptyset <_r A <_r C$ *then* $\{x|A <_r \nu x\}$ *is* $\Sigma_2^{-1,\emptyset^2}$*-complete.*

(ii) If $\emptyset <_r A_0 <_r A_1 <_r A_2 < \cdots$ *then the sets* $\{x|A_0 \leq_r \nu x \wedge A_1 \not\leq_r \nu x\}$, $\{x|(A_0 \leq_r \nu x \wedge A_1 \not\leq_r \nu x) \vee A_2 \leq_r \nu x\}, \ldots$ *are m-complete respectively in* $\Sigma_2^{-1,\emptyset^2}, \Sigma_3^{-1,\emptyset^2}, \ldots$

Theorems 4.4.2 and 4.4.3 are applicable also to the structures $(K_n; \leq_m)$ and $(K_n; \leq_{sm})$ from the previous section. To the last structure we can also apply Theorem 4.4.4 and get the following result in which $\emptyset$ and C denote respectively the least and the greatest elements of $(K_n; \leq_{sm})$.

Corollary 5.5.4. *For all* $A, B \in K_n$ *we have:*

(i) If $\emptyset <_{sm} B <_{sm} C$ *then* $\{x|\nu x <_{sm} B\}$ *is* $\Sigma_2^{-1,\emptyset^2}$*-complete.*

(ii) If $\emptyset <_{sm} A$ *or* $B <_{sm} C$ *then* $\{x|A <_{sm} \nu x <_{sm} B\}$ *is* $\Sigma_2^{-1,\emptyset^2}$*-complete.*

Theorem 4.5.4 may be applied to the classification of some index sets in the structure $(\mathbf{R}_m; \leq)$ of c.e. m-degrees. The structure $\mathbf{R}_m$ has a natural numbering $\nu_n = deg(W_n)$. For simplicity we think that all recursive sets form a single m-degree. We classify index sets of some principal ideals $\hat{\mathbf{a}} = \{\mathbf{x}|\mathbf{x} \leq \mathbf{a}\}$ of $\mathbf{R}_m$.

Let $\mathcal{L}_0$ be the category of the Σ_3^0-distributive lattices with 0, and let $\mathcal{L}_0^1$ be the category of Σ_3^0-distributive lattices with 0 and 1. For an object $\mathbf{B}$ of $\mathcal{L}_0$ let

$\mathbf{B}^*$ be obtained from $\mathbf{B}$ by joining the greatest element. Note that from a given index of $\mathbf{B}$ in the acceptable numbering of $\mathcal{L}_0$ one can effectively compute an index of $\mathbf{B}^*$ in the acceptable numbering of $\mathcal{L}_0^1$. In [Se88a] it was noted that from the characterization of the principal ideals of $\mathbf{R}_m$ in [La72] it follows that from a given (index of) object $\mathbf{A}$ of $\mathcal{L}_0^1$ one can effectively compute a degree $\mathbf{a} \in \mathbf{R}_m$ such that $\hat{\mathbf{a}}$ and $\mathbf{A}$ are isomorphic as upper semilattices with 0 and 1.

The next result from [Se96] follows now from Theorem 4.5.4.

Corollary 5.5.5. *(i) The sets $\{n|\hat{\nu}_n \in LO\}$ and $\{n|\exists l \leq k(\hat{\nu}_n \simeq l+2)\}$ are Π_4^0-complete.*

(ii) The set $\{n|\hat{\nu}_n\, is\, finite\}$ is Σ_5^0-complete.

(iii) The set $C = \{n|\hat{\nu}_n \simeq \omega + 1\}$ is Π_6^0-complete.

(iv) The set $\{n|\hat{\nu}_n \in WO\}$ is Π_1^1-complete.

Many other facts on index sets of c.e. m-degrees may be deduced from results of Section 4.5. We formulate one such example taken from [Se91]. For a class $\mathcal{X}$ of BA's, let $s(\mathcal{X}) = \{n|(\hat{\nu}_n; \subseteq) \in \mathcal{X}\}$. Let $\mathcal{A}_k^n, \mathcal{B}_k^n, \mathcal{A}^n$ be the classes of BA's defined in Section 4.5.

Corollary 5.5.6. *For any $n > 0$, the index sets $s(\mathcal{A}_0^n), s(\mathcal{A}_{k+1}^n), s(\mathcal{B}_k^n), s(\mathcal{A}^n)$ are m-complete respectively in $\Sigma_{4n+3}^0, \Pi_{4n+4}^0, \Pi_{4n+5}^0, \Pi_{4n+6}^0$.*

Now consider index sets of some classes of hyperhypersimple sets. For a class $\mathcal{X}$ of BA's, let $h(\mathcal{X}) = \{n|L(W_n) \in \mathcal{X}\}$ and $h^*(\mathcal{X}) = \{n|L^*(W_n) \in \mathcal{X}\}$, where $L(W_n)$ is the lattice of c.e. supersets of W_n, and $L^*(W_n)$ is the quotient of $L(W_n)$ modulo finite sets. From Theorems 4.5.5, 4.5.6 and Lachlan's theorems on hyperhypersimple sets it is not hard to deduce the following results from [Se90].

Theorem 5.5.7. *(i) The index sets $h^*(\mathcal{A}_0^n), h^*(\mathcal{A}_{k+1}^n), h^*(\mathcal{B}_k^n), h^*(\mathcal{A}^n)$ are m-complete respectively in $\Sigma_{4n+3}^0, \Pi_{4n+4}^0, \Pi_{4n+5}^0, \Pi_{4n+6}^0$.*

(ii) For any finitely axiomatizable class $\mathcal{X}$ of BA's, the index set $h^*(\mathcal{X})$ is m-complete in one of levels of the long difference hierarchy.

In [Le89, Se90, Se91] one could find many other similar results for the index sets $h(\mathcal{X})$ and $h^*(\mathcal{X})$. We do not know the answer to the following interesting question: is the index set of any elementarily definable in $\mathcal{E}$ class of hyperhypersimple sets m-complete in a level of the long difference hierarchy (or the fine hierarchy)? From Fact 2 on page 151 of [HN98] it follows that this class of index sets is much bigger that the class from statement (ii) of the last theorem (e.g., it contains $h^*(\mathcal{X})$ for the classes of BA's defined as in Section 4.5 with Frechet derivative or the I-derivative in place of the Tarski-Ershov derivative).

We conclude this section by citing a result from [Se92, Se96a] giving a complete classification of index sets of definable predicates for a natural positive structure. For a numbered structure $\mathbf{A} = (A; \alpha)$ and a predicate $P(v_1, \dots, v_n)$ on A, let $\alpha^{-1}(P) = \{\langle x_0, \dots, x_k \rangle | P(\alpha x_0, \dots, \alpha x_k)\}$ be the α-index set of P. *A definable index set* of $\mathbf{A}$ is index set $\alpha^{-1}(P)$ of a predicate P elementarily definable in $\mathbf{A}$.

Theorem 5.5.8. *Any definable index set in the universal positive BA is m-complete in one of levels Σ_α, Π_α, $\Delta_{\lambda+1}$ ($\alpha, \lambda < \varepsilon_0$, λ is limit) of the fine hierarchy, and all the possibilities are realized. There is an algorithm which computes from a formula defining predicate P the level of the fine hierarchy where $\alpha^{-1}(P)$ is m-complete.*

The universal positive BA is thus an example of an *easy* structure (m-degrees of its definbale index sets are completely understood and effectively computable). There are also well known examples of *hard* structures, in which the standard model of arithmetic is interpretable without parameters (the structures of m-degrees of their definable index sets are close to the arithmetical m-defrees, see [Se96a] for more details). But there are also natural structures outside these two classes. As follows from results of [HN98], the lattice $\mathcal{E}$ of c.e. sets is such a structure.

We do not know whether any definable index set of the structure $\mathcal{E}$ is m-complete in one of levels of the fine hierarchy. Undecidability of $Th(\mathcal{E})$ implies that, in the case of positive answer to this question (which seems not very plausible), the level cannot be effectively computed. In [Se91] it was shown that for any level of the fine hierarchy one can find a definable predicate whose index set is complete in this level.

6. Conclusion

We hope that this survey may convince the reader that the theory of positive structures is already a rather matured discipline with some general theory, deep results on concrete classes of structures and with many interesting applications. It has its own flavour (as compared say with the well-developed theory of constructive structures). At the same time, the theory is still in its beginning and there are a lot of interesting questions for future work.

Acknowledgement. I am grateful to Yu. L. Ershov and S. S. Goncharov for useful remarks and bibliographical hints.

References

[Ar89] M. M. Arslanov. Completeness in the arithmetical hierarchy and fixed points. *Algebra i Logika*, 28 (1989), 3—18 (in Russian, English translation: *Algebra and Logic*, 28 (1989), p.1—9.).

[BT80] J. A. Bergstra and J. V. Tucker. A characterization of computable data types by means of a finite, equational specification method. Lec. Notes Comp. Sci., v. 85 (1980), 76—90.

[Ba74] W. Baur. Rekursive Algebren mit Kettenbedingungen. *Z. Math. Logik und Grundl. Math.*, 20, No 1 (1974), 37—46.

[BD70] I. Bucur and A. Deleany. *Introduction to the Theory of Categories and Functors.* Wiley, London, 1970.

[CN99] D. Cenzer and A. Nies. Initial segments of the lattice of Π^0_1-classes (to appear).

[CR98] D. Cenzer and J. B. Remmel. Π^0_1-classes in mathematics. In [EGNR98], v. 2, 623—822.

[CK73] C. C. Chang and H. J. Keisler. *Model Theory.* North Holland, Amsterdam, 1973.

[Co81] P. M. Cohn. *Universal Algebra.* Reidel, Dordrecht, 1981.

[De70] S. D. Denisov. On m-degrees of recursively enumerable sets. *Algebra i Logika*, 9 (1970), p.422—427 (in Russian, there is an English translation).

[Dob98] V. P. Dobritsa. Computable classes of constructive models, in [EGNR98], v. 1, 183—234.

[Dow98] R. G. Downey. Computability theory and linear orderings, in [EGNR98], v. 2, 823—976.

[DN98] R. G. Downey and A. Nies. Undecidability results for low-level complexity degree structures. *J. Comput.Systems Sci.*, 60 (2000).

[Dv84] S. G. Dvornikov. The lattice of degrees of separability. *Abstracts of Seventh all-Union Conf. in Math. Log.*, Novosibirsk (1984), p.55 (in Russian).

[EM85] H. Ehrig and B. Mahr. *Fundamentals of Algebraic Specification 1.* Springer, Berlin, 1985.

[EM90] H. Ehrig and B. Mahr. *Fundamentals of Algebraic Specification 2.* Springer, Berlin, 1990.

[Er68] Yu. L. Ershov. Numbered fields. *Proc. of 3-d Int. Congress for LMPS*, North Holland, Amsterdam, 1968, 31—34.

[Er69] Yu. L. Ershov. Completely numbered sets. *Sib. Mat. Zhurn.*, 10 (1969), 1048—1064 (in Russian, there is an English translation).

[Er77] Yu. L. Ershov. *Theory of numberings.* Nauka, Moscow, 1977 (in Russian).

[Er80] Yu. L. Ershov. *Decidability Problems and Constructive Models.* Nauka, Moscow, 1980 (in Russian).

[EGNR98] Yu. L. Ershov, S. S. Goncharov, A. Nerode and J. B. Remmel, eds. *Handbook of Recursive Mathematics*, v. 1 and 2, Elsevier, Amsterdam, 1998.

[EG99] Yu. L. Ershov, S. S. Goncharov. *Constructive Models.* Plenum, New York, 1999.

[FS56] A. Fröhlich and J. C. Shepherdson. Effective procedures in field theories. *Philos. Trans. London Roy. Soc.*, 248, No 950 (1956), 407—432.

[Go83] S. S. Goncharov. Universal recusively enumerable boolean algebras. *Sib. Math. J.*, 24, No 6 (1983), 36—43 (in Russian, there is an English translation).

[Go85] S. S. Goncharov. Models of data and languages for their description. *Computing Systems*, v. 107 (1985), 52—70 (in Russian).

[Go96] S. S. Goncharov. *Countable Boolean Algebras and Decidability.* Plenum, New York, 1996.

[Gor99] V. A. Gorbunov. *Algebraic Theory of Quasivarieties.* Plenum, New York, 1999.

[Gr71] G. Grätzer. *Lattice theory.* Freeman, San Francisco, California, 1971.

[Ha93] T. Hammond. Nonisimorphism of lattices of recursively enumerable sets. *J. Symbol. Log.*, 58, No 4 (1993), 1177—1189.

[HN98] L. Harrington and A. Nies. Coding in the partial order of enumerable sets. *Advances in Math.*, 133, No 1 (1998), 133—162.

[He84] E. Herrmann. The undecidability of the elementary theory of the lattice of recursively enumerable sets. *Proc. 2-nd Frege Conf. at Schwerin, GDR*, 20 (1984), 66—72.

[JLSS89] C.G. Jockusch jr., M. Lerman, R. I. Soare and R. M. Solovay. Recursively enumerable sets modulo iterated jumps and extensions of Arslanov's completeness criterion. *J. Symbol. Logic*, 54 (1989), p.1288—1323.

[Kal71] S. Kallibekov. On index sets of m-degrees. *Sib. Mat. Zhurn.*, 12 (1971), p.1292–1300 (in Russian, there is an English translation).

[Kal73] S. Kallibekov. On tt-degrees of recursively enumerable sets. *Mat. Zametki*, 14, No 5 (1973), p.421—426 (in Russian, there is an English translation).

[Kas89] N. Kh. Kasymov. Positive algebras with congruences of finite index. *Abstracts of Int. Conf. on Algebra (Theory of models and algebraic systems)*, Institute of Math., SD USSR, Novosibirsk, 1989, 57.

[Kas92] N. Kh. Kasymov. Positive algebras with Nöther congruence lattices. *Sib. Mat. Zhurn.*, 33, No 2 (1992),181—185 (in Russian, there is an English translation).

[Kas96] N. Kh. Kasymov. Recursively separable numbered algebras. *Uspechi Mat. Nauk*, 51 (1996), 145—176 (in Russian, there is an English translation).

[Kh88] N. G. Khisamiev. Arithmetical hierarchy of abelian groups. *Sib. Mat. Zhurn.*,29, No 6 (1988), 144—159 (in Russian, there is an English translation).

[Kh98] N. G. Khisamiev. Constructive abelian groups. In [EGNR98], v. 2, 1177—1232.

[KP67] S. Kripke and M. B. Pour-El. Deduction preserving recursive isomorphism between theories. *Fund. Math.*, 61 (1967), 141—163.

[Ku56] A. V. Kuznetsov. On equality problems and functional completeness for algebraic systems. *Trans. 3-d All-Union Math. Congress*, Moscow, 2 (1956), 145—146 (in Russian).

[Ku58] A. V. Kuznetsov. Algorithms as operations in algebraic systems. *Uspechi Mat. Nauk*, 13, No 3 (1958), 240—241 (in Russian).

[La72] A. H. Lachlan, Recursively enumerable many–one degrees, *Algebra i Logika*, 11, No 3 (1972), 326—358.

[La87] A. H. Lachlan. A note on positive equivalence relations. *Z. Math. Logik Grundl. Math.*, 33 (1987), 43—46.

[Le87] S. Lempp. Hyperarithmetical index sets in recursion theory. *Trans. Amer. Math. Soc.*, 303, No 2 (1987), 559—584.

[Ll84] J. M. Lloyd. *Foundations of Logic Programming*. Springer, Berlin, 1984.

[Ma61] A. I. Mal'cev. Constructive algebras, *Uspechi mat. nauk*, 16, No 3 (1961) 3—60 (in Russian, English translation in: The Metamathematics of Algebraic Systems (North Holand, Amsterdam) 1971, p. 148—214).

[Ma73] A. I. Mal'cev. *Algebraic Systems*. Springer, Berlin, 1973.

[Ma63] A. I. Mal'cev. Completely numbered sets. *Algebra i Logika*, 2 (1963), p.4—29 (in Russian).

[Ma70] A. I. Mal'cev. *Algorithms and recursive functions.* Groningen, Wolters–Noordhoff, 1970.

[MP70] D. Martin and M. B. Pour-El. Axiomatizable theories with few axiomatizable extensions. *J. Symbol. Logic*, 35 (1970), 205—209.

[McK45] J. G. G. McKinsey. The decision problem for some classes of sentences without quantifiers. *J. Symbol. Log.*, 8, No 3 (1945), 61—76.

[MN77] G. Metakides and A. Nerode, Recursively enumerable vector spaces. *Annals Math. Logic*, 11, No 2 (1977) 147—172.

[MS85] F. Montagna and A. Sorbi. Universal recursion–theoretic properties of c.e. preordered structures. *J. Symbol. Logic*, 50 (1985), 397—406.

[Mo52] A. Mostowski. On models of axiomatic systems. *Fund. Math.*, 39 (1952), 133—158.

[Mo53] A. Mostowski. On a system of axioms which has no recursively enumerable arithmetic model. *Fundam. Math.*, 40 (1953), 56—61.

[Ni97] A. Nies. Intervals of the lattice of computably enumerable sets and effective boolean algebras. *Bull. Lond. Math. Society*, 29 (1997) 683-692.

[Ni98] A. Nies. *Coding methods in computability theory and complexity theory*. Habilitation thesis, Heidelberg, 1998.

[Ni00] A. Nies. Effectively dense boolean algebras and their applications. *Trans. Amer. Math. Soc.*, 352, No 11 (2000), 4989-5012.

[Nu74] A. T. Nurtazin. *Computable classes and critetia of autostability*. PhD Thesis, Alma-Ata, 1974 (in Russian).

[O86] S. P. Odintsov. Lattices of recursively enumerable subalgebras of a recursive algebra. *Algebra i Logika*,25, No 6 (1986), 631—642 (in Russian, there is an English translation).

[O98] S. P. Odintsov. Generally constuctive boolean algebras. In [EGNR98], v. 2, 1319—1354.

[OS89] S. P. Odintsov and V. L. Selivanov. Arithmetical hierarchy and ideals of numbered boolean algebras. *Sib. Mat. Zhurn.*, 30, N 6 (1989), 140—149 (in Russian, there is an English translation).

[Pe00] M. G. Peretyat'kin. Finitely axiomatizable theories and Lindenbaum algebras of semantic classes. *Abstracts of AMS-SIAM Conf. on Computability and applications in Boudler*, (to appear in Contemporary Math., 2000).

[Ra60] M. O. Rabin, Computable algebra, general theory and theory of computable fields. *Trans. Amer. Math. Soc.*, 95, No 2 (1960) 341—360.

[Re78] J. B. Remmel. Recursively enumerable Boolean algebras. *Annals Math. Log.*, 15, No 1 (1978), 75—107.

[Re89] J. B. Remmel. Recursive Boolean algebras. In: *Handbook of Boolean algebras*, v. 3, J. D. Monk et.al. (eds.), North Holland, Amsterdam, 1989, 1097—1165.

[Ro67] H. Rogers jr. *Theory of recursive functions and effective computability*. New York, McGraw-Hill, 1967.

[Se82] V. L. Selivanov. On the structure of degrees of generalized index sets. *Algebra i Logika*, 21, N 4 (1982), 472–491 (in Russian, there is an English translation).

[Se83] V. L. Selivanov. Hierarchies of hyperarithmetical sets and functions, *Algebra i Logika*, 22, No 6 (1983) 666—692 (in Russian, English translation in: Algebra and Logic, 22, No 6 (1983) 473—491).

[Se87] V. L. Selivanov. Index sets of factor-objects of the Post numbering. In: Proc. of the Int. Conference in Foundations of Computation Theory in Kazan. *Lecture Notes in Computer Science*, v. 278. Berlin: Springer, 1987, 396–400.

[Se88] V. L. Selivanov. Index sets of factor–objects of the Post numbering, *Algebra i Logika*, 27, No 3 (1988) 343—358 (in Russian, English translation in: Algebra and Logic, 27, No 3 (1988) 215—224).

[Se88a] V. L. Selivanov. On algorithmic complexity of algebraic systems. *Mat. Zametki*, 44, No 6 (1988) 823—832 (in Russian, English translation in: Math. Notes, 44, No 5—6 (1988) 944—950).

[Se89] V. L. Selivanov. Applications of precomplete numberings to tt-type degrees and to index sets. *Algebra i Logika* Logic, 12, N 1 (1989), 165—185 (in Russian, there is an English translation).

[Se89a] V. L. Selivanov. *Hierarchical classification of arithmetical sets and index sets*. Habilitation thesis, Institute of Mathematics SD RAS, Novosibirsk, 1989, 170 p. (in Russian).

[Se90] V. L. Selivanov. Index sets of classes of hyperhypersimple sets. *Algebra i Logika*, 29, N 2 (1990), 220–240 (in Russian, there is an English translation).

[Se91] V. L. Selivanov. Fine hierarchies and definable index sets, *Algebra i Logika*, 30, No 6 (1991) 705—725 (in Russian, English translation in: Algebra and Logic, 30, No 6 (1991) 463—475).

[Se91a] V. L. Selivanov. Universal Boolean algebras with applications. *Abstracts of Int. Conf. in Algebra*, Novosibirsk, 1991, 127 (in Russian).

[Se92] V. L. Selivanov. Computing degrees of definable classes of sentences. In: Proc. of Int. Conference in Algebra in Honor of A. I. Malcev in Novosibirsk. Contemporary Mathematics, v. 131, part 3 (1992), 657–666.

[Se92a] V. L. Selivanov. *Hierarchies, Numerations, Index Sets*. Handwritten notes, 1992, 290pp.

[Se94] V. L. Selivanov. Precomplete numerations with applications, Preprint No 13, 1994, the University of Heidelberg, Chair of Mathematical Logic, 59 pp. (see Proceedings of Int. Conference in Honor of A. I. Maltsev, Novosibirsk, 1999 for a shorter version).

[Se95] V. L. Selivanov. Fine hierarchies and Boolean terms. *The Journal of Symbolic Logic*, 60, N 1 (1995), 289–317.

[Se96] V. L. Selivanov. On recursively enumerable structures. *Annals of Pure and Applied Logic*, 78 (1996), 243–258.

[Se96a] V. L. Selivanov. Fine hierarchy and definability in the Lindenbaum algebra. In: *Logic: from foundations to applications, Proceedings of the Logic Colloquium-93 in Keele*. Oxford, 1996, 425–452.

[So87] R. I. Soare. *Recursively Enumerable Sets and Degrees*. Springer, Berlin, 1987.

[T94] J. J. Thurber. Recursive and recursively enumerable quotient boolean algebras. *Arch. math. Logic*, 33 (1994), 121—129.

[Vi80] A. Visser. Numberings, λ-calculus and arithmetic. In: *To H.B.Curry: Essays on combinatory logic, λ-calculus and formalism*, New York, Acad. Press,1980, p.259—284.

LOCAL PROPERTIES OF THE NON-TOTAL ENUMERATION DEGREES

Boris Solon*
Ivanovo State University
Ivanovo
Russia 153038
solon@icti.ivanovo.su

Abstract The non-total e-degrees are examined as a special subclass of the enumeration degrees.

Keywords: Enumeration reducibility; enumeration degrees.

Some of the motivations for studying enumeration reducibility are described as follows by Cooper [Coo90]:

> In a computation using auxiliary informational inputs one can think of the external resource making itself available in different ways. One way is via an oracle as in Turing reducibility, where information is supplied on demand without any time delay. Alternatively, the Scott graph model for lambda calculus suggests a situation where new information, only some of it immediately related to the current computation, is constantly being generated (or enumerated) over a period of time in an order which is not under the control of the computer. For some purposes, such as in classifying the relative computability of total function without any time restriction, it makes no difference whether oracles or enumerations supply auxiliary informational inputs. But this is not generally the case in situations involving partially accessible information . . .

The former model of computation based on oracles leads to the notion of Turing reducibility between sets, and the latter model based on enumeration of the information leads to enumeration reducibility between sets. Turing reducibility and its related degree structure have been studied since 1936, and have a rich and extensive theory. Enumeration reducibility, which emerged in 1955, has been an active field of research in the last decade.

*This work was supported by INTAS-RFBR grant No. IR-97-139.

Before discussing this reducibility further, we need to review the notation and terminology used in this article. That used is generally similar to that of the monograph [Rog67].

Let ω denote the set of positive integers; $A, B, C, Q, \ldots$ (with or without indices) are variables for subsets of ω; D, F, E (with or without indices) will be used to denote finite subsets of ω. Let $\{W_n\}_{n\in\omega}$, $\{\phi_n\}_{n\in\omega}$ be Gödel numberings of all computably enumerable (c.e.) sets and all partial computable (p.c.) functions, respectively. Let $\{W_n^s\}_{s\in\omega}$ give finite computable approximations to the c.e. set W_n. As usual, D_u is the finite set with canonical index u; $\langle k, l\rangle$ is the Cantor number of the ordered pair (k, l). If t is the Cantor number of an ordered pair (k, l) then $\langle t\rangle_1 = k$ and $\langle t\rangle_2 = l$. Let $\langle x, D\rangle = \langle x, u\rangle$ where $D = D_u$. We denote by $|A|$ the cardinality of the set A. Sometimes we will write $|A| = \infty$ if A is an infinite set and $|A| < \infty$ if A is a finite set. As usual, let $K = \{x : x \in W_x\}$ and $\overline{K} = \omega \setminus K$.

Given a partial function $\alpha : \omega \to \omega$ let $\delta\alpha$ and $\rho\alpha$ be the domain and the range of α, respectively, and let $\tau\alpha = \{\langle x, \alpha(x)\rangle : x \in \delta\alpha\}$ be the graph of α. Sometimes, instead of $x \in \delta\alpha$, we will write $\alpha(x)\downarrow$, and instead of $x \notin \delta\alpha$, we will write $\alpha(x)\uparrow$. We restrict the use of the symbols f, g to denote *total* functions, i.e. functions such that $\delta f = \delta g = \omega$. As usual ϕ_n is the partial computable (p.c.) function with index n. A set A is said to be *single-valued* if $A = \tau\alpha$ for some function α.

In order now to justify the formal definition of enumeration reducibility we make more precise the intuition behind it. Let a procedure be given by which, intuitively, B is enumeration reducible (or, e-reducible) to A (denote this fact by $B \leq_e A$), i.e. there is a uniform algorithm for obtaining the enumeration of the members of B from any given enumeration of A. Consider all finite sequences of distinct integers. Let us say that an integer x is *caused* by a sequence $\langle y_1, \ldots, y_k\rangle$ if the procedure lists x as an output after $y_1, \ldots, y_k$ are given in order as inputs, but before any other input is requested. By examining all finite sequences, we can effectively enumerate the set

$$\{\langle x, u\rangle : D_u \text{ consists of the members of a finite sequence of distinct integers that cause } x\}.$$

Let this set be W_z. As the set B is listed regardless of the order in which A is presented, for every $x \in \omega$ we have that

$$x \in B \iff \exists u(\langle x, u\rangle \in W_z \wedge D_u \subseteq A).$$

Conversely, for any z it is possible to organize a procedure for given sets A and B which intuitively enumeration reduces B to A. For this we simultaneously begin to enumerate W_z and request inputs. If we find that an $\langle x, u\rangle$ has already appeared in the set W_z, for which D_u is contained in the inputs already obtained then we add the integer x to the output list. It is obvious that this procedure yields the same set of outputs (up to their order), regardless of the order in which the set of inputs is presented. Thus any z and any A generate a unique set B such that $B \leq_e A$ via z, namely,

$$B = \{x : \exists u(\langle x, u\rangle \in W_z \wedge D_u \subseteq A)\}.$$

Hence each $z \in \omega$ determines a total mapping $\Phi_z : 2^\omega \to 2^\omega$ given by

$$\Phi_z(A) = \{x : \exists u(\langle x, u\rangle \in W_z \wedge D_u \subseteq A)\},$$

which is called the *enumeration operator with index z*.

The concept of an enumeration operator, formally defined in [FR59], substantially coincides with the concept of a one-place computable operation in the sense of Uspensky [Usp55]. It is possible to give the general definition of a computable operation in the following form:

Definition 1. *Let $2^{<\omega}$ be the set of all finite subsets of ω. The mapping $\psi : 2^{<\omega} \to 2^\omega$ is called computable if the set $\{\langle u, x\rangle : x \in \psi(D_u)\}$ is computably enumerable. The mapping $\psi : 2^\omega \to 2^\omega$ is called a computable operation if its restriction to $2^{<\omega}$ is computable and it is monotone (i.e. $A \subseteq B \to \psi(A) \subseteq \psi(B)$) and continuous (i.e. $x \in \psi(A) \to \exists D(|D| < \infty \wedge D \subset A \wedge x \in \psi(D))$).*

As usual, let

$$A \equiv_e B \iff A \leq_e B \wedge B \leq_e A,$$

and let $d_e(A) = \{B : B \equiv_e A\}$ be the *e-degree* of the set A, and let

$$d_e(A) \leq d_e(B) \iff A \leq_e B$$

be the partial ordering of e-degrees. If two e-degrees **a** and **b** are incomparable with respect to $\leq$ then we denote this fact by $\mathbf{a} \mid \mathbf{b}$; we also write $\mathbf{a} < \mathbf{b}$ if $\mathbf{a} \leq \mathbf{b}$ and $\mathbf{b} \not\leq \mathbf{a}$. For functions α, β we will write $\alpha \leq_e A$ or $\alpha \leq_e \beta$ if $\tau\alpha \leq_e A$ or $\tau\alpha \leq_e \tau\beta$, respectively. We denote by $\mathbf{D}_e$ the set of e-degrees, partially ordered by the relation $\leq$; we also let $\mathbf{D}_e(\leq\mathbf{a})=\{\mathbf{x}\in\mathbf{D}_e\colon \mathbf{x}\leq\mathbf{a}\}$, $[\mathbf{a},\mathbf{b}]_e=\{\mathbf{x}\in\mathbf{D}_e\colon \mathbf{a}\leq\mathbf{x}\leq\mathbf{b}\}$, $(\mathbf{a},\mathbf{b})_e=\{\mathbf{x}\in\mathbf{D}_e\colon \mathbf{a}<\mathbf{x}<\mathbf{b}\}$. It is well known (see [Cas71]) that $\mathbf{D}_e$ forms an upper semilattice, which is not a lattice, with least element $\mathbf{0}_e$, given by $\mathbf{0}_e=\{W_n : n \in \omega\}$. If $\mathbf{a}=d_e(A)$ and $\mathbf{b}=d_e(B)$, then the supremum of the e-degrees **a** and **b** is equal to $\mathbf{a}\cup\mathbf{b}=d_e(A \oplus B)$ where $A \oplus B = \{2x : x \in A\} \cup \{2x+1 : x \in B\}$. We denote by $\mathbf{a}\cap\mathbf{b}$ the infimum of the e-degrees **a** and **b**. As was shown in [Cas71], there are e-degrees **a** and **b** for which there is no infimum $\mathbf{a}\cap\mathbf{b}$ in $\mathbf{D}_e$. So the binary operation $\cap$ is partial.

We denote by $\mathbf{D}_T$ the upper semilattice of Turing degrees (or, simply, *degrees*). As for all A, B,

$$A \leq_T B \iff c_A \leq_e c_B$$

there is an isomorphic embedding $\epsilon : \mathbf{D}_T \to \mathbf{D}_e$, $\epsilon(d_T(A)) = d_e(c_A)$, which preserves the supremum operation.

An e-degree **a** is said to be *total*, if it contains a total function. It is obvious that **a** is total iff there is $B \in \mathbf{a}$ such that $B \equiv_e c_B$. We denote by **Tot** the set of all total e-degrees ordered by $\leq$, i.e. $\mathbf{Tot} = \{d_e(\tau c_A) : A \subseteq \omega\}$. It is rather obvious that **Tot** is an upper subsemilattice of $\mathbf{D}_e$ by virtue of the fact that $d_e(\tau c_A) \cup d_e(\tau c_B) = d_e(\tau c_{A \oplus B})$. Thus $\epsilon \mathbf{D}_T = \mathbf{Tot}$ and **Tot** is isomorphic to $\mathbf{D}_T$.

We note that the embedding ϵ need not preserve infima of T-degrees. Indeed in [Roz81] it was shown that

$$(\forall \mathbf{i} \in \mathbf{D}_e)(\exists \mathbf{a}, \mathbf{b} \in \mathbf{Tot})(\mathbf{a}|\mathbf{b} \wedge \mathbf{i} = \mathbf{a} \cap \mathbf{b}).$$

Let now **i** be given. As was noted above, in this case there are sets $A \in \mathbf{a}$ and $B \in \mathbf{b}$ such that $c_A \in \mathbf{a}$ and $c_B \in \mathbf{b}$, then $\epsilon(d_T(A)) = \mathbf{a}$ and $\epsilon(d_T(B)) = \mathbf{b}$. If we suppose that **i** is a non-total e-degree, then $\mathbf{i} \neq \epsilon(\mathbf{x})$ for every T-degree **x**. So even if the infimum $d_T(A) \cap d_T(B) = d_T(C)$ exists, then its image under the embedding $\epsilon(d_T(C)) = d_e(\tau c_C)$ cannot be equal to **i**.

In $\mathbf{D}_T$ the *jump* of a set $A \subseteq \omega$ is a set A', which decides the halting problem for procedures which are computable relative to the set A: "Given any $n \in \omega$, does $n \in W_n^A$ or not?". In $\mathbf{D}_e$ the jump of A must answer questions of the form: "Given any $n \in \omega$, does $n \in \Phi_n(A)$ or not?" or "Given any $n, x \in \omega$, does $x \in \Phi_n(A)$ or not?". In other words the e-jump of the set A in $\mathbf{D}_e$ is the least set (with respect to $\leq_e$) to which the characteristic function of every set e-reducible to A is e-reducible, i.e. $\mathbf{J}(A)$ is the e-jump of A iff

$$\forall B((B \leq_e A \to c_B \leq_e J(A)) \wedge \forall C((B \leq_e A \to c_B \leq_e C)) \to J(A) \leq_e C).$$

A formal definition of the e-jump of A is given in [McE85]. Let $A^e = \{x : x \in \Phi_x(A)\}$, and define the *e-jump* of the set A to be the set

$$\mathbf{J}(A) = \tau c_A \equiv A^e \oplus \overline{A^e}.$$

Denote the e-jump of the e-degree $\mathbf{a} = d_e(A)$ by $\mathbf{a}' = (d_e(A))' = d_e(\mathbf{J}(A))$. (One notes that the jump of a T-degree in $\mathbf{D}_T$ is denoted by the same symbol $'$. This will be justified below, and the exact meaning of the symbol $'$ will be clear from the context.)

Let $\mathbf{D}_T^{c.e.}$ be the set of all c.e. T-degrees. Denote by $\mathbf{D}_e^{co-c.e.}$ the set of all co-c.e. e-degrees, i.e. $\mathbf{D}_e^{co-c.e.} = \{d_e(A) : \overline{A} \text{ c.e.}\}$. It is clear that $\mathbf{D}_T^{c.e.} \subseteq \mathbf{D}_T(\leq \mathbf{0}'_T)$, and, as was shown in [Coo84], $\epsilon(\mathbf{D}_T^{c.e.}) = \mathbf{D}_e^{co-c.e.} \subset \mathbf{D}_e(\leq \mathbf{0}'_e)$. Using terminology from the arithmetical hierarchy it is possible to say that for any e-degree **a**,

$$\mathbf{a} \in \mathbf{D}_e^{co-c.e.} \iff \mathbf{a} \in \Pi_1 \iff \mathbf{a} = \epsilon(\mathbf{b})$$

for some c.e. T-degree **b**.

Let $\mathbf{0}'_e = d_e(\overline{K}) = d_e(\mathbf{J}(\emptyset))$. It is well known that $\mathbf{a} \in \Sigma_2 \iff \mathbf{a} \leq \mathbf{0}'_e$ and $\mathbf{a} \in \Delta_2 \iff A \leq_T K$. It is easy to verify that $\mathbf{Tot}(\leq \mathbf{0}'_e) \subseteq \Delta_2 \subseteq \Sigma_2$, where $\mathbf{Tot}(\leq \mathbf{0}'_e)$ is the set of total e-degrees $\leq \mathbf{0}'_e$. In [CC88] it is shown that $\Sigma_2 \setminus \Delta_2 \neq \emptyset$. In fact, in [CC88] an e-degree $\leq \mathbf{0}'_e$ is constructed which is incomparable with all intermediate e-degrees in Δ_2. E-degrees belonging to the class $\Sigma_2 \setminus \Delta_2$ are called *properly* Σ_2 *e-degrees.* In [CC88] it is proved also that every high e-degree **h** bounds a properly Σ_2 e-degree which is incomparable with all e-degrees $\mathbf{a} \in \Delta_2$ with $\mathbf{0}_e < \mathbf{a} < \mathbf{h}$. In [BCS97] it is proved that for every e-degree $\mathbf{a} < \mathbf{0}'_e$ there is an e-degree **b** such that $\mathbf{a} \leq \mathbf{b} < \mathbf{0}'_e$ and all e-degrees **c** with $\mathbf{b} \leq \mathbf{c} < \mathbf{0}'_e$ are properly Σ_2 and so are non-total e-degrees. That the class of Δ_2 e-degrees extends the class of Π_1 e-degrees follows from Yates's result, [Yat65], giving the existence of a T-degree **b** belonging to Δ_2, which is incomparable with any intermediate c.e. T-degree, together with some properties of the isomorphic embedding ϵ. In other words, the e-degree $\mathbf{a} = \epsilon(\mathbf{b}) \in \Delta_2 \setminus \Pi_1$, and is incomparable with all Π^0_1 e-degrees apart from $\mathbf{0}_e$ and $\mathbf{0}'_e$. We note that such an e-degree **a** is total. In [Coo90] there was constructed a non-total (in fact, quasi-minimal) e-degree belonging to Δ_2.

In [McE85] it was shown that the embedding $\epsilon : \mathbf{D}_T \to \mathbf{D}_e$ is monotone relative to the jump of degrees. Thus the embedding ϵ realizes an embedding of the algebraic structure $\langle \mathbf{D}_T, \leq, \cup, ' \rangle$ (where $\leq$ is the partial ordering relation, $\cup$ is the binary supremum operation, and $'$ is the T-jump operation in $\mathbf{D}_T$) into the algebraic structure $\langle \mathbf{D}_e, \leq, \cup, ' \rangle$ where $\leq, \cup, '$ are the corresponding relation and operations in $\mathbf{D}_e$, and also $\epsilon(\mathbf{D}_T) = \mathbf{Tot}$.

The first serious question about e-reducibility was solved by Medvedev [Med55], who showed that $\mathbf{D}_e \setminus \mathbf{Tot} \neq \emptyset$. Medvedev constructed a partial function $\psi : \omega \to \omega$ which is not computable and for every total function f if $f \leq_e \psi$ then f is a computable function. Case, [Cas71], called Medvedev's e-degrees *quasi-minimal.* Note that if we consider e-reducibility on subsets of ω then it is reasonable to give the definition of a quasi-minimal set in the following form: a non-c.e. set A is *quasi-minimal* if $\forall f (f \leq_e A \to f \text{ is computable})$. From the definition it follows that if **a** is a quasi-minimal e-degree then it is non-total and $\mathbf{D}_e(\leq \mathbf{a}) \cap \mathbf{Tot} = \{\mathbf{0}_e\}$. We denote by **Q** the set of all quasi-minimal e-degrees.

The existence of quasi-minimal sets can be proved with the help of the embedding ϵ. Let **a** be a minimal T-degree, that is a T-degree satisfying

$$\mathbf{0}_T < \mathbf{a} \wedge (\forall \mathbf{x} \in \mathbf{D}_T)(\mathbf{x} < \mathbf{a} \to \mathbf{x} = \mathbf{0}_T).$$

The existence of minimal T-degrees was first proved in [Spe56]. We consider the e-degree $\mathbf{b} = \epsilon(\mathbf{a})$. It is clear that $\mathbf{b} \neq \mathbf{0}_e$. In [Gut71] it was shown that $\mathbf{D}_e$ does not have minimal elements so there is an e-degree $\mathbf{c} \neq \mathbf{0}_e$ such that $\mathbf{c} < \mathbf{b}$. If we assume that **c** is a total e-degree then there is a set $C \in \mathbf{c}$ such that $C \equiv_e c_C$. We consider $d_T(C)$. As $\epsilon(d_T(C)) = \mathbf{c}$ so $d_T(C) < \mathbf{a}$ and so $d_T(C) = \mathbf{0}_T$ (i.e. C is a computable set). Then **c**= $\epsilon(\mathbf{0}_T)$ =$\mathbf{0}_e$, which contradicts our premise. Thus **c** is a non-total e-degree. From the previous argument it follows that all e-degrees except $\mathbf{0}_e$ which are strictly smaller than **b**= $\epsilon(\mathbf{a})$ (the e-degree **b** is total) where **a** is a minimal T-degree, are non-total and moreover are quasi-minimal. In addition, using a Baire category argument one can be justified in concluding that there are quasi-minimal e-degrees which do not lie under images, via ϵ, of minimal T-degrees. In [Sac61] it was proved that there are minimal T-degrees below $\mathbf{0}'_T$ hence for every such minimal T-degree $\mathbf{a}<\mathbf{0}'_T$ there is a countable set of quasi-minimal e-degrees below $\epsilon(\mathbf{a})$ $<\mathbf{0}'_e$. Thus $\mathbf{Q}\cap\Sigma_2 \neq \emptyset$, moreover $|\mathbf{Q}\cap\Sigma_2| = \aleph_0$.

In [Joc68] it was introduced the notion of a semirecursive set: $A \subseteq \omega$ is *semirecursive* if there is a computable function $\lambda x.\lambda y f(x,y)$ for which

(i) $f(x,y) \cap \{x,y\} \neq \emptyset$,

(ii) $\{x,y\} \cap A \neq \emptyset \rightarrow f(x,y) \in A$ for all $x,y \in \omega$.

In the same place it was proved that every nonzero T-degree contains a semirecursive set A such that A and $\overline{A}$ are both not c.e. and $d_e(A)|d_e(\overline{A})$. In [ACK01] it was shown that in this case the e-degree $d_e(A)$ is quasi-minimal (and $d_e(\overline{A})$ is quasi-minimal too). As $\tau c_A \equiv_T A \equiv_T \overline{A}$, $d_e(\tau c_A) = d_e(A) \cup d_e(\overline{A})$ and $d_e(A) \cap d_e(\overline{A})$ =$\mathbf{0}_e$ (see also [ACK01]) we have that for every nonzero total e-degree one can embed a diamond into $\mathbf{D}(\leq\mathbf{a})$ the bottom of which is $\mathbf{0}_e$, the top is **a**, and the two remaining points are the quasi-minimal e-degrees $d_e(A)$ and $d_e(\overline{A})$. From here in particular we get the following result (first proved in [Sor90]):

Proposition 2. $\mathbf{D}_e(\leq\mathbf{a})\cap\mathbf{Q}\neq \emptyset$ *for any e-degree* $\mathbf{a}\neq\mathbf{0}_e$.

An e-degree **a** is called *splitting* if there are e-degrees **b** and **c** such that $\mathbf{0}_e < \mathbf{b} < \mathbf{a}$, $\mathbf{0}_e < \mathbf{c} < \mathbf{a}$ and $\mathbf{a}=\mathbf{b}\cup\mathbf{c}$. Thus any total e-degree is splitting in the quasi-minimal e-degrees. At the same time, by properties of the embedding ϵ and the fact there are T-degree that are not splitting, it is obvious that there are total e-degrees which are not splitting in the total e-degrees.

Nonzero e-degrees **a** and **b** form a *minimal pair* if $\mathbf{a}\cap\mathbf{b}=\mathbf{0}_e$. From previously mentioned results it follows that any total nonzero e-degree bounds a minimal pair which is formed by quasi-minimal e-degrees.

As we can see from the above mentioned results all natural classes of e-degrees contain both total and non-total degrees. The class $\Sigma_2 \setminus \Delta_2$ of properly Σ_2 e-degrees consists of non-total e-degrees only, moreover as we can see from [CC88], among properly Σ_2 e-degrees there exist quasi-minimal e-degrees. In this paper we study local properties of non-total e-degrees.

We deal first with a characteristic property of non-total e-degrees. We denote by **Tot** the set $\mathbf{Tot}_A = \{d_e(\tau f) : \rho f = A\}$ where $|A| > 1$. The following lemma is of independent interest.

Lemma 3. *If* $A \leq_e B$ *and* $d_e(B) \in\mathbf{Tot}$ *then* $d_e(B) \in\mathbf{Tot}_A$.

Proof. Let g be a total function such that $\tau g \equiv_e B$. As $A \leq_e B$ there is an e-operator Φ such that $A = \Phi(\tau g)$. If we give as inputs to Φ the elements of the set τg in the standard order $\langle 0, g(0)\rangle, \langle 1, g(1)\rangle, \ldots$ then as output we will obtain the elements of the set A in some (fixed) order. We define:

$h(0) =$ [the first element of A in the output of Φ];
$h(1) =$ [the second element of A in the output of Φ];
$\ldots$;
$h(n) =$ [$n+1$-th element of A in the output of Φ];
$\ldots$.

It is clear that $h \leq_e g$ and $\rho h = A$. By the assumption that $|A| \geq 2$, we may suppose that $h(0) \neq h(1)$.

We define the function f:

$$f(0) = h(0), \ldots, f(g(0)+1) = h(0),\ f(g(0)+2) = h(1).$$

Thus the value of $g(0)$ is coded with the help of the function f:

$$g(0) = \mu_x[f(x) = h(1)] - 2.$$

We continue our definition of the function f. Let $x_1 = g(0)+3$, set

$$f(x_1) = h(0), \ldots, f(x_1 + g(1) + 1) = h(0),\ f(x_1 + g(1) + 2) = h(1).$$

Thus the value of $g(1)$ is coded with the help of the function f:

$$g(1) = \mu_{x > x_1}[f(x) = h(1)] - (x_1 + 2).$$

If $h(2) \notin \{h(0), h(1)\}$ then define $f(x_1 + g(1) + 3) = h(2)$ and set $x_2 = x_1 + g(1) + 4$. Otherwise, set $x_2 = x_1 + g(1) + 3$. Next we define

$$f(x_2) = h(0), \ldots, f(x_2 + g(2) + 1) = h(0),\ f(x_2 + g(2) + 2) = h(1).$$

If $h(3) \notin \{h(0), h(1)\}$ then define $f(x_2 + g(2) + 3) = h(3)$ and set $x_3 = x_2 + g(2) + 4$. Otherwise, set $x_3 = x_2 + g(2) + 3$ and so on.

From the definition we can see that

(i) $d_e(\tau f) \in \mathbf{Tot}_A$.

(ii) $f \leq_e g$ as $h \leq_e g$ and the construction is computable in an enumeration of the set τg.

(iii) $g \leq_e f$ as was proved in the course of the definition of the function f.

Thus, $d_e(B) \in \mathbf{Tot}_A$, as was to be proved. □

Define $\mathbf{Tot}_A(B) = \{d_e(\tau f) : \rho f = A \wedge B \leq_e f\}$. Obviously, for all sets A with $|A| > 1$, and all B,

$$\mathbf{Tot}_A(B) \subseteq \mathbf{Tot}_A,$$

and if $B \leq_e A$ then $\mathbf{Tot}_A(B) = \mathbf{Tot}_A$.

Theorem 4. *For any sets* A, $|A| > 1$, *and* B

$$B \leq_e A \iff \mathbf{Tot}_A \subseteq \mathbf{Tot}_A(B).$$

Proof. Suppose that $\mathbf{Tot}_A \subseteq \mathbf{Tot}_A(B)$ and let $A \leq_e f$. Then by Lemma 3 there is g such that $g \equiv_e f$ and $\rho g = A$. Then by the assumption $g \in \mathbf{Tot}_A(B)$. Hence $B \leq_e g$, and thus $B \leq_e f$. By a result in [Roz78] it follows that $A \leq_e B$. □

Corollary 5. $\mathbf{D}_e(>\mathbf{a})$ *does not have a least element for any e-degree* $\mathbf{a}$.

Proof. If $\mathbf{a}$ is a total e-degree then, as was shown in [Gut71], we have that

$$\forall \mathbf{b}[\mathbf{a} < \mathbf{b} \rightarrow \exists \mathbf{c}[\mathbf{a} < \mathbf{c} < \mathbf{b}]]. \tag{1}$$

If we assume that $\mathbf{D}_e(>\mathbf{a})$ has a least element $\mathbf{b}$ then

$$\mathbf{a} < \mathbf{b} \wedge \forall \mathbf{c}[\mathbf{a} < \mathbf{c} \rightarrow \mathbf{b} \leq \mathbf{c}]. \tag{2}$$

In this case, as follows from (1), for the e-degree $\mathbf{b}$ there is an e-degree $\mathbf{c}_0$ such that $\mathbf{a} < \mathbf{c}_0 < \mathbf{b}$, which contradicts the assumption (2). Thus if $\mathbf{a}$ is a total e-degree then $\mathbf{D}_e(>\mathbf{a})$ does not have a least element.

Let $\mathbf{a}$ be a non-total e-degree, and let us assume that $\mathbf{b}$ is the least element of $\mathbf{D}_e(>\mathbf{a})$. In this case we have

$$\mathbf{D}_e(\geq \mathbf{a}) = \{\mathbf{a}\} \cup \mathbf{D}_e(\geq \mathbf{b}).$$

As $\mathbf{a}$ is a non-total e-degree then by Lemma 3 we have

$$\mathbf{Tot}_B = \mathbf{Tot}_B(A),$$

and by Theorem 4 in this case we have that $B \equiv_e A$, which contradicts our assumption. Thus, for any e-degree $\mathbf{a}$ the cone $\mathbf{D}_e(>\mathbf{a})$ does not have a least element. □

Corollary 6. *An e-degree* $\mathbf{a} = d_e(A)$ *is non-total iff* $\mathbf{Tot}_A$ *does not have a least element.*

Proof. $\Rightarrow$: Let $\mathbf{a} = d_e(A)$ be a non-total e-degree, i.e. for any total function f such that $\rho f = A$, $f \not\leq_e A$. We assume that there is a total function g such that $\rho g = A$ and $d_e(\tau g)$ is the least element of $\mathbf{Tot}_A$. This means that

$$\forall f(d_e(\tau f) \in \mathbf{Tot}_A \rightarrow g \leq_e f).$$

In other words, from any enumeration of the set A one can effectively get an enumeration of A in some fixed order. In this case, as was shown in [Cas71], the e-degree $d_e(A)$ is total and we obtain the contradiction. Thus, $\mathbf{Tot}_A$ does not have a least element.

$\Leftarrow$: Let $\mathbf{a} = d_e(A)$ be a total e-degree. Then from any enumeration of A one can effectively get an enumeration of A in a fixed order $A = \{a_n : n \in \omega\}$. Consider the total function $g(n) = a_n$, for all $n \in \omega$. It is clear that $d_e(\tau g) \in \mathbf{Tot}_A$ and $\tau g \equiv_e A \leq_e f$ for any total function f with $\rho f = A$. This means that $d_e(\tau g)$ is the least element of $\mathbf{Tot}_A$. □

The existence of non-total e-degrees which are not quasi-minimal was first proved in [Sas75]. To this end, the notion of a quasi-minimal e-degree was relativized to the notion of quasi-minimality relative to an e-degree $\mathbf{c}$=$d_e(C)$. A set A is called *C-quasi-minimal*, (and the e-degree $\mathbf{a} = d_e(A)$ is called *$\mathbf{c}$-quasi-minimal*) if $C <_e A$ and for any total f

$$f \leq_e A \rightarrow f \leq_e C.$$

We denote by $\mathbf{Q_c}$ the set of $\mathbf{c}$-quasi-minimal e-degrees. It is clear that $\mathbf{Q}_{\mathbf{0}_e} = \mathbf{Q}$. In [Sas75] it was proved that $\mathbf{Q_c} \neq \emptyset$ for every e-degree $\mathbf{c}$. In [Sol99] it was shown that for any e-degree $\mathbf{c}$, $\mathbf{Q_c}$ contains an uncountable antichain and an uncountable ascending chain.

In the following theorem it is shown that the partially ordered set $\mathbf{Q_c}(< \mathbf{a})$, consisting of the $\mathbf{c}$-quasi-minimal e-degrees smaller than a total e-degree $\mathbf{a}$>$\mathbf{c}$, has a sufficiently "rich" structure.

Theorem 7. *If* $\mathbf{c}$<$\mathbf{a}$ *and* $\mathbf{a}$ *is a total e-degree then in* $\mathbf{Q_c}(<\mathbf{a})$ *one can isomorphically embed any partially ordered set whose cardinality is at most countable.*

Proof. Let $\mathbf{a}$ and $\mathbf{c}$ be e-degrees satisfying the condition of the theorem, i.e. $\mathbf{c}$<$\mathbf{a}$ and $\mathbf{a}$ is a total e-degree and let $C \in \mathbf{c}$, $A \in \mathbf{a}$, with $\overline{A} \leq_e A$. We denote by $\Phi_l^s(M)$ the set $\Phi_l^s(M) = \{x : \exists u(\langle x, u\rangle \in W_l^s \wedge D_u \subseteq M)\}$. Let $C = \Psi(A)$ for some e-operator Ψ, and let $C^s = \Psi^s(A)$ for every s. In the proof we will exploit the idea of a *counter* from [Ers77].

We define for any $z, k, l \in \omega$ and any computable set $R \subseteq \omega$ the functions

$$\begin{aligned}\lambda s.f_{k,l}(z,R,s) = \max\{y : y \geq z \wedge \forall y'[z \leq y' < y \rightarrow \\ [y' \in A \iff \langle k,l,y'\rangle \in \Phi_l^s(C^s \oplus R)]]\}\end{aligned}$$

and

$$\lambda s.F_{k,l}(z,R,s) = \max\{f_{k,l}(z,R,s') : s' \leq s\}.$$

It is obvious that $\lambda s.f_{k,l}(z,R,s)$ and $\lambda s.F_{k,l}(z,R,s)$ are total functions which are computable relatively to the set A, and $\overline{A} \leq_e A$ implies that $f \leq_e A$ and $F \leq_e A$. As the function $\lambda s.F_{k,l}(z,R,s)$ is monotonically nondecreasing then there exists (finite or infinite) the limit, $\lim_s F_{k,l}(z,R,s) = F_{k,l}(z,R)$.

We now prove some lemmas from which Theorem 7 will follow.

Lemma 8. *For any $k, l, z \in \omega$ and any computable set $R \subseteq \omega$, $F_{k,l}(z, R) < \infty$.*

Proof of Lemma 8. As $A \not\leq_e C$ there is $y_0 \geq z$ such that

$$\neg[y_0 \in A \iff \langle k, l, y_0\rangle \in \Phi_l(C \oplus R)].$$

We can find a big enough $s_0 \in \omega$ such that for all $s \geq s_0$ we have

$$\langle k, l, y_0\rangle \in \Phi_l^s(C^s \oplus R) \iff \langle k, l, y_0\rangle \in \Phi_l(C \oplus R);$$

but then $f_{k,l}(z, R, s) \leq y_0 + 1$ for all $s \geq s_0$ and hence

$$F_{k,l}(z, R, s) \leq \max\{F_{k,l}(z, R, s_0), y_0 + 1\},$$

which proves the lemma. □

We now introduce notation which we will use in the proof of the theorem. For any $M \subseteq \omega$ and $k \in \omega$ we write: $M^k = \{x : x \in M \wedge \exists l, y[\langle k, l, y\rangle = x]\}$ and $\tilde{M}^k = \{x : x \in M \wedge \forall l, y[\langle k, l, y\rangle \neq x]\} = M \setminus M^k$.

Lemma 9. *There is a set $B \leq_e A$ such that $C \oplus B$ is C-quasi-minimal and $B^k \not\leq_e C \oplus \tilde{B}^k$ for all $k \in \omega$.*

Proof of Lemma 9. We construct the set B via a construction which is computable in A. In the course of the construction at every step we will place the symbol "+" near finitely many integers in ω; in addition the set B will consist of the integers near which we have placed the symbol "+" at some step of the construction. We denote by B_+ the set of integers which have "+" at a given step of the construction. The construction is organized in such a way that $|B_+| < \infty$, as the step increases the set B_+ increases under inclusion, and $\lim_s B_+ = B$.

Given a pair of integers (k, l) in the construction we will use two movable $\langle k, l\rangle$-labels: a *lower label* $b(k, l)$ and an *upper label* $h(k, l)$. Both $\langle k, l\rangle$-labels can move only along their *path* $T_{\langle k,l\rangle} = \{\langle k, l, y\rangle : y \in \omega\}$. The integers associated to the lower $\langle k, l\rangle$-label and to the upper $\langle k, l\rangle$-label at a given point of the construction will be denoted by $\langle k, l, b(k, l)\rangle$ and $\langle k, l, h(k, l)\rangle$, respectively. In the course of the construction labels may move higher along their path $T_{\langle k,l\rangle}$, i.e. the movable values of $b(k, l)$ and $h(k, l)$ do not decrease, and moreover $b(k, l) \leq h(k, l)$ at all steps of the construction. (Note that the labels and their associated integers are denoted by the same symbols. Whether these symbols denote labels or their associated integers will be clear from the context.) The integer $\langle k, l, y\rangle$ is called *free* at a given point of the construction if $\langle k, l, y'\rangle \notin B_+$ for all $y' \geq y$. Let

$$\tilde{T}_{\langle k,l\rangle} = \{\langle k', l', y\rangle : k' \neq k \wedge \langle k, l\rangle < \langle k', l'\rangle\}$$

and

$$\tilde{T}^*_{\langle k,l\rangle} = \{\langle k', l', y\rangle : \langle k, l\rangle < \langle k', l'\rangle\}.$$

It is clear that $\tilde{T}_{\langle k,l\rangle} \subseteq \tilde{T}^*_{\langle k,l\rangle}$ for all $k, l \in \omega$.

We now move on to the description of the construction of the set B. At the beginning of the construction let us put $B_+ = \emptyset$, and $b(k,l) = h(k,l) = 0$ for all $k, l \in \omega$.

Step 2s. This step is divided into two stages. At each of these two stages, we perform, one after another, substeps t for $t = 0, 1, \ldots, s$. The substeps relative to the first stage are called *p-substeps*, and the ones relative to the second stage are called *n-substeps*. For $s = 0$ we do not go through any p-substeps.

p-substep t. Let $t = \langle k, l\rangle$. Check whether the following condition is satisfied

$$F_{k,l}(b(k,l), \tilde{B}^k_+ \cup \tilde{T}_{\langle k,l\rangle}, s) > F_{k,l}(b(k,l), \tilde{B}^k_+ \cup \tilde{T}_{\langle k,l\rangle}, s-1). \qquad (1)$$

If (1) is true then for all k', l' such that $\langle k', l'\rangle > \langle k, l\rangle$ we set both $\langle k', l'\rangle$-labels near the least element belonging to the set of free integers of their paths $T_{\langle k',l'\rangle}$, and not lying below $h(k', l')$; the label $h(k,l)$ is set near the integer $\langle k, l, F_{k,l}(b(k,l), \tilde{B}^k_+ \cup \tilde{T}_{\langle k,l\rangle}, s)\rangle$. If (1) is not true then we move to the next step.

After the end of the p-substep s we move on to the second stage, and perform one after another the n-substeps t for $t = 0, 1, \ldots, s$.

n-substep t. Let $t = \langle k, l\rangle$. We do the following operations one after another:

$(n.t.1)$ For each integer $y \in A$ such that $b(k,l) \leq y \leq h(k,l)$ and $\langle k, l, y\rangle \notin B_+$ we set the sign "+" near the integer $\langle k, l, y\rangle$ (and shall say that the integer y *puts* "+" near the integer $\langle k, l, y\rangle$).

$(n.t.2)$ For all y such that $b(k,l) \leq y \leq h(k,l)$, we check whether the following condition is satisfied:

$$\langle k, l, y\rangle \in \Phi^s_l(C^s \oplus (\tilde{B}^k_+ \cup \tilde{T}_{\langle k,l\rangle})) \setminus \Phi^s_l(C^s \oplus \tilde{B}^k_+).$$

For any such y which satisfies this condition we find the least integer $\langle\langle k, l, y\rangle, u\rangle \in W^s_l$ such that $D_u \subseteq C^s \oplus (\tilde{B}^k_+ \cup \tilde{T}_{\langle k,l\rangle})$, we set "+" near all integers from the set $\{x : 2x+1 \in D_u\}$ which do not have this sign (and we shall say that the integer y $\langle k, l\rangle$-*induces* the sign "+" near the corresponding integers). Next, for each k', l' such that $\langle k', l'\rangle > \langle k, l\rangle$ we set both $\langle k', l'\rangle$-labels near the least free integer from the path $T_{\langle k',l'\rangle}$ which does not lie below $h(k', l')$. We move to the next substep.

Step 2s+1. Let $s = \langle k, l\rangle$. We check whether the following condition is satisfied:

$$\Phi^s_l(C^s \oplus (B_+ \cup T^*_{\langle k,s\rangle})) \text{ is not single-valued.} \qquad (2)$$

If (2) is true then let D^* be the finite set with least canonical index among all finite sets D such that $D \subseteq C^s \oplus (B_+ \cup T^*_{\langle k,s\rangle})$ and $\Phi^s_l(D)$ is not single-valued. We set "+" near all integers from the set $\{x : 2x+1 \in D^*\}$, which do not have this sign (in this case we shall say that we *l-append* the sign "+" to the corresponding integer x). Next for each k', l' such that $\langle k,l\rangle < \langle k',l'\rangle$ we set both $\langle k',l'\rangle$- labels near the least free integer from the path $T_{\langle k',l'\rangle}$, which does not lie below $h(k',l')$. We move to the next step.

If the condition (2) is not true then we move to next step. The description of the construction is now complete.

Let $n = \langle k,l\rangle$. We now prove some statements from which Lemma 9 will follow.

1°. Each $\langle k,l\rangle$-label moves only finitely many times.

2°. On the path $T_{\langle k,l\rangle}$ there is only a finite number of "+".

3°. $\langle k,l\rangle$-inductions take place only finitely many times.

4°. there is only a finite number of l-appendings of "+".

5°. $B^k \neq \Phi_l(C \oplus \tilde{B}^k)$.

6°. $\tau g = \Phi_l(C \oplus B) \rightarrow g \leq_e C$ for every total function g.

The proofs of these statements will be by induction on n, assuming that for all $n' < n$ the given statements are true. Note that for $n = 0$ all statements are true by definition. (We will assume that $W_0 = \emptyset$.)

1°. Let $s_o \geq n$ be such that, beginning with step $2s_0$, for all $n' < n$ there are no reallocations of n'-labels, no n'-inductions, no n'-appendings, and on the paths $T_{n'}$ no new signs "+" appear. Let B_0 be the set of integers which have "+" by the beginning of step $2s_0$. Then for all $s \geq s_0$ we have

$$\tilde{B}^k_+ \cup \tilde{T}_{\langle k,l\rangle} = \tilde{B}^k_0 \cup \tilde{T}_{\langle k,l\rangle} \tag{3}$$

and

$$\tilde{B}^k_+ \cup \tilde{T}^*_{\langle k,l\rangle} = \tilde{B}^k_0 \cup \tilde{T}^*_{\langle k,l\rangle} \tag{4}$$

Moreover $b(k,l) = b_0(k,l)$ where $\langle k,l,b_0(k,l)\rangle$ is the integer to which is associated the lower label $b(k,l)$ at step s_0. In this case at the p-substep $\langle k,l\rangle$ of any step $2s$ for $s \geq s_0$ we have

$$F_{k,l}(b(k,l), \tilde{B}^k_+ \cup \tilde{T}_{\langle k,l\rangle}, t) = F_{k,l}(b_0(k,l), \tilde{B}^k_0 \cup \tilde{T}_{\langle k,l\rangle}, t)$$

for all $t \in \omega$ and hence at the steps $2s$, $s \geq s_0$ the upper label $h(k,l)$ can go up only if the function $F_{k,l}(b_0(k,l), \tilde{B}^k_0 \cup \tilde{T}_{\langle k,l\rangle}, s)$ increases its value. By Lemma 8 this can happen only finitely many times and after that the label $h(k,l)$ stops forever near the integer $\langle k,l,h_0(k,l)\rangle$, where $h_0(k,l) = F_{k,l}(b_0(k,l), \tilde{B}^k_0 \cup \tilde{T}_{\langle k,l\rangle})$. This proves statement 1°.

2°. Let $Y = \{y : b_0(k,l) \leq y \leq h_0(k,l)\}$. As at the steps $2s$, $s \geq s_0$ the sign "+" can appear on the path $T_{\langle k,l\rangle}$ only when some integer y puts "+"

near an integer which lies above $h(k, l)$, we have that 2° follows from 1°. At steps $2s + 1$, $s \geq s_0$ by the inductive assumption that no n'-appending for any $n' < n$ takes place any more, and the remark that $\tilde{T}_{\langle k,l \rangle} \subseteq \tilde{T}^*_{\langle k,l \rangle}$, we easily see that no more signs "+" appear near any integers which lie above $h(k, l)$. Thus statement 2° is proved.

3°. Note that if the integer y causes a $\langle k, l \rangle$-induction at some step $2s$ then after this at all steps $2s'$, $s' > s$, we will have $\langle k, l, y \rangle \in \Phi_l^{s'}(C^s \oplus \tilde{B}^k_+)$. So the integer y can no longer cause any $\langle k, l \rangle$-induction. As at steps $2s$, $s \geq s_0$, the integers from the set Y can only cause $\langle k, l \rangle$-induction, and moreover each of them at most once. Statement 3° immediately follows.

4°. Note that an l-appending of the sign "+" may take place at a given step $2s + 1$ where $s = \langle k, l \rangle$, may not take place at this step but may take place at some later step, or may never take place. It is clear that in the last case the statement 4° is proved.

We assume that an l-appending of the sign "+" took place at the step $2s_1 + 1$, $s_1 \geq s_0$, $s_1 = \langle k_1, l \rangle$, and we l-appended the sign "+" to the integers from the set $\{x : 2x + 1 \in D^*\}$, where the finite set D^* was chosen at step $2s_1 + 1$ at which condition (2) was satisfied. As a result, at all subsequent steps $2s + 1$, $s \geq s_1$, we will have $D^* \subseteq B_+$. If condition (2) is satisfied for $s = \langle k, l \rangle$, $k \geq k_1$, we get that $\Phi_l^s(C^s \oplus (B_+ \cup \tilde{T}^*_{\langle k,l \rangle})$ is not single-valued, and moreover $D^* \subseteq C^{s_1} \oplus B_+ \subseteq C^s \oplus (B_+ \cup \tilde{T}^*_{\langle k,l \rangle})$. In this case we l-append the sign "+" to the integers from the set $\{x : 2x + 1 \in D\}$, where the set D was chosen at step $2s + 1$ and has canonical index which is smaller than or equal to the canonical index of the set D^*.

Thus l-appendings of the sign "+" take place only finitely many times for each $l \in \omega$ and statement 4° is proved.

5°. Let $2s_1$, $s_1 \geq s_0$, be a step such that at any step $s \geq s_1$ there is no reallocation of the upper labels $h(k, l)$, no $\langle k, l \rangle$-inductions take place, no l-appendings take place, and such that for all $y \in Y$

$$\langle k, l, y \rangle \in \Phi_l(C \oplus (\tilde{B}^k_0 \cup \tilde{T}_{\langle k,l \rangle})) \iff \langle k, l, y \rangle \in \Phi_l^{s_1}(C^{s_1} \oplus (\tilde{B}^k_0 \cup \tilde{T}_{\langle k,l \rangle})). \tag{5}$$

Let B_1 be the set of integers which have "+" at the beginning of step $2s_1$. Because of the operations $(n.t.1)$ which are performed at the n-substeps t of the steps $2s$, for all $y \in Y$ we have

$$\langle k, l, y \rangle \in B^k \iff y \in A. \tag{6}$$

As there are no more $\langle k, l \rangle$-inductions and l-appendings at any step $2s_1$, by (3) for all $y \in Y$ we have

$$\langle k, l, y \rangle \in \Phi_l^{s_1}(C^{s_1} \oplus (\tilde{B}^k_0 \cup \tilde{T}_{\langle k,l \rangle})) \to \langle k, l, y \rangle \in \Phi_l^{s_1}(C^{s_1} \oplus \tilde{B}^k_1).$$

But $C^{s_1} \oplus \tilde{B}_1^k \subseteq C \oplus \tilde{B}^k$. So for all $y \in Y$ we have

$$\langle k, l, y\rangle \in \Phi_l^{s_1}(C^{s_1} \oplus (\tilde{B}_0^k \cup \tilde{T}_{\langle k,l\rangle})) \to \langle k, l, y\rangle \in \Phi_l(C \oplus \tilde{B}_1^k). \quad (7)$$

On the other hand from $\tilde{B}^k \subseteq \tilde{B}_0^k \cup \tilde{T}_{\langle k,l\rangle}$ it follows that for all $y \in Y$

$$\langle k, l, y\rangle \in \Phi_l(C \oplus \tilde{B}^k) \to \langle k, l, y\rangle \in \Phi_l(C \oplus (\tilde{B}_0^k \cup \tilde{T}_{\langle k,l\rangle})),$$

from which by (5) it follows that for all $y \in Y$

$$\langle k, l, y\rangle \in \Phi_l(C \oplus \tilde{B}^k) \to \langle k, l, y\rangle \in \Phi_l^{s_1}(C^{s_1} \oplus (\tilde{B}_0^k \cup \tilde{T}_{\langle k,l\rangle})),$$

which together with (7) proves the equivalence

$$\langle k, l, y\rangle \in \Phi_l(C \oplus \tilde{B}^k) \iff \langle k, l, y\rangle \in \Phi_l^{s_1}(C^{s_1} \oplus (\tilde{B}_0^k \cup \tilde{T}_{\langle k,l\rangle})), \quad (8)$$

for all $y \in Y$.

For the proof of statement 5° we show that there exists $y \in Y$ such that

$$\neg[\langle k, l, y\rangle \in B^k \iff \langle k, l, y\rangle \in \Phi_l(C \oplus \tilde{B}^k)].$$

Indeed, if we suppose otherwise then by (6) and (8) we have

$$\forall y[y \in Y \to [y \in A \iff \langle k, l, y\rangle \in \Phi_l^{s_1}(C^{s_1} \oplus (\tilde{B}_0^k \cup \tilde{T}_{\langle k,l\rangle}))]]$$

from which it follows that

$$F_{k,l}(b_0(k,l), \tilde{B}_0^k \cup \tilde{T}_{\langle k,l\rangle}, s_1) \geq f_{k,l}(b_0(k,l), \tilde{B}_0^k \cup \tilde{T}_{\langle k,l\rangle}, s_1)$$
$$\geq h_0(k,l) + 1 > F_{k,l}(b_0(k,l), \tilde{B}_0^k \cup \tilde{T}_{\langle k,l\rangle}),$$

and this is not possible. Statement 5° is thus proved.

6°. Let $\tau g = \Phi_{l_0}(C \oplus B)$. We prove that for all $s \geq s_0 = \langle k_0, l_0\rangle$, $s = \langle k, l_0\rangle$, $k \geq k_0$,

$$\Phi_{l_0}^s(C \oplus (B_0 \cup \tilde{T}^*_{\langle k,l_0\rangle})) \text{ is single-valued,}$$

where B_0 is the set of the integers which have "+" at step $2s_0 + 1$, and s_0 is as in the proof of statement 5°. We assume that is not true i.e. for some $s_1 = \langle k_1, l_0\rangle$

$$\Phi_{l_0}^{s_1}(C \oplus (B_0 \cup \tilde{T}^*_{\langle k_1,l_0\rangle})) \text{ is not single-valued.}$$

In this case there is a finite set $D \subseteq C \oplus (B_0 \cup \tilde{T}^*_{\langle k_1,l_0\rangle})$, such that $\Phi_{l_0}^{s_1}(D)$ is not single-valued. We choose $s \geq s_1$, $s = \langle k, l_0\rangle$, such that

$$D \subseteq C^s \oplus (B_0 \cup \tilde{T}^*_{\langle k,l_0\rangle}) \subseteq C^s \oplus (B_+ \cup \tilde{T}^*_{\langle k,l_0\rangle}).$$

In this case at step $2s+1$ the condition (2) is satisfied and we $\langle k, l_0\rangle$-append a sign "+", which is not possible by the choice of step $2s_0+1$. Thus for all $s = \langle k, l\rangle$ the set $\Phi^s_{l_0}(C \oplus (B_0 \cup \tilde{T}^*_{\langle k,l_0\rangle}))$ is single-valued. Note that in this case the set $\Phi_{l_0}(C \oplus (B_0 \cup \tilde{T}^*_{\langle k_0,l_0\rangle}))$ is single-valued too.

Thus from the inclusion $C \oplus B \subseteq C \oplus (B_0 \cup \tilde{T}^*_{\langle k_0,l_0\rangle})$ we have

$$\tau g = \Phi_{l_0}(C \oplus B) \subseteq \Phi_{l_0}(C \oplus (B_0 \cup \tilde{T}^*_{\langle k_0,l_0\rangle}))$$

from which it follows that $\tau g = \Phi_{l_0}(C \oplus (B_0 \cup \tilde{T}^*_{\langle k_0,l_0\rangle}))$, i.e. $g \leq_e C \oplus (B_0 \cup \tilde{T}^*_{\langle k_0,l_0\rangle})$. And since $B_0 \cup \tilde{T}^*_{\langle k_0,l_0\rangle}$ is a computable set, it also follows that $g \leq_e C$, as was to be proved.

Thus all foregoing statements 1° – 6° are proved for all $n = \langle k, l\rangle$, and this completes the proof of Lemma 9. □

Lemma 10. *For any c.e. set W the e-degree $d_e(C \oplus \bigcup_{k\in W} B^k)$ is C-quasi-minimal and $d_e(C \oplus \bigcup_{k\in W} B^k) \leq$* **a**.

Proof of Lemma 10. Let $f \leq_e C \oplus \bigcup_{k\in W} B^k$. As for all $k \in \omega$

$$\forall x[x \in B^k \iff x \in B \wedge \exists l, y[\langle k, l, y\rangle = x]],$$

we have that $B^k \leq_e B$ for all $k \in \omega$, and hence $\bigcup_{k\in W} B^k \leq_e B$. Thus $f \leq_e C \oplus B$, and by Lemma 9 the set $C \oplus B$ is C-quasi-minimal, so $f \leq_e C$.

All conditions which are verified during the construction and all performed operations are computable in the set A, and as $\overline{A} \leq_e A$ then they are computable in any enumeration of A. Thus $C \oplus B \leq_e A$ (more exactly, $C \oplus B <_e A$ by non-totality of $d_e(C \oplus B)$) and $C \oplus \bigcup_{k\in W} B^k \leq_e A$. The lemma follows. □

Lemma 11. *The mapping $\epsilon(W) = d_e(C \oplus \bigcup_{k\in W} B^k)$ isomorphically embeds the lattice of the c.e. sets into* $\mathbf{Q_c}(\leq\mathbf{a})$.

Proof of Lemma 11. From Lemma 10 it follows that if W is a c.e. set then $\epsilon(W) \in \mathbf{Q_c}(\leq\mathbf{a})$. We now verify that for any c.e. V and W one has

$$V \subseteq W \iff (C \oplus \bigcup_{k\in V} B^k) \leq_e (C \oplus \bigcup_{k\in W} B^k).$$

Let $V \subseteq W$. Then

$$C \oplus \bigcup_{k\in V} B^k = (C \oplus \bigcup_{k\in W} B^k) \cap (\omega \oplus \{\langle k, y\rangle : k \in V \wedge y \in \omega\}) \leq_e (C \oplus \bigcup_{k\in W} B^k).$$

Let now

$$(C \oplus \bigcup_{k\in V} B^k) \leq_e (C \oplus \bigcup_{k\in W} B^k).$$

Assume that $V \not\subseteq W$ and let $m \in V \setminus W$. Then $\{m\} \subseteq V$ and $W \subseteq \omega \setminus \{m\}$, from which it follows that

$$B^m \leq_e (C \oplus \bigcup_{k\in\{m\}} B^k) \leq_e (C \oplus \bigcup_{k\in V} B^k) \leq_e (C \oplus \bigcup_{k\in W} B^k)$$
$$\leq_e (C \oplus \bigcup_{k\in\omega\setminus\{m\}} B^k) = C \oplus \tilde{B}^m,$$

which contradicts Lemma 9. Thus, $V \subseteq W$ and the lemma is proved. □

We can now complete the proof of the theorem. In the paper [Goe76] it was shown that there exists a partially ordered subset Re of the lattice of c.e. sets such that any partially ordered set, which is at most countable, can be isomorphically embedded into Re. In Lemma 11 it is shown that Re can be isomorphically embedded into $\mathbf{Q}_c(\leq \mathbf{a})$. □

As the following theorem shows, the class of non-total e-degrees is not exhausted by the set of **c**-quasi-minimal e-degrees.

Theorem 12. *For any set C if there is a total function f for which $C \leq_e \tau f <_e \mathbf{J}(C)$ and $\mathbf{J}(\tau f) \equiv_e \mathbf{J}(C)$, then there is a set $A \subseteq \omega$ such that*
(i) $C <_e A <_e \mathbf{J}(C)$;
(ii) $d_e(A)$ is the non-total e-degree;
(iii) A is not a B-quasi-minimal set for any $B <_e A$.

Proof. Let g, h (with indices or not) be variables which range over the set of total functions and let D be a variable for finite sets. We construct a set A such that

$$\forall h(h \leq_e A \rightarrow \exists g(g \leq_e A \wedge g \not\leq_e h)). \quad (1)$$

We verify that in this case $d_e(A)$ is a non-total e-degree. We assume that $d_e(A)$ is total and $\tau h \in d_e(A)$ then $h \leq_e A$ and $A \leq_e h$. Let g be a total function the existence of which follows from (1) relatively to this function h. Then $A \leq_e h \rightarrow g \leq_e A \leq_e h$, so $g \leq_e h$ which contradicts (1). Thus if the condition (1) is satisfied then $d_e(A)$ is a non-total e-degree.

Let f be a total function for which $C \leq_e f$ and $\mathbf{J}(\tau f) \equiv_e \mathbf{J}(C)$. Let $C \leq_e f <_e g_0 <_e g_1 <_e \cdots <_e \mathbf{J}(C)$ be a sequence of total functions. Let

$$G_s = \{\langle s, n, g_s(n)\rangle : n \in \omega\};$$
$$\tilde{M}_s = \{\langle m, x, y\rangle : m > s \wedge m, x, y \in \omega\}.$$

It is clear that $\tau g_s \equiv_e G_s$ and $\tilde{M}_s$ is computable for all $s \in \omega$. With the help of the following construction we form a sequence of sets $A_0 \subseteq A_1 \subseteq \cdots \subseteq A_s \subseteq$ such that

$$C \leq_e A_s \leq_e G_s <_e \mathbf{J}(C) \quad (2)$$

for all $s \in \omega$.

Step 0. We set

$$A_0 = G_0.$$

As $C \leq_e G_0 <_e \mathbf{J}(C)$ then $C \leq_e A_0 \leq_e G_0 <_e \mathbf{J}(C)$.

Step s+1. We verify whether the following condition is satisfied:

$$\exists D(D \subset \tilde{M}_s \wedge \Phi_s(A_s \cup D) \text{ is not single-valued}). \quad (3)$$

If (3) is true then set

$$A_{s+1} = A_s \cup G_{s+1} \cup D^*,$$

where D^* is the finite set with the least canonical index among all finite sets D which satisfy (3). If (3) is not true then set

$$A_{s+1} = A_s \cup G_{s+1}.$$

As $A_s \leq_e G_s$, one can test (3) computably in $\mathbf{J}(G_s) \equiv_e \mathbf{J}(\tau g_s) \equiv_e \mathbf{J}(C)$. Hence the set A_{s+1} which was constructed at step $s+1$ satisfies property (2):

$$C \leq_e A_{s+1} \leq_e G_{s+1} <_e \mathbf{J}(C).$$

From the construction one can see that $A_s \subset A_{s+1}$ for all $s \in \omega$. We define $A = \bigcup_{s\in\omega} A_s$. As $C \leq_e A_s <_e \mathbf{J}(C)$ for all $s \in \omega$, we have that $C \leq_e A \leq_e \mathbf{J}(C)$.

We now prove two lemmas from which the theorem will follow.

Lemma 13. $g_s \leq_e A$ *for all* $s \in \omega$.

Proof of Lemma 13. Given any enumeration of A we enumerate all elements from A of the form $\langle s, x, y\rangle$ with the help of this enumeration. The set $G^s = \{\langle x, y\rangle : \langle s, x, y\rangle \in A\}$ differs from τg_s by at most the finitely many elements which were put into A_i at steps $i \leq s$. In this case $G^s \equiv_e \tau g_s \equiv_e G_s$ and

$$g_s \leq_e G_s \leq_e G^s \leq_e A,$$

hence $g_s \leq_e A$ for all $s \in \omega$. □

Lemma 14. $A_s \leq_e g_s$ *for all* $s \in \omega$.

Proof of Lemma 14. As $g_0 <_e g_1 <_e \cdots <_e g_s$ then $\bigcup_{i\leq s} G_i \leq_e G_s$. From the construction one can see that

$$A_s = \bigcup_{i\leq s} G_i \cup F$$

for some finite set F so $A_s \leq_e g_s$ for all $s \in \omega$. □

We now complete the proof of the theorem. From the construction it follows that $C \leq_e A \leq_e \mathbf{J}(C)$. If we assume that $A \leq_e C$ then from Lemma 13 it follows that $g_s \leq_e A \leq_e C \leq_e f$ which contradicts the assumption $f <_e g_s$ (i.e. $g_s \not\leq_e f$). Moreover from (ii) it follows that the e-degree $d_e(A)$ is non-total so $C <_e A <_e \mathbf{J}(C)$ and the requirement (i) is satisfied.

To show (ii), we prove that (1) holds. Let $h \leq_e A$ be a total function, i.e. $\tau h = \Phi_s(A)$ for some $s \in \omega$. Let us consider step $s+1$, we show that in this case the condition (3) is not satisfied. Indeed, if (3) is satisfied then the set $\Phi_s(A_s \cup D)$ is not single-valued for some finite set $D \subset \tilde{M}_s$, then the finite set D^* which was chosen in this case is a subset of the set A, and so the equality $\tau f = \Phi_s(A)$ is impossible. Thus the condition (3) is not satisfied so

$$\forall D(D \subseteq \tilde{M}_s \to \Phi_s(A_s \cup D) \text{ is single-valued}). \tag{4}$$

Obviously, in this case the set $\Phi_s(A_s \cup \tilde{M}_s)$ is single-valued. Thus, we have that $\tau h = \Phi_s(A) \subseteq \Phi_s(A_s \cup \tilde{M}_s)$ and $\tau h = \Phi_s(A_s \cup \tilde{M}_s)$ by the totality and single-valuedness of τh, i.e. $h \leq_e A_s \cup \tilde{M}_s \leq_e A_s \leq_e g_s$.

We assume that $g_{s+1} \leq_e h$, then $g_{s+1} \leq_e h \leq_e g_s$ which contradicts the choice of the sequence of the functions $\{g_s\}_{s\in\omega}$ where $g_{s+1} \not\leq_e g_s$ for all $s \in \omega$. Thus,

$$h \leq_e A \to \exists s(g_s \leq_e A \wedge g_s \not\leq_e h),$$

and the condition (1) is satisfied. We noted above that in this case $d_e(A)$ is a non-total e-degree and the requirement (ii) is satisfied.

Finally, we prove that A is not a B-quasi-minimal set for any $B <_e A$. We assume that this is false and there is a set B such that $B <_e A$ and

$$\forall h(h \leq_e A \to h \leq_e B).$$

From Lemma 13 it follows that $g_s \leq_e A$ for all $s \in \omega$, so $g_s \leq_e B$ for all $s \in \omega$. As

$$G^s = \{\langle x, y\rangle : \langle s, x, y\rangle \in A\} \leq_e A$$

for all $s \in \omega$, and the set τg_s differs from the set G^s by at most finitely many elements (G^s can be a set which is not single-valued), then $G^0 <_e G^1 <_e \cdots \leq_e B$, and as $G_s \equiv_e \tilde{G}^s = \{\langle s, x, y\rangle : \langle x, y\rangle \in G^s\}$ then $\tilde{G}^0 <_e \tilde{G}^1 <_e \cdots \leq_e B$. It is clear that $A = \bigcup_{s\in\omega} \tilde{G}^s$, so $A \leq_e B$, which contradicts our assumption and thus (iii) is satisfied.

The theorem is now completely proved. □

Corollary 15. *For every total e-degree* $\mathbf{c}$=$d_e(C)$ *there exists an e-degree* $\mathbf{a}$=$d_e(A)$ *such that* $\mathbf{c} < \mathbf{a} < \mathbf{c}'$, *and* $\mathbf{a}$ *is a non-total not* $\mathbf{c}$*-quasi-minimal e-degree.*

Proof. From the construction of the proof of Theorem 12 it follows that $C <_e A$ so for $B = C$ we prove our statement. □

We conclude by observing that a complete description of the non-total e-degrees is still lacking.

References

[AL98] Ahmad, S. and Lachlan, A. H., *Some special pairs of Σ_2^0 e-degrees properties of Σ_2^0 -enumeration degrees*, Math. Logic Quarterly **44** (1998), 431–444.

[ACK01] Arslanov, M., Cooper, S. Barry and Kalimullin, I. S., *Splitting properties of total e-degrees*, to appear.

[BCS97] Bereznyuk, S., Coles, R. and Sorbi, A., *The distribution of properly Σ_2^0 e-degrees*, J. Symb. Logic, 65, n. 1, 2000, pp. 2856297.

[Cas71] Case, J., *Enumeration reducibility and partial degrees*, Ann. of Math. **2** (1971), 419–439.

[CC88] Cooper, S. B. and Copestake, C. S., *Properly Σ_2 e-degrees*, Z. math. Logik Grundlag. Math. **34** (1988), 491–522.

[Coo84] Cooper, S. B.,*Partial degrees and the density problem. Part 2: The enumeration degrees of the Σ_2^0 sets are dense*, J. Symb. Logic., 49, 1984, pp. 503–513

[Coo90] Cooper, S. B., *Enumeration reducibility, nondeterministic computations and relative computability of partial functions*, Recursion Theory Week (Oberwolfach, 1990), Lecture Notes in Mathematics (ed. K. Ambos-Spies, G. Müller, and G. E. Sacks) Vol. 1432, pp. 57–110

[Ers77] Ershov, Yu. L., *Theory of numberings*, "Nauka", Moscou, 1977

[FR59] Friedberg, R. and Rogers, H., *Reducibility and completeness for sets of integers*, Zeit. Math. Log. Grund. Math. **5** (1959), 117–125.

[Goe76] Goetse, B. G., *The structure of the lattice of recursive sets*, Z. math. Log. und Grundl. Math. **22** (1976), 187–191.

[Gut71] Gutteridge, L., *Some results on e-reducibility*. Ph. D. Diss. Simon Fraser University, Burnaby 1971.

[Joc68] Jockusch, C. G., Jr., *Semirecursive sets and positive reducibility*, Trans. Amer. Math. Soc. **131** (1968), 420–436.

[McE85] McEvoy, K., *Jumps of quasi-minimal enumeration degrees*, J. Symb. Logic **50** (1985), 839–848.

[Med55] Medvedev, U. T., *Degrees of difficulty of the mass problem*, Dokl. Akad. Nauk SSSR **104** (1955), 501–504.

[Myh61] Myhill, J., *Note on degrees or partial functions*, Proc. of the Amer. Math. Soc. **12** (1961), 519-521.

[Rog67] Rogers, H. Jr, *Theory of Recursive Functions and Effective Computability*, McGraw-Hill, New York, 1967

[Roz78] Rozinas, M. G., *The semilattice of e-degrees*, Recursive function (E.A. Polyakov, Editor, Ivanovskii Gosudarstvennyi Universitet, Ivanovo, 1978, pp.71684.

[Roz81] Rozinas, M. G., *About infimum in the semilattice of e-degrees*, VINITI Dep. **3564-81** (1981)

[Sac61] Sacks, G. E., *A minimal degree less than* $\mathbf{0}'$, Bull. Amer. Math. Society **67** (1961), 416-419.

[Sas75] Sasso, L. P.. *A survey of partial degrees*, J.Symb. Logic **40** (1975), 130–140.

[Sol99] Solon, B., *e-reducibility and the problem of the nontotal property of e-degrees*, Recursion Theory and Complexity (Kazan, 1997), de Gruyter Series Logic Appl. No. 2, de Gruyter, Berlin, 1999, pp. 173–191.

[Sor90] Sorbi, A., *On some filters and ideals of the Medvedev lattice*, Arch. Math. Logic, 30, 1990, pp.29648.

[Spe56] Spector, C., *On degrees of recursive unsolvability*, Ann. of Math. **64** (1956), 581–592.

[Usp55] Uspensky, V. A., *About computable operations*, Dokl. Akad. Nauk SSSR **103** (1955), 773–776.

[Yat65] Yates, C. E. M., *Three theorems on the degrees of recursively enumerable sets*, Duke Math. J. **32** (1965), 461–468.

Index

X-Rogers semilattice of $\mathcal{A}$, 15
L-identity, 323
L-quasiidentity, 323
L-rule, 323
L-structure
  extensional, 323
  intensional, 323
  positive, 323
m-complete, 4
m-reducibility, 1, 290
Q-reducibility, 290
  bounded, 291
  bounded strong, 291
  strong, 291
R-reducibility, 4
sm-reducibility, 342
tt-complete, 7–8
wtt-complete, 3
wtt-reducibility, 1
Aberth O., 94
Abstract data type, 326
  specifiable by identities, 326
Ambos-Spies K., 270, 273, 275–276, 280, 282, 303
Analytical hierarchy, 324
Appel K., xvi, 166
Arithmetical hierarchy, 324
Arslanov
  Completeness Criterion, 2
  Cupping Theorem, 282
Arslanov M.M., xi, xv, xvii, 1, 4, 277, 282–284, 292, 356
Automorphism base, 282
Bārzdiņš J., 215–216, 240
Bārzdiņš
  Conjecture, 217–218, 220, 224–225, 244
Badaev S.A., xi, 17, 37, 43, 45–47, 71–73, 75, 79–80, 84
Banach S., 94
Bauer A., 98, 102
Beigel R., 161, 164, 168, 171–172, 179–180, 188
Belegradek O.V., 290
Bi interpretability conjecture, 149
Bickford M., 277–278, 284
Bishop E., 94
Blum L., 221, 224, 245
Blum M., 221, 224, 245, 290, 295–296
Brattka V., xi, 93, 95, 100, 106, 116, 119, 123, 128–129, 131–133
Bukharaev N.R., xvii
Bulitko V.K., 2, 9
Cai J., 164
Cannonito F.B., 253
Case J., 217, 220, 226–227, 239, 241–242, 353, 359
Ceĭtin G., 104
Chaos theory, 141
Chervonenkis A.Y., 171
Cholak P., 275, 281
Church's thesis, 138
Church A., 137–138, 148, 151
Clarke S., 171
Computably bounded set, 167
Computably enumerable sets
  congruence relations, 197
  factor lattices, 197
  lattice, 197
Continuity of Capping Theorem, 280
Continuity of Cupping Theorem, 280
Cooper
  Automorphism Theorem, 282
  Minimal Pair Theorem, 271
  Splitting Theorem, 283
Cooper-Li Continuity Theorem, 280
Cooper-Li Splitting Theorem, 284
Cooper-Yates Noncupping Theorem, 271
Cooper S.B., xi, 137, 153, 157, 270–272, 277–284, 316, 321, 351, 354, 356
Copestake C.S., 355
Dark matter, 155
Decoherence, 141, 156
Definability, 141–143
  and invariance, 141
  in Nature, 155, 159
  mathematical, 141
  Turing, 149, 155
Degree
  contiguous, 313
  cuppable, 270
  deep, 278

deep$_n$, 278
low$_n$ cuppable, 274
noncappable, 270
promptly simple, 273
strongly noncappable, 270
Degtev A.N., xvi, 310
Dekker J.C.E., 187
Denisov S.D., xvi, 26, 308
Determinism, 158
Diagonally non-recursive sets, 167
Diagonally noncomputable functions, 1
Diamond base, 276
Ding D., 271, 284
Dobritsa Y.P., 290
Downey Diamond Theorem, 283
Downey R.G., xvi, 260–261, 272, 275, 281, 283, 303, 306, 310, 331
Duhem-Quine thesis, 156
Dushnik-Miller Theorem, 261
E-degree, 353
properly Σ_2, 355
quasi-minimal, 355
total, 354
E-jump, 354
Effectively separable, 109
Effectively speedable set, 296
Effectively subcreative, 298
Einstein-Podolsky-Rosen thought experiment, 154
Emergence, 157
Enumerability class, 164
Enumeration
operator, 353
reducibility, 352
Epistemological relativism, 158
Ershov
difference hierarchy, 324
lattice, 329
Ershov Yu.L., xv–xvii, 12, 15, 19–20, 22–24, 40, 46–47, 49, 54, 72, 80, 95, 321, 329, 335, 337, 346
Extensive set, 167
Feiner L., 86, 331, 341
Fejer P., 275
Fenstad J.E., 94
Feynman R.P., 143
Finitism, 143
Finkelstein M., 253
Fisher P., 303
Fixed-point free function, 1, 3, 9
Fröhlich A., 321
Frechet
derivative, 332
ideal, 332
Friedberg R.M., xv, 197, 268
Friedman H., 94
Fulk M., 215, 217, 244
Function learning, 215
Gödel's Incompleteness Theorem, 141
Gödel K., 137–138
Gasarch W., xi, 161, 180
Gauge theory, 156
Generalised Low Splitting Theorem, 277
Gill J.T., 290–291, 295, 297, 314
Giorgi M.B., 280
Gold E.M., 215–216, 218
Goncharov S.S., 11–13, 16–17, 37, 43, 45–47, 54, 71–73, 79–80, 85–86, 321, 327, 346
Groszek M., 275
Grzegorczyk A., 94, 127
Gutteridge L., 355
Hanf W., 339
Harrington, 277
cap and cup theorem, 274
nonsplitting base, 272
Harrington L., 62, 213, 271–272, 279–280, 283–284, 331, 341
Hauck J., 94
Hay L., 161, 261, 311
Hemachandra L.A., 164
Hemmerling A., 129
Herrmann E., xii, 197–198, 205
Hertling P., 100, 129
High/low hierarchy, 270
Hooke R., 140
Hyperdegree, 150
Hyperhypersimple set, 291
Hypersimple
finitely strongly, 292
Immanence in Nature, 155
Incomputability
in mathematics, 143, 146–147
in Nature, 138–139, 143, 147, 158
Inflation, 156
Invariance, 142
Ishmukhametov Sh.T., xvii, 284
Jahn M.A., 301
Jain S., xii, 215, 217–218, 223, 226–227, 229, 235, 242, 244
Jiang Z., 283
Jockusch-Li-Yang Join Theorem, 278
Jockusch C.G., xv, 166–167, 174, 273–274, 278–279, 356
Kalimullin I.Sh., xii, 6, 249, 356
Kaufmann S., 239, 241
Khisamiev N.G., 331
Khoussainov B.M., 326
Khutoretsky A.B., 46
Kinber E., 239, 241
Kipling R., xv
Kleene S.C., 137–138, 148
Klein F., 217
Ko K.-I., 94
Kobzev G.N., xvii, 310
Koz'minyh V.V., 249, 251–252, 255–256

Kreisel G., 147
Kreitz C., 94–95, 100
Kummer M., 173, 239, 241
Kuznetsov A.V., 321
Lachlan
  Nonbounding Theorem, 272
  Nondiamond Theorem, 270
  nonsplitting base, 277
  nonsplitting pair, 277
  Nonsplitting Theorem, 272, 306
Lachlan A.H., 2, 12, 40, 56, 62, 75, 84, 87, 197, 210, 269, 272, 276, 279–280, 283, 293, 314, 339
Lacombe D., 94, 104, 108, 127
Ladner R.E., 310
LaForte G., 283, 303, 306
Laplace P.-S., 138
Lavrov I.A., 26, 54
Learnable (class of functions), 218
Learning
  BC-, 219
  confident, 219
  conforming, 219
  consistent, 219
  Ex-, 219
  hyperrobust, 232
  reliable, 219
  robust, 223
  uniformly robust, 226
Leibniz G.W., 154, 217
Lempp S., xii, 259, 261, 270, 272, 275, 278, 283, 339
Li-Yi Cupping Theorem, 283
Li A., xii, 267, 271–274, 276–278, 280–281, 283–284
Linear ordering
  computable, 260
  interval (or convex subset) of, 260
  self-embedding of, 260
  strongly η-like, 260
Link
  hard, 273
  soft, 273
Lu H., 284
Maass W., 272
Macintyre A., 290
Major subdegree, 279
Mal'cev Stability Theorem, 123
Mal'cev A.I., 12, 19, 21, 26, 46, 80, 321
Many worlds, 156
Marchenkov S.S., xvii–xviii, 84, 290, 301, 308
Markov A.A., 94
Marques I., 290, 295–297
Martin D.A., 167
Mazur S., 94
McCoy C.F.D., xii, 259
McEvoy K., 354
McLaughlin T.G., 166
McNicholl T., 188
Medvedev U.T., 355
Metric space
  computable, 104
  effective, 104
  recursive, 104
  separable, 104
Miller D., 271
Miller W., 167
Mills C.F., 277–278, 284
Minicozzi E., 221, 225
Minimal Pair Theorem, 269
Minimal program, 297
  modulo h, 297
Morozov A.S., xii, 259
Morris P.H., 290–291, 295, 297, 314
Moschovakis Y.N., 94, 104
Mostowski A., 94, 321
Muchnik A.A., xv, xvii, 268
Myhill J., 314
Nerode A., 321
Newton I., 137, 139–140
Nies A., 62, 275, 303, 306, 341
Non-locality, 141, 154
Non-Union Theorem, 227
Nondensity Theorem of n-c.e. Degrees, 284
Nonspeedable set, 296
Nowhere simple set, 293
Num, 216
Numbering
  $\mathbf{X}$-minimal, 15
  Σ^0_{n+1}–computable, 13
  $\mathbf{X}$-universal in $\mathrm{Com}^0_{n+1}(\mathcal{A})$, 15
  h-2-complete, 329
  acceptable, 326
  arithmetical, 13, 80
  complete, 11, 324
  computable, 11
  Friedberg, 72
  precomplete, 19, 324
  splittable, 22
  uniformly $\mathbf{X}$-complete, 38
  universal h-2-complete, 330
  universal, 11
Occam's Razor, 156
Odifreddi P., 2, ix, xiii, xviii, 137–138, 250
Odintsov S.P., 331–332
Omanadze R.Sh., xiii, xvii, 289, 292–295, 298–299, 301–303, 307–308, 311–315
Ott M., 232–233, 236, 242
Ouspensky V., xvii
Owings J., 161, 171–172, 174
PA sets, 167
Peretyat'kin M.G., 327, 339
Permutation
  computable, 249

primitive recursive, 249
Pilot wave, 156
Pitt L., 240
Podnieks K., 220
Podzorov S.Yu., xiii, 45, 62, 73, 75
Polynomial time computably enumerable (p.t.c.e.), 251
Positive Q-model
component of, 327
universal, 327
Post E.L., xv, 138, 148, 268, 289–290
Pour-El M.B., 94, 107, 139
Presentation
scientific, 151
Prestructure, 118
Primitive recursively enumerable (p.r.e.), 251
Qian L., 284
Quantum theory, 141
Quantum
ambiguity, 154
structure, 154
theory, 154
Quasivariety, 323
Rabin M.O., 321
Rasiowa H., 94
Recursive closure schemes, 114
Recursive retraction, 121
Recursively approximable real numbers, 104
Reductionism, 153
Relative Nonbounding Theorem, 273
Representation
Cauchy, 103
of a set, 95
product, 98
sequence, 98
Retraceable set, 292
Rice H.G., 94
Richards J.I., 94, 107, 139
Robinson Low Splitting Theorem, 271
Robinson J., 253
Robinson R., 94
Robinson R.W., 271
Rogers
semilattice, 45–46, 62, 80
isomorphism type of, 85
minimal cover in, 71
minimal elements in, 71
strong minimal cover in, 71, 75
Rogers H., 12, xv, 46, 80
Rosenstein J.G., 261
Sacks
Splitting Theorem, 269
coding stategy, 269
Density Theorem, 283
preservation, 269
Splitting Theorem, 283
Sacks G.E., 268–269, 283
Sasso L.P., 310, 356, 359
Sauer N., 171
Schröder M., 101–102, 133
Seetapun Continuity Theorem, 280
Seetapun D., 280
Selective sets, 166
Selivanov V.L., xiii, xvi, 46, 321, 327–332, 334, 336–337, 339, 342–346
Semi–computable
lower, 103
upper, 103
Semi-located
lower, 108
upper, 108
Semirecursive set, 301, 356
Sharma A., 242
Shelah S., 171
Shepherdson J.C., 94, 110, 321
Shoenfield's Conjecture, 269
Shoenfield J.R., 269, 279
Shore R.A., xvi, xviii, 268, 270, 272–275, 277, 279, 282, 293
Simpson S.G., 261
Singularities in Nature, 155–156
Skordev D.G., 94
Slaman-Soare Extensions of Embeddings Theorem, 273
Slaman T.A., 275–279, 283
Smith C., 217, 220, 223, 235, 240, 244
Soare R.I., xv–xvi, 12, 46, 80, 88, 139, 149, 162, 167, 174, 262, 270, 273, 279–280, 283, 295, 297, 299
Solomon D.R., xiii, 259, 339
Solon B., xiii, 351, 359, 368
Solov'ev V.D., 291, 293
Sorbi A., xiii, 11–13, 16, 43, 45–47, 79, 356
Special f-partial ordering, 334
effectively dense, 334
effectively upward dense, 334
Speedable set, 296
Speedable
bounded strongly effectively, 301
strongly effectively, 298
Spreen D., 104
Spriggs J., 171
Standard model, 156
Stephan F., xiii, 173, 180, 185, 188, 215, 224, 226–227, 232, 236, 242
Structure, 119
Δ_1^0-, 324
complete, 120
constructivizable, 325
effective, 121
effectively broad, 334
natural, 120
numbered negative, 323
numbered positive, 323

- recursive, 121
- representation of, 120

Subcreative set, 296
Superstring theory, 156
Tarski-Ershov
- derivative, 332
- ideal, 332

Tarski A., 337
Tennenbaum S., 289
Troelstra A., 98
Truth-table, 1
- completeness, 1
- reducibility, 1

Tucker J., 94–95
Turing
- rigidity, 149
- automorphism, 150
- degree, 150, 268
- jump, 268
- model, 147–148, 152, 154, 157
- reducibility, 268
- rigidity, 149
- universe, 148, 154

Turing A.M., ix, 93–94, 137–138, 147–148, 151, 153
Uspensky V.A., 353
V'jugin V.V., 54
Vapnik V.N., 171
Walk S., 275, 281
Wang D., 273
Watson P., 283
Weak Density Theorem, 283
Weihrauch K., 94–95, 100, 104, 133
Weyl H., 217
Wiehagen R., 217, 222–223, 226–227, 235, 244
Wu G, 273
Yang D., 276
Yang Y., 274, 278
Yates C.E.M., 269–271, 278–279, 282, 355
Yi X., 277, 283–284
Yu L., 271
Zelmanov E., xvii
Zeugmann T., 224, 237, 240, 244
Zhang Z., 273
Zucker J., 94–95

MIX
Papier aus verantwortungsvollen Quellen
Paper from responsible sources
FSC® C105338

If you have any concerns about our products,
you can contact us on
ProductSafety@springernature.com

In case Publisher is established outside the EU,
the EU authorized representative is:
Springer Nature Customer Service Center GmbH
Europaplatz 3, 69115 Heidelberg, Germany

Printed by Libri Plureos GmbH
in Hamburg, Germany